Developmental and Cell Biology Series

The aim of the series is to present relatively short critical accounts of areas of developmental and cell biology where sufficient information has accumulated to allow a considered distillation of the subject. The fine structure of cells, embryology, morphology, physiology, genetics, biochemistry and biophysics are subjects within the scope of the series. The books are intended to interest and instruct advanced undergraduates and graduate students and to make an important contribution to teaching cell and developmental biology. At the same time, they should be of value to biologists who, while not working directly in the area of a particular volume's subject matter, wish to keep abreast of developments relevant to their particular interests.

RECENT BOOKS IN THE SERIES

FUNGAL MORPHOGENESIS

DAVID MOORE

PUBLISHED BY THE PRESS SYNDICATE OF THE UNIVERSITY OF CAMBRIDGE
The Pitt Building, Trumpington Street, Cambridge CB2 1RP, United Kingdom

CAMBRIDGE UNIVERSITY PRESS
The Edinburgh Building, Cambridge CB2 2RU, UK http: //www.cup.cam.ac.uk
40 West 20th Street, New York, NY 10011-4211, USA http: //www.cup.org
10 Stamford Road, Oakleigh, Melbourne 3166, Australia

© First published 1998

Printed in the United States of America

Typeset in Times Roman and Friz Quadrata [KW]

A catalog record for this book is available from
the British Library

Library of Congress Cataloging-in-Publication Data
Moore. D. (David). 1942–
Fungal morphogenesis / David Moore.
p. cm. — (Developmental and cell biology series)
Includes bibliographical references (p.) and index.
ISBN 0-521-55295-8 (hb)
1. Fungi—Morphogenesis. I. Title. II. Series.
QK601.M648 1998
571.8′295—dc21

97-43011
CIP

Living is easy with eyes closed,
Misunderstanding all you see.

Strawberry Fields
John Lennon (1966)

Contents

Preface

This book is aimed at all biologists. Certainly, I started out with the intention to write a biological text rather than a mycological one because I believe the fungi are too important to remain in an intellectual ghetto in some faintly plant-like place which most people visit rarely, and then with unease.

Throughout, I have attempted to blend together physiological, biochemical, structural and molecular descriptions within an evolutionary framework, combining the older literature with the most recent. Without attempting a comprehensive description of fungi, I hope that I have provided sufficient information about fungal biology to give the general reader a rounded view of the mycological context within which fungal morphogenesis is played out without obscuring the broader biological significance. If I have got the balance right, the reader with knowledge of basic biology should not need to bring any other knowledge with him or her, nor need to refer elsewhere, in order to appreciate fungal morphogenesis.

The first chapter aims to give an overview of the evolutionary origins of fungi and the central role they played (and still play) in the evolution of life on Earth. The second chapter introduces hyphal growth, the essence of the fungal lifestyle, and identifies features which are crucial aspects of morphogenesis. Chapter 3 summarises fungal primary and secondary metabolism, necessary here because adaptation of primary metabolism and exploitation of secondary metabolism are both critical to fungal morphogenesis. In Chapter 4 the impact of physiology on morphogenesis is discussed, the genetic components of differentiation and morphogenetic change being dealt with in Chapter 5. The development of form and structure is the main theme of a lengthy Chapter 6, and the ideas developed here are brought together and summarised in the final Chapter 7.

I would like to thank Peter Barlow for suggesting the book in the first place and for helpful comments on the manuscript and the proof. Most of the planning, and then the writing, of this volume were done during two extended visits I was able to make to the Chinese University of Hong Kong. My sincere thanks are due to the Leverhulme Trust for the award of a Research Grant which enabled the first of these, in 1995, and also to the Royal Society for award of a Kan Tong Po Visiting Professorship which enabled my second (writing) visit to the Department of Biology at CUHK in 1996. I wish also to extend my thanks to the School of Biological Sciences and The University of Manchester for leave of absence on these occasions and to the staff in Manchester for managing to get along without me. I greatly appreciate the hospitality of the Department of Biology at CUHK, the CUHK Guesthouse system and Shaw College and particularly thank Professors Samuel Sun and Norman Woo for all they did to facilitate my visits. Special thanks are reserved for Professor Siu Wai Chiu for her constant encouragement and help. She also commented on early drafts of the manuscript and produced many of the photographic illustrations. Thanks for everything, Suzie! I also thank Ms Carmen Sánchez and Dr Halit Umar for providing me with previously unpublished photographs, Rebecca Jane Moore for advice on organic chemistry and for drawing most of the structural formulae and Sophie Anne Moore for help with index preparation. I offer my deepest appreciation to my wife and daughters for tolerating my eccentric behaviour whilst writing this book.

1

Fungi: a Place in Time and Space

The basic shape, form and structure of an organism (whether fungal, plant or animal) does not arise all at once. Rather, the shape and form emerge as a result of a sequence of developmental adjustments. Each of these is usually irreversible within its morphogenetic sequence although often reversible by some gross disturbance; for example, differentiated cells being put into tissue culture, nuclear and cell transplants, regeneration after injury, etc. The whole process in which the final organisation and pattern of the organism is established is termed 'morphogenesis'.

The most extensive research on the topic has been done with animals and from this a vocabulary has been established which describes morphogenetic events without pre-judging the mechanisms which may be involved (Slack, 1991). It is evident that as the embryonic organism develops towards adulthood, each intermediate state represents a reduction in developmental potential compared with the previous state. Each adjustment (or developmental 'decision') is made by cells already specified by earlier adjustments to belong to a particular developmental pathway. Consequently, developmental decisions are made from among progressively smaller numbers of alternatives until the particular structure to which the cell will contribute is finally determined. Classic embryological transplantation experiments revealed these states. Where the explant differentiated to a state representative of its old position then it was said to have been determined prior to transplantation. If it developed in accord with its new position, then it had not been determined, but may have been specified.

Within the developing tissues, cells embark on particular routes of differentiation in response to the playing out of their intrinsic genetic programme, in response to external physical signals (light, temperature, gravity, humidity), or in response to chemical signals from other regions of the developing structure. These chemicals may be termed organisers, inducers or morphogens, and seem to inhibit or stimulate entry to particular states of determination. Chemical signals may contribute to a *morphogenetic field* around a structure (cell or organ) which permits continued development of that structure but inhibits formation of another structure of the same type within the field.

All of these phenomena contribute to the *pattern formation* which characterises the 'body plan' which is created by the particular distribution of differentiated tissues in the structure (organ or individual). Pattern formation depends on *positional information*, which prompts or allows the cell to differentiate in a way appropriate to its position in the structure. Positional information is usually thought to be conveyed by concentration gradients of one or more morphogens emitted from one or more spatially distinct organisers. The responding cell senses the concentration of the morphogen and initiates a differentiation programme appropriate to the physical position at which that morphogen concentration is normally found. In essence, the cell 'triangulates' on the incoming signals and adjusts its morphogenetic response in accord with its position relative to the controlling organisers. Populations of cells which respond like this are said to show *regional specification*. The operation can be divided into (i) an *instructive* process which provides positional information, and (ii) an *interpretive* process in which the receiving cell or tissue responds.

The basic rules of pattern formation seem to be that regional specification (directed by organisers producing morphogens) occurs first, regulating gene activity in ways specifically geared to morphogenesis so that particular cells are first specified (a state which is still flexible) and then determined (a state which is inflexible) to their differentiated fates. Cell differentiation is a consequence of these events – cells which are either specified or determined are not necessarily morphologically different from their neighbours or predecessors (the morphological change may occur much later, or the differentiation may involve change only in molecular or metabolic attributes).

The vocabulary outlined above highlights the major events contributing to animal development and is useful, too, in descriptions of plant morphogenesis. One might ask why it is featured so early in a book about the development of fungal structures. Unfortunately, fungal development has been rather ignored as a topic in its own right. The great majority of the published research on fungal morphogenesis has been done with taxonomic intentions. It has great value for its descriptive and comparative content, but precise developmental accounts are extremely rare and experimental approaches rarer still. The dearth of research on experimental fungal developmental biology forces us to seek parallels between fungi and other eukaryotes so that we can make use of the conceptual framework which has been established, in embryology, cell and evolutionary biology. It is not a negative comparison because it can reveal common strategies and conserved pathways as well as alternative approaches, providing insight into the response of very different living organisms to the need to solve the same sorts of structural and morphological problems.

But just as an organism develops and evolves, so the science of studying organisms develops and evolves, but along a tortuous route which includes the turns and roadblocks of misinterpretation and misconception. By making bold comparisons across the boundaries between the major eukaryotic kingdoms we can learn from past mistakes rather than repeat them.

This urge to compare must be tempered with full appreciation that fungi have attributes which are unique to them which must affect their developmental mechanisms. Fungi are 'modular organisms', like clonal corals and vegetatively-propagated plants, among others (Harper *et al.*, 1986; Carlile, 1995). In modular organisms growth is repetitive and a single individual (though the definition of 'individual' is open to debate) will have localised regions at very different stages of development (Andrews, 1995). The constituent cells are generally considered to be totipotent (able to follow any pathway of differentiation), because a mycologist would expect to be able to produce a tissue (mycelial) culture in a culture dish from a fragment removed from a mature, fully differentiated structure, like a mushroom fruit body, collected from the field. This cannot be done routinely with animals, and most plants demand far more stringent *in vitro* growth media and conditions than do most fungi. This behaviour reflects the nutrient-absorptive fungal lifestyle, but it also says something about the control of fungal development because even highly differentiated fungal cells will revert readily to vegetative growth if they are explanted to a (relatively simple) nutrient medium (see section 6.5).

This is not to say that fungal cell differentiation is any less sophisticated or complex than is found in animals and plants, but fungi can vary the timing, extent, and mode of differentiation in response to external signals, interconverting growth forms and reproductive phases of their life cycle in ways which make them supremely adaptable to challenging conditions. This results in a morphological plasticity which surpasses that of other organisms and provides an intellectual challenge in terms of developmental biology, taxonomy and genetics (Watling and Moore, 1994).

It is still often necessary to remind people that fungi are not plants. Even mycologists are not immune to an occasional lapse, maybe referring to 'saprophytic' (rather than saprotrophic) fungi in the heat of an argument, but though this does great disservice to the study of fungi as a unique kingdom (Hawksworth, 1995) it is usually just a slip of the tongue.

More dangerous is the fact that there are still a great many people whose education was completed before the revolution in systematics in the mid-1960s, or who were taught by teachers whose education was completed before then, and who are consequently firmly convinced that fungi are plants – peculiar plants, perhaps, but plants nevertheless. Any such idea is completely wrong because plants, animals and fungi are now considered to be three quite distinct kingdoms of eukaryotic organisms (Whittaker, 1969; Margulis, 1974; Cavalier-Smith, 1981, 1987).

Arranging organisms into kingdoms is a matter of systematics – an agreed scheme of categorisation – but this arrangement is mirrored in current ideas about the early evolution of eukaryotes (Fig. 1.1).

Whilst there is debate over the most likely sequence of early evolutionary events, all of the schemes argue that the major kingdoms separated from one another at some unicellular level. If it really is the case that the last common ancestor of plants, animals and fungi was a unicell, then these kingdoms have independently evolved all of the mechanisms which they currently use to organise populations of cells into multicellular organisms.

1.1 Fungal lifestyle

It is worth dwelling on evolutionary aspects, because the three main eukaryote kingdoms are very different from one another in ways that

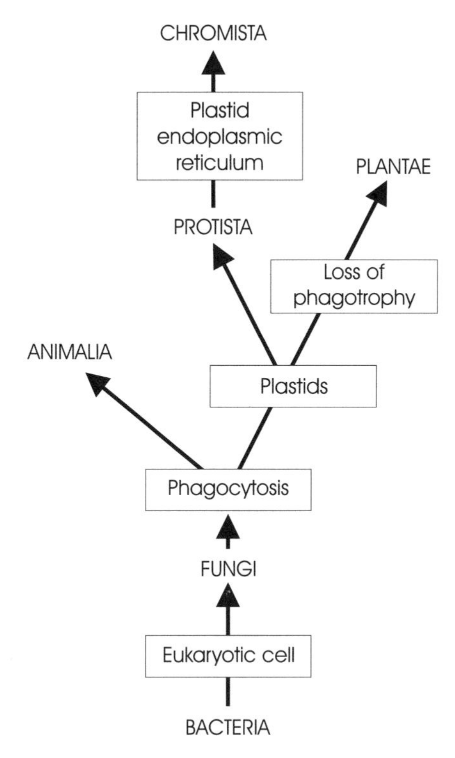

Figure 1.1 Simplified, elementary phylogeny and six-kingdom classification of living organisms. The major evolutionary events are shown in the boxes. The kingdom Protista as shown here is a heterogeneous group which would include myxomycetes and most organisms normally understood to be protozoa. Kingdom Chromista includes a range of golden and brown algae and diatoms (but green and red algae are included in kingdom Plantae) as well as fungus-like phyla interpreted as having lost plastids secondarily, including the Oomycota, Hyphochytriomycota and Labyrinthulomycota. Redrawn from Cavalier-Smith (1981).

are crucial to determining shape and form. A key part of the original definition of the kingdoms (Whittaker, 1969) was their mode of nutrition (plants use radiant energy, animals engulf food, fungi absorb nutrients), and this apparently simple basis for systematic separation then reveals other differences in structure and lifestyle which are correlated with nutrition (presence of chloroplasts, internalised digestion, export of digestive enzymes, etc.).

Approaching consideration of metabolic processes from an evolutionary viewpoint is appropriate since the mode of nutrition has always been a major characteristic in schemes of classification. The photosynthetic plants have always been clearly distinguishable from ingesting animals;

but only recently have the fungi been properly placed in the scheme of living things in a way which clearly recognises the fundamentally different mode of nutrition they employ. In placing the fungi in an entirely separate kingdom, Fungi, Whittaker (1969) emphasises that the "...nutritive mode and way of life of the fungi differ from those of the plants. Fungi characteristically live embedded in a food source or medium, in many cases excreting enzymes for external digestion, but in all cases feeding by absorption of organic food from the medium. Their organisation, whether mycelial, chytrid, or the unicellular of yeasts, is adapted to this mode of nutrition."

This, then, is how fungi have evolved to grow: effectively embedded in a substratum which they digest by the excretion of appropriate enzymes. The smaller molecules produced by the activity of those enzymes are the nutrients which can be absorbed across the plasmalemma. In this the fungi are distinct from both plants and animals for, as Whittaker (1969) also points out, these latter groups have internalised the absorption process; animals by the process of ingestion, but plants too, by the elaboration of membranes around their photosynthetic organelles. Among eukaryotes only the fungi (though this is a character they share with bacteria) must digest their substrates externally prior to absorption of the smaller molecules of which the substrates are composed. There are ecological and structural, as well as biochemical, consequences of this. A protozoan or metazoan can immediately capture a morsel of food by ingesting it into a food vacuole or digestive tract where it can be converted to its components without fear of loss to competing organisms. A fungus may be capable of digesting the same food source, but must perform most of that digestion externally with the valuable products of the digestion being open to absorption by competitors until they can be internalised by the fungus. This may have influenced the evolution of extracellular mucilages, cell walls, membrane components and digestive enzymes so that the fungus can improve its competitive effort by exerting effective control over the environment in the immediate vicinity of its cell surface.

The kingdom Fungi encompasses a tremendously diverse and enormously versatile range of organisms. It is unlikely that there is a compound, organic or inorganic, on the planet that some fungus cannot utilise, transform, modify or otherwise deal with. It could reasonably be argued that lignin is the most exotic and biologically demanding potential nutrient. Lignocellulose constitutes about 95% of terrestrial biomass (Janshekar and Feichter, 1983), and except for the activities of the very few organisms which are able to degrade the lignin component, the rest of us would be overwhelmed! But there are examples of some extreme fungal abilities. *Phanerochaete chrysosporium*, one of the key organisms in studies of lignin breakdown, has also been shown to degrade PCBs the polychlorinated biphenyls used as electrical insulators which are such persistent pollutants (Bumpus *et al.*, 1985; Eaton, 1985). This, and other fungi, can also degrade pesticides (Kumar *et al.*, 1996), packaging materials (Pagga *et al.*, 1995) and other xenobiotics (Shah *et al.*, 1992; Singleton, 1994), abilities which make them ideal candidates as contributors to the microbial degradation of industrial wastes as a cost-effective method of removing such pollutants from the environment by a process now known as bioremediation (Alexander, 1994).

Some species of the yeast *Candida* can use n-alkanes as sole carbon source, and specifically modify the structure of their walls to enable them to do so (Kappeli *et al.*, 1978). Hydrocarbons, and other organic vapours, in the atmosphere can also be scavenged and used for growth (Mirocha and DeVay, 1971; Tribe and Mabadeje, 1972). At the other end of the molecular spectrum, fungi can fix CO_2 from the atmosphere (Tabak and Bridge-Cooke, 1968), some being able to use CO_2 as sole carbon source indefinitely (Mirocha and DeVay, 1971), and others are claimed to fix more CO_2 when illuminated (Hilgenberg and Sandmann, 1977; and see Wainwright, 1988). Although these are some of the most unusual (and debatable) metabolic situations, it is becoming increasingly evident that some soil fungi can grow under oligotrophic conditions; i.e. in a purely mineral medium (Wainwright *et al.*, 1994).

The genetic apparatus of the cell encapsulates its form and nature in an informational archive, which is expressed through the metabolic activities of the cell. Through its metabolism the cell interacts with its environment and neighbouring organisms; metabolism provides for the energy requirements of the cell and satisfies the demands of its biosynthetic machinery for the precursors of those polymers whose assembly creates, maintains and modifies the physical form of the cell itself.

The metabolic apparatus is both powerhouse and workshop, and is the working, responsive interface with whatever may be ‘outside’ the cell. The key word here is ‘responsive’, for metabolism is continually changing to accommodate changing circumstances both within and outside the cell. It must be appreciated that the biochemical transformations occurring in a cell at any one time are only a small subset of those that are possible. The adjustments and changes between those subsets emphasise different aspects of metabolism suiting the prevailing conditions and often occurring for reasons of economy. Indeed, politico-economic phrases like cost-effective, return on the investment made, and profit-and-loss relation can all be applied effectively in discussion of metabolic regulation. The 'choice' (not a conscious choice, of course) between alternative metabolic processes is always made on the basis of economy of effort, because in competitive evolutionary terms greater economy of effort is a selective advantage. Of course, this does not mean that the most energetically economic path is always taken, though it often is. It does mean that the advantage which accrues to the organism must be worth the costs incurred. In some cases the advantage is in successful completion of a developmental pathway and intermediary metabolism is adapted to some morphogenetic purpose, rather than to a purely nutritional one (examples in Chapters 3 and 4).

1.2 The essential nature of fungi

Other differences between the eukaryotic kingdoms, not obviously correlated with mode of nutrition, include the way in which multicellular structures can be organised. In animals, even lower animals, the movement of cells and cell populations plays a crucial role in development, so cell migration (and everything that controls it) is a key feature of all aspects of animal morphogenesis. Plant cells, on the other hand, are encased in walls and have little scope for movement relative to each other. Changes in shape and form in plants are dependent upon control of the orientation and position of the mitotic division spindle because the new cell wall which will separate the parental cell into two daughter cells arises from the phragmosome at the equator of the mitotic division spindle. Consequently, the orientation and position of the dividing parental nucleus will determine the orientation and position of the daughter cell wall.

Fungi are also encased in walls; but their basic structural unit, the hypha, has two peculiarities which mean that fungal morphogenesis must be totally different from plant morphogenesis. These are that hyphae grow only at their tips and that cross-walls form only at right angles to the long axis of the hypha. The consequence is that fungal morphogenesis depends on the placement of hyphal branches. To proliferate a hypha must branch, and to form a structure the position at which the branch emerges and its direction of growth must be controlled (see Chapter 2).

1.3 Evolutionary origins

Most aspects relating to the origins and subsequent evolution of fungi are impossible to establish from any fossil record, so ideas and concepts must be gleaned from other sources (Rayner *et al.*, 1987). Key events in fungal evolution probably took place in the early Palaeozoic or late Precambrian (see Table 1.1), and the likelihood of finding definitive fossil evidence for them is small (Sherwoodpike, 1991). Fungal spores are often well-preserved, but in addition, fungi characteristically interact with other organisms, particularly plants, and the fossil record can provide some information about these (Taylor and Osborn, 1996; and see below). However, the general lack of convincing fossil records of fungal origin means that comparative analyses of molecular sequences (protein and nucleic acid) provide the strongest evidence about the nature of evolutionary relationships between existing groups of organisms. Hendriks *et al.* (1991) examined the small ribosomal subunit RNA sequence of 58 eukaryotes. Evolutionary trees constructed from the data showed a nearly simultaneous radiation of metazoa, the red alga *Porphyra umbilicalis*, the sporozoa, the higher fungi, the ciliates, the green plants, plus some other groups. Higher fungi formed a monophyletic cluster when all alignment positions were used to construct the evolutionary tree. Although the red alga and fungi seem to diverge at nearly the same evolutionary time, no evidence could be detected to support the idea that higher fungi and red algae might have shared a common origin.

There are probably at least 1.5 million species of fungi in the world today (Hawksworth, 1991) and if Whittaker's paper made 1969 a memorable year by establishing their true importance in terms of their rank in the scheme of things, 1993 seems to have been another golden year, this time for publications establishing fungal relationships.

Margulis (1992) had already pointed out that the field of systematic biology had been reorganised with logical, technical definitions for each of the three major kingdoms of eukaryotes (Mycota, 'true' fungi; Plantae, bryophytes and tracheophytes; and Animalia) "This classification scheme requires changes in social organisation of biologists, many of whom as botanists and zoologists, still behave as if there were only two important kingdoms (plants and animals)." So 1993 seems rather special as a year in which papers appeared with the titles 'Monophyletic origins of the Metazoa – an evolutionary link with fungi' (Wainright *et al.*, 1993) and 'Animals and fungi are each others closest relatives – congruent evidence from multiple proteins' (Baldauf and Palmer, 1993). Wainright *et al.* (1993) analysed small subunit ribosomal RNA sequences and deduced

Table 1.1. *A version of the geological time-table. Inspired by a figure in J. A. Moore (1993).*

Million years ago	Eon	Era	Period	Description
13,000				Universe forms and expands.
4,600	Hadean			Earth forms by accretion and was probably under repeated bombardment by meteorites, each cataclysmic event destroying any prior chemical evolution.
3,800	Archean			Oldest known rocks, meteorite bombardment declines sufficiently for evolution to progress. Autotrophic carbon fixation dated to 3800 million years ago (on the basis of carbon isotope ratios in sedimentary organic carbon). Oldest prokaryotic fossils about 3500 million years old.
2,500	Proterozoic			Prokaryotes abundant. Atmospheric oxygen about 0.2%. Eukaryotes appear during this Eon, about 2000 million years ago. Last common ancestor of fungi, animals and plants about 1000 million years ago. Plants diverge first. Fossil algae known about 1000 million years old. Fungi and animals shared a common ancestor more recently than either did with plants. Metazoans emerge about 800 to 1000 million years ago.
670	Phanerozoic	Palaeozoic	Ediacaran	Some fossils interpreted to be of lichen origin. Oldest known metazoans (coelenterates, annelids, arthropods).
570			Cambrian	Atmospheric oxygen reaches 2%. Trilobites and brachiopods abundant. All metazoan phyla present.
510			Ordovician	Terrestrial fungi diverge from chytrids about 550 million years ago. Earliest vertebrates appear.
435			Silurian	Possible fungal fossils. First fishes with jaws. Animals and plants invade land. Atmospheric oxygen reaches 20%.

405		Devonian	Mycorrhizas evident in plant fossils 400 million years old. Ascomycetes separate from basidiomycetes. Land plants and land arthropods abundant. First insects. First amphibians. Continents move towards one another.
355		Carboniferous	Widespread forests of primitive plants eventually form coal deposits. Reptiles appear at the end of the period.
290		Permian	Land masses form a single continent. Mammal-like reptiles. Frigid conditions and massive extinctions at the end of the period.
250	Mesozoic	Triassic	First dinosaurs. Cycads and conifers abundant. Continents move apart.
205		Jurassic	Fungal spores in amber about 225 million years old. Basidiomycete radiation begins, poroid and agaricoid forms evident. The genera *Penicillium* and *Aspergillus* may have appeared about 200 million years ago. Dinosaurs abundant. First birds and mammals.
135		Cretaceous	Fossilized fungal remains known from this period. Major radiation of flowering plants. Amber 90 to 94 million years old contains gilled mushrooms strongly resembling current species. Dinosaurs become extinct; this and many other extinctions possibly associated with meteor impact at the end of the period.

(*cont.*)

Table 1.1. (*cont.*)

Million years ago	Eon	Era	Period	Description
65		Cenozoic	Tertiary	Continued diversification of birds, mammals and flowering plants. Grasses and grazing mammals abundant. The genus *Homo* appears at the end of the period.
1.6			Quaternary	Humans widespread. Large mammals become extinct. Glaciation in northern hemisphere.
0				Present day. Mycorrhizas essential to all terrestrial plants. Wood decay fungi essential to lignin degradation. Fungi essential to many daily human activities, social and commercial. Fungi rule!

that animals and fungi shared a unique evolutionary history, their last common ancestor being a flagellated protist similar to present-day choanociliates (Fig. 1.2). Current theorising about the early evolution of metazoans also places their origin about 800 million to 10^9 years ago from pre-existing unicellular organisms which were most probably flagellates (Denis, 1995). This paper is also interesting in that it is argued that the earliest metazoans arose not by aggregation of solitary cells, but by subdivision of multinucleated cells – in fungi, too, it is assumed that subdivision of the mycelial hyphae into compartments by septa is a later development.

Baldauf and Palmer (1993) compared sequences from 25 proteins, constructing phylogenetic trees for four of the most frequently sequenced proteins (actin, α-tubulin, β-tubulin and elongation factor 1α). All four trees placed animals and fungi together as a monophyletic group to the exclusion of plants and a range of protists. The authors concluded that "This congruence among multiple lines of evidence strongly suggests, in contrast to traditional and current classification, that animals and fungi are sister groups while plants constitute an independent evolutionary lineage."

Two other publications in 1993 reported similar conclusions. Hasegawa *et al.* (1993) also analysed amino acid sequence data of elongation factor-1α and concluded that kingdom Fungi is the closest relative of kingdom Animalia (they also concluded that the ancestor of the cellular slime mould, *Dictyostelium discoideum*, had not diverged from the line leading to Plantae–Fungi–Animalia before these three kingdoms separated). Van de Peer *et al.* (1993) used small ribosomal subunit RNA sequences but concluded that animals, green plants, and fungi form monophyletic groups which seem to have originated nearly simultaneously, certainly over a relatively short time interval.

Since 1993 there has been ample reinforcement of these ideas. There may still be debate over the exact sequence of branching of the three kingdoms; but there is unanimity over the fact that the three kingdoms are entirely separate and became so at a very early stage in evolution. The evidence for these generalisations now comes from a very wide range of protein and nucleotide sequences.

Study of the phylogenetic relationship among the kingdoms Animalia, Plantae, and Fungi using 23 different protein species and three different methods produced a maximum-likelihood tree ((A,F),P) in which Plantae (P) is an outgroup to an Animalia (A)Fungi (F) clade (a clade is a group of organisms that share a common ancestor). On this evidence, kingdom Animalia is more closely related to Fungi than to Plantae (Nikoh *et al.*,

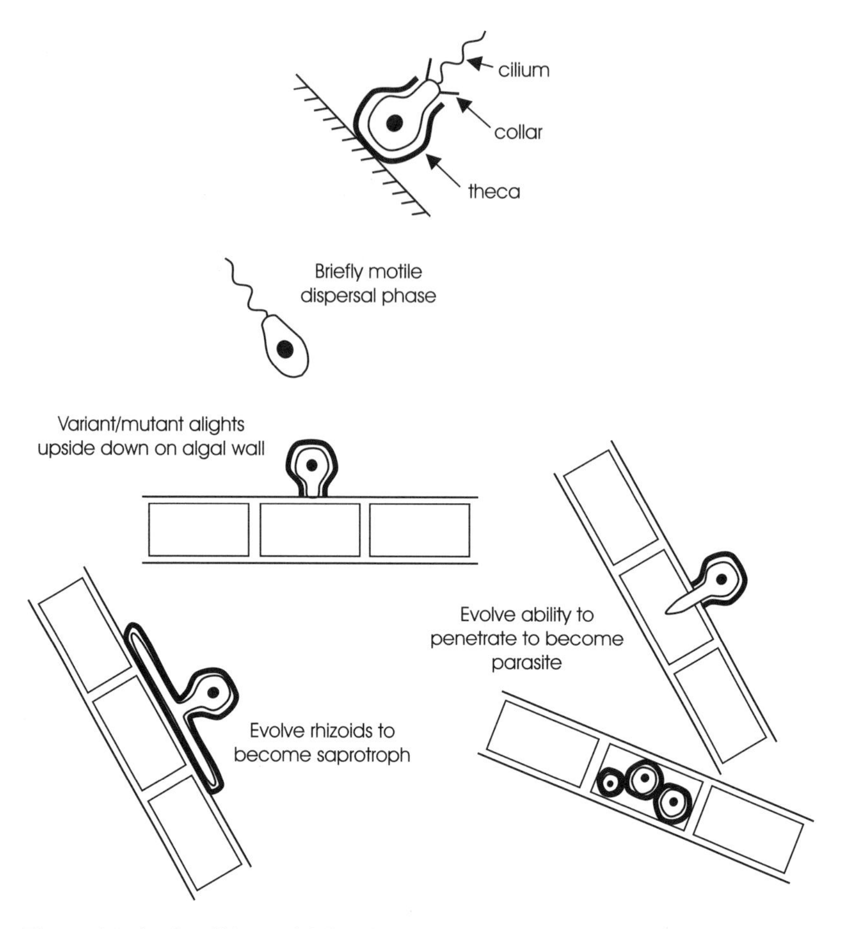

Figure 1.2 A plausible model for the origin of fungi from a sessile choanociliate with a chitinous theca (protective capsule). Present-day sessile (epiphytic) choanociliates phagocytose bacteria drawn onto the cytoplasmic collar by the current produced by the beating of the cilium. The idea illustrated here suggests that fungal organisms could have arisen if the dispersal stage attached its apical rather than basal end to the algal filament when it alighted. Survival would then depend on evolution towards saprotrophism or parasitism. Modified after Cavalier-Smith (1987).

1994). Gupta (1995) examined 31 heat-shock protein sequences and found closer relationships between plant and animal sequences than between either of them and the fungi, concluding that the relationship among the animal, plant and fungal kingdoms remained unresolved. Kumar and Rzhetsky (1996) studied 214 sequences of the nuclear small-subunit ribosomal RNA gene and also concluded that the branching order of eukaryotic lineages may be difficult to resolve, though they did indicate that animals and true fungi are indeed closer to each other than to any other major group in the eukaryote phylogenetic tree.

A possible reason for uncertainty in determining phylogenetic branching order is the different rates of change in protein sequences during evolution of the different kingdoms. Doolittle *et al.* (1996) examined amino acid sequence data from 57 different enzymes, calculating that plants, animals, and fungi last shared a common ancestor about 10^9 years ago. With regard to these protein sequences, plants were slightly more similar to animals than were the fungi, though phylogenetic analysis of the same sequences indicated that fungi and animals shared a common ancestor more recently than either did with plants, the greater difference resulting from the fungal lineage changing faster than the animal and plant lines over the last 965 million years.

Analyses of sequences of the two largest subunits of RNA polymerase II suggested that plants and animals share a last common ancestor that excluded fungi, the lineage of which originated earlier (Sidow and Thomas, 1994). On the other hand, a phylogenetic tree based on 65 actin protein sequences shows that green plants, fungi and animals are monophyletic groups; and that the animal and fungal lineages shared a more recent common ancestor than either did with the plant lineage; it also indicated that oomycetes are not related to higher fungi. In contrast to small-subunit rRNA trees, this tree shows that slime moulds diverged after the plant lineage. The slower rate of evolution of actin genes of slime moulds relative to those of plants, fungi, and animals might be responsible for this incongruent branching (Drouin *et al.*, 1995). Karlin and Ladunga (1994) also detected variable evolutionary rates during the metazoan radiations, but claimed there had been slower changes in the fungal lineage, using a method based on relative abundances of short oligonucleotides in large DNA samples. Paquin *et al.* (1995) based their phylogeny on mitochondrial NAD5 (a subunit of the NADH dehydrogenase) protein sequences. The NAD5 tree indicates an early divergence of the Chytridiomycetes, appearance of Zygomycetes prior to the divergence of Ascomycetes and Basidiomycetes, and that Oomycetes are clearly unrelated to fungi. In addition, this analysis predicted a common ancestor of fungi and animals, to the exclusion of green algae and plants. Riley and Krieger (1995) studied the strongly conserved cyclin-dependent kinase (CDK) and produced dendrograms consistent with rRNA-phylogenetic inference of early divergence of the eukaryotic lineages.

There is still uncertainty in the exact sequence of divergence of the major kingdoms, probably because of the effect of variable rates of evolution between the different groups (and, indeed, between the different molecules analysed). However, uncertainty over this point of detail should not obscure the major message being delivered which is that the major kingdoms have been separated for an enormous length of time. Another important date arising from the analyses of Doolittle *et al.* (1996) is that eukaryotes and eubacteria last shared a common ancestor about 2×10^9 years ago. To see the significance of these dates, it is important to realise that there is evidence for the activities of living organisms in terrestrial rocks which are 3.5×10^9 years old. Consequently, there was a period of 1.5×10^9 years during which early living things evolved before the eukaryotes diverged from the prokaryotes. This was followed by 10^9 years of eukaryote evolution before plants, animals and fungi began to diverge from each other (presumably at some unicellular grade of organisation). Finally, the kingdoms have been separate for the past 10^9 years.

1.4 Evidence from fossils

Fossilised fungal spores in deposits up to 60 or so million years old can be found fairly readily. Many are highly distinctive and extremely similar to present-day species (e.g. see Kalgutkar and Sigler, 1995). Such fossils give a strong impression that even the most distinctive fungal morphologies are maintained for long periods of time. Pirozynski (1976) put it this way: "...evidence accumulates to support the long-held view that the history of fungi is not marked by change and extinctions but by conservatism and continuity."

There is plausible evidence for all major groups of extant fungi existing during the Palaeozoic era (Sherwoodpike, 1991). The fossil record indicates morphological conservatism and establishment quite early in evolution of a spectrum of intimate associations between fungi and vascular plants. Almost all terrestrial plant species of today form mutualistic associations with soil fungi, their roots being infected with mycorrhizal fungi which contribute to their mineral nutrition and can benefit the plants in a variety of other ways (Read, 1991). This mutualistic association would have eased, if not solved, many of the difficult problems encountered by the first land plants during the Silurian epoch, 430 million years ago (Letacon and Selosse, 1984). It is most probable that the initial exploitation of the terrestrial environment by plants depended on the establishment of mutualistic associations; between fungi and algae (mycophycosymbiosis and lichens) on the one hand, and between fungi and vascular plants (mycorrhizas) on the other. Marks (1991) suggested that mycorrhizas could have developed as soil microbes started to colonise rhizospheres and rhizoplanes. Some of the rhizoplane microbes probably developed a parasitic lifestyle, eventually giving rise to minor pathogens of roots, root necrotrophs and a very small and highly specialised group of root biotrophs. Several examples of mycoparasitism have been described in specimens from the Lower Devonian Rhynie chert (Hass *et al.*, 1994), and the same fossil material shows that vesicular-arbuscular mycorrhizas occurred early in evolutionary history. Remy *et al.* (1994) and Taylor *et al.* (1995) described arbuscles morphologically identical to those of living arbuscular mycorrhizas in plant specimens from these 400 million-year-old fossils. Such mycorrhizas are now widespread and widely distributed (Allen, 1996). Ectomycorrhizas seem to be another early type from which others developed as intercellular invasion gave way to intracellular infection. On this basis, the ericoid and orchid mycorrhizas could be regarded as the most advanced. Under normal conditions, these associations are still necessary to the majority of land plants. Several associations have probably arisen independently in different groups or families. Certainly, there is evidence from analysis of small subunit ribosomal DNA sequences for at least five independent origins of the lichen habit in different groups of ascomycetes and basidiomycetes (Gargas *et al.*, 1995).

Fossilised fungal remains have been identified in middle to late Cretaceous deposits from Hokkaido, Japan in association with ferns, gymnosperms and angiosperms (Nishida, 1991). Earlier fossils are less easy to identify as being definitely fungal in origin. Fragmented filaments in Scottish Silurian and Lower Devonian 'phytodebris' deposits have been interpreted as being of fungal origin (Wellman, 1995). These fragments were similar to material described from contemporaneous strata from elsewhere, indicating that the organisms from which they derived were geographically widespread.

Oldest of all the fossils so far found seem to be the Ediacaran fossils (from about 650 million years ago). These are impressions in quartz sandstones and have been thought to be fossils of worms or jellyfish. However, Retallack (1994) claims to have falsified this notion with an argument based on the presumed compaction resistance of the object which made the impression, and suggests that lichens are a better candidate for this type of preservation. He also points out that microscopic tubular structures and darkly pigmented cells in permineralised late Precambrian fossils from Namibia and China are also compatible with interpretation as lichens.

Amber preserves soft-bodied organisms and various microorganisms, including fungi. Fungal spores, have been found in amber 220–230 million years old. Many of these microorganisms can be assigned to present-day groups (Poinar *et al.*, 1993). Perhaps most remarkable of all such specimens reported so far, are the remains of two gilled mushrooms in 90 – 94 million year old amber, which bear a strong resemblance to the existing genera *Marasmius* and *Marasmiellus* (Hibbett *et al.*, 1995). The close similarity between the fossil specimen and existing genera again emphasises the trend in fungal evolution to conserve form and morphology over very long periods of time. So before the age of mammals, when dinosaurs still ruled the Earth, there existed mushrooms almost the same as those existing today.

Saprotrophic, parasitic and biotrophic interactions between fungi and plants are clearly evident in fossils which represent some of the earliest land plants. Lichens and, most especially, mycorrhizas are key among these and emphasise the importance of fungi in shaping the ancient ecosystem (Taylor and Osborn, 1996). The impression given by the ancient origin and wide geographic prevalence of mycorrhizas is that the invasion of the land by plants was only possible because of their association with fungi. Current knowledge of existing mycorrhizas indicates that bacteria may have been involved, too. There is evidence that the symbiotic establishment of mycorrhizal fungi on plant roots is affected in various ways by the other microorganisms of the rhizosphere. Most especially by some bacteria which consistently promote mycorrhizal development, leading to the concept of 'mycorrhization' helper bacteria (Garbaye, 1994a,b). Remarkably, Bianciotto *et al.* (1996) showed that the cytoplasm of the arbuscular mycorrhizal fungus *Gigaspora margarita* harbours a bacterial endosymbiont, so the mycorrhizal system in this case comprises plant, fungal and bacterial cells.

1.5 Origin of development

As well as being concerned with the phylogenetic origins of the organisms, it is appropriate to enquire about the origins of multicellular development within those organisms. The main eukaryotic divergences probably occurred about 10^9 years ago. Yet only three (animals, fungi and plants) of the 20 or so eukaryotic lineages have evolved complex multicellular development. Broadly speaking, this (at least in plants and animals because evidence for the fungi does not exist) emerged about 600 million years ago. Most of the cellular machinery that is necessary for development appears early in eukaryotic evolution, and is expressed by many lineages. This machinery includes a common system of gene regulation and structure (introns and exons, *trans*-acting regulatory elements whose proteins modify *cis*-acting sites to regulate activity of adjacent structural gene sequences), the ability of cells (and regions of cells) to differentiate to specific functions, the cytoskeletal architecture, signal transduction mechanisms and cell-to-cell signalling. If these components of complex development occurred so early and are shared so widely, why are animals, fungi and plants the only lineages to assemble the components into a morphogenetic phenomenon? What event or events enabled the emergence of complex development?

The answer for animals seems to be the acquisition of basement membranes and associated traits (gap junctions, cellular adhesion molecules, collagen, laminin and integrins) which permit the formation of epithelial cell sheets, and a sufficient variety of cell types to furnish the tissues of complex developmental process (Erwin, 1993). Erwin (1993) favours a correlated progression model (derived from Thompson, 1988) in which sequential innovations permit the developmental process to emerge. So, for Metazoa, the sequence of innovations might have been: collagen, basement membranes, gap junctions, synaptic connections, then production of additional cell types (and a corresponding increase in complexity of the regulatory circuits which control them). Erwin (1993) argues convincingly that plant development has many similarities with metazoan development and implies that the same model might be applicable. However, he seems to exclude the fungi in the sentence "...we have two lineages (metaphytes and metazoans) which exhibit a progressive increase in the number of cell classes and in both morphologic and taxonomic diversity." A significant purpose of this book is to enquire whether kingdom Fungi might be a third lineage to which these arguments might apply (see Chapter 7).

1.6 Evolution within kingdom Fungi

In the true fungi, phylogenetic relationships inferred from 18S ribosomal DNA sequence data agree with those derived from morphology. In particular they support the traditional fungal subdivisions Ascomycotina and Basidiomycotina. Consideration of morphology alone produces conflict when it is impossible to decide the difference between features which might be either simple (and therefore ancient) or reduced (and therefore recent). Sequence data help in resolution of some such conflicts. For example, phylogenetic trees from rDNA indicate that those morphologically simple ascomycetes classified as yeasts are polyphyletic (which is probably also true of basidiomycetous yeasts (Prillinger *et al.*, 1991)), and that forcible spore discharge was lost convergently from three lineages of ascomycetes producing flask-like fruiting bodies (Berbee and Taylor, 1992). Berbee and Taylor (1993) constructed a relative timescale for the origin and radiation of major lineages of the true fungi, using 18S ribosomal RNA gene sequence data calibrated with fossil evidence (see also Berbee and Taylor, 1995). These analyses indicated that the terrestrial fungi diverged from the chytrids approximately 550 million years ago. After plants invaded the land approximately 400 million years ago, ascomycetes split from basidiomycetes. Mushrooms, many ascomycetous yeasts, and common moulds in the existing genera *Penicillium* and *Aspergillus* may have evolved after the origin of angiosperm plants and in the last 200 million years. Swann and Taylor (1993) evaluated sequences of the 18S rRNA gene from nine basidiomycetes and obtained a phylogenetic tree in which the basidiomycetes fell into three major lineages; the Ustilaginales (smuts), simple septate basidiomycetes, and hymenomycetes. Later, Swann and Taylor (1995) used the 18S ribosomal RNA sequences to establish a new class level taxonomy for basidiomycetes consisting of the Urediniomycetes, Ustilaginomycetes, and Hymenomycetes. Other characters such as cellular carbohydrate composition, major ubiquinone system, 5S rRNA sequence, basidium morphology, and septum and spindle pole body ultrastructure are also incorporated into the scheme. Combining analysis of light microscopic and ultrastructural characters with the nucleotide sequence from the 5′ end of the nuclear large subunit ribosomal RNA gene, McLaughlin *et al.* (1995) found that both morphological and molecular characters supported similar phylogenetic conclusions. The Uredinales were shown to be advanced, arising from the simple-septate Auriculariales; some characters that they share with the ascomycetes resulting from convergent evolution. The simple-septate Auriculariales consisted of more than one clade, and the related gasteroid *Pachnocybe ferruginea* exhibited numerous derived light microscopic characters, including holobasidia (the cell in which meiosis occurs is not septate).

DNA sequences of nuclear small subunit rRNA genes have been used to determine phylogenetic relationships among basidiomycetes, ascomycetes and chytridiomycetes (the three major classes of organisms most widely considered to be true fungi). A significant point at issue was the relationship between the two classes of non-flagellated fungi (ascomycetes and basidiomycetes) and the flagella-bearing chytridiomycetes. The results indicate that basidiomycetes and ascomycetes are the most closely related, but chytridiomycetes did group with these higher fungi rather than with the protists. *Neocallimastix*, a eukaryote lacking mitochondria which is an economically important anaerobic rumen microorganism, showed closest molecular affinities with the chytridiomycete fungi of the order Spizellomycetales (Bowman *et al.*, 1992).

Using the primary structure sequence of the small rRNA subunit, Wilmotte *et al.* (1993) produced a phylogenetic tree which confirmed the early divergence of the zygomycetes and the classical division of the higher fungi into basidiomycetes and ascomycetes. The tree divided basidiomycetes into true basidiomycetes and ustomycetes. Within the ascomycetes, the major subdivisions of hemiascomycetes and euascomycetes were recognised by the tree, but *Schizosaccharomyces pombe* did not cluster with the hemiascomycetes, to which it is assigned in classical taxonomic schemes, but formed a distinct lineage. Among the euascomycetes, the plectomycetes and the pyrenomycetes were distinguished. Within hemiascomycetes, the polyphyletic nature of genera like *Pichia* and *Candida* and of families like the Dipodascaceae and the Saccharomycetaceae were evident. More detailed phylogenetic analyses with sequences of the small subunit ribosomal DNA (Spatafora, 1995) detected four major groups of filamentous ascomycetes: group 1, pyrenomycetes (Hypocreales, Microascales, Diaporthales, Sordariales) and loculoascomycetes (Pleosporales); group 2, operculate discomycetes (Pezizales); group 3, inoperculate discomycetes (Geoglossaceae); and group 4, plectomycetes (Eurotiales, Onygenales) and loculoascomycetes (Chaetothyriales). Well-supported clades, which correspond to groupings based on fruit body morphology, were resolved.

Other significant observations in the literature include studies of evolutionary relationships of *Lentinus* to the Tricholomataceae and Polyporaceae using restriction analysis of nuclear-encoded ribosomal RNA genes (rDNA) which suggests that species placed in the genus *Lentinus* today evolved to this morphology from different routes and, therefore, that spore-bearing gills (or lamellae) have arisen repeatedly by convergent evolution (Hibbett and Vilgalys, 1991, 1993). It should also be emphasised that the phrase ‘true fungi’ has been used above. This implies that there are some organisms which are not considered ‘true’ fungi. There are three groups of organisms which are traditionally studied by mycologists which are considered to be so phylogenetically remote from kingdom Fungi as to be placed in a different kingdom. These are the Hyphochytriomycota, Labyrinthulomycota and the Oomycota which are all placed in the kingdom Chromista (see discussions in Hawksworth *et al.*, 1995). The Oomycota includes many economically important parasites and the water moulds *Achlya* and *Saprolegnia*. These two genera have been the subjects of some of the most elegant cell biological experiments over many years. Some of the results obtained will be discussed in this book with a view to incorporating them into our understanding of fungi. It is essential, though, to remember two things. First, that there is an enormous phylogenetic gulf between these water moulds and true fungi. Second, that the Oomycota are in no sense ancestral to true fungi; they represent a *different,* not an earlier, level of organisation.

1.7 Horizontal transfer of genetic information

What has been said and implied in the above discussion about evolutionary relationships assumes a gradual change in DNA coding sequences from generation to generation. This concept is of transmission of genetic information in a 'vertical' direction, from one generation to the next. There is evidence, however, for horizontal transmission of genetic information. This is transmission of genetic information from one organism to another within the same generation.

One of the first possible examples concerned the biosynthetic pathways of β-lactam compounds (penicillins, cephalosporins and cephamycins). These are found in a wide range of microorganisms, including fungi, actinomycetes and Gram-negative bacteria. As molecular information became available, comparisons of gene sequences (particularly the genes encoding isopenicillin N-synthetase) and gene organisation in these different microorganisms revealed them to be so similar that the idea that they evolved independently seemed less likely than the proposition that in some way they might have been transferred from bacterial β-lactam producers to filamentous fungi about 850 to 950 million years ago (Turner, 1992; Buades and Moya, 1996). Other examples include: evidence that an intron in an angiosperm mitochondrial gene arose recently by horizontal transfer from a fungal donor (Vaughn *et al.*, 1995); horizontal transfer of a mitochondrial plasmid from the discomycete *Ascobolus immersus* to the pyrenomycete *Podospora anserina* (Kempken, 1995); and the distribution of the transposable element *mariner* which has been found in many species of Drosophilidae (dipteran flies), several other groups of Arthropods, and in Platyhelminthes and a phytopathogenic fungus (Capy *et al.*, 1994). Metzenberg (1990) points out that the hallucinogen bufotenin is produced by the toad *Bufo* and in a species of the mushroom *Amanita*. All of these observations raise the question of whether the similar or identical sequences which exist in very different organisms are relics of some ancient origin, recent acquisitions by horizontal transfer or instances of parallel evolution. Where phylogenetic studies confirm a relationship based upon horizontal transfer, a powerful mechanism is revealed. It is a mechanism which enables a cell to recruit to its own use new gene sequences defining metabolic and regulatory functions. Importantly, the sequences may have evolved in response to selection pressures to which the recipient may never have been exposed. One imagines, however, that the less spectacular evolution of transcriptional regulatory proteins by the recruitment of functional domains provided by metabolic enzymes remains the more likely way by which a cell might acquire new and interesting regulators (Hawkins *et al.*, 1994).

1.8 Comparing and combining

Discussions of fungal developmental biology must combine together results of work done on different fungi and even draw in those derived from work with plants and animals. It is right to ask how much reliance can be placed on such comparisons and combinations.

Of particular interest is the analysis of Taylor *et al.* (1993) of the phylogenetics of those model organisms (*Saccharomyces*, *Aspergillus*, and *Neurospora*) which are widely used because they are so convenient to maintain under laboratory conditions. These organisms fit well within the robust phylogeny of ascomycetes that wider phylogenetic analysis has established. The authors predict that truly comparative molecular biological studies of fungal development will be available shortly because the prerequisites have been completed. These include a well-supported phylogeny, sophisticated molecular techniques, including transformation by complementation and gene disruption, and morphological developmental pathways that are simpler than those of plants or animals.

This analysis of Taylor *et al.* (1993) allows for some confidence when combining observations made with the frequently used model organisms. However, because of the lack of specifically-fungal research in many areas it is often necessary to make use of work done with animals and plants to try to complete the fungal picture, so it is worth asking what sort of background of similarity there may be to provide some comfort for such an approach. The basic cytoplasmic and cytoskeletal machinery is thought to be of ancient origin (Perasso and Barointourancheau, 1992) and common to the major kingdoms; as is the overall genetic architecture (Jimenez-Garcia *et al.*, 1989).

The Golgi body of eukaryotic cells consists of one or more stacks of flattened saccules (cisternae) and an array of fenestrae and tubules continuous with the peripheral edges of the saccules. All Golgi bodies are involved in the production and translocation of proteins and membranes within the cell or to the cell exterior. Rudimentary Golgi bodies, consisting of tubular vesicular networks, have been identified in fungi and may represent an early stage of Golgi body evolution (Mollenhauer and Morre, 1994). It seems paradoxical that a group of organisms so dependent for everyday activities on protein export should have a 'rudimentary Golgi body', but perhaps this is yet another example of the extreme conservatism of fungal evolution – if it works, don't fix it.

Controlled organelle movement is an important feature of many aspects of development. Kinesins comprise a conserved family of molecular motors for organelle transport that have been identified in various animal species. An equivalent cytoplasmic motor in *Neurospora crassa* has been cloned and characterised (Steinberg and Schliwa, 1995). It proved to be a distant relative of the family of conventional kinesins, and is thought to have diverged early in the evolution of this family of motors.

Another important aspect of fungal developmental biology is the potential involvement of hormones and growth factors in the signalling which controls hyphal growth and tissue formation. At the time of writing, no fungal growth hormones are known to science and the evidence for their occurrence at all is fragmentary and unconvincing (Novak Frazer, 1996; and see Chapter 4). In contrast, animal cytokines, hormones and other growth regulators on the one hand, and plant auxins, cytokinins, gibberellins and similar growth regulators on the other hand are so well understood that they are the basis of numerous commercial products, many of which we use in everyday life.

Interestingly, a number of fungi synthesise steroids that are hormonally active in animals (Lenard, 1992; Agarwal, 1993); even the cellular effect of such steroids in their native organism appears to be comparable to that in mammals and is mediated via receptors organised in a manner similar to that seen in animals. Agarwal (1993) suggests that the ancestry of ligand-induced transactivation via zinc finger proteins might be more ancient than the early Cambrian burst of metazoan evolution 500 million years ago. A plausible explanation might lie in the observation that mammalian integral membrane receptors are homologous to metabolite transporters specific for amino acids, polyamines, and choline, which catalyse solute uniport, solute/cation symport, or solute/solute antiport in yeast, fungi, and eubacteria (Reizer *et al.*, 1993). These observations imply that cell surface receptor proteins in mammals are transport proteins that share a common origin with transport proteins of single-celled organisms. These ancient homologies make the absence of readily identifiable fungal hormones extremely difficult to understand. This situation becomes even more paradoxical when one considers the ways in which pathogenic fungi of today manipulate their hosts by synthesising and deploying host hormones. Production of gibberellin by *Gibberella fujikuroi* is a classic example among plant pathogens (Cerdá-Olmedo *et al.*, 1994), but the entomopathogenic fungus *Entomophthora muscae* which infects and kills domestic flies (*Musca domestica*) produces pheromones which attract other flies, especially males, towards dead, infected flies (Moller, 1993): come hither and be infected!

2

Hyphal growth

Despite the fungal mode of nutrition figuring prominently in originally defining kingdom Fungi (Chapter 1), the fundamental aspect of cell biology which sets the fungi off from other major kingdoms is the apical growth of hyphae. Extension growth of the hypha is limited to the apex. Growth at the tip is what makes new fungal hypha, but there is more to development than growth at hyphal tips can achieve alone. The vegetative fungal mycelium is an exploratory, invasive organism. Its component hyphae are regulated to grow outwards into new territory and consequently possess controls which ensure that hyphae normally grow away from one another to form the typical 'colony' with an outwardly-migrating growing front (see Carlile, 1995). Tissue development requires that different hyphae cooperate in an organised way. For tissue to be formed the invasive outward growth pattern of the vegetative mycelium must be modified so that independent hyphal apices grow towards each other, allowing their hyphae to branch and differentiate in a cooperative fashion.

Since the developing structures (spore-forming fruiting bodies, for instance) are actually produced by the vegetative mycelium, these changes in growth pattern must be localised, and must be a response to regulatory processes which are imposed upon the vegetative mycelium. Another aspect of localisation is that tissue formation demands that the continuous tube of hypha produced by the growing apex is divided up into cells or compartments by the formation of cross-walls (usually called septa). This enables differentiation to be localised, offering the possibility that adjacent compartments might follow different pathways of differentiation, and even be of different size.

2.1 Fungal cells

The above paragraph includes the phrase 'cells or compartments' because mycologists over the years have been very sensitive to the question of whether fungi have cells, and how fungal cells and their interactions compare with those of plant and animal cells. Most lower fungi (e.g. *Mucor*) have coenocytic hyphae; but they do not form multicellular structures. Hyphae of fungi which do exhibit complex developmental pathways form septa at regular intervals, but the septa usually have a pore (more or less central) which may be elaborated with the parenthesome apparatus in basidiomycetes or be associated with Woronin bodies in ascomycetes (Markham, 1995). The pore is what worries people about the definition of fungal cells, because the implication carried with the word 'pore' is that all of the cytoplasm of a hypha is in continuity even though it might be subdivided by the septa into compartments.

It must be admitted that movement of cytoplasm and organelles through septa has often been described and is frequently easy to demonstrate. But it is also clearly the case that the movement or migration of cytoplasmic components between adjacent cells is under very effective control. There are instances in which nuclei move freely, but mitochondria do not, and others in which rapid migration of vacuoles is not accompanied by migration of any other organelle. Some biochemical experiments have even demonstrated that different sugars can be translocated in opposite directions in a hypha at the same time. There are also numerous examples available where grossly different pathways of differentiation have been followed on the two sides of what appear (to the electron microscope) to be open septal pores. Clearly, whatever the appearance, the hypha can be separated into compartments whose interactions are carefully regulated and which can exhibit contrasting patterns of differentiation. There may still be a semantic argument for preferring 'compartment' to 'cell', but from this point on I will take the pragmatic view that if it looks like a cell and if it behaves like a cell, then I will call it a cell. But please don't forget that every fungal cell *is* just a segment of a tubular hypha!

Griffin *et al.* (1974), pointed out that in mycelial fungi, branch formation (by increasing the number of growing points) is the equivalent of cell division in animals, plants and protists; and the kinetic analyses of Trinci (1974; Fiddy and Trinci, 1976) show clearly that fungal filamentous growth can be interpreted on the basis of a regular cell cycle. These and subsequent observations will be discussed in detail later in this chapter (section 2.5). For the moment, these observations contribute to the conclusion that fungi can quite reasonably be considered to be cellular organisms producing differentiated tissues composed of cells which are the progeny of an initial cell or cell population which is induced to start multiplication and differentiation.

Ross (1979) indicates that almost without exception, multicellular organisms are constructed from a basic unit which is the uninucleate cell, and that, fern and moss gametophytes apart, those cells are diploid. Among the fungi, multicellular structures are formed only by the ascomycetes and basidiomycetes and the key evolutionary step allowing this seems to be formation of the dikaryon – a hyphal compartment containing two complementary haploid nuclei. In ascomycetes, ascogenous hyphae are dikaryotic and organised multicellular structures (the ascomata) often form around the asci. In the classic description of basidiomycete organisation the entire secondary mycelium is dikaryotic, and the fruit body is composed of dikaryotic cells. Ross concludes that the dikaryon is a truly cellular system and that attainment of the ability to produce this grade of cell organisation either allows, or is at least coincident with the step-over into the ability to form multicellular structures.

I subscribe to this argument to an extent, but dikaryosis cannot have a causal connection with the ability to assemble multicellular structures because there are too many fundamental exceptions. The hyphae which make up the ascomata are not themselves dikaryotic; they arise from the maternal mycelium and are heterokaryotic. Also, the classic description of basidiomycete organisation, though it appears in most text books, does not describe all, and may not even describe the majority, of basidiomycete developmental strategies. Nevertheless, evolution of some sort of stable nuclear-cytoplasmic relationship from the potentially chaotic heterokaryotic condition is a likely candidate as another prerequisite in the evolution of fungal complex development.

In my view the crucial evolutionary step (in all eukaryote lineages) which permits organised multicellularity is the development of mechanisms for dividing a cell, together with a mechanism for controlling the placement of the plane of cell division in particular relation to the orientation of nuclear division. Common to normal morphogenesis in animals and plants alike is the concept of cellular polarity and the developmental consequences of precise positioning of the plane of cleavage (in animals) or wall formation at the cell plate or phragmoplast (in plants; Samuels *et al.*, 1995). The classic examples of embryology in both groups of organisms include instances of asymmetric divisions partitioning 'stem' cells in ways which result in the daughter cells expressing some sort of differentiation (not necessarily immediately expressed) relative to one another.

The apical growth characteristic of the fungal hypha is *the* prime attribute of fungi and is, of course, an extreme cellular polarity (Gow, 1995a). An instructive comparison which should bring out this point about the orientation of cell division is the way the filamentous mode of growth, effectively generating a one-dimensional (linear) structure, is converted to a two- or three-dimensional structure in different eukaryote kingdoms. A good example from the plant kingdom is the development of the protonema following germination from spores of bryophytes and pteridophytes. Here, a filament of cells in a single file is formed as a result of the constraint of new cell walls to an orientation perpendicular to the long axis of the protonema. This growth pattern persists for a time. Eventually, and usually in the apical cell of the protonema, the division plane becomes reoriented so that the new cell walls are formed obliquely or parallel to the long axis of the protonema filament. As a result, a flat plate of cells (the gametophyte prothallus) is formed. This transition, analysed in detail by Miller (1980), epitomises the importance in plant morphogenesis of the mitotic orientation. This strategy is totally alien to the fungal approach to solution of the same problem.

Cross-walls in fungal hyphae are rarely formed other than at right angles to the long axis of the hypha. Except in cases of injury or in hyphal tips already differentiated to form sporing structures, hyphal tip cells are not subdivided by oblique cross-walls, nor by longitudinally oriented ones. Even in fission yeast cells which are forced to produce irregular septation patterns under experimental manipulation, the plane of the septum was always perpendicular to the plane including the longest axis of the cell (Miyata *et al.*, 1986a). In general, then, the characteristic fungal response to the need to convert the one-dimensional hypha into a two-dimensional plate or three-dimensional block cannot depend on a different geometrical arrangement of the septum. The only solution open to the fungal hypha is the formation of branches. Casual observation shows that branching patterns alter greatly.

For example in the gill tissue of the *Coprinus cinereus* mushroom, successive branches may emerge at intervals of less than 5μm, whereas the same dikaryon, growing as a vegetative mycelium on the surface of agar medium, forms branches at average intervals of 73 μm (Horner and Moore, 1987). The kinetics of branch formation in fungal tissues of different degrees of complexity would be a promising line of enquiry (see Chapter 4), but at this point I wish to emphasise that whilst plant morphogenesis depends on placement of the cross-wall, in fungal hyphae the general inability to vary the orientation of the septum means that the position of the new branch tip must be specified. The new apex must extend sufficiently to form an apical compartment before the parental cytoplasm can be subdivided.

The septum in the branch will still be formed at right angles to the long axis of the branch, but its orientation relative to the parent hypha will depend entirely on the positioning of the branch apex (which, of course, is established some time prior to septum formation). Consequently, there are two fundamental processes involved in construction of fungal multicellular structures; the first is the origin of the branch (its appropriate placement and orientation on the parent hypha) and the second is the direction of growth of the new hyphal apex which is created by the branching event.

The former process seems to be the formal equivalent of determination of morphogenetic growth by orienting the plane of division and the new cross-wall as is seen in plants, and the latter has much in common with the morphogenetic cell migrations that contribute to development of body form and structure in animals. Viewed in this light, therefore, the fungal kingdom is seen as employing morphogenetic processes which have affinities with both of the other major kingdoms; presumably using in combination ancestral capabilities which have been adapted differently, each to the exclusion of the other, in animal and plant kingdoms.

2.2 Hyphal tip extension

Extension of the cell wall at the hyphal apex is the most striking characteristic of the fungi and possible mechanisms to account for it have been discussed since the end of the nineteenth century. As knowledge of the chemical and ultrastructural nature of the fungal cell wall has accumulated the discussions have become more detailed and, hopefully, closer to the truth. But throughout this time the concern has been essentially the same: how can wall extension be achieved without jeopardising the integrity of the existing hypha? It is quite clear that new wall material must be delivered to the apex and inserted into pre-existing wall (note that throughout this discussion the phrase 'new wall material' should be interpreted to include *everything* that is necessary to extend the tip: membrane, periplasmic materials and all wall layers). The models which have been suggested over the years to account for this process differ in how they account for this being achieved, but underlying them all is the recognition that the act of inserting new wall material could itself weaken the wall. The potential validity of the models must therefore be judged not only on how they provide for wall synthesis, but on how they safeguard the integrity of the hypha whilst wall synthesis is in progress.

2.2.1 Structure of the hyphal wall

I do not wish to describe in detail the chemical structure of hyphal walls throughout the fungi. In particular, the structure of phycomycete walls is not considered relevant at the moment as these simple fungi do not undertake complex developmental pathways. However, a brief, and very generalised overview of ascomycete and basidiomycete wall structure is pertinent to provide some context for discussion of wall synthesis (see Sentandreu *et al.*, 1994; Gooday, 1995a,b,c).

The wall is a sophisticated cell organelle. It defines the volumetric shape of the cell, provides osmotic and physical protection and, together with the plasma membrane and periplasmic space, influences and regulates the influx of materials into the cell. However, it is also able to control the environment in the immediate external vicinity of the cell membrane, and it represents the interface between the organism and the outside world. This is an active interface, since the interaction of the organism and the outside world (and the latter will include other cells) is subject to modulation and modification. The fungal cell wall is metabolically active, interactions between its components occur to give rise to the mature cell wall structure. So the wall must be understood to be a dynamic structure which is subject to modification at various times to suit various functions. Besides enclosing and supporting the cytoplasm, those functions include selective permeability, as a support for immobilised enzymes and cell-cell recognition and adhesion.

The wall is a multilayered complex of polysaccharides, glycoproteins and proteins. The polysaccharides are glucans and mannans and include some very complex polysaccharides (like gluco-galacto-mannans). In hyphae the major component of the wall, and certainly the most important to structural integrity, is chitin though this is frequently crosslinked to other wall constituents, particularly a $\beta(1\longrightarrow3)$-glucan (Alfonso *et al.*, 1995), the terminal reducing residue of a chitin chain being attached to the non-reducing end of a $\beta(1\longrightarrow3)$-glucan chain by a $(1\longrightarrow4)$ linkage (Kollar *et al.*, 1995). Synthesis of $\beta(1\longrightarrow3)$-glucan is achieved by a membrane-bound, GTP-stimulated $\beta(1\longrightarrow3)$-glucan synthase (Mol *et al.*, 1994; Polizeli *et al.*, 1995).

Chitin is the most important structural component of fungal walls. It is a $\beta(1\longrightarrow4)$-linked homopolymer of *N*-acetylglucosamine (Fig. 2.1). Chitin is synthesised by the enzyme chitin synthase which adds two molecules of uridine diphospho-*N*-acetylglucosamine (UDPGlcNAc) to the existing chitin chain in the reaction:

$$(\mathrm{GlcNAc})_n + 2\mathrm{UDPGlcNAc} \longrightarrow (\mathrm{GlcNAc})_{n+2} + 2\mathrm{UDP}$$

Most fungi have several chitin synthase genes, though the purpose of the redundancy is uncertain. Chitin synthases require phospholipids for activity and are normally integral membrane-bound enzymes. The enzyme protein spans the membrane, accepting substrate monomers from the cytoplasm on the inner face of the membrane and extruding the lengthening chain of chitin through the outer membrane face and into the wall (Gooday, 1995b). The essential chitin subunit is a microfibril made up of a number of individual chitin molecules which are probably held together by hydrogen bonding. Gooday (1995b) suggested that the microfibrils might be assembled by the joint action of several chitin synthases grouped in enzyme complexes which might contain other enzymic and/or regulatory polypeptides. Although during normal apical growth of the hypha the incorporation of newly synthesised chitin is limited to the hyphal apex, there is evidence that inactivated chitin synthases are widely distributed in the plasmalemma. One piece of evidence for this is that inhibition of protein synthesis with cycloheximide in *Aspergillus nidulans* resulted in chitin synthesis occurring uniformly over the hypha (Katz and Rosenberger, 1971; Sternlicht *et al.*, 1973; and see Gooday, 1995b for other examples). This observation implies that protein synthesis is required to maintain chitin synthases which are already in place in the membrane in an inactive state. Thus, inactivated chitin synthase activity appears to be an intrinsic property of the plasmalemma, the enzyme being activated somehow specifically at the hyphal apex and at sites where branch formation is initiated.

Chitin is important at particular sites in yeast walls, although the major structural component in these organisms is a fibrillar inner layer

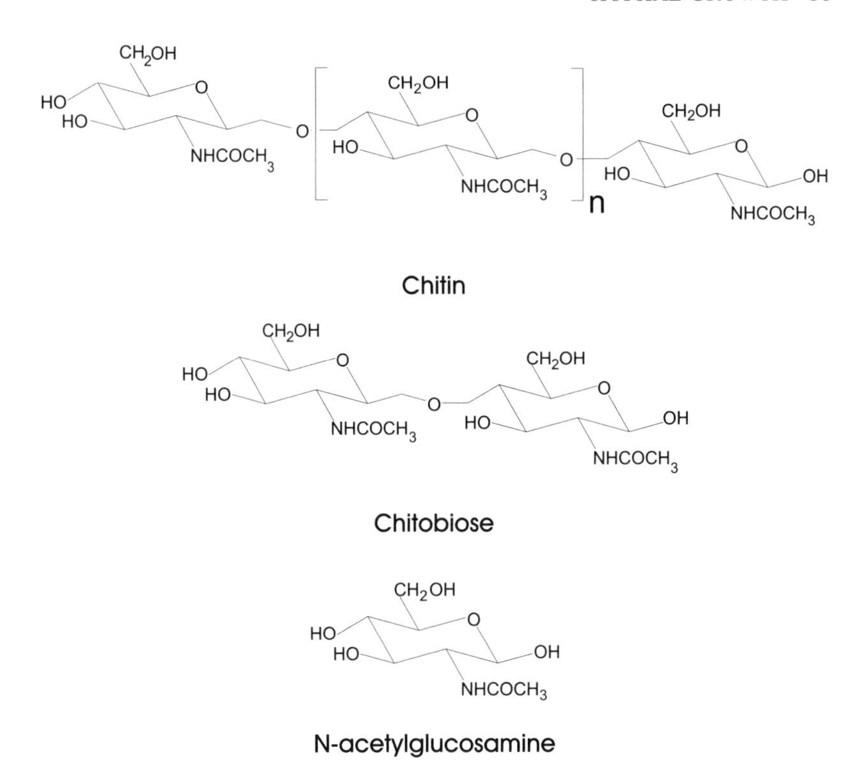

Fig. 2.1 Structural formulae of chitin and its constituents.

of β-glucan. The glucan has secreted mannoproteins attached to it, and this introduces the idea that the protein components of fungal walls are of considerable importance, too. Indeed, recent work has revealed a class of glycosyl phosphatidylinositol (GPI)-anchored, serine/threonine-rich wall proteins which are thought to be anchored in the plasma membrane, spanning the wall with extended glycosylated structures which protrude out into the surrounding medium (Stratford, 1994). Mannoproteins play an essential role in cell wall organisation, and there is evidence for the formation of covalent bonds between these molecules and the structural polymers (glucans and chitin) outside the plasma membrane (Gozalbo *et al.*, 1993). Mannoproteins were secreted from the beginning of the regeneration of walls by protoplasts of *Candida albicans* and were incorporated into the regenerating walls by covalent bonding (Elorza *et al.*, 1994). Some of the proteins identified in walls have enzymic activities associated with them. These include α-glucosidases (Reiser and Gasperik, 1995) and β-glucosidases, the latter being bound to the wall by a heteroglycan composed of mannose, galactose, glucose, and glucuronic acid (Rath *et al.*, 1995). Angiolella *et al.* (1996) showed that an enolase protein of *Candida albicans* was associated with the glucans of the inner layers of the wall, for example, and Campo-Aasen and Albornoz (1994) found an alkaline phosphatase in the cell wall of the yeast phase of *Paracoccidiodes brasiliensis*. Other obvious components are proteins involved in cell-to-cell recognition like the products of mating type factors.

Agglutination (cell-to-cell adhesion) between cells of different mating type in *Saccharomyces cerevisiae* depends on interaction of two cell surface glycoproteins, the gene products of components of the mating type factors (see section 5.2.2.1) which are inserted into the outer coat of the cell wall (Cappellaro *et al.*, 1994). The cell wall form of α-agglutinin is covalently linked to β(1⟶6)-glucan, this extracellular crosslinkage being dependent on prior addition of a GPI anchor to α-agglutinin, though the fatty acid and inositol components of the anchor are lost before the reaction with the glucan (Lu *et al.*, 1995).

The outer surface of walls is usually found to be composed of a layer of rodlets (e.g. Bobichon *et al.*, 1994; Gerin *et al.*, 1994). These are presumably composed of proteins which modify the biophysical properties of the wall surface appropriately. Rodlets do occur on wettable surfaces, but on hydrophobic surfaces (like spores and aerial mycelium) hydrophobin (see section 5.4.4) occurs in the outermost rodlet layer (Hazen and Glee, 1994). Bidochka *et al.* (1995b) found another major species of cell wall protein (cwp1) in cell walls of *Beauveria bassiana* aerial or submerged conidia which occurs primarily next to the rodlet layer.

The above is a highly generalised and abbreviated, even idiosyncratic, description of wall structure. The justification for an absence of much specific detail is that fungal walls are remarkably variable in their detailed aspects of structure. For example, Calonje *et al.* (1995) were able to detect significant differences in the gross chemical composition of the walls of four commercial strains of the cultivated mushroom *Agaricus bisporus*. Differences detected included overall composition of the wall and in the polysaccharide structures. In the face of that sort of varietal difference, attempting to describe the structure of the wall of a particular species would be a pointless exercise.

The justification for such a generalised and abbreviated description is that the intention here is to present a concept of the fungal wall, rather than a specifically detailed description, because our next task is to conceptualise how that wall might be assembled during extension growth of the hypha. In brief, then, the concept is this:

- the main structural substance of the wall is provided by polysaccharides, mostly glucan, and in filamentous forms the shape-determining component is chitin;
- the various polysaccharide components are linked together by hydrogen bonding and by covalent bonds;
- a variety of proteins/glycoproteins contribute to wall function, some of these are structural, some are enzymic, and some vary the biological and biophysical characteristics of the outer surface of the wall;
- proteins may be anchored in the plasma membrane, covalently bonded to wall polysaccharides or more loosely associated with the wall;
- the wall is a dynamic structure which is modified (a) as it matures, and/or (b) as part of hyphal differentiation, and/or (c) on a short-term basis to react to changes in physical and physiological conditions.

Finally, by definition, the wall is extracellular, its entire structure lies outside the plasma membrane, so all additions to its structure must be externalised through the membrane before the wall can be restructured and some of the chemical reactions which link wall components together are extracellular reactions.

2.2.2 Growth at the tip: the problem

A fungal hypha is part of a closed hydraulic system which is under pressure. The osmotic influx of water, due to the difference in water activity between the inside and the outside of the semipermeable plasma membrane, attempts to increase the cytoplasmic volume but is counteracted by the wall pressure due to the mechanical strength of the wall outside the plasma membrane. The difference between these two forces is the turgor pressure which is the resultant 'inflation pressure' which keeps the hypha inflated. An interesting thing about pressure is that in a closed vessel the pressure is the same over the whole of the inside wall surface. This applies whatever the shape of the vessel; irrespective of shape the wall pressure is uniform. The only way that pressure can be varied locally is to enclose part of the pre-existing vessel wall. But then, of course, there would be two vessels, to each of which the previously-stated rule of uniformity would apply, but which could now have different internal pressures. The pressurised vessel will remain intact for as long as the mechanical strength of its wall is sufficient to resist the internal pressure.

In the context of the hypha, providing the mechanical force which the wall can exert is equal to or greater than the force exerted by turgor, it will remain intact. However, if turgor exceeds the breaking strain of the wall *at any point* then the wall will rupture (possibly explosively), and the cell will die. The problem we face in understanding apical hyphal growth is that the structure of the wall needs to be weakened to allow insertion of new wall material to the continuously elongating tip. How can that be achieved without exploding the tip?

Most of the foregoing two paragraphs is theoretical, it may be plumbing theory, but it is theory, nevertheless. It can be argued that with the expenditure of sufficient energy the hypha could 'change the rules' and avoid these physical consequences of its plumbing. Obviously, that *does* happen. The question is how? There is no doubt that hyphal tips can be caused to burst by experimental treatments. Classic experiments of this sort were carried out by Robertson (1965a,b, 1968) who flooded hyphal tips growing on agar with distilled water and found that some tips burst, others ceased extension growth, but many eventually resumed growth after a period of adaptation. Application of sugar analogues which are thought to interfere with synthesis of wall polysaccharides also causes tip bursting in filamentous hyphae (Moore, 1969a, 1981b) and yeasts (Megnet, 1965). Thus, it is generally thought that, in the true fungi, turgor is controlled to contribute to tip extension by driving the tip forward and shaping it by plastic deformation of the newly-synthesised wall (Ray *et al.*, 1972; Wessels, 1993). I emphasise 'in the true fungi' here because the oomycetes *Achlya bisexualis* and *Saprolegnia ferax* can grow in the absence of turgor (Kaminskyj *et al.*, 1992; Money and Harold, 1992, 1993).

2.2.3 Growth at the tip: towards a solution

We saw in Chapter 1 that the oomycete lineage probably diverged early from that which gave rise to kingdom Fungi. Their cell biology today probably reflects their adaptations to their specific lifestyles. They are *not* representative of true fungi but their behaviour patterns *may* reflect mechanisms which the true fungi could have further adapted to their own specific circumstances. Whatever the truth of the matter, consideration of the behaviour of these fungi is instructive in indicating what can be achieved without implying that it is a generally applicable pattern.

Achlya bisexualis and *Saprolegnia ferax* do not regulate turgor, their response to the addition of nutrients which raise the osmotic pressure of the medium is to produce a more plastic wall and continue to grow. Near-normal hyphae are formed by *S. ferax* in the absence of a measurable turgor pressure (Money, 1994; Harold *et al.*, 1995). Harold *et al.* (1995) claimed that hyphal extension in oomycetes is similar to pseudopod extension in animal cells, in that polymerisation of the actin cytoskeletal network at the apex of the hypha plays an indispensable role in the absence of turgor, actin polymerization becoming the main driving force for extension. When turgor pressure is high, though, the consequent hydrostatic pressure probably stresses the wall to expand and admit new wall material. The latter is likely to be the normal circumstance for a water mould living in a dilute aqueous medium. In this case the actin network is thought to mechanically reinforce the hyphal apex (Jackson and Heath, 1990; Heath, 1995b).

Kaminskyj and Heath (1995) have contributed to this idea of the actin cytoskeleton being involved in tip growth of the oomycete hypha by demonstrating that *S. ferax* possesses a polypeptide homologous to the highly conserved cytoplasmic domain of the cytoplasmic anchor integrin. The polypeptide was involved in cytoplasm–cell wall attachment, in patches enriched in actin and integrin. These attachment patches were more abundant towards the tip and were reduced in number when the growth rate was reduced. *Saprolegnia* also has a spectrin homologue with a similar distribution to that of integrin, though it did not participate in cytoplasm-wall anchoring. Spectrins are actin-binding proteins which are thought to control the stability and flexibility of membranes (Hitt and Luna, 1994), so the *Saprolegnia* spectrin homologue is presumed to contribute to the strength of the plasma membrane in the vicinity of the integrin anchor.

Heath (1994, 1995a,b) has blended these observations into a model which envisages that oomycete hyphal tip expansion is regulated (restrained under normal turgor pressure and protruded under low turgor) by a peripheral network of F-actin-rich components of the cytoskeleton that is attached to the plasmalemma and the cell wall by integrin-containing linkages. A virtue of the latter point is that it places control in the cytoplasm where it is accessible to normal intracellular regulatory processes. Heath (1995a) places great emphasis on the F-actin system, envisaging that it also functions in cytoplasmic and organelle motility, control of plasmalemma-located stretch-activated Ca^{2+}-transporting ion channels, vesicle transport and exocytosis. He suggests that Ca^{2+} may be involved in regulating the system, a feedback regulatory mechanism possibly being provided by a gradient of the stretch-activated channels with the highest concentration being at the hyphal tip.

This model is not obviously applicable to the true fungi if it is interpreted as postulating involvement of the actin cytoskeleton in order to drive tip extrusion. This is unnecessary because the true fungi can use the hydrostatic pressure of their regulated turgor to 'inflate' a plastic hyphal tip. However, this is not the only virtue of Heath's model, because it enables us to see a possible mechanism to solve the potential problem that the extending hyphal tip might not be strong enough to resist turgor during the process of wall synthesis.

There are data which indicate that the cytoskeleton, in this case both microfilaments and microtubules, is involved in the control of polarity in the hyphal growth phase of *Candida albicans*, the two cytoskeletal components cooperating in the establishment of polarity which leads to hyphal outgrowth from the yeast phase (Akashi *et al.*, 1994). Further, actin plaques have been observed in association with septum formation during yeast cell division (Robinow and Hyams, 1989; Mulholland *et al.*, 1994). Thus an actin cytoskeleton with firm attachments through the membrane to the growing wall appears to be involved at both the tip and the septal regions of wall synthesis. In the case of septum formation, mechanical resistance to turgor does not arise, so an obvious interpretation of the actin network is that it functions in transport, directing the thousands of vesicles containing wall precursors to the growing sites for terminal exocytosis (Johnson *et al.*, 1996). This seems to be supported by observations that temperature-sensitive mutants of *Saccharomyces cerevisiae* which have a dysfunction of the actin cytoskeleton at their restrictive temperature form an aberrant wall over the whole surface of the isodiametrically-growing cell at the restrictive temperature (Gabriel and Kopecka, 1995).

Johnson *et al.* (1996) concluded that actin at the fungal hyphal tip functions only in vesicle traffic control and biosynthesis, with no strengthening function at all. Yet actin microfilaments are tension elements. Actin filaments anchored to the wall by integrin-like molecules through the membrane near to growth sites could certainly serve to direct microvesicles to their target, but in addition they would be providing tension anchors to supplement the mechanical strength of the synthesis-weakened, plastic cell wall. By being involved in both the directional control of the synthesis precursors *and* in mechanical support such structures would be ideally placed to serve also as sensors of the local mechanical strength of the wall and thereby act as regulators of the amount of new synthesis required to restore wall integrity. This would satisfy Heath's (1995a) requirement that control of synthesis of the extracellular wall is reserved to the intracellular environment. This idea is also attractive in the sense that it can be readily appreciated how it might have arisen as an evolutionary development of some ancestral version of the 'cytoskeletal extrusion' mechanism which is used in present day oomycetes.

Thus, I conclude that the F-actin cytoskeleton at the hyphal apex: (i) is anchored through the membrane to the wall at its growing points; (ii) thereby reinforces the wall so that its components can be partially disassembled for new material to be inserted; (iii) directs precursors to those points; (iv) acts as a strain gauge, adjusting the traffic of precursors in step with local mechanical requirements (Fig. 2.2). Importantly, the microtubule components of the cytoskeleton do not seem to be major contributors to apical growth. Apical growth of *Schizophyllum commune* hyphae continued for several hours after drug-induced disruption of cytoplasmic microtubules (Raudaskoski *et al.*, 1994).

2.2.4 Strategies for synthesis

Two major models have been proposed to explain the mechanism of hyphal tip growth in the mycelial fungi, both envisage the tip wall being enlarged as the result of fusion with the plasma membrane of vesicles (chitosomes) carrying precursors and enzymes, particularly chitin synthase (Sietsma *et al.*, 1996), so externalising their contents.

2.2.4.1 The hyphoid model

Bartnicki-Garcia *et al.* (1989) suggested a model in which the rate of addition to any part of the wall depends on its distance from an autonomously moving vesicle supply centre (VSC) which is presumed to be a representation of the Spitzenkörper, a vacuolar organelle which is usually found in the vicinity of the hyphal apex. The model originates from an interpretation of the particular shape of the hyphal tip as a 'hyphoid' curve (as opposed to being hyperbolic or hemispherical), the hyphoid equation then being elaborated into a mathematical model which assumes that wall-building vesicles emanate from the VSC. The VSC is an organiser from which vesicles move radially to the hyphal surface in all directions at random. Forward migration of the VSC generates the hyphoid shape (Bartnicki-Garcia *et al.,* 1989; 1995a,b). Computer modelling suggests that the position and movement of the VSC determines the morphology of the fungal cell wall. The model both mimics observations made on living hyphae and predicts observations which

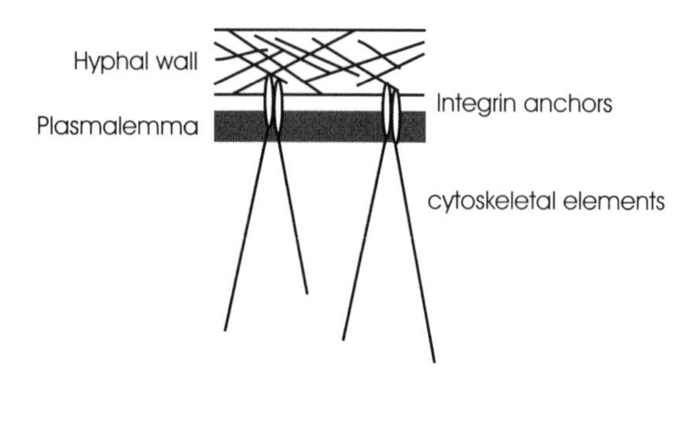

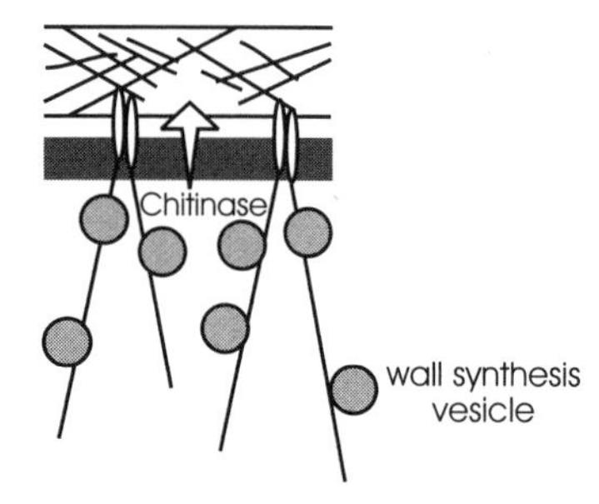

Fig. 2.2 Opportunities for outside-in signalling in fungal wall growth. The top diagram shows how integrins spanning the membrane and specifically bound to components of the wall matrix on the outside of the membrane might effectively connect the wall to the cytoskeleton within the membrane. The lower diagram indicates that turgor imbalance resulting from weakening of the wall through chitinase (or other lytic enzyme) activity would consequently stress the cytoskeleton and trigger the channelling of wall synthetic microvesicles to the locality. The cytoskeletal attachments would protect the wall against catastrophic failure.

were subsequently confirmed, making the hyphoid model and its VSC concept a very plausible hypothesis to explain hyphal morphogenesis.

Enzymically, the hyphoid model derives from earlier ideas which postulated a delicate balance between wall synthesis and wall lysis during wall growth (Bartnicki-Garcia, 1973; Bartnicki-Garcia and Lippman, 1969, 1972). Explicitly, fusion of vesicles with the plasma membrane externalises their content of lytic enzyme (endoglucanase, perhaps chitinase) hydrolyses structural glucan molecules in the existing wall, mechanical stretching pulls the broken molecules apart, then resynthesis occurs either by insertion of oligoglucan or by synthetic extension of the divided molecule. The resynthesised molecules have the same mechanical strength as before, but have been lengthened and the tip has grown.

2.2.4.2 The steady state or 'soft spot' model

This model assumes that turgor pressure stretches the wall at the hyphal tip where it is still plastic and, in addition, that the synthesis vesicles fuse with the membrane only if they reach parts of the wall that are sufficiently new to be still plastic (Wessels and Sietsma, 1981; Wessels, 1993; Sietsma and Wessels, 1994). Thus new wall will be incorporated preferentially into the most recently synthesised wall and this cooperative insertion of newer wall into new wall is the steady state synthesis to which the name of the model refers. Turgor stretches the new, plastic, wall thin, but then it is rethickened and restored by vesicular exocytosis of proteins and polysaccharides as wall precursors. The preferential targeting of these vesicles to the most recently synthesised wall means that, once established, the growing point will be maintained (steady state, again). Pre-softening to generate the plastic stretch-thinning in the first place occurs through endolytic cleavage (by chitinase?) though in this model such an enzyme activity is employed only briefly to initiate the growth mode but not to sustain it.

Stretching of the wall and addition of new wall material from the cytoplasmic side occur maximally at the extreme tip. Newly added wall components are chitin and β-(1⟶3)-glucan molecules. With time, these two polymers interact to form covalent linkages and to cross-link with proteins. At the extreme tip the wall is minimally cross-linked and supposed to be most plastic. Subapically, wall added at the apex becomes stretched and partially cross-linked while new wall material is added from the inside to maintain wall thickness. Wall material at the outside is always the oldest. Cross-linking increases progressively from the tip and as 'wall hardening' proceeds the wall hardly yields to turgor pressure and stretching and synthetic activity decline. If the tip ceases to extend for any reason (e.g. change in turgor) the steady state breaks down, newer wall is not added to new wall and cross-linking between the wall polymers spreads into the apical dome and over the whole apex. From this stopped state, a fresh round of endolytic cleavage would be necessary to restart tip growth.

2.2.4.3 Hybrid models and modifications

Both Koch (1994) and Johnson *et al.* (1996) express the view that there may be a role for a combination of the VSC and steady state models working together. Based on a detailed analysis of the enzymological requirements, together with an experimental test (Johnson *et al.*, 1995) which indicated that fission yeast (*Schizosaccharomyces pombe*) cell walls have an out-to-in gradient of hardening, Johnson *et al.* (1996) describe a hybrid model in which they retain the concepts of exocytosis of wall components and relevant enzymes from secretory vesicles at the tip, the out-to-in gradient of cross-linking (hardening) between wall polysaccharides and regulation involving stretch receptors (Wessels, 1993) of the steady state model. They reject the steady state model's reliance on transient endoglucanase activity to establish a pre-softened tip. From the enzymological basis of the VSC model they retain continuous endoglucanolytic activity and resynthesis either by insertion of monomers or oligomers across the break, or by linear synthetic addition to the new, endoglucanase-generated ends of the polysaccharide molecules (Johnson *et al.*, 1996). The new hybrid model accents the need for controlled endolytic activity and resynthesis of polysaccharide at the tip (derived from the VSC model) and the need for the cross-linking of polymers to 'harden' the maturing wall (from the steady state model).

Gooday (1995b,c) and Gooday and Schofield (1995) suggested a modification to the existing models to account for the fact that inactive chitin synthase is distributed throughout the hyphal plasma membrane and activated only at the apex (and at incipient branch sites). They argue that physical stressing of the membrane as a consequence of turgor pressure may activate the chitin synthase locally. Experiments showed that cells subjected to osmotic stress had raised native chitin synthase activities. Membrane stress may cause a conformational change in molecules of chitin synthase in the membrane, or cause changes in the interactions between chitin synthase and associated polypeptides, leading to activation.

Evaluation of the VSC model as it might apply to *Saprolegnia ferax* led Heath and van Rensburg (1996) to the conclusion that the model as proposed by Bartnicki-Garcia *et al.* (1989, 1995a,b) may not be applicable to oomycete hyphal tips. Heath and van Rensburg (1996) developed a mathematical model which could generate biologically-accurate shapes for hyphal tips on the assumption of a tip-high gradient of vesicle fusions with the plasma membrane without the need to specify that the vesicles originate from a point source (which would be the VSC). By observation of living hyphal tips, they also showed that tracks of vesicles presumed to be moving to contribute to apical extension of the hypha did not plot to a position which might represent a VSC. Indeed, apart from an abundance of vesicles close to and parallel with the plasma membrane, no consistent pattern in vesicle distribution could be detected. Heath and van Rensburg (1996) emphasise that it is the gradient in the fusion of wall vesicles with the plasma membrane which is most important in determining the shape of the apex. Their 'vesicle fusion gradient' (VFG) model which results has no cytological structure analogous to the VSC to direct this gradient, but the authors point out that targeting of vesicles to specific endomembrane structures in animal cells is known to be extremely precise and they postulate that peripheral actin microfilaments and a membrane skeleton of the plasma membrane (Heath, 1995a,b) may be involved. These experiments and arguments clearly show that the VSC model is not the only explanation for hyphal tip growth, despite its impressive ability to mimic hyphal growth forms and morphogenesis in computer visualisations (Bartnicki-Garcia *et al.*, 1995a,b). The crucial contribution of this evaluation of the VSC model is that it has clearly focused attention on the fact that the vesicle fusion gradient matters more than the way in which the gradient is achieved. If we again think of *Saprolegnia* as representing an alternative adaptation of some ancestral condition, perhaps the point of difference which has emerged is that the ancestral hyphal tip may have been forced to rely on membrane architecture and the actin cytoskeleton to both extrude the tip and construct its wall. As true fungi emerged they developed the ability to modulate and use cell turgor as a hydraulic tool. This may have released the membrane structure and cytoskeleton from responsibility for driving the apex forwards, allowing them to adapt, enhancing their vesicle control and regulatory capability, possibly culminating in the VSC/Spitzenkörper arrangement.

2.2.5 Growth at the tip: a consensus model of tip growth

The hybrid model of Johnson *et al.* (1996) is comprised of an attractive combination of enzymological features. Associating chitin synthase activity with stretch receptors (Gooday, 1995a,b; Gooday and Schofield, 1995), involving membrane architectural proteins (Heath, 1995a) and recognising the primacy of vesicle gradients (Heath and van Rensburg, 1996) all bring new elements into regulation of wall synthesis. When these are combined with other components of the two prime models (Bartnicki-Garcia *et al.*, 1989; Wessels and Sietsma, 1981; Wessels, 1993) a consensus model emerges which could claim to be the best available explanation of hyphal tip growth. This model has the following features.

1. *The enzymological features of the hybrid model of Johnson* et al. *(1996) are adopted.*

I believe it is quite clear that glucanases must have a central role in fungal wall growth. This has not been definitively demonstrated but there are many suggestive observations, such as those of Cole *et al.* (1995)

which show that wall hydrolases (glucanase, chitinase) play key roles in wall development in the pathogen *Coccidioides immitis*. Equally, there seems to be little reason to doubt that cross-linking, and particularly in this context cross-linking to proteins, is involved in wall maturation. Sentandreu *et al.* (1995) describe formation of the cell wall of *Candida albicans* following synthesis of its individual components as a two-step extracellular assembly in which a viscoelastic composite is formed:

1. noncovalent interactions occur between mannoproteins and other wall components;
2. the initial network is consolidated by formation of covalent cross-linkages among the wall polymers.

The authors suggest that in both processes specific proteins, carrying out part of what they call a morphogenetic code, may regulate the final yeast or mycelial morphology. Experiments have shown that some mannoproteins form supramolecular complexes, being secreted independently but released together from cell walls digested by hydrolases. Transglutaminase may be involved in the formation of covalent bonds between different cell wall proteins during the final assembly of the mature cell wall (Ruiz-Herrera *et al.*, 1995a).

The description given by Sentandreu *et al.* (1995) of wall formation in *Candida albicans* highlights the point that the final structure of the wall may not be made immediately. There is evidence that low-molecular-mass oligomers made of neutral sugars, GlcNAc and amino acids may contribute to a core scaffold of the cell wall of the fungus (Ruiz-Herrera *et al.*, 1994). However, it is also important to stress that wall construction must be a dynamic process. The wall which is first made is not simply 'hardened' and left at that. Rather it is remodelled and adapted to suit the immediate conditions. Most excreted proteins leave hyphae at the growing apices. Sietsma *et al.* (1995) suggest that proteins excreted at the apex may pass through the wall by being carried with the flow of wall material (bulk flow) rather than through pores in the wall. Among proteins that can be permanently retained by the wall are the hydrophobins that self-assemble at the outer wall surface when confronted with a hydrophilic-hydrophobic interface. These have been shown to mediate both the emergence of aerial hyphae and the attachment of hyphae to hydrophobic substrates (by electrostatic interaction) and wall surface properties have been shown to change with time following biochemical reorganisation of the cell wall (Tronchin, *et al.*, 1995). Changes in wall structure correlate with morphogenetic changes (Wessels, 1994b; Yabe *et al.*, 1996).

2. Cytoplasmic vesicles and vacuoles are assumed to be crucial to the growth of hyphal apices (Grove, 1978; Bartnicki-Garcia, 1990; Markham, 1995) and to be responsible for delivering the enzymes and substrates needed for wall construction.

The demonstration that a mutant of *Ustilago maydis* defective in apical extension growth lacks a component of clathrin-coated vesicles (Keon *et al.*, 1995) shows that vesicle fusion is a component of apical growth. Organelle movements seem to depend on microtubules in *Neurospora crassa* (Steinberg and Schliwa, 1993) and in *Schizophyllum commune* the microtubular cytoskeleton is needed for maintaining the direct extension of the leading hyphae at the colony edge though microtubules are not the major elements in hyphal extension growth (Raudaskoski *et al.*, 1994). The cytoskeleton is clearly used differentially by different organelles; nuclear migration and hyphal tip extension growth are coordinated and interrelated, but independent processes (Aist, 1995).

Possibly related to organelle movements and the supply of vesicles to the tip is the 'pulsed growth' of hyphal tips of seven different species of fungi recorded by López-Franco *et al.* (1994) using video analysis and image enhancement. In all fungi tested, the hyphal elongation rate fluctuated continuously with more or less regular intervals of fast and slow growth. The authors suggest that the observations are consistent with a causal relationship between fluctuations in the overall rate of secretory vesicle delivery or discharge at the hyphal apex and the fluctuations in hyphal elongation rate. A problem with video observations is that the video raster and pixel (picture element) structure of the electronically-observed image can impose pulsations upon smooth movements. As the edge of the moving object moves from one pixel to the next there is a defined time interval during which no observation is possible, yet the eye, and computer-aided image enhancements, can either compensate for this or amplify it depending on circumstances which have nothing to do with the moving object itself. This effect can readily generate pulsations where they do not exist (Hammad *et al.*, 1993c). The simple test of the validity of pulsations in video measurements which was recommended by Hammad *et al.* (1993c) does not seem to have been applied by López-Franco *et al.* (1994). Stepwise changes in elongation rate of hyphae of *Streptomyces* have been recorded using photographic methods not prone to this particular artifact (Kretschmer, 1988) so the phenomenon *could* be more general. However, the 'high technology' of their approach itself raises doubts about the validity of the interpretations of López-Franco *et al.* (1994) because their video enhancement process depends upon undisclosed and undefined computer routines.

Thus, cyclical variations in vesicle delivery *may* cause pulsations in apex extension rate. However, there is no need to doubt the general ability of hyphae to supply vesicles either in the number or at the rate that the VSC model would require. Motile vacuolar systems have been observed in a range of different fungi, rapidly moving extensive vacuolar 'trains' along hyphal tip cells (Shepherd *et al.*, 1993a,b; Rees *et al.*, 1994) which could easily supply the vesicles used for apical extension growth.

3. The actin cytoskeleton is assumed to be involved directly in hyphal tip growth in a number of ways.

In animal cells focal contacts are specialised membrane domains where bundles of actin filaments terminate at integrin-mediated attachments to the extracellular matrix. This consensus model of tip growth assumes that similar structures, 'wall contacts', at the hyphal apex would permit the actin cytoskeleton to be directly linked, through the plasma membrane, to components of the wall. Again in animals, integrins form a family of cell adhesion receptors which shows a range of specificities for the extracellular molecules to which they will bind (Kühn and Eble, 1994). It is therefore feasible to anticipate that such wall contacts might also be specific for different wall components or even fragments of wall components. Wall contacts are anticipated to function:

1. as additional mechanical support to the wall (and via spectrins to the membrane) to compensate for loss of wall strength due to enzymatic cleavage of wall components as new wall materials are incorporated;
2. as detectors of mechanical strain which might then regulate enzyme activity (Gooday and Schofield, 1995) or modulate the flow of vesicles to the local region;
3. as feedback detectors of the progress of wall synthesis. In this respect it should be noted that, again in animal systems, interaction between integrins and the extracellular matrix generates signals which cause phosphorylation of intracellular signal transduction pathways (Williams *et al.*, 1994). This is known as 'outside-in signalling'. Such a mechanism at the fungal wall would enable the progress of wall synthesis to be reported to the synthesis machinery allowing it to modulate vesicle supply in quantity and/or kind;
4. as vesicle traffic directors and regulators; close to the membrane the actin filaments will be directing vesicles with great specificity,

> in response to the outside-in signalling information; more distant filaments will be marshalling and collecting vesicles into the supply pathways which result in the required overall vesicle fusion gradient. In many cases vesicle supply will be channelled through a vesicle supply centre/Spitzenkörper structure; in other cases this will not be within the capabilities of the particular cell biology concerned and vesicle supply may be decentralised. In polarised epithelial cells of animals sequential transport of vesicles on both microtubules (over long distances and to the vicinity of the target) and actin filaments (which complete terminal delivery); the spectrin-based cytoskeleton may then be involved in protein sorting in the plane of the membrane (Mays *et al.*, 1994).

Firm mechanical connection between the body of the hypha and its wall synthesis apparatus is implied by observations of spiral and helical growth. Spiral growth, where the pattern of lead hyphae in the colony spirals from the centre rather than being strictly radial has often been observed on the surface of solid media but is not seen in liquid medium, in aerial hyphae or in hyphae penetrating agar (Ritchie, 1960; Madelin *et al.*, 1978). The proportion of spiral hyphae varies with the gel strength of the agar (Trinci *et al.*, 1979) and is best observed when growth is sparse because negative autotropic hyphal interactions obscure spiral growth in dense colonies. Spiralling is thought to result from rotation of the apex during extension growth causing the hyphal tip to roll over the surface. Tip rotation has not been observed in vegetative hyphae but occurs in sporangiophores of *Phycomyces blakesleeanus* and *Mucor mucedo*, and conidiophores of *Aspergillus giganteus* (Gooday and Trinci, 1980). A more extreme example seems to be that, when grown on a range of surfaces in conditions favouring hyphal growth, hyphae of *Candida albicans* grew in a right-handed helical fashion, though in liquid media the hyphae grow straight (Sherwood-Higham *et al.*, 1994). Spiral and helical growth show that the hypha can interact directly with its environment and that the structures are sufficiently rigid so that mechanical work can be done as a consequence of the wall synthesis.

There are numerous other examples of environmental interaction through the wall which may be the equivalent of outside-in signalling, especially in exchange of signals between plant pathogenic fungi and their hosts (e.g. Podila *et al.*, 1993; Clay *et al.*, 1994; Clement *et al.*, 1994; Jelitto *et al.*, 1994; M. C. Heath, 1995; Green *et al.*, 1995; 1996). Bircher and Hohl (1997) have identified some glycoproteins at the apex of the germ tube of germinating zoospores of *Phytophthora palmivora* which regulate adhesion by sensing the hydrophobicity of the contact surface. This particular phenomenon relates to smooth surfaces only, but surface topology and chemicals on the surface of the host plant can trigger germination of fungal spores and differentiation of the germ tubes into appressoria (infection structures). Some fungal genes are uniquely expressed during appressorium formation as a result of induction by topography and surface chemicals. Penetration of the cuticle by the fungus is assisted by fungal cutinase, small amounts of which are carried by spores of virulent pathogens, but contact with the plant surface releases cutin monomers which then trigger cutinase gene expression (Kolattukudy *et al.*, 1995). In the *Colletotrichum lindemuthianum*/bean interaction, the extracellular matrices associated with conidia, germ tubes, and appressoria differ in composition, with extracellular glycoproteins organised into specific regions of the fungal cell surface. In addition, the plasma membrane of appressoria is differentiated into distinct domains. In the *Erysiphe pisi*/pea interaction extrahaustorial membrane contains both specific components and glycoproteins in common with the host plasma membrane (Green *et al.*, 1995; 1996).

2.3 Septation

The hyphal growth form of filamentous fungi is an adaptation to the active colonisation of solid substrata. By hyphal extension and regular branching the fungal mycelium can increase in size without disturbing the cell volume/surface area ratio so that metabolite and end-product exchange with the environment can involve translocation over very short distances (Rayner *et al.*, 1995b). Growth of the mycelium is regulated, of course, and Bull and Trinci (1977) identified three mechanisms involved in regulating the growth pattern of undifferentiated mycelia. These are: the regulation of hyphal polarity, the regulation of branch initiation, and the regulation of the spatial distribution of hyphae. Fungal hyphae are variable between species, but generally speaking the hyphal filament, when separated into compartments by cross-walls, has an apical compartment which is perhaps up to ten times the length of the intercalary compartments. The septa which divide hyphae into cells may be complete (imperforate), penetrated by cytoplasmic strands, or perforated by a large central pore. The pore may be open (and offer little hindrance to the passage of cytoplasmic organelles and nuclei), or may be protected by a complex cap structure derived from the endoplasmic reticulum (the dolipore septum of basidiomycetes). In ascomycetes, which characteristically lack the parenthesome apparatus, the pore may be associated with Woronin bodies (Markham, 1994, 1995; Heath, 1995b). Septal form may be modified by the hyphal cells on either side of the septum, and may vary according to age, position in the mycelium, or position in the tissues of a differentiated structure (Gull, 1976, 1978). These features make it clear that the movement or migration of cytoplasmic components between neighbouring compartments is under effective control. So, the cellular structure of the hypha extends, at least, to its being separated into compartments whose interactions are carefully regulated and which can exhibit contrasting patterns of differentiation.

Early research showed that septation in the main hypha is in some way defined by the position of the dividing nucleus (Talbot, 1968; Girbardt, 1979), though branch formation does not seem to be dependent on orientation of the nuclear division spindle. There also seems to be some sort of 'memory' in yeasts, for both *Saccharomyces cerevisiae* and *Schizosaccharomyces pombe* cells re-initiate growth after division in the region where the cell extended in a previous cycle. Johnson *et al.* (1996) draw attention to observations of Marks *et al.* (1987) that some fission yeast cells had an actin ring, presumably apposed to the plasmalemma, that marked the last pre-mitotic position of the dividing nucleus. Johnson *et al.* (1996) suggest that vestiges of actin, or some actin-associated structure, from earlier cell cycle events might remain to mark the site for subsequent events which themselves involve actin assemblies (Jackson and Heath, 1990; Prosser and Trinci, 1979). Spindle pole bodies, which are fungal organelles equivalent to the centrosome, appear to regulate the polymerisation of both the spindle and astral microtubules, and, consequently, are the major effectors of mitosis and septation (Butt *et al.*, 1989; Kamada *et al.*, 1989a,b, 1993; Morris and Enos, 1992; Clutterbuck, 1994; Harris *et al.*, 1994; Heath, 1995b). These may be the agents which identify and act upon any cytoskeletal 'memory' of past events.

Primary septa in fungal hyphae are formed by a constriction process in which a belt of microfilaments around the hyphal periphery interacts with microvesicles and other membranous cell organelles (Girbardt, 1979; Heath, 1995b). Girbardt (1979) emphasised the correspondence between fungal septation and animal cell cleavage (cytokinesis). In filamentous fungi cytokinesis is equivalent to formation of cross-walls, that is the primary septa, which occurs close to the site of nuclear division in many organisms (Valla, 1984). In dikaryotic mycelia of clamp connection-forming basidiomycetes there are two classes of postmitotic nuclear migration. Immediately after the completion of mitosis, the first three sibling nuclei migrate. This is followed by the migration of the fourth nucleus from the clamp cell into the subterminal hyphal segment (see section 5.2.2.3). Different fungi may vary in the fine details of cytokinesis. In *Schizophyllum commune*, the septum separating the progeny nuclei in the parent hypha had formed before formation of the septum delimiting the clamp cell whereas in *Pleurotus ostreatus*, the formation of the clamp septum preceded that of the septum in the parent hypha (Yoon and Kim, 1994). Coupling between karyokinesis and cytokinesis varies from a rather loose association in hyphae where multinucleate cells are formed to a strict coupling during spore formation when uninucleate spores are produced (Trinci, 1978; Doonan, 1992). Coupling is also evident during conidiospore germination in *Aspergillus nidulans*, in which a uninucleate G1-arrested conidium underwent its first nuclear division as the spore swelled to signal germination (Morris and Enos, 1992). The second nuclear division was generally coincident with germ tube emergence; by the completion of this division (the four nucleus stage) the fungus had already established a polarised growth axis. Then the most apical nucleus normally entered mitosis first and a wave of intranuclear mitotic activity passed down the hypha (Morris and Enos, 1992), followed by formation of septa. Analysis of some *sep* mutants of *Aspergillus nidulans* (mutants identified as failing to form septa, i.e. defective in cytokinesis) showed that in addition to the septal defect, they arrested at the third nuclear division stage and had aberrant, elongated and multilobed, nuclear morphology (Harris *et al.*, 1994). Therefore, this class of mutants suggests the existence of a regulatory mechanism to ensure the continuation of nuclear division following the initiation of cytokinesis during spore germination. Unfortunately, the early events in spore germination cannot be generalised in fungi as, in contrast to the *A. nidulans*, mechanism, in the basidiomycete *Volvariella*, a germ tube emerges before the basidiospore nucleus migrates to initiate its first mitotic division (Chiu, 1993).

Nuclear migration features in most division processes in fungi and the machinery for nuclear movement seems to depend on the cytoskeleton structure (McKerracher and Heath, 1986; Osmani *et al.*, 1990; Clutterbuck, 1994). Treating a diploid fungus with an antimicrotubular drug, such as benomyl, induces haploidisation (Anderson *et al.*, 1984; Büttner *et al.*, 1994). Kamada *et al.* (1989a,b, 1993) induced and isolated benomyl-resistant (*ben*) mutants which were defective in structural genes for α- or β-tubulin in *Coprinus cinereus*. These *ben* mutants were blocked in transhyphal migration of nuclei in dikaryosis (formation of dikaryotic hyphae with clamp cells, see section 5.2.2.3), and they were used to demonstrate that microtubules participate in the pairing and positioning of the two conjugate nuclei in dikaryotic cells (Kamada *et al.*, 1993). However, although they were defective in nuclear migration in hyphae, they were not affected in migration of nuclei into developing spores. This indicates that the tubulin-mediated movement system is not the sole mechanism for translocation of nuclei. The involvement of actin, the other major cytoskeleton component, in nuclear migration has also been demonstrated (Grove and Sweigard, 1980; Doonan, 1992; Harris *et al.*, 1994). Immunohistochemical labelling of actin filaments reveals a close colocalisation with the multiple nuclei in stem hyphae of *Flammulina* (Monzer, 1995). The contracting actin microfilament band, the ‘septal band’, which appeared early at the position of a presumptive septum (Girbardt, 1979; Doonan, 1992; Kobori *et al.*, 1992) seems to be required, as application of the inhibitor cytochalasin A not only disrupted actin organisation but also prevented septum formation (Grove and Sweigard, 1980; Kobori *et al.*, 1992). Gene disruption experiments further demonstrated the requirement for myosin, various phosphatases and actin-associated proteins for cytokinesis, cytoskeleton–cell wall attachment and for nuclear migration as well as motility (Harris *et al.*, 1994; Kaminskyj and Heath, 1995). In *Aspergillus nidulans*, the *nudA* gene, which is involved in nuclear migration, encodes a cytoplasmic dynein protein. Cytoplasmic dynein has been implicated in nuclear division in animal cells, and is concentrated at the growing tip of the *Aspergillus* hyphae. Mutants in which *nudA* has been disrupted are able to grow slowly and undergo nuclear division, suggesting that there are alternative motor proteins for both nuclear migration and nuclear division (Xiang *et al.*, 1995).

Proteins homologous to septins have been identified as being involved with actin in septation in *Aspergillus nidulans* (Momany *et al.*, 1995). The actin septal bands which appear around the periphery of the hypha and locate the position of the septum appear to act as sites for chitin deposition (Momany and Hamer, 1996); this is similar to the situation during cytokinesis in *Schizosaccharomyces pombe* (Frankhauser and Simanis, 1994). There is a protein in *S. pombe* which is required for proper formation of the mitotic spindle during nuclear division (karyokinesis) as well as the actin ring and septum. The gene product can induce septum formation in the absence of mitosis when overexpressed (Ohkura *et al.*, 1995). Septins are a family of proteins which are involved in cytokinesis in animal cells and in bud site selection in budding yeast cells (Sanders and Field, 1994). The septins were first recognised in yeast but are now known to be present also in other fungi, insects, and vertebrates. Despite the apparent differences in cytokinesis in animals and fungi, septins

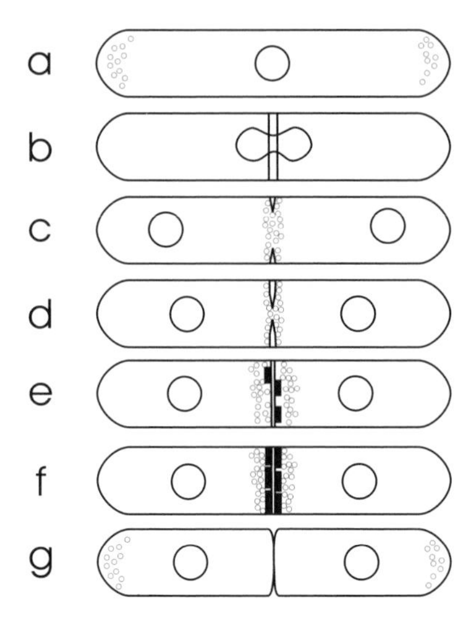

Fig. 2.3 Septum formation and cytokinesis in *Schizosaccharomyces pombe*. Fission yeast cells grow mainly by extension at both tips (a), actin is located at the growing ends of the cell (shown here as small circles) and a sheaf of microtubules runs from end to end. When mitosis is initiated, actin is relocated to form an equatorial ring of filaments around the nucleus (b), the position of which anticipates the position of the septum. As mitosis is completed the primary septum originates in the periplasmic space and grows into the cell (c and d). As synthesis of the septum proceeds, actin distribution changes from a filamentous ring to clusters of 'vesicles' or 'dots' (c to e) which are located where septum material is deposited. Secondary septa are formed either side of the primary (e and f), and the latter is removed to bring about cell separation (g). Redrawn after Frankhauser and Simanis (1994).

appear to be essential for this process in both cell types, and also appear to be involved in various other aspects of the organisation of the cell surface (Simanis, 1995; Longtine *et al.*, 1996) including spore formation in *Saccharomyces cerevisiae* (Fares *et al.*, 1996).

Taken together, the observations from the 1970s to the present day echo the ancient phylogenetic relationships by emphasising the remarkable similarity between cell cycle events in animals and fungi. Although the picture is not yet complete, the indications are that septation in fungi involves chitin deposition in a ring defined by a pre-formed ring of actin microfilaments (Fig. 2.3). There is evidence for cytoskeleton-related functions being shared by karyokinesis and cytokinesis and these may form the basis of a structural memory which allows septa to be laid down in positions defined by a nuclear division which occurred some time before.

2.4 Branching

Hyphal branches may be described as primary (which subtend no branches and arise directly from the main hypha), secondary (which subtend a primary branch), tertiary (which subtend a secondary branch), etc. In many biological and non-biological branching systems there is an inverse logarithmic relation between the number of branches belonging to a particular order and order number (Park, 1985a). Such a relation describes the situation where maximum surface area can be colonised with minimum total length of filament (Rayner *et al.*, 1995a). For the theoretical mycelial fungus, therefore, the relation would optimise the efficiency of the mycelium in colonising substrata while minimising the amount of biomass required to do so. The relation has been shown to hold in the majority of fungal species examined (Park, 1985b). A minority showed proportionality between the logarithm of mean branch length and order number, or an arithmetic relationship, deviations from the theoretical expectation being thought to represent stages in colony development and/or mycelial differentiation which is discussed further below. With regard to the relation between *length* of branches and their hierarchical order, Park (1985a) suggested that where growth is not limited the mean length of branches will increase arithmetically with order number. On the other hand, when growth is limited, by space or nutrient limitation or production of staling compounds, hyphal extension rate will decline and a logarithmic relationship may be expected. This sort of regular change in branching structure may have significance in differentiation.

Relationships of this sort in mycelia specifically suit the fungal mycelial colony morphology to exploration of the substratum and colonisation of the substrates it contains. The formative rules which determine the mycelial growth form are the very rules which the organism must vary in order to create tissues and organise them into multicellular organs. It is worth dallying a while, therefore, to make some sweeping generalisations in an attempt to describe development of a 'typical' mycelium. Most of what follows is derived from Prosser (1983, 1993; 1995a,b), Trinci (1984), Trinci and Cutter (1986), Jennings (1986), Gow (1995a) and Rayner *et al.* (1995a,b).

When provided with appropriate incubation and nutritional conditions, fungal spores form a germ tube which initially increases in length exponentially; after this initial phase a constant linear extension rate is achieved and branches are formed, either dichotomous or lateral. In their early growth, hyphal branches repeat this pattern: an initial exponential rate of extension being replaced by a constant linear extension rate.

Measured in terms of total mycelial length, growth is exponential as a result of exponential branch formation. Total mycelial length and the total number of hyphal apices (all lead hyphae and branches) both increase exponentially at the same specific rate. This rate is equivalent to the specific growth rate of the organism under the same growth conditions in liquid medium, where biomass can be measured directly as dry weight (Trinci, 1974).

Trinci (1974) showed that a new branch is initiated when the mean volume of cytoplasm per hyphal tip (the hyphal growth unit) exceeds a particular critical value. For a range of fungi the hyphal growth unit increased following spore germination but then exhibited a series of damped oscillations tending towards a constant value (Trinci, 1974). Such constancy demonstrates that over the mycelium as a whole, and not just in single hyphae, the number of branches is regulated in accord with increasing cytoplasmic volume. The hyphal growth unit (Caldwell and Trinci, 1973) is the population parameter which describes this. It is calculated as the ratio of total mycelial length to total number of hyphal tips, and can therefore be interpreted as the average volume of cytoplasm necessary to support the extension growth of a single average hyphal apex (Robinson and Smith, 1979).

In early phases of growth, branches usually subtend an angle of approximately 90° to the long axis of the parent hypha, but hyphae tend to avoid their neighbours (negative autotropism) and to grow radially away from the centre of the colony. Thus, a circular colony is formed eventually, with radially directed hyphae, approximately equally spaced, and extending at the margin at a constant rate. Colony differentiation may result from production of a secondary metabolite or 'staling compound' at the colony centre, where growth is reduced, which acts as a morphogen, diffusing towards the colony margin and causing reduced rate of branch initiation, increased extension rate and negative chemotropism. Equally, the behaviour may be a response to depletion of nutrients below a developing colony. Robson *et al.* (1987) showed that the glucose concentration remaining in the agar fell sharply behind the advancing colony margin in Petri dish cultures of *Rhizoctonia solani*. Though these experiments are usually described in terms of 'substrate utilisation by colonies growing on agar and the nutritional limitations to growth at the colony centre' (Prosser, 1995a), there was no evidence of glucose limitation in these cultures. I prefer to see these experiments as further evidence of the ability of fungal mycelia to *internalise* nutrients very rapidly (see introduction to Chapter 4). Nevertheless, whether internalised or utilised, glucose (together, presumably, with other nutrients) is removed rapidly from beneath the colony. Consequently, the advancing hyphal tips are always likely to experience a nutrient gradient offering better supplies ahead than behind. This may be a component of the regulatory circuit which causes the outward migration and substratum exploration by hyphal apices.

As the colony circumference increases the apices of some branches catch up with their parent hyphae to maintain hyphal spacing at the colony margin. This occurs either by relaxation of controls on extension rate of branches as they become further separated from main hyphae or as a result of simple variability in extension rate (Hutchinson *et al.*, 1980). A specific example of change in hyphal behaviour during colony development is provided by the analysis of mycelial differentiation in *Neurospora crassa* by McLean and Prosser (1987). Up to about 20 h of growth, all hyphae in mycelia of this fungus have similar diameters, growth zone lengths and extension rates and all branches are at an angle of 90° to the parent hypha. After about 22 h growth branch angle decreases to 63°, hyphal extension rates and diameters increase and a hierarchy is established in which main hyphae are wider and have greater extension rates than their branches; the ratios between diameters of leading hyphae, primary branches and secondary branches being 100:66:42 and between extension rates being 100:62:26. Similar ratios have been found in other *Neurospora* strains and in other fungi (Prosser, 1983, 1995a).

When growth becomes restricted at the centre of the colony, exponential growth is confined to its margin in what is called the peripheral growth zone. The width (w) [better, depth] of this remains constant as the colony expands. The radial rate of colony expansion (Kr) is determined solely by growth in the peripheral growth zone and is related to the specific growth rate in liquid culture (μ) by the relation $Kr = \mu w$ (Trinci, 1971). The peripheral growth zone is virtually an epidermal layer. If we view the colony as it appears on the surface of solid medium in a culture dish as a section through a spherical structure, then the peripheral growth zone overlies a region in which growth has ceased or been reduced, which is depleted of external resources and is consequently dependent on translocation of nutrients from other regions and on turnover of existing biomass. Branches are narrower in this region, hyphal growth direction is more variable and fusions between vegetative hyphae are common. The tissue beneath the peripheral growth zone 'epidermis' has differentiated. For the colony as a whole, the peripheral growth zone is a ring of active tissue at the colony margin which is responsible for expansion of the colony. At the level of the individual hypha, the peripheral growth zone corresponds to the volume of hypha contributing to extension growth of the apex of that hypha (the hyphal growth unit).

Because true extension growth is absolutely limited to the hyphal tip, the whole morphology of the hypha depends on events taking place at its apex (Grove, 1978). It follows from this that the pattern of hyphae in a mycelium, which is largely a consequence of the distribution of hyphal branches, depends on the pattern of formation of the hyphal tips which initiate those branches. It seems now to be generally accepted that the materials necessary for hyphal extension growth are produced at a constant rate (related to the specific growth rate) throughout the mycelium and are transported towards the tip of the growing hypha. Among the materials taking part in this polarised transport are the cytoplasmic vesicles, which are thought to contain wall precursors and the enzymes needed for their insertion into the existing wall to which reference is made above. Growth of the hypha involves the integration of all of the cellular growth processes contributing to karyokinesis, cytokinesis and extension growth so as to produce an ordered sequence of events contributing to a duplication cycle which is exactly analogous to the cell cycle of uninucleate cells (Trinci, 1979).

Trinci (1974, 1978, 1979) has argued that lateral branches are formed at locations where wall synthetic vesicles (and other components) affect the rigidified wall of the hypha so as to produce a new 'hyphal tip'. What specifies the site of the branch initiation is not entirely clear, though Trinci (1978, 1979) pointed out that in fungi which form septa, branching is usually associated with their presence though not invariably so. In *Geotrichum candidum*, a close relationship exists between septation and branching; branches are formed at a specific time after septation and are positioned immediately behind septa. In most other fungi the septation/branching relationship is less obvious and branch positioning is more variable, *Aspergillus nidulans* is a well-researched example of this. Septal-association of branching may be another expression of the putative memory-marker from cytokinesis suggested by Johnson *et al.* (1996).

It was mentioned at the beginning of this chapter that the regulatory target in fungal morphogenesis is the machinery which is involved in generating a new apical growth centre (to become the hyphal tip of the new lateral branch) and the determination of its position, orientation and direction of outgrowth from the parent hypha, because it is these factors which establish the developing tissue patterning in any fungal multicellular structure. Origin of the branch seems to be the formal equivalent of determination of morphogenetic growth by orienting the plane of division and the new cross-wall, as occurs in plants, and directional growth of the new hyphal apex has much in common with the morphogenetic cell migrations that contribute to development of body form and structure in animals (Moore, 1984a).

Cytoplasmic vesicles are crucial to the growth of hyphal apices so it is reasonable to expect that distribution of microvesicles may be closely connected with branch initiation. Localised accumulation of microvesicles caused by disturbance to their flow may be a cause of branch initiation (Trinci, 1978) but this seems to be an excessively simplistic interpretation of what must be a complex and sophisticated transport and targeting process.

For some time, differential ion fluxes were thought to be involved in determining branching. Application of electrical fields will affect the site of branch formation and the direction of hyphal growth in young mycelia of several fungi (McGillivray and Gow, 1986). An ion current caused by influx of protons (as an amino acid symport) characterised the hyphal tip of *Achlya* (Kropf *et al.*, 1983, 1984) and a new zone of proton influx often preceded and predicted the emergence of a branch. However, these phenomena seem to be more related to nutrient uptake than tip growth (Harold and Caldwell, 1990; de Silva *et al.*, 1992; Gow, 1994) in the organisms concerned, and are not representative of true fungi (Harold, 1994). Levina *et al.* (1995) demonstrated that although growing hyphae of *Neurospora crassa* contained a tip-high gradient of cytoplasmic Ca^{2+} (absent in non-growing hyphae) there was no corresponding gradient of either spontaneous (K^{+}-inward) or stretch activated (Ca^{2+}-inward) ion channels. The ion channels were not essential for tip growth and the authors concluded that the Ca^{2+} gradient was generated from internal stores.

Stretch activation, but specifically of the enzyme chitin synthase, was suggested by Gooday (1995a, b) and Gooday and Schofield (1995) to be a means of activating the enzyme only at the apex and incipient branch sites. However, the description given of the supposed mechanism is limited and physically implausible as a means of defining a branch position. Gooday (1995b) states that. "The sites of chitin synthesis, i.e. apex and branch points, are where the cell's turgor pressure may stretch the membrane outwards against the overlying plastic wall." which I think makes it clear that stretch-activation of chitin synthase cannot be the initiating event. Until the wall is weakened (made plastic by enzymatic cleavage of some of its polymers) the hydraulic turgor pressure will be uniform over the entire wall. Thus, although elevated chitin synthesis has been associated with branching of *Gaeumannomyces graminis* (Yarden and

Russo, 1996), if chitin synthase is stretch-activated it must be a secondary consequence of some primary branch-initiation process.

What that process might be is not at all clear at the moment. Inhibitors of phosphoinositide turnover were found by Hosking *et al.* (1995) to act as paramorphogens, inhibiting hyphal extension and increasing hyphal branching of *Neurospora crassa* without affecting specific growth rate. The authors suggest that this indicates that phosphoinositides have a role in branching. Calcium-signalling also seems to be involved in branch initiation in strain A3/5 of *Fusarium graminearum* (Robson *et al.*, 1991a), which is the strain used to produce Quorn mycoprotein, and cAMP added to the medium acted as a paramorphogen by causing significant decreases in both mean hyphal extension rate (E) and hyphal growth unit length (G, see below), that is, cAMP caused mycelia to branch profusely (Robson *et al.*, 1991b). In view of the rather startling concentrations of cAMP used (up to 50 mM) the true significance of these last experiments is uncertain, although Pall and Robertson (1986) suggested a role for cAMP in internal control of branching patterns as reflected in hyphal diameter and extension rate. It is significant, however, that in this organism there is evidence that hyphal extension and branch initiation are regulated independently (Wiebe *et al.*, 1992).

However, the only hint of a possible mechanism for branch initiation is in the reports implicating heat-shock proteins in branching (Silver *et al.*, 1993). Heat-shock proteins are polypeptides which interact with other proteins. They are 'molecular chaperones' which bind to and stabilise other proteins to prevent incorrect intermolecular associations, then aid their correct folding by releasing them in a controlled manner. It is feasible that branch initiation requires assisted conformational alterations of wall proteins or that heat-shock proteins assist in the delivery of a branch-initiating polypeptide to the appropriate position. Although heat-shock proteins are induced in many organisms as they enter into morphogenetic pathways (Plesofsky-Vig, 1996), heat-shock proteins themselves do not normally have the specificity which is likely to be required for exact positioning of branch initiation sites during, say, the formative stages in production of a fungal tissue.

Very little is known about the control of growth direction of hyphae and hyphal branches, though they often tend to avoid each other. Minimum distance of approach, due to negative autotropism, of two hyphae was 30, 27 and 24 μm respectively in *N. crassa*, *Aspergillus nidulans* and *Mucor hiemalis* (Trinci *et al.*, 1979). Similar data were obtained by Hutchinson *et al.* (1980) but they concluded that hyphal distribution in colonies of *Mucor mucedo* was not due to such avoidance responses because the majority of hyphae in the colony were spaced further apart than this. Oomycetes show chemotropic growth towards amino acids (Musgrave *et al.*, 1977; Manavathu and Thomas, 1985) which may result from the apical location of the amino acid symport (Kropf *et al.*, 1983, 1984). Robinson and Bolton (1984) showed that negative autotropism in hyphae of *Saprolegnia ferax* was affected by amino acid concentration. The hyphae would approach each other to within touching distance in media containing a high ($0.47\,g\,l^{-1}$) concentration but maintained a minimum 28 μm separation when the amino acid concentration was $0.025\,g\,l^{-1}$. In this case negative autotropism may result from a positive chemotropism towards a higher concentration of amino acids, but tropism towards chemical nutrients is rare in fungi. A more general explanation for negative autotropism may be in the reaction of hyphal tips to gradients in oxygen concentration which surround hyphae (Robinson, 1973a,b). This is still a chemotropism, of course, and the hyphal tip is assumed to be positively chemotropic for oxygen so that a hyphal apex approaching a second growing hypha will turn away in order to move up the oxygen concentration gradient. The mechanism suggested by Robinson (1973b) that a gradient of respiratory activity across a hyphal tip could lead to increased production and concentration of vesicles at the side of the hypha subjected to a higher oxygen concentration. He then argued that chemotropism towards oxygen could result in increased extensibility of the wall in this region, leading to bulging and consequent extension towards the higher oxygen concentration. The extension zone has been demonstrated to be the region giving rise to tropic responses in *Phycomyces* sporangiophores (Trinci and Halford, 1975), but the tropic responses of such structures make it clear that increased wall extensibility on the side receiving the signal leads to bending *away* from the signal (i.e. a negative tropism). Robinson's argument is not, therefore, consistent with current models of wall growth. Nevertheless, in *Geotrichum candidum* branches originate from sides of hyphae facing the higher oxygen concentration so a powerful influence of oxygen cannot be excluded. Equally, redistribution of vesicles does influence the direction of branch growth. During initial outgrowth of clamp connections of *Schizophyllum commune* vesicles are displaced in the direction of curvature of the clamp cell soon after its emergence (Todd and Aylmore, 1985), whilst computer modelling of the VSC combined with observations made on Spitzenkörpers in living hyphae show that vesicle distributions certainly can explain hyphal morphogenesis (Bartnicki-Garcia *et al.*, 1989; 1995a,b).

Unfortunately, the information we have about branch initiation is too sparse for a model to be constructed. Also, it all derives from experiments with vegetative hyphae in which controls of karyokinesis, cytokinesis and branch initiation may well be relaxed. In the tissues of fungal multicellular structures the constituent cells are generally smaller and less vacuolated than typical hyphal cells, features often associated with rapidly dividing cells in animals and plants. In the fungi, a consequence of rapid karyogamy and frequent branching might be that a much closer correlation is maintained between karyokinesis, cytokinesis and branching. Furthermore, a highly hydrated extracellular matrix, comprised predominantly of glucans, fills the interhyphal spaces of most fungal multicellular structures and may be important in providing an environment in which morphogenetic control agents can interact. It seems almost inevitable that different tissues will secrete different polysaccharides and/or glycoproteins to provide a specific local environment for the hyphae. In such a case reactions may be like those in animal systems where response of a cell to a growth factor is influenced by the extracellular matrix within which the interaction takes place (Nathan and Sporn, 1991) and the extracellular matrix regulates transcription directly, probably through integrin-mediated signalling (Damsky and Werb, 1992; Hynes, 1992; Streuli, 1993). If fungi have extended their homologies with animals to this level, the possibility exists that a hypha may directly influence the gene expression of its neighbours through the extracellular matrix molecules it secretes.

2.5 Growth kinetics

The basic features of the fungal duplication cycle, the hyphal growth unit and peripheral growth zone, are numerical factors which can be established quite readily, thereby providing a quantitative, and potentially objective, basis for description of fungal growth. Prosser (1993) gives a clear example of this in a discussion of paramorphogens, which are metabolic inhibitors which affect hyphal morphogenesis by causing a highly branched, restricted-extension growth morphology (Moore, 1981b). Prosser (1993) shows that this (and other effects) can be explained by the intimate relationship between hyphal extension rate, branch initiation and specific growth rate. When specific growth rate is unaltered but extension rate is decreased, newly synthesised biomass must be redirected towards increased branch formation. Similarly, any increase in specific growth rate which is not reflected by corresponding increase in extension rate will also increase branching. The relationship is expressed in the equation $E = \mu G$, where E is the mean extension rate, μ the specific growth rate and G is the hyphal growth unit length, which decreases as branch formation increases. This relationship was discussed by Bull and Trinci (1977) and Prosser (1983). Prosser (1993) illustrates it with the example of L-sorbose inducing formation of densely branched, slowly-extending colonies in *Neurospora crassa* as distinct from the diffuse rapidly-spreading colonies normally observed during growth of this organism. Sorbose does not affect specific growth rate but results in reduced hyphal growth unit length and reduced hyphal extension rate which are expressed as increased branch formation. At a concentration of 20 g l^{-1}, hyphal growth unit length was decreased from 323 to 40 μm while colony radial growth rate decreased from 1001 μm h^{-1} to 100 μm h^{-1} (Trinci and Collinge, 1973). This sort of alteration in morphology may be what occurs in the formation of differentiation structures which involve hyphal aggregation and increased branching. Metabolic inhibitors often produce cell morphologies which mimic those encountered as differentiated cells in multicellular structures (Moore and Stewart, 1972; Moore, 1981b). The basic effect of sorbose in *N. crassa* may be inhibition of wall synthesising enzymes leading to reduction in extension rate, with increased branch formation a secondary consequence of the '$E = \mu G$' rule. Similar phenotypes have been observed in mutants in which hyphal growth and branching have been altered. Colonial mutants of *N. crassa* and other fungi have the same colonial morphology, again due to an increase in branching frequency per unit length of hypha but with specific growth rate equal to that of the wild-type strain (Trinci, 1973, 1994; Trinci *et al.*, 1994; Wiebe *et al.*, 1993, 1996).

To extend this quantitative approach to description of hyphal growth, various authors have presented descriptions of fungal growth in mathematical models (Prosser 1982, 1983, 1993, 1995b). Prosser and Trinci (1979) constructed a model in which vesicles were assumed to be produced at a constant rate in distal hyphal regions, to be transported to the tip at a constant rate, where they accumulated and fused with existing wall and membrane to give hyphal extension. The assumption that accumulation of vesicles behind septa resulted in formation of lateral branches, which then extended like their parent hypha, incorporated the duplication cycle. Predictions of the model (changes in total mycelial length, number of branches and interbranch distances) compared well with observations of growth of *Aspergillus nidulans* and *Geotrichum lactis*.

Edelstein (1982) and Edelstein and Segel (1983) extended the model of Prosser and Trinci (1979) by considering average properties of hyphae and emphasising the role of the hyphal tip in controlling and regulating mycelial growth, hyphal death, hyphal fusion and different forms of branching. These authors described growth in terms of two basic properties: hyphal density (hyphal length per unit area, symbolised by ρ), and tip density (the number of tips per unit area, symbolised by n). Their basic model consists of two partial differential equations relating changes in these two variables.

$$\frac{\partial \rho}{\partial t} = nv - d(\rho) \quad (1)$$

$$\frac{\partial \mathrm{n}}{\partial t} = \frac{\partial nv}{\partial x} + \sigma(\rho, n) \quad (2)$$

Equation 1 indicates that hyphal density will increase at a rate equal to the product of tip density and tip extension rate (v), which is assumed constant; nv is the tip flux. Hyphal density decreases through hyphal death (d) which is described as a function of ρ and a rate constant for autolysis (σ). Changes in tip density depend on tip flux within a region (the first term in equation 2) and increase due to branching and/or decrease due to tip death or anastomosis, all of which are represented by σ. Substitution for σ by suitable functions allows the model to describe dichotomous and lateral branching:

- dichotomous branching is represented by $\alpha_1 n$, where α_1 is the product of the rate of branching and the number of daughters produced per tip;
- lateral branching is represented by α_2, this being the number of branches produced per unit length of hypha in unit time.

Other functions describe tip to hypha and tip to tip anastomoses and tip death due to atrophy or overcrowding. This powerful model was also extended to consider uptake of a growth limiting nutrient and its redistribution within the mycelium. Experimental data on colony growth of *Sclerotium rolfsii* agreed qualitatively with predictions of the model (Edelstein *et al.*, 1983). The model offers a much fuller description of colony growth than previously. A variety of other mathematical models have been published describing aspects of hyphal growth (Yang *et al.*, 1992; Viniegra-Gonzalez *et al.*, 1993; Nielsen, 1993), spore germination (Bosch *et al.*, 1995), the growth of fungal biomass (Georgiou and Shuler, 1986) and the mycelial aggregates (pellets) which form in many fermenter cultures (Pirt, 1966; Edelstein and Hadar, 1983; Prosser and Tough, 1991; Meyerhoff *et al.*, 1995; Tough *et al.*, 1995; Paul and Thomas, 1996).

All of these models are useful. At the very least the intellectual discipline required to construct (and then to understand) such models concentrates the mind upon aspects of hyphal behaviour which might otherwise have been neglected. Some of these mathematical models are extremely valuable and have shown themselves capable of remarkably accurate simulations of some aspects of fungal growth. But the problem remains that rate constants and other parameters required for quantitative simulations derived from the models cannot yet be measured reliably. In some cases ignorance is due to technical problems which might be solvable as measurement techniques improve in the future. Unfortunately, in other cases our ignorance is due to the fact that mycologists in general are not quantitative biologists. Most mycologists are still more likely to observe events and attempt verbal descriptions illustrated with hand-drawn 'interpretive' (subjective) diagrams than to enumerate, make kinetic measurements and establish Cartesian relationships for objective comparison of structures. As a result we are woefully ignorant of the most basic anatomy, aspects of hyphal relationships, branching patterns, growth directions of apices, etc., even in simple colonial patterns (Petri dish cultures and fermenter pellets). The idea that cells in more complex fungal structures like fruit bodies could actually be counted and measured has occurred to regretfully few people. The few brave souls who are attempting a mathematical description of fungal growth and development are treading a lonely path at the moment.

2.6 Dynamic boundaries

Hyphae differentiate almost as soon as they are formed, whether as branches from main hyphae or as germ-tubes emerging from spores, as soon as the first septum is formed the apical cell is differentiated as the site of extension growth and the subapical cell as its supplier and provider. Viewed in this microscale, it is not surprising to find that the mycelium differentiates on a microscale, different parts of a single mycelium (or colony) may express different phenotypes from among those accessible to the genotype within that environment. Phenotypes may differ in terms of cell size and proportions, branch type and frequency, wall structure and pigmentation, distribution of aerial, surface and submerged hyphae, spore production, organelle content and distribution, and metabolic processes both primary and secondary. It is important to emphasise that although many, perhaps most, of the studies of this topic have used *in vitro* experiments in Petri dishes, or at best microcosms which attempt to simulate more natural conditions, there is evidence for mycorrhizal hyphae changing from their normal branching pattern and apical dominance to differentiate locally in a new irregular, septate branching pattern with reduced interhyphal spacing in response to the presence of the roots of host plants (Giovannetti *et al.*, 1993).

Differentiation of primary metabolism is technically difficult to demonstrate on a small scale, such as that representing differentiation at the cellular level. On the larger mycelial or tissue scale, factors which have been shown to be related to differentiation include various aspects of carbohydrate metabolism (Moore and Ewaze, 1976; Tan and Moore, 1994, 1995), nitrogen metabolism (Moore, 1984a), polyamine metabolism (Ruiz-Herrera, 1994), and phosphatase distribution (Jones *et al.*, 1995). Some of these aspects of metabolism will be discussed further in Chapter 3. Secondary metabolites are mostly low molecular weight natural products forming families of chemically-related compounds that are synthesised after active growth has ceased, often at the same time as differentiated morphological features (Bennett, 1995). Secondary metabolites are widely exploited as drugs, toxins and flavourings, but in few cases is their function known, though they are thought to benefit their producers as poisons that protect them against competitors, predators or parasites. They are produced from universally present precursors (most often acetyl-CoA, amino acids or shikimate) by specific enzymes thought to have originated, from different ancestral enzymes at different times in evolution (Cavalier-Smith, 1992). Secondary metabolism has a characteristic relationship with differentiation which is expressed even in fermenters, where the products are not formed until the primary growth phase of fungal pellets is over (Michel *et al.*, 1992; and see section 3.13.3). Secondary metabolism, like morphological differentiation, occurs after the primary growth phase has secured sufficient reserves to support morphogenetic processes (see section 3.13).

The ways in which mycelia differentiate morphologically have been recorded in a very extensive literature, some of which can be found in Burnett and Trinci (1979), Frankland *et al.* (1982), Boddy and Rayner (1983a,b), Jennings and Rayner (1984), Dowson *et al.* (1986), Sharland and Rayner (1986) and Trinci *et al.* (1994). The accepted interpretation is that fungal mycelia can express a range of alternative phenotypes that enable them to explore (rapid extension of sparsely-branched hyphal systems), assimilate (locally enhanced branching), conserve (cells differentiated to store metabolites), and redistribute (via strands and similar hyphal systems adapted to translocation) resources in ecological niches which are heterogeneous in both space and time. Although the exact causes and mechanisms are unknown, partial or complete occlusion of septal pores may occur during normal ageing and development, possibly as an aspect of stress management (Markham, 1992), with septal pore occlusions developing in response to severe trauma, such as tip bursting or severance of the hypha, but also as a reaction to less disruptive stress, such as nutrient deprivation. Complex structures often develop in the septal pores which are believed to partly occlude the pore, acting as subcellular sieves to permit selective passage of ions, molecules and small cytoplasmic components, whilst regulating or preventing the migration of large organelles (Markham, 1994, 1995).

Rayner (1992) and Rayner *et al.* (1994, 1995a,b) formulate an argument that the different local phenotypes result from changes in the hyphal envelopes (membranes and walls) which affect the flow and passage of molecules. They stress the involvement of epigenetic processes which, once initiated, may be beyond immediate genetic control. Later, Rayner (1997) developed these ideas derived from observations with fungi into a more widely-ranging discussion which relates reactions across the dynamic boundary between a living organism and its surroundings to the diversity in behaviour expressed by the organism.

It begins to become evident from discussions of this sort that an important component of the differentiation of hyphal cells is that the differentiation is a reaction to the local environment. Response to the environment (and the word 'environment' must be taken to encompass other cells, the substrate, the substratum, the gaseous atmosphere and physical environmental conditions) has a large measure of control over the initiation of differentiation. The genetic constitution of the cell (or perhaps the subset of genes which are currently being expressed) determine the nature of the differentiation of which the cell is capable, but the local environment determines where, whether, and to what degree that differentiation will occur.

Rhythmic, or cyclical, growth of colonies on solid medium is an excellent example of this. Regular concentric banding of colonies grown *in vitro* is seen quite often. It results from regular changes in hyphal extension rate and branch formation as hyphae react to local conditions as part of an endogenous or externally-regulated circadian clock (Bell-Pedersen *et al.*, 1996). Lysek (1984) described the rhythmic growth or banding of the clock-mutant of *Podospora anserina* as resulting from increased branching of hyphae growing on the agar surface eventually causing staling and limiting further extension on the surface. On the other hand, submerged hyphae do not increase branching and, escaping the staling limitation, continue to extend and reach the surface some distance beyond the staled surface mycelial front (Fig. 2.4). Emergence of the submerged hyphae prevents further growth of the old surface mycelium but produces a new generation of surface hyphal tips which go through the same process of branching, staling and growth limitation. Repetition of the cycle gives rise to zones of alternately dense and sparse surface mycelium which are visible as a regular series of bands on the surface. Reduced extension rates and increased branching are accompanied by increased oxygen uptake and exposure to light. Lysek (1984) points out that rhythmic growth is a differentiation process which separates hyphae with different functions and properties in space and time. Extending hyphae, exploring for new substrates, are separated spatially from staling hyphae which may differentiate into sporing and resting structures. Lysek (1984) suggests that membrane structure and/or permeability play a role in these sorts of reactions, which anticipates the type of interpretation developed by Rayner *et al.* (1995b).

Among other examples of mycelial growth rhythms are the concentric rings and radial zonations which can be induced in the band-mutant of *Neurospora crassa* (Deutsch *et al.*, 1993) by such treatment as change of certain salt concentrations in the medium. Deutsch *et al.*, (1993) take their analysis beyond simple observation by using a 'cellular automaton model' which mimics growth, branching and differentiation of a fungal mycelium. They postulate a homogeneous microscopic hyphal branching pattern in which inhomogeneities, like spore differentiation, are induced by extracellular activator and inhibitor molecules. Regalado *et al.* (1996) developed a similar mathematical model in an attempt to assess the role of exogenous and endogenous factors in generating the heterogeneity which is so characteristic of fungal mycelia. Their analysis shows that nutrient availability is a decisive factor in generating mycelial heterogeneity. Similar postulates can be made (and maybe similar mechanisms actually exist) about the hyphal growth and branching patterns which result in formation of fungal tissues in more complex structures. When such tissues are studied *in vitro* the differentiated cells may or may not form and/or continue their differentiation in a normal fashion (Chiu and Moore, 1988a,b, 1990c; Butler, 1992a,b; see section 6.5). The interpretations placed upon these observations are that the cells differentiate in response to signals they receive from their immediate environment within the tissue and that maintenance of their state of differentiation requires continual reinforcement from those signals, whether these are nutritional

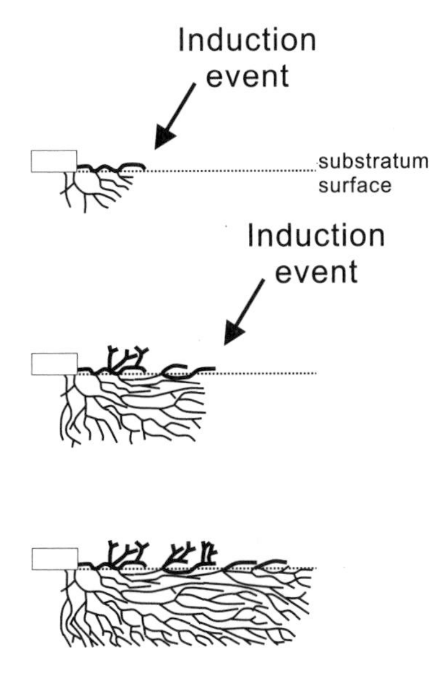

Fig. 2.4 Schematic illustration of the origin of rhythmic growth. Top: aerial and submerged hyphae grow away from the inoculum (rectangle) but in response to an induction event (may be an endogenous physiological cycle or external signal, like light or temperature fluctuation) surface growth is stopped. The check to surface extension prompts aerial branching and allows submerged hyphae to grow beyond the surface mycelium, branches from submerged hyphae re-establishing surface growth. A second induction event initiates another cycle of aerial branching after the check to extension on the surface (bottom diagram). Modified after Lysek (1984).

or purely informational. These aspects will be further discussed in Chapter 6, but it is worth emphasising here that it is again the cell surface, membrane and wall, which is the focus of the control mechanisms which are envisaged.

Arguably, therefore, differentiation starts with the wall and membrane because creation of a cell-specific barrier is a crucial step. Secondary or complex wall structures are commonly formed, reaching their highest expression, perhaps, in the greatly thickened and often highly sculptured walls of spores, even those of relatively simple fungi like *Rhizopus sexualis* (Hawker and Beckett, 1971). There are several examples in which the phenotypes of differentiated cells of fungal mycelia have been shown to be similar or identical to those of cells normally found in multicellular structures. Waters *et al.* (1972, 1975a), for example, describe two different cell types with heavily secondarily-thickened walls in mycelia of *Coprinus cinereus* which were otherwise characteristic of the inner and outer layers of sclerotia of that organism. In the extreme case, in *Hymenochaete corrugata* a cell type which seems adapted to serve as a protective 'epidermis' of the sclerotium may be formed on the mycelium in plates much larger than any normal sclerotium (Sharland *et al.*, 1986). Such observations might imply that fungi have a relatively restricted range of differentiation pathways open to them.

On the other hand, more fundamental changes are evident in some of the spore formation processes in which normal apical wall building mechanisms are highly adapted. In these cases there seems to be no shortage of developmental diversity. Many deuteromycete fungal conidiospores are produced in chains as the result of proliferation processes in which new wall is formed in a highly organised pattern at specific sites distant from the apex (Minter *et al.*, 1983). In some cases the original hyphal apex is lost with the first (i.e. top) spore, and is replaced by a new apex at the top of the conidiogenous cell, this process being repeated to produce the chain of spores. In another case, continuous upward production of wall for successive spores is the result of wall synthetic activity in a localised collar, or ring, in the upper region of the conidiogenous cell (Hawes and Beckett, 1977a,b); a mechanism known as 'ring wall building'. Clearly, wall synthesis at the hyphal apex is far from the end of the story. Further synthesis of new wall as well as modification of existing wall is a frequent occurrence. When and where it occurs is under exquisite control. Given that there are so many instances in which fungal wall synthesis is positionally and temporally regulated to produce regular change in morphology, one wonders why so much attention has been (and continues to be) lavished on the mundane activities located at the apex of the vegetative hypha.

Chemical modifications to the wall, in particular in chitin content, have also been associated with differentiation, and different chitin synthases may be involved in different stages of development (Motoyama *et al.*, 1994). Yarden and Russo (1996) show that hyphal branching of *Gaeumannomyces graminis* in culture is accompanied by localised elevation of chitin synthesis. During stem elongation of *Coprinus cinereus* the polysaccharide composition of the cell walls changes drastically (Kamada and Takemaru, 1983); at least part of the change being a matter of the left- and right-handed helicity of the chitin microfibrils in the walls (Kamada *et al.*, 1991; Kamada and Tsuru, 1993). Other very obvious aspects of differentiation-related modification of wall structure are the formation of spore appendages and extracellular adhesives involved in attachment of spores and hyphae to surfaces (Jones, 1994; Nicholson, 1996), and modification of the wall surface with hydrophobins, for example in ectomycorrhizas where they influence the colonisation of host roots (Tagu *et al.*, 1996).

In *cho*C mutants of *Aspergillus nidulans*, which are deficient in choline synthesis, direct linear correlations were observed between phosphatidylcholine content and chitin synthase activity and between the latter and hyphal extension rate (Binks *et al.*, 1991). Clearly, there is a close relationship between the chemical and biophysical constitution of the wall and the state of differentiation of the hyphal cell and the structure and nature of the wall reflects membrane composition (Gooday, 1995a). Differentiation of the plasma membrane of *Phytophthora palmivora* for proton pumping controlling internal pH during differentiation of zoospore has been demonstrated (Grzemski *et al.*, 1994). Further differentiation is implied by the range of different microvesicles which have been identified during zoosporogenesis in *Phytophthora cinnamomi* (Hyde *et al.*, 1991; Chambers *et al.*, 1995; Dearnaley *et al.*, 1996). Possibly the most remarkable recent report linking the membrane to differentiation is that of Belozerskaya *et al.* (1995) which claims that hyphae contain 600 to 800 μm long electrically interconnected assemblies comprising cells with different physiological functions. They claim that the electrical interconnections are involved in genetic control. In particular, a general photoregulatory mechanism was postulated to control light-induced changes in the electrophysiological properties of the hypha as well as gene induction. The main function of the electrical responses of the membrane to illumination being synchronisation of the operation of individual cells which are otherwise physiologically heterogeneous. Electric currents are ionic flows which are well documented in and around fungal hyphae (Harold, 1994), but taken together with observations of current surges, like action potentials, in hyphae in response to presentation of food sources (Olsson and Hansson, 1995), the idea of electrical signalling within and between hyphae looks very plausible. Such an extreme view, virtually postulating a fungal nervous system, needs very much more research before it can be taken seriously. However, it has attractions in explaining some observations which have been made of 'response at a distance'. For example Kuhad *et al.* (1987) found that cAMP (0.8 fmol) added to specific sites in Petri dish cultures of a mature dikaryotic colony of *Coprinus cinereus* caused a two-fold increase in the amount of glycogen over the whole culture within 19 h. The accumulation of glycogen occurred uniformly over the whole culture despite the non-uniform method of administration of the active chemical. Diffusion of a molecule about the size of dibutyryl-cAMP would achieve a radius of only about 12 mm from the point of administration within

24 h, yet by this time even the most distant sampling positions (about 40 mm from the point of administration of cAMP) were showing statistically significantly elevated glycogen contents. The observation implies rapid transmission of some sort of signal to orchestrate a homogeneous response across the mycelium and this may have some relation to the electrical interconnections postulated by Belozerskaya *et al.* (1995).

3

Metabolism and biochemistry of hyphal systems

In this chapter I present an account of the ways in which fungal hyphal systems obtain, absorb, metabolise, reprocess and redistribute nutrients. The description is relatively brief and more details can be found in texts on fungal physiology, such as Jennings (1995).

This material is relevant here for a number of reasons. First, these metabolic and biochemical activities of fungi provide the background and context within which their differentiation and morphogenesis occur. In discussing differentiation and morphogenesis there will be frequent need to refer to biochemical processes and it is useful to have those details readily to hand for reference. Second, fungi adapt the metabolism which normally serves their vegetative hyphal growth phases in specific ways to provide for and support their morphogenesis. Consequently, description of the basic metabolism is a valuable preparation for understanding the ways in which it is adapted. Third, description of basic metabolism provides another opportunity to make comparisons with the other eukaryotic kingdoms and to show, again, how kingdom Fungi make use of advanced and sophisticated mechanisms in the management of their cell biology.

Almost all of the processes described in this chapter will appear in some guise again in discussions of fungal morphogenesis in later chapters.

3.1 Nutrients in nature

Texts on fungal physiology usually present lists of chemicals that fungi can utilise when added to culture media. Like attempting to study the nutrition of *Homo sapiens* by noting the range of fast food outlets on the local high street, such an approach merely illustrates the range of nutrients the organism can be forced to endure. It reveals little about the natural substrates which have formed part of the evolutionary biology of the organism.

There are three major fungal nutritional modes: probably the majority are *saprotrophs* for which the substrates are dead organic materials not killed by the fungus itself (Cooke and Rayner, 1984). *Necrotrophs* invade living tissues which they kill and then utilise, whereas *biotrophs* exploit host cells which remain alive. In the latter case one might expect that though local digestion of host tissue may be necessary for penetration or establishment of the pathogen, only simple nutrients would be removed from the host because of the damage which would be inflicted on the host by large-scale digestion of polymeric cell constituents. Biotrophs may be host-specific, but saprotrophs and necrotrophs generally have a very large range of habitats open to them, and in the majority of these polymeric sources of nutrients predominate.

This predominance of polymers as sources of nutrients is obviously true for such materials as herbaceous plant litter, wood and herbivore dung, but it also applies to the soil.

The bulk of plant litter consists of plant cell walls and consequently contains large amounts of cellulose, hemicellulose and lignin, even though the cytoplasm of the dead plant cells will contribute lipids, proteins and organic phosphates to the remains. Wall components amount to 90% of the dry weight of wheat straw, for example (Chang, 1967). Wood, since it derives from secondary wall growth, is especially rich in wall polymers and correspondingly relatively poor in other potential nutrients, particularly nitrogen and phosphorus (Swift, 1977). Digested litter (a euphemism for herbivore dung) is, on the other hand, relatively enriched in nitrogen, vitamins, growth factors and minerals, since in passage through the intestine it accumulates the remains of bacteria, protozoa and other microorganisms (Lodha, 1974). The composition of animal tissue varies enormously according to the particular organ system considered, but in nature most animal remains will be eaten by animal scavengers too rapidly for any microbes to be able to compete, so the microorganisms will be left with the parts – 'skin, gristle and bone' – that other organisms cannot reach.

In soil, nitrogen exists largely in the form of organic compounds; the proportion of nitrogen occurring as ammonium, nitrate or nitrite rarely exceeds 2%, although there may be a higher proportion of clay-fixed ammonium in some soils (Bremner, 1967). Inorganic nitrogen compounds only predominate in agricultural soils which are repeatedly dosed with chemical fertilisers. Nitrite is not usually detectable and nitrate content is usually very low in natural soils because these salts are so readily leached out by rain, so in most cases exchangeable-ammonium and the organic nitrogen provide saprotrophs with the most readily-available sources of this element. In most surface soils, 20–50% of total nitrogen occurs in proteinaceous form, and 5–10% as combined and complexed amino sugars. The amino sugars also contribute, of course, to the carbohydrate component of the soil, which represents 5–16% of the total organic matter. Here, again, though most soil carbohydrate is in polymeric form. Monosaccharides represent less than 1% of the carbohydrate but cellulose can account for up to 14% of total carbohydrate and chitin must also be well represented in view of the occurrence of amino sugars. Ericoid mycorrhizas are able to use chitin as a sole source of nitrogen (Leake and Read, 1990). About 50–70% of total phosphorus in soil is organic, mostly as phosphate esters related to or derived from compounds like nucleic acids, inositol phosphates and phospholipids.

Although inorganic forms of sulfur may accumulate in some soils (e.g. as calcium and magnesium sulfates in arid regions; calcium sulfate co-crystallised with calcium carbonate in calcareous soils) there is little inorganic sulfur in the surface horizons of soils in humid regions. Organic sulfur occurs in the form of methionine and cystine (and derivatives), and sulfate esters, including sulfated polysaccharides and lipids.

That fungi can utilise the polymeric nutrient sources which are in most ready supply can be demonstrated indirectly by experimentally monitoring growth on, or degradation of, particular materials. For example, supplementation of mushroom compost with vegetable oils and linoleic and oleic acid esters causes growth stimulation of the cultivated mushroom *Agaricus bisporus* (Wardle and Schisler, 1969), which must mean that, at least in pure culture, the organism is able to utilise supplied lipids. Similarly, though lignin is considered to be highly resistant to degradation, about two-thirds of the lignin present in compost when first inoculated with *A. bisporus* has disappeared by the time the crop is picked. *Agaricus bisporus*, shares with two other litter-degrading mushrooms, *Coprinus cinereus* and *Volvariella volvacea*, the ability to use protein as efficiently as the sugar glucose as a sole source of carbon, and can use protein, additionally, as a source of nitrogen and sulfur (Kalisz *et al.*,

1986). *A. bisporus* and a wide range of other filamentous fungi (Fermor and Wood, 1981; Grant *et al.*, 1986) have been shown to be able to degrade dead bacteria and to utilise them as sole source of carbon, nitrogen and phosphorus.

For most fungi in most circumstances, therefore, the initial nutritional step is the excretion of enzymes able to convert polymers to the simple sugars, amino acids, carboxylic acids, purines, pyrimidines, etc, the cell can absorb. In doing this they obviously contribute to recycling and, mineralisation of nutrients.

3.2 Extracellular polymer-degrading enzymes

Besides lignin (see below), the bulk of plant cell biomass consists of the polysaccharides cellulose, hemicelluloses, and pectins in varying proportions depending on the type of cell and its age. Plant biomass does not consist of neatly isolated packets of polysaccharide, protein and lignin; these three (and other materials) are intimately mixed together, so that it is better to think of the degradation of lignocellulosic and/or lignoprotein complexes. A typical agricultural residue, like cereal straw or sugar cane bagasse, contains 30–40% cellulose, 20–30% hemicellulose and 15–35% lignin. Organisms may differ in their ability to degrade components of this mixture. On this sort of basis wood-decay fungi have been separated into white-rot, brown-rot and soft-rot species. The white-rot fungi (about 2000 species, mostly basidiomycetes) can metabolise lignin, on the other hand, brown-rot fungi (about 200 basidiomycete species) degrade the cellulose and hemicellulose components without much effect on the lignin. Soft-rot species (mostly soil-inhabiting ascomycetes and deuteromycetes) have rather intermediate capabilities, being able to degrade cellulose and hemicellulose rapidly, but lignin only slowly. These differences in behaviour are a reflection of the different enzymes produced by these organisms and serve to emphasise that the organisms must digest complexes of potential nutrient sources and assemble panels of different enzymes to do so. For ease of presentation here, however, I must consider degradation of specific compounds separately.

3.2.1 Polysaccharide degradation

Polysaccharides are polymers of monosaccharides in which the constituent sugars are connected with glycosidic bonds. There is a considerable variety of polysaccharides, both because of the number and variety of available sugars and because of the diversity of bonding possibilities between different carbon atoms of the adjacent sugar residues. There is a matching variety of enzymes, hydrolases or glucosidases, capable of hydrolysing this range of glycosidic links. Enzymes responsible for polymer degradation (any polymer, not just polysaccharide) may employ one of two strategies of attack. They may attack randomly, effectively fragmenting the polymer molecule into a number of oligomers, these are the endo-enzymes, or they may approach terminally, digesting away monomers or dimers, the exo-enzymes.

Cellulose is the most abundant organic compound on Earth and accounts for over 50% of organic carbon; about 10^{11} tons are synthesised each year. It is an unbranched polymer of glucose in which adjacent sugar molecules are joined by β-1⟶4 linkages (Fig. 3.1); there may be from a few hundred to a few thousand sugar residues in the polymer molecule, corresponding to molecular weights from about 50,000 to approaching 1 million. Breakdown of cellulose is chemically straightforward, but is complicated by its physical form, which is still not completely understood. Mild acid hydrolysis of cellulose releases soluble sugars, but does not go to completion; oligomers of 100–300 glucose residues remain. The fraction which is readily hydrolysed is called amorphous cellulose while that which is resistant to acid is called crystalline cellulose. Since it influences chemical breakdown, the conformation and three-dimensional structure of cellulose must influence cellulolytic enzyme activity.

The cellulolytic enzyme (cellulase) complex of white-rot fungi like *Phanerochaete chrysosporium* and deuteromycetes like *Trichoderma reesei* consists of a number of hydrolytic enzymes: endoglucanase, exoglucanase and cellobiase (which is a β-glucosidase) which work synergistically and are organised into an extracellular multienzyme complex called a cellulosome (Lemaire, 1996). Endoglucanase attacks cellulose at random, producing glucose, cellobiose (a disaccharide) and some cellotriose (a trisaccharide). Exoglucanase attacks from the non-reducing end of the cellulose molecule, removing glucose units; it may also include a cellobiohydrolase activity which produces cellobiose by attacking the non-reducing end of the polymer. Cellobiase is responsible for hydrolysing cellobiose to glucose. Glucose is, thus, the readily-metabolised end-product of cellulose breakdown by enzymatic hydrolysis.

In addition to catalytic regions, many cellulolytic enzymes contain domains not involved in catalysis, but participating in substrate binding, multi-enzyme complex formation, or attachment to the cell surface. These domains assist in the degradation of crystalline cellulose by securing the

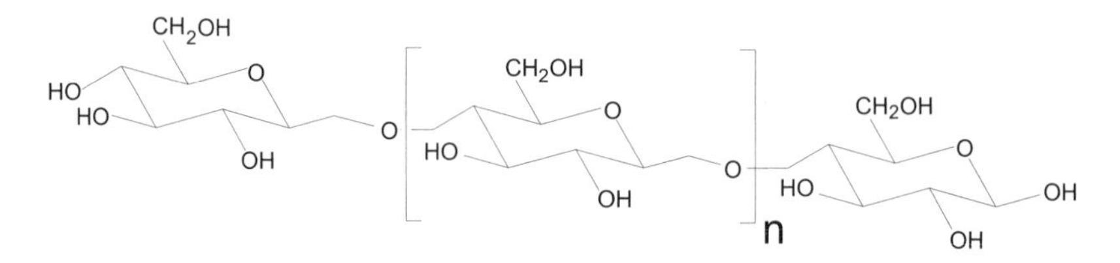

Figure 3.1 Structural formula of cellulose.

enzymes to the substrate, by focusing hydrolysis on restricted areas in which the substrate is synergistically destabilised by multiple cutting events, and by enabling recovery of degradation products by the producing organism (Beguin and Aubert, 1994; Radford *et al*., 1996).

When grown on cellulose, the white-rot fungi like *Phanerochaete chrysosporium* produce two cellobiose oxidoreductases; a cellobiose: quinone oxidoreductase (CBQ) and cellobiose oxidase (CBO). Cellobiose oxidase is able to oxidise cellobiose to the δ-lactone, which can then be converted to cellobionic acid and then glucose + gluconic acid; cellobiose d-lactone can also be formed by the enzyme cellobiose: quinone oxidoreductase. Similar cellobiose-oxidising enzymes, capable of utilising a wide variety of electron acceptors, have been detected in many other fungi, though their role is uncertain. These enzymes are probably of most significance in regulating the level of cellobiose and glucose, the accumulation of which can inhibit endoglucanase activity. The role originally ascribed to CBQ was as a link between cellulose and lignin degradation. Cellobiose oxidase also reduces Fe(III) and together with hydrogen peroxide, generates hydroxyl radicals. These radicals can degrade both lignin and cellulose, possibly indicating that cellobiose oxidase has a central role in degradation of wood by wood-degrading fungi (Ander, 1994). However, most evidence available so far indicates that the presence of CBO/CBQ with lignin peroxidases and laccases actually reduces the rate of oxidation of lignin degradation products (Eriksson *et al*., 1993).

Brown-rot fungi use a rather different initial cellulolytic system to the hydrolytically-based one employed by the white-rots. Brown-rot fungi are able to depolymerise cellulose rapidly and virtually completely. Even cellulose deep within the walls and protected by lignin polymers is prone to attack. The process seems to depend on H_2O_2 (secreted by the fungus) and ferrous ions in the wood oxidising sugar molecules in the polymer, thereby fragmenting it and leaving it open to further attack by hydrolytic enzymes. Interestingly, it has been suggested that the oxalate crystals which coat so many fungal hyphae are responsible for reducing the ferric ions normally found in wood to ferrous ions, so potentiating oxidative cleavage of the cellulose. Although the white-rot fungi produce H_2O_2 for lignin degradation they do not secrete oxalate and therefore fail to depolymerise cellulose oxidatively.

Hemicellulose is a name which covers a variety of branched-chain polymers containing a mixture of various hexose and pentose sugars, which might also be substituted with uronic and acetic acids. The main hemicelluloses found in plants are xylans (1⟶4-linked polymers of the pentose sugar xylose), but arabans (polyarabinose), galactans (polygalactose), mannans and copolymers (e.g. glucomannans and galactoglucomannans) are also encountered. The major angiosperm hemicellulose is a xylan with up to 35% of the xylose residues acetylated, and it is also substituted with 4-*O*-methylglucuronic acid in dicotyledonous plants. Enzymes responsible for hemicellulose degradation are named according to their substrate specificity; for example, mannanases degrade mannans, xylanases degrade xylans, etc. As xylans predominate in plant walls, more is known about xylanases.

Xylanases can be induced by their substrate, the response being for the fungus to produce a complex of enzymes rather than a single one. The complex consists of at least two endoxylanases and a β-xylosidase. The endoxylanases degrade xylan to xylobiose and other oligosaccharides while the xylosidase degrades these smaller sugars to xylose. Some arabinose is also formed, showing that the xylanase complex is able to hydrolyse the branch points in xylan.

Pectins consist of chains of β-1⟶4 linked galacturonic acids, in which about 20 to 60% of the carboxyl groups are esterified with methanol. They occur primarily in the middle lamella between plant cells. As this represents only a small proportion of the plant wall they are correspondingly of little importance as a component of plant litter. However, extensive breakdown of the middle lamella of living plants is brought about by necrotrophic parasites. Pectinases, therefore, are of great importance during fungal invasion of plant tissue (Byrde, 1982).

Polygalacturonases and pectin lyases attack the true pectins, while arabanases and galactanases degrade the neutral sugar polymers associated with them. These activities have drastic effects on the structural integrity of the tissues which may extend to death of the cell due to osmotic stresses imposed by damage to the wall. It seems likely that the products of pectinase activity will be absorbed as nutrient by the fungus, but these enzymes are better considered to be part of the machinery by which plant defences are breached than as being concerned primarily with nutrient supply.

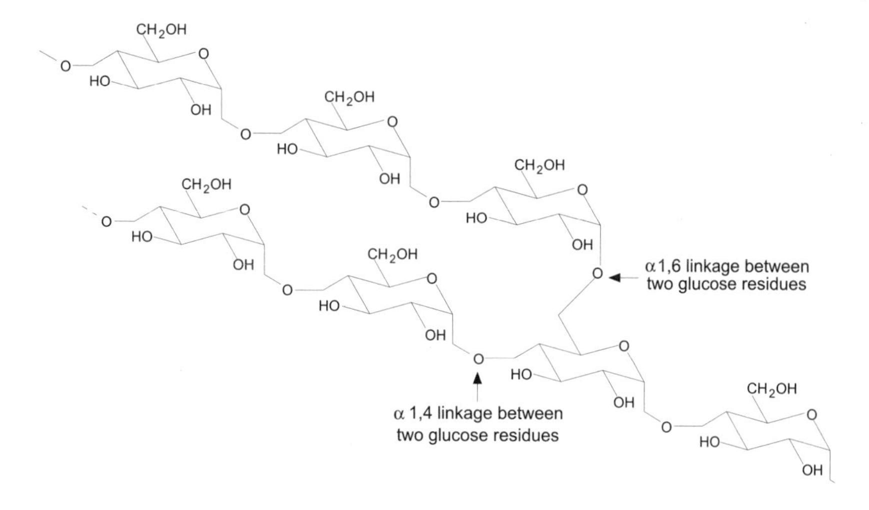

Figure 3.2 Structural formula of amylopectin.

Chitin, in which the repeating unit is the same as that in cellulose except that the hydroxyl group at C-2 is replaced by an acetamido group (Fig. 2.1), is the second most abundant polymer on Earth as it occurs in the exoskeletons of arthropods and, of course, in fungal cell walls. Polysaccharides which contain amino sugars or their derivatives are called mucopolysaccharides. Chitin is degraded by chitinase, a glucan hydrolase which attacks the β-1⟶4 glycosidic bonds, eventually producing the disaccharide chitobiose which is then converted to the monosaccharide *N*-acetylglucosamine by chitobiase. Chitinase may also be involved in fungal wall synthesis (see Chapter 2).

Starch, the major reserve polysaccharide of plants, contains glucose polymers with α-1⟶4 glycosidic bonds. Amylose is constituted of long unbranched chains, whereas amylopectins (comprising 75–85% of most starches) have branch points formed from α-1⟶6 glycosidic bonds (Fig. 3.2). Starch degrading enzymes include: α-amylases, which are endoamylases acting on 1⟶4 bonds and bypassing the 1⟶6 bonds; β-amylases, which are exoamylases producing the disaccharide maltose by splitting alternate 1,4 bonds until they reach a 1,6 branch point (which they cannot bypass); amyloglucosidases (or glucoamylases), which can act on both 1⟶4 and 1⟶6 bonds, seem to occur almost exclusively in fungi; debranching enzymes (e.g. pullulanase) which sever 1⟶6 bonds; α-glucosidases which hydrolyse 1⟶4 glycosidic linkages in disaccharides and oligosaccharides, producing glucose as the end-product of starch breakdown (Radford *et al.*, 1996).

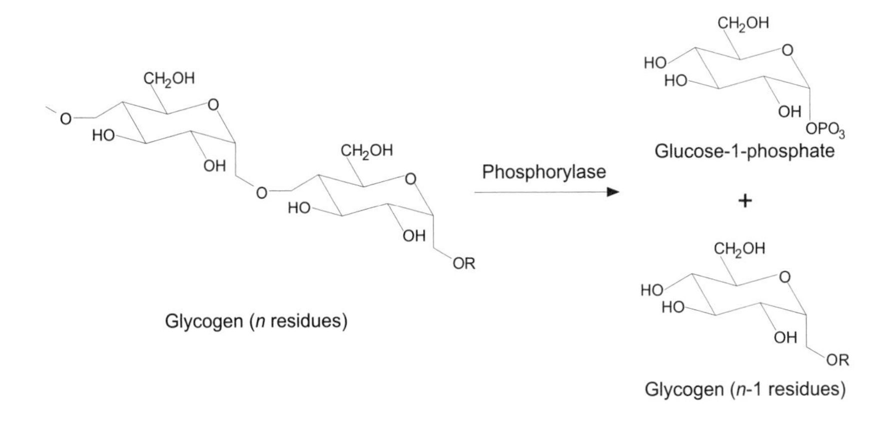

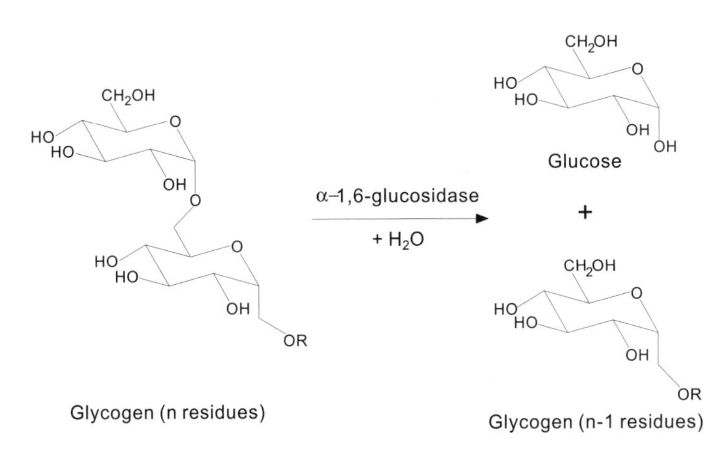

Figure 3.3 Intracellular degradation of glycogen. The top panel shows glycogen phosphorylase activity severing one of the 1⟶4 linkages. The bottom panel shows the 'debranching activity' of the glucosidase hydrolysing a 1⟶6 linkage.

Glycogen is very similar to starch, being a branched polymer composed of glucose residues linked by α-1⟶4 glycosidic bonds; about every tenth residue is involved in a branch formed by α-1⟶6 glycosidic bonds. It is the polysaccharide reserve found in animal tissues, and in the fungi themselves. Most fungi are likely to encounter glycogen in their surroundings, as they are likely to be surrounded by dead and dying fungal cells. Intracellularly, glycogen is degraded by a phosphorylase which releases glucose 1-phosphate for metabolic use (Fig. 3.3), with the aid of a transferase and α-1⟶6-glucosidase (activities of a single polypeptide) to deal with the branches. Extracellularly, glycogen is probably degraded by components of the amylase enzyme complex.

3.2.2 Lignin degradation

Lignins are high-molecular-weight, insoluble polymers which have complex and variable structures, being composed essentially of many methoxylated derivatives of benzene, especially coniferyl, sinapyl, and coumaryl alcohols, the proportions of these three differing between angiosperms and gymnosperms (Fig. 3.4). The ability to degrade lignin is limited to a very few microorganisms, including a range of basidiomycetes, some ascomycetes and a few bacteria. Consequently, lignin is extremely resistant to microbial degradation itself and can protect other polymers from attack. It has been suggested that lignin may be degraded non-enzymically, being oxidised by a chemically-produced 'activated oxygen'; at least part of the argument being that lignins are so variable in chemical structure that any enzymes concerned would have to be non-specific. However, even the information about enzymes seems to point towards something like a 'combustion' process (Reid, 1995).

Most research has been concentrated on two ligninolytic organisms; the white-rot basidiomycete fungus *Phanerochaete chrysosporium* (= *Sporotrichum pulverulentum*) and the actinomycete bacterium *Streptomyces viridosporus*. *P. chrysosporium*, in axenic culture, is able to mineralise lignin completely to CO_2 and water and it is now quite clear that both organisms do produce extracellular enzymes which are specifically and necessarily involved in lignin degradation, the best characterised of which are laccase, lignin peroxidases and manganese peroxidases.

Lignin peroxidase (ligninase) is the key lignin-degrading enzyme of white-rot fungi. *P. chrysosporium* produces a family of lignin peroxidases, which are extracellular glycosylated heme proteins, as major components of its lignin-degrading system. There can be as many as 15 lignin peroxidase isozymes, ranging in molecular mass from 38,000 to 43,000, the spectrum of isozymes produced depending on culture conditions and strains employed. Manganese-dependent peroxidases are a second family of extracellular heme proteins produced by *P. chrysosporium* that are also believed to be important in lignin degradation by this organism (Elisashvili, 1993; Reddy and Dsouza, 1994; Broda *et al.*, 1996; Cullen and Kersten, 1996).

The lignin-degradative system of *P. chrysosporium* appears after cessation of primary growth (i.e. it is an aspect of the secondary metabolism of

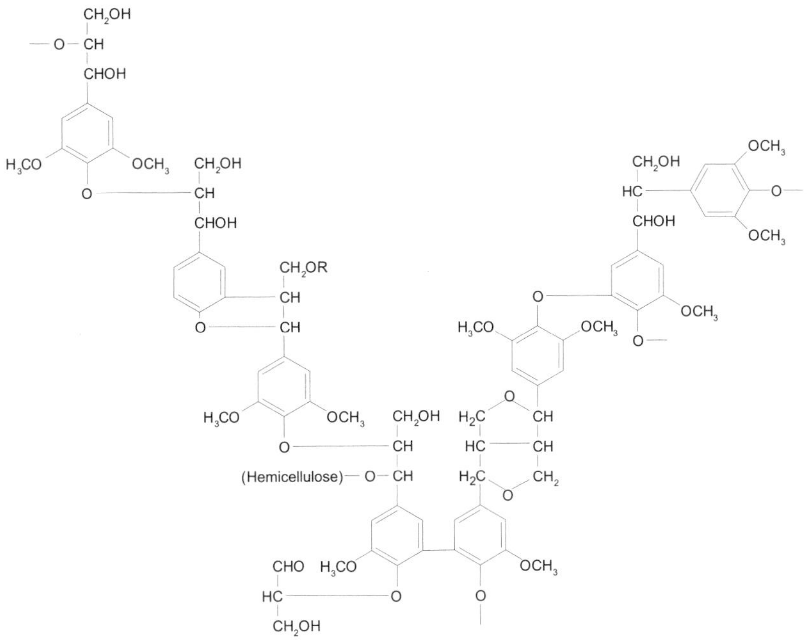

Figure 3.4 Schematic formula of angiosperm lignin.

the organism) and can be induced by nitrogen starvation. Intracellular cAMP levels appear to be important in regulating the production of lignin peroxidases and manganese peroxidases, though production of the former is affected more than that of the latter. When the fungus is grown in low-nitrogen medium there is an increase in H_2O_2 production by cell extracts which correlates with the appearance of ligninolytic activity; experimental destruction of H_2O_2 by adding the enzyme catalase strongly inhibits lignin breakdown. Thus, evidence for involvement of H_2O_2 in lignin degradation by *P. chrysosporium* is conclusive. So it seems that activated oxygen is involved in degrading lignin, but is held in the active site of a specific extracellular enzyme, the lignin peroxidase. The initial step involves oxidation by one electron and produces unstable intermediates which can undergo a wide range of subsequent oxidative reactions (Kirk *et al.*, 1990; Hatakka, 1994; Cullen and Kersten, 1996).

Although few fungi produce ligninolytic enzymes, a much wider range excrete laccases as extracellular enzymes. These are copper-containing oxygenases which are able to oxidise *o*- and *p*-phenols and are required for the metabolism of lignin degradation-products. They are particularly interesting as their appearance or disappearance in fungal cultures has been correlated with sexual and asexual reproduction in a number of cases. Thus, during mycelial growth of the cultivated mushroom, *Agaricus bisporus*, a large proportion of the compost lignin is degraded and correspondingly high activities of laccase are recorded. This one enzyme can amount to 2% of the total fungal protein (Wood, 1980a). Yet, as the culture forms fruit bodies laccase activity is rapidly lost, initially by inactivation and subsequently by proteolysis (Wood, 1980b). The activity of a manganese-dependent lignin-degrading peroxidase has also been monitored from the time of colonisation of the compost through the development of fruit bodies and was found to be correlated with the laccase, suggesting that both enzymes have significant roles in lignin degradation by this fungus (Bonnen *et al.*, 1994). The pattern of behaviour of these two enzymes illustrates both the changing nutritional demands of fungal mycelia as they process through successive developmental phases and the ability of the mycelium to act on its environment to satisfy those demands.

3.2.3 Protein degradation

Proteinases are peptide hydrolases; a group of enzymes which hydrolyse the peptide bonds of proteins and peptides, cleaving the substrate molecule into smaller fragments and, eventually, into amino acids. This is a complex group of enzymes, varying greatly in physicochemical and catalytic properties. Proteolytic enzymes are produced intra- and extracellularly, playing important roles in regulatory processes of the cell as well as contributing to nutrition through degradation of protein food sources. Intracellular proteolysis seems to be the responsibility of large multicatalytic complexes of proteinases which are called proteasomes. 20S proteasomes are cylindrical particles found in the cytoplasm and nucleoplasm of all eukaryotes. They are composed of a pool of 14 different subunits (M_r 22–25 kDa) arranged in a stack of 4 rings. 26S proteasomes are larger, comprising a 20S proteasome as core particle with additional subunits complexed at the ends of the 20S cylinder. Proteasomes are needed for stress-dependent and ubiquitin-mediated proteolysis. They are involved in the degradation of short-lived and regulatory proteins and are, therefore, important for cell differentiation, adaptation to environmental changes, and control of the cell cycle (Hilt and Wolf, 1995).

Extracellular proteinases are involved mainly in the hydrolysis of large polypeptide substrates into the smaller molecules which can be absorbed by the cell. Extracellular proteinases are produced by many species of fungi but most is known about protein utilisation and proteinase production by *Aspergillus* species and *Neurospora crassa*. The basidiomycetes *Agaricus*, *Coprinus*, and *Volvariella* have been shown to be able to use protein as a sole source of carbon about as efficiently as they can use the sugar glucose, and can also use protein as a source of nitrogen and sulfur (Kalisz *et al.*, 1986, 1987, 1989). The real value of protein as a nutrient for fungi became evident only comparatively recently. Mycorrhizal fungi have been shown to use protein as a source of both nitrogen and carbon (Bajwa and Read, 1985; Spinner and Haselwandter, 1985) and some ectomycorrhizas supply nitrogen derived from proteins in soil to their higher plant symbionts (Abuzinadah and Read, 1986a, 1986b; Abuzinadah *et al.*, 1986; Read *et al.*, 1989; Read, 1991). Protein is probably the most abundant nitrogen source available to plant-litter-degrading organisms in the form of plant protein, lignoprotein and microbial protein. Many pathogenic microorganisms secrete proteinases which are involved in the infection process and some, including the apple pathogenic fungus *Monilinia fructigena*, are known to utilise host proteins for nutrition. The virulence of a few pathogenic fungi is correlated with their extracellular proteinase activity. Several species release specific proteinases which can hydrolyse structural and other proteins resistant to attack by most other proteinases, such as insect cuticles (Samuels and Paterson, 1995: St Leger, 1995). The dermatophytes *Microsporum* and *Trichophyton* produce collagenases, elastases and keratinases.

Enzymes which degrade proteins form two major groups: peptidases and proteinases (Kalisz, 1988). Exopeptidases remove terminal amino acids or dipeptides and are subdivided according to whether they act at the carboxy terminal end of the substrate protein (carboxypeptidases); the amino terminal end (aminopeptidases), or on a dipeptide (dipeptidases). Proteinases cleave internal peptide bonds; they are endopeptidases. The proteinase catalytic mechanism can be determined indirectly by study of response to inhibitors which react with particular residues in the active site of the enzyme. This leads to subclassification into four groups.

- Serine proteinases are the most widely distributed group of proteolytic enzymes. They have a serine residue in the active site, are generally active at neutral and alkaline pH, and show broad substrate specificities.
- Cysteine proteinases occur in few fungi though extracellular cysteine proteinases have been reported in *Microsporum* sp.,

Aspergillus oryzae, and *Phanerochaete chrysogenum* (*Sporotrichum pulverulentum*).

- Aspartic proteinases show maximum activity at low pH values (pH 3 to 4) and are widely distributed in fungi.
- Metalloproteinases have pH optima between 5 and 9 and are inhibited by metal-chelating reagents,such asethylenediamine tetraacetic acid (EDTA). In many cases the EDTA-inhibited enzyme can be reactivated by zinc, calcium or cobalt ions. Metalloproteinases are widespread, but only a few have been reported in fungi and most of these are zinc-containing enzymes, with one atom of zinc per molecule of enzyme.

3.2.4 *Lipases and esterases*

Lipases and esterases catalyse the hydrolysis of esters made between alcohols and organic ('fatty') acids. They generally have low specificity and any lipase will hydrolyse virtually any organic ester, though different esters will be acted upon at different rates. The main factors influencing what specificity is expressed are the lengths and shapes of hydrocarbon chains either side of the ester link. The term esterase is generally applied to enzymes 'preferring' short carbon chains in the acyl group and these are the enzymes which have so often been examined for electrophoretic variants in studies of population genetics. The lipases 'proper' tend to favour long carbon chains in the acyl group. Their substrates include fats, the lipid components of lipoprotein and the ester bonds in phospholipids.

Extracellular lipase production has been detected in *Agaricus bisporus* during degradation of bacteria (Fermor and Wood, 1981). In fermenter cultures most of the lipase is produced in the stationary phase (i.e. is a secondary metabolic activity) and regulation of lipase production in *Rhizopus* is very much affected by carbon and nitrogen sources in the medium and by the oxygen concentration (Mukhamedzhanova and Bezborodov, 1982; Guiseppin, 1984).

3.2.5 *Phosphatases and sulfatases*

Phosphatases are also esterases, acting on esters of alcohols with phosphoric acid. They are enzymes of comparatively low specificity but fall into groups depending on their activity as phosphomonoesterases, phosphodiesterases and polyphosphatases. The phosphomonoesterases are further distinguished according to their pH optima as alkaline or acid phosphatases. For example, phosphatases are among the extracellular enzymes produced by *Agaricus bisporus* during growth on compost and must be important, therefore, in the nutrition of such litter degrading fungi. Sulfatases act on sulfate esters in the same way that phosphatases act on phosphate esters. They may be important in recovery of sulfate from the sulfated polysaccharides which are found in soils.

3.3 Production, location, regulation and use of degradative enzymes

3.3.1 Production

Extracellular enzymes are produced within the cell but act outside it. Consequently, they must be secreted across the plasmalemma. The indications are that the processes involved in protein translocation across membranes are very similar in all eukaryotes. Polypeptides destined for secretion are identified by short amino terminal transient 'signal' sequences which consist of uninterrupted stretches of at least six hydrophobic amino acid residues. The signal sequence is in the first part of the polypeptide to be synthesised on the ribosome and, as its hydrophobicity confers an affinity for the lipid environment of a membrane bilayer, the signal sequence 'targets' the ribosome producing it onto the endoplasmic reticulum membrane. The translocation machinery includes a targeting system on the side of the membrane where the polypeptide is being synthesised, an oligomeric transmembrane channel, a translocation motor powered by the hydrolysis of nucleotide triphosphate, and a protein folding system on the far side of the membrane (Edwardson and Marciniak, 1995; Schatz and Dobberstein, 1996). The signal peptides are cleaved off by proteolysis during membrane passage, yielding active polypeptides which can be delivered to their site of action in membrane-bound vesicles.

Fungal compartments housing Golgi functions in secretory transport do not resemble the interphase Golgi apparatus of other eukaryotes, but are more fragmentary. Nevertheless, the overall process is similar to the mammalian secretion scheme and morphologically distinct populations of transport microvesicles are formed from endoplasmic reticulum and Golgi cisternae. Thus, transfer of proteins between different compartments in the cell and between the endoplasmic reticulum and the plasmalemma involves coated vesicles which bud off from the endoplasmic reticulum, migrate to the plasmalemma with which they fuse, externalising their contents. Components of the membrane-associated protein translocation machinery have been recognised in the yeast, *Saccharomyces cerevisiae*. Similarly, coated vesicles have been identified in *S. cerevisiae*, *Neurospora crassa* and the rust fungus *Uromyces phaseoli*. The protein clathrin, which has a major role in constructing a basket-like coat around these transport vesicles has been extracted from *S. cerevisiae*, and has been associated with vesicles containing secretable enzymes. Gene products involved in sorting vacuolar protein, which are responsible for the recognition, packaging, and vesicular transport of proteins to the vacuole of yeast are homologous with those of other eukaryotic cells. There is no doubt that the characteristic protein export mechanisms of eukaryotes operate in fungi, but with some modification, presumably to compensate for the poorly-developed Golgi complex. For example, in fungi there is evidence that the nuclear envelope lumen houses certain functions normally associated with the endoplasmic reticulum and some steps of outer-chain glycosylation may occur in microvesicles during transport.

3.3.2 Location

Some of the extracellular enzymes which are produced are soluble and are freely dispersed in fluid films surrounding the hyphae, but others are fixed in space by being bound to the hyphal wall, extracellular matrix or even to the substrate itself. The exo- and endoglucanases of the cellulase complex strongly absorb to native cellulose; cellobiase does not absorb at all. The fungus excretes the enzymes into the substratum where the glucanases bind to the insoluble native cellulose to produce cellobiose which is hydrolysed in the soluble phase by cellobiase. The producing organism must then absorb the glucose end-product. Some lipases act preferentially at a water/lipid interface. In contrast, other extracellular enzymes seem to remain associated with the envelope of the producing cell.

Many cell-wall bound enzymes have been reported in fungi, including glucosidase of *Trichoderma reesei*, trehalase, invertase and proteinases of *Neurospora*, and cellulases of *Volvariella volvacea*. Enzymes have also been found inside layers of extracellular polysaccharide-rich sheaths outside the cell wall.

This sort of natural immobilisation of enzymes offers advantages to fungi. Wall-located enzymes would only degrade substrate in the immediate vicinity of the cell, so ensuring that the organism producing the enzyme has some advantage in the competition with surrounding organisms for the nutrients produced by the enzyme activity. In this way the fungus can exert a degree of control over its immediate environment. For basidiomycetes there is evidence that wall-bound proteinases may be released as their substrate becomes depleted so that they can scavenge further afield in the substratum. Other extracellular enzymes produced by their parent organisms may be among the proteins they scavenge: proteolysis is one of the processes causing decline in extracellular laccase when fruit bodies form in cultures of *Agaricus bisporus* (Wood, 1980b).

3.3.3 Regulation

Production of most enzymes is regulated according to the need for the enzyme activity. The cellulase enzyme complex, for example, is inhibited and probably repressed by the end-products of its activity, i.e. cellobiose and, especially, glucose. On the other hand, enzyme synthesis can be induced, and known inducers of cellulases include cellulose, cellobiose, sophorose and lactose. Induction of cellulases *in vivo* appears to be due to soluble products generated from cellulose by cellulolytic enzymes synthesised constitutively at a low level. These products are thought to be converted into true inducers by transglycosylation reactions.

In *Aspergillus* species, proteinase production is controlled by derepression; the *Neurospora crassa* proteinase is controlled by induction and repression. In neither case is proteinase produced in the presence of ammonia; which appears, therefore, to be the primary source of nitrogen. In sharp contrast, production of extracellular proteinases in the basidiomycetes is regulated mainly by induction; as long as substrate protein is available the proteinases are produced, even in the presence of adequate alternative supplies of ammonia, glucose and sulfate (Kalisz *et al.*, 1986). So in this case the protein might be presumed to be the 'first choice' substrate.

3.4 The menu of basic nutrients

With the prevalence of glucose polymers in the natural world it is inevitable that this monosaccharide proves to be the most widely utilised carbon source. Other monosaccharides will serve too, since most can be readily incorporated into metabolism after minor isomeric transformation. Among nitrogen sources, nitrate will be rarely encountered, ammonium could be met with in quantity either as in its clay-bound form or as an 'excretion' product of microorganisms using protein as a carbon source. However, in most locations it seems likely that fungi will find amino sugars and, especially, amino acids to be quantitatively the most available source of nitrogen and to represent good return on any energy invested in uptake as they would serve as carbon sources also. This sort of consideration is likely to explain the fact that the biomass yields of certain basidiomycete fungi were 4 – 7 times greater from an ammonium-medium containing both glucose and protein as carbon sources than from media containing either one of these nutrients as sole carbon source (Kalisz *et al.*, 1986).

3.5 The wall and membrane as barriers

3.5.1 Barriers to the extracellular environment

The plasma membrane is the boundary between what constitutes the cell and what constitutes the environment. In models of the origin of life, formation of such a membrane represents a crucial step. Only if the cell is completely separated from its environment can the fundamental chemistry of living things be made to work. The essential function of the plasma membrane is protection. Its purpose is to prevent leakage of cellular material to the environment and to prevent intrusion of environmental molecules into the cell.

The wall, too, serves a protective function, but a primarily mechanical one. However, it is essential to appreciate that wall and plasma membrane together form a physicochemical mechanism to regulate entry into the cell of that all-pervasive environmental molecule, water. Unlike other parts of the cell, for example, enzymes, mitochondria, even mitotic division spindles, isolated fungal walls have chemical and physical properties but exhibit no function similar to their natural one. Also, the other part of the mechanism, fungal protoplasts, which (in theory, at least) are intact fungal cells from which the wall has been removed (Peberdy and Ferenczy, 1985), are unable to regulate water flow and must be stabilised by suspension in abnormally high concentrations of osmotically active solutes. Evidently, the native function is lost when the two components of the mechanism are separated. The water relationships of fungal cells will be discussed towards the end of this section; prior to that, attention will be focused on the relationship of aqueous solutes with the wall and membrane.

3.5.2 Transfer across the plasma membrane

The plasma membrane is a lipoidal layer which effectively separates the aqueous 'bubble' of the cell from its aqueous surroundings. Obviously, this separation is not complete or absolute. The cell must exchange chemicals with the environment, extruding excretion products and absorbing nutrients. However, only molecules which dissolve readily in lipid are able to penetrate the membrane without assistance. Since the vast majority of molecules the cell needs to transfer across the membrane are hydrophilic rather than lipophilic, plasma membranes have evolved a range of associated transfer systems (see below) which permit selective communication between the two sides of the membrane. This selectivity permits the cell to exercise considerable control over its interaction with the environment.

3.5.3 Barriers within the cell: compartments

Prokaryotic cells are simply single compartments lacking internal membranous subdivision and bounded by the plasma membrane. In eukaryotic cells, on the other hand, membranous structures abound. What has been said above about the plasma membrane acting as a barrier to the free movement of molecules between the aqueous environment of the cell and the aqueous external environment applies equally to intracellular membranes. These, too, separate a volume of an aqueous solution (in this case 'cytoplasm') from its surroundings (also 'cytoplasm') so that the compartment they delimit can, again through the selective transfer of molecules, undertake chemical reactions which are not possible elsewhere in the cell. Even the membrane itself may be a compartment, in the sense that it can serve as a structural support for enzyme trains whose successive reactions have to be coordinated by being spatially fixed relative to one another.

Some of these intracellular compartments (mitochondria, vacuoles, nuclei) are visible with the light microscope, others require electron microscopy (endoplasmic reticulum, Golgi-like cisternae, cytoplasmic vesicles) and still others can only be inferred from indirect experimentation. They all permit selective localisation of enzymes, substrates and/or products; providing the means for attaining reaction conditions (high metabolite concentration, unusual pH or ionic concentrations, for example) which cannot be maintained, or perhaps cannot be permitted, through the whole volume of the cell. Such compartmentalisation also has implications for control of metabolism as, for example, two processes with a common substrate requirement may be regulated independently by having their enzyme systems in different membrane-bound compartments so that transfer of the substrate into one or other of the compartments determines which process is carried out (Jennings, 1995).

Much of the metabolic integration which is responsible for the rich variety of eukaryotic biochemistry is dependent on intracellular compartmentalisation. Unfortunately, because it does occur within the structure of the living cell it is extremely difficult to investigate and very little is known about it.

3.6 The flow of solutes

3.6.1 Solute behaviour in solutions

In an unbounded solution molecules of solute can move within the solution in two ways. Whole volumes of solution may be transported from place to place, taking solute molecules with them. This is bulk flow or mass flow and results from such things as convection flows and other large scale disturbances within the solution. As far as living organisms are concerned, bulk flow may be achieved through cytoplasmic streaming, transpiration streams and similar processes. It is likely to be associated more with distribution of materials (whether in solution or not) over large cellular or intercellular dimensions than with transfer across the lesser dimension of a lipid bilayer membrane.

The second mode of solute movement is diffusion, where random thermal motion at the molecular level causes all solute molecules to move continuously. If the solution is completely homogeneous then any molecules which move out of a particular unit volume will be replaced by an identical number moving into that unit volume and, under all but the most stringent experimental conditions, the exchange of solute molecules will not be detectable. On the other hand, if there is a concentration gradient within the solution there will be a net flow of solute molecules from the high concentration end of the gradient, towards the low concentration end. Note that this gradient can be a chemical gradient of uncharged molecules (e.g. a sugar), an electrical gradient of a charged ion (e.g. K^+) or a combination of the two. This diffusion process is extremely relevant to the behaviour of every cell, since there is likely to be a concentration gradient across the plasma membrane for just about every solute of importance to the cell.

3.6.2 Transport systems

To traverse the biological membranes a solute must leave the aqueous phase for the lipoidal environment of the membrane, traverse that, and then re-enter the aqueous phase on the other side of the membrane. Unaided simple diffusion of molecules across biological membranes depends considerably on their solubility in lipids.

There are exceptions to this generalisation, though, as some small polar molecules (such as water) enter cells more readily than would be expected from their solubility in lipid. They behave as though they are traversing the membrane by simple diffusion through gaps or pores which are transiently generated by random movements of the acyl chains of the membrane phospholipids. Since transfer of these materials (like that of molecules which are soluble in the hydrophobic lipid bilayer of the membrane, such as O_2 and CO_2) depends on diffusion, their rate of movement is proportional to the concentration differential on the two sides of the membrane, and the direction of movement is from the high to the low concentration side. No metabolic energy is expended and no specific membrane structures are involved in this mode of transfer, but net transfer ceases when the transmembrane concentrations are equalised.

In some, though probably rare, cases these concentration gradients across membranes will be oriented so that solute flows in the direction required by the cell and at a rate suitable for its metabolism. However, in most cases the gradient will be adverse, or the concentration differential so small that the rate of diffusion is inadequate. Furthermore, only a minority of compounds pass through biological membranes *in vivo* by simple diffusion, and the vast majority of metabolites that the cell needs to absorb or excrete are too polar to dissolve readily in lipid and too large in molecular size to make use of transient pores. To cope with these circumstances the membrane is equipped with solute transport systems. This argument applies to intracellular membranes bounding compartments within the cell as well as to the plasma membrane (see Garrill, 1995 for discussion of tonoplast and mitochondrial transport).

The essential component of any transport system is a transporter molecule, a protein which spans the membrane and assists transfer of the metabolite across the lipid environment of the membrane. With both passive and active transporters, substrate translocation depends on a conformational change in the transporter such that the substrate binding site is alternately presented to the two faces of the membrane. These transporters are glycoproteins of around 500 amino acids arranged into three major domains: 12 α-helices spanning the membrane, a highly charged cytoplasmic domain between helices 6 and 7, and a smaller external domain, between helices 1 and 2, which bears the carbohydrate moiety. Sequence homology between the N- and C-terminal halves of the protein suggest that the 12 α-helix structure has arisen by the duplication of a gene encoding a 6-helix structure. Ion channels are different as their polypeptide subunits form a β-barrel containing a pore. A loop of the polypeptide is folded into the barrel and amino acids of this loop determine the size and ion selectivity of the channel. This transporter alternates between open and closed conformations (Jennings, 1995).

If the transfer is passive, that is without a requirement for metabolic energy, then the transport process is described as facilitated diffusion. Such a process still depends upon a concentration differential existing between the two sides of the membrane, transfer occurring 'down the gradient' (towards the compartment which has the lower concentration). However, transfer is much faster than would be predicted from the solubility of the metabolite in lipid, the high rate of transfer depending on the fact that the transporter and the transporter–metabolite complex are highly mobile in the lipid environment of the membrane (see Fig. 3.9). The major difference from simple diffusion is that facilitated diffusion exhibits saturation kinetics and (usually) high substrate specificity. Showing saturation kinetics means that as the concentration of the metabolite being transported is increased, the rate of transport increases asymptotically towards a theoretical maximum value at which all the transporters are complexed with the metabolite being transported (i.e. transporters are saturated).

Facilitated diffusion can transport a specific substrate very rapidly; but can only equalise the concentrations of the transported metabolite on the two sides of the membrane. Yet in many cases the cell needs to transfer a metabolite against its concentration gradient. The prime example will be where the cell is absorbing a nutrient available at only a low concentration; if growth of the cell is not to be limited by the external concentration of the nutrient, the cell must be able to accumulate the nutrient to concentrations greater than those existing outside. In which case an adverse gradient of concentration will have to be established and maintained. Neither simple diffusion nor facilitated diffusion can do this; to achieve it the cell must expend energy to drive the transport mechanism. Such a process is called active transport.

Active transport is a transporter-mediated process in which movement of the transporter–substrate complex across the membrane is energy dependent. The transporter exhibits the same properties as a facilitated transport transporter (saturation kinetics, substrate specificity, sensitivity to metabolic inhibitors). In addition to these properties, active transport processes characteristically transfer substrate across the membrane against a chemical and/or electrochemical gradient, and are subject to inhibition by conditions or chemicals which inhibit metabolic energy generation.

The mechanism is often a co-transport in which the movement of an ion down its electrochemical gradient is coupled to transport of another molecule against its concentration gradient. When the ion and the transported substrate move in the same direction the co-transporter is called a symport, whereas transporters which transport the two in opposite directions are termed antiporters. The electrochemical gradients, most usually of protons or K^+ in fungi, are created by ion pumps in which hydrolysis of ATP phosphorylates a cytoplasmic domain of the ion channel. Consequential conformational rearrangement of the protein then translocates the ion across the membrane and reduces the affinity of the binding site to release the ion at the opposite membrane face. Dephosphorylation restores the pump to its active conformation (and may translocate another ion or molecule in the opposite direction).

3.7 Transport strategy

Complex interactions occur in transport of anions, cations and non-electrolytes; interactions which may depend on metabolic, chemical, biophysical and/or electrochemical relationships between a number of different molecular species and with the rest of metabolism. There are indications of what might be called transport strategy in operation in most cells. Single uptake systems are rarely encountered; dual or multiple systems are the norm, the different components being suited to different environmental conditions the organism may encounter. Multiple uptake systems for a single substance are found widely in plants as well as fungi. They inevitably result in complex uptake kinetics which might be indicative of physically separate transport transporters, each showing Michaelis–Menten kinetics (like the glucose transporters in *Neurospora*), or of single molecules exhibiting kinetics modulated by their environment (like the glucose transporter in *Coprinus;* Fig. 3.5). Whatever the physical basis, the regulatory properties of the components of such 'families' of transport processes appear to be interlinked to ensure that nutrient uptake is maintained at a reliable level whatever the variation in substrate availability in the environment.

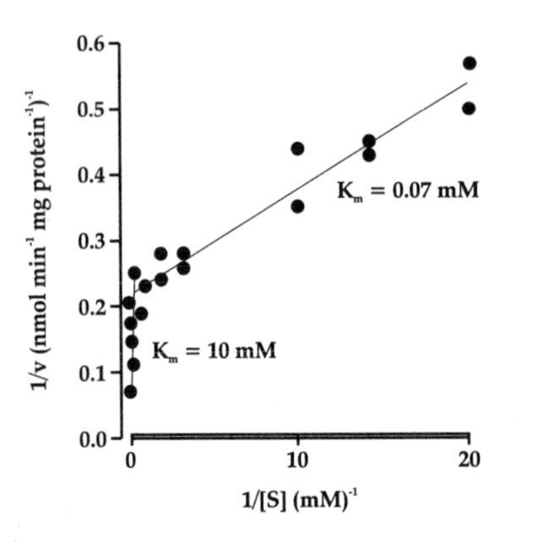

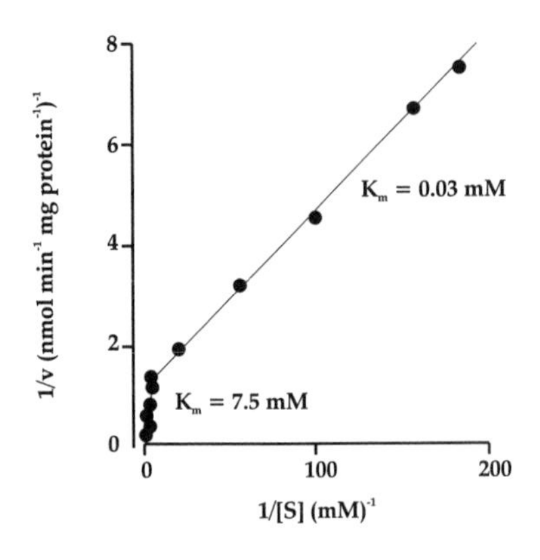

Figure 3.5 Initial rates of uptake of two glucose analogues as a function of concentration in *Neurospora* (left) and *Coprinus* (right). Note the biphasic nature of the plots, indicating that there are at least two kinetic entities in each case, one with a much higher substrate affinity than the other. Data taken from Neville *et al.* (1971), and Moore and Devadatham (1979).

Probably the most important generalisation that can be made about transport processes, though, is that for almost all of them the active extrusion of protons from the fungal cell seems to be of significance. The proton gradient thereby established provides for uptake of sugars, amino acids and other nutrients by proton co-transport down the gradient, and is directly involved in cation transport (like the K^+/H^+ exchange or antiport).

Thus, each fungus possesses multiple uptake systems for most nutrients but the same basic process (active H^+ extrusion) provides the energisation for most if not all. An aqueous suspension of yeasts may be a fairly homogeneous population, expressing relatively uniform transport properties, but a yeast colony on a solid medium, and most definitely a mycelial fungus on any solid substratum will inevitably be expressing different transport processes in different regions of the mycelium. This conclusion stems from consideration of such data as those dealing with the amino acid composition of suspensions of yeast cells. Recall that the composition of the amino acid pool varies with age, with the growth rate, and with the composition of the medium. Many of these observed differences in amino acid pool reflect differences in transport processes. Applying these observations to mycelia of filamentous fungi, different hyphal cells in a filamentous mycelium will be expressing different transport capabilities because of inevitable differences in age, growth rate and composition of the substratum on which the hypha is growing.

What is true for amino acid uptake is probably also true for other nutrients and it is quite evident that uptake systems must be regulated not only by reference to their own substrates but in relation to other substrates and other metabolic conditions as well. The majority of the transport proteins which have been characterised can be grouped into families which appear to be conserved between different organisms. As more genomes are sequenced, computer analysis of many transport proteins has become feasible, allowing deduced protein sequences of unknown function to be characterised by sequence similarity with known membrane transport proteins (Andre, 1995).

A crucial point, which has not yet been taken into account, is that the sorts of transport systems so far considered will inevitably alter the solute concentrations of the cell and thereby influence the movement of that all-pervading nutrient, water.

3.8 Water relationships

Water, of course, is the solvent within which the majority of life processes are played out, but it is also a significant (even if often overlooked) component of innumerable biochemical processes (Ayres and Boddy, 1986). For example, every hydrolytic enzyme reaction uses a molecule of water, every condensation reaction produces a molecule of water and respiration of 1 g of glucose produces 0.6 g of water. The water relationships of the fungal cell are an important aspect of its overall economy.

Water availability is determined by its potential energy – referred to as the water potential, symbolised by the Greek letter psi (Ψ). Zero water potential is the potential energy of a reference volume of free, pure water. The water in and around living fungal cells will have positive or negative potential energy relative to that reference state, depending on the effect(s) of osmotic, turgor, matrix and gravitational forces. Water will flow spontaneously along a water potential gradient, from high to low potentials, though in the normal state for most fungi this will mean from a negative to a more negative potential. The lower the water potential the less available is the water for physiological purposes and the greater is the amount of energy that must be expended to make the water available.

On the face of it, two things need to be considered. One is some sort of compensation for change in the solute relationships of the cell resulting from uptake of some substrate – such a process would further reduce the potential of the cell water and increase the tendency of external water to influx. The other is to provide the cell with a means to regulate its water uptake even though the external water potential is uncontrollable. In fact, of course, these are just two facets of the same problem. In either case the fungus must cope with water potential stress and the evidence indicates that solute transport systems provide the mechanism which permits this. The internal maintenance of turgor pressure by movement of water across the membrane is related to transport of ions across the membrane and to the breakdown of macromolecules and biosynthesis of solutes. Inorganic ions usually make the greatest percentage contribution to the osmotic potential of the protoplasm. The main ions involved are K^+ and Na^+, with Cl^- being moved to balance the cation content. Some organic solutes also make major contributions, including glycerol, mannitol, inositol, sucrose and proline (Eamus and Jennings, 1986).

The most immediate response to water potential stress is change in cell volume by the rapid flow of water into or out of the cell. The consequent change in turgor affects the cell membrane permeability and electrical properties so that the cell can restore the volume by transporting ions or other solutes across the membrane and/or by synthesising solutes or by obtaining them by degrading macromolecules. Response to water potential stress can be extremely rapid. Experimentally this is particularly evident in fungal protoplasts, the size of which alters soon after change in the solute concentration of the suspending medium. Such behaviour attests to the ready permeability of the cell membrane to water even though it is quite obvious that water molecules are unable to penetrate the lipid (hydrophobic) environment of the membrane. The route taken by water molecules has not been established but they are thought to penetrate by migration through pores or channels in proteins which span the membrane. These proteins may have other functions; although no such research has been done with fungi, there is evidence that in the rabbit corneal endothelium water molecules flow preferentially through a glucose transporter protein. The managed flow of water, coordinated with control of the wall synthetic apparatus, must be a prime factor in controlling the inflation of fungal cells which is responsible for so many of the changes in cell shape which characterise fungal cell differentiation. Furthermore, since most of the increase in size during maturation of many fruit bodies and related structures is itself an expansion due to inflation of the constituent cells (rather than an increase in cell number), management of water movement is an equally important aspect of gross morphogenesis.

Turgor also contributes to flow along the fungal hypha. As this is a filamentous structure, flow of water and solutes within the hypha (i.e. translocation) is of enormous importance. Although our current view of apical growth requires that fungi can organise rapid translocation and specific delivery of microvesicles containing enzymes and substrates for wall synthesis, the evidence indicates that more general water flow along the hypha is driven by a turgor gradient and that solutes are translocated by this turgor-driven bulk flow. Translocation of nutrients of all sorts in this manner is of crucial importance to morphogenesis because it must be the main way in which developing multicellular structures, such as a fruit body developing on a vegetative colony, are supplied with nutrients and water. Translocation is ably discussed by Jennings (1995; see his Chapter 14) and the mechanism can best be illustrated by quoting his description of the way in which *Serpula lacrymans* (the major timber decay organism in buildings in northern Europe) translocates carbohydrate:

> Mycelium attacks the cellulose in the wood, producing glucose, which is taken into the hyphae by active transport. Inside the hypha, glucose is converted to trehalose, which is the major carbohydrate translocated. The accumulation of trehalose leads to the hypha having a water potential lower than outside. There is a flux of water into the hyphae and the hydrostatic pressure so generated drives the solution through the mycelium. The sink for translocated material is the new protoplasm and wall material produced at the extending mycelial front. The mechanism of translocation in *S. lacrymans* is thus the same as that now accepted for translocation in the phloem of higher plants, namely osmotically driven mass flow (Jennings, 1995; p. 459).

This description could be paraphrased to apply it to other circumstances by, for example, featuring nutrients other than carbohydrates (but note that the molecule which is translocated may be a different species from those which are initially absorbed or finally used) and/or alternative sinks, such as fruit bodies or particular tissues in fruit bodies. Importantly, this bulk flow does not have to be unidirectional *within a tissue*. Because the tissue is comprised of a community of hyphae, different hyphae in that community may be translocating in different directions. Carbon is translocated simultaneously in both directions along rhizomorphs of *Armillaria mellea* (Granlund *et al.*, 1985). In mycorrhizas, carbon sources from the host plant and phosphorus absorbed by the hyphae from the soil must move simultaneously in opposite directions. Indeed, the flow of carbon in mycorrhizas must be fairly complex as carbon can be transferred between two different plants which are connected to the same mycorrhizal system (Watkins *et al.*, 1996). Cairney (1992) has proposed that contrary flows in individual hyphae may be achieved if there is a spatial separation between nutrient flow through the cytoplasm and a bulk flow of water in the reverse direction through the hyphal walls. Although much remains to be learned, there is clear evidence that nutrients (including water in that category) can be delivered over long distances through hyphal systems, and that the flows can be managed and targeted to specific destinations.

3.9 Intermediary metabolism

Metabolism is the sum total of the chemical changes which occur in a fungus; changes which are both destructive (catabolism) and constructive (anabolism). The different metabolic processes are interlinked in a very complex manner in a living cell, but for study and description metabolism must be subdivided and categorised. It is important to recognise that these subdivisions are not absolute in terms of cellular biochemistry, but are extremely useful in terms of its interpretation. It is also important to note that much of the biochemistry described in this chapter is common to all free-living organisms. In the sections which follow, the general metabolism is described first and then any particular mycological points are emphasised.

A fungus grows at its maximum rate (μ_{max}) when all the nutrients it requires are freely available. If a nutrient is in short supply, growth rate will be reduced to that which can be supported by the amount of the limiting nutrient which is available. Metabolic pathways which characteristically operate when a fungus is growing at or near its maximum rate are described as primary pathways; secondary pathways become operational (or amplified) when growth rate is limited in some way to a level below the maximum (Bu'Lock, 1967).

The fundamental function of metabolism is the utilisation of nutrients to form ATP, reduced nucleotide coenzymes (NADH and NADPH) and the compounds which serve as precursors of cellular components, especially macromolecular constituents. This section will be concerned with aspects of metabolism responsible for formation of substrates which serve biosynthetic pathways and compounds providing energy and reducing power – the so-called primary or intermediary metabolism.

3.10 Carbon metabolism

The major source of energy and reducing power is the catabolism of carbohydrate. Although other carbon-containing compounds can be utilised for these purposes by most cells, the full sequence of enzymic processes are conventionally represented as involving the controlled release of energy by the use of atmospheric oxygen to convert glucose to CO_2 and water. This overall process is described as respiration and its chemically balanced (or stoichiometric) summary equation is:

$$C_6H_{12}O_6 + 6O_2 \longrightarrow 6CO_2 + 6H_2O + \text{energy}$$

Note that this equation does not even begin to describe the biochemical mechanisms which achieve the indicated chemical transformation, but it does emphasise that for the conversion of each mole of glucose, six moles of oxygen must be absorbed from the atmosphere, and six moles of CO_2 and six moles of water appear within the cell and 2900 kJ of free energy (i.e. energy capable of doing some work) are released. To put this into more readily grasped units, respiration of 1 g of glucose uses 1.07 g (747 cm^3) oxygen and produces 1.47 g (747 cm^3) CO_2 and 0.6 g water, releasing 16.1 kJ of energy.

This basic chemistry can be demonstrated by simply combusting glucose under suitably controlled conditions. Obviously, the living cell cannot do this, but instead uses a sequence of enzymically controlled reactions. These are conveniently divided into three phases or subpathways: glycolysis, the tricarboxylic acid (TCA) cycle, and oxidative phosphorylation. The indications are that these pathways are much the same in fungi as they are in other organisms (Blumenthal, 1965, 1968; Lindenmeyer, 1965; Niederpruem, 1965; Cochrane, 1976; Watson, 1976; Fothergill-Gilmore, 1986; Jennings, 1995; Van Laere, 1995).

3.10.1 Glycolysis: conversion of glucose to pyruvate

The word glycolysis describes the conversion of glucose to pyruvate without implying a particular pathway. In fact, there are three enzymic pathways which might be used, though one does tend to predominate.

The *Embden–Meyerhof–Parnass (EMP) pathway* is the major pathway in most species; it comprises nine enzymic steps, all of which occur in the cytoplasm (Fig. 3.6). The net outcome of the reactions summarised in Fig. 3.6 is that one molecule of glucose is converted to two molecules of pyruvic acid plus two molecules of ATP and two molecules of $NADH_2$. Thus, the energy yield ($2ATP + 2NADH_2$, since the latter do represent potential chemical work), is rather small; the main function of the EMP pathway being conversion of glucose to pyruvate for processing in the TCA cycle.

A commonly encountered alternative glycolytic pathway is the *pentose phosphate pathway (PPP)*, also called the hexose monophosphate path-

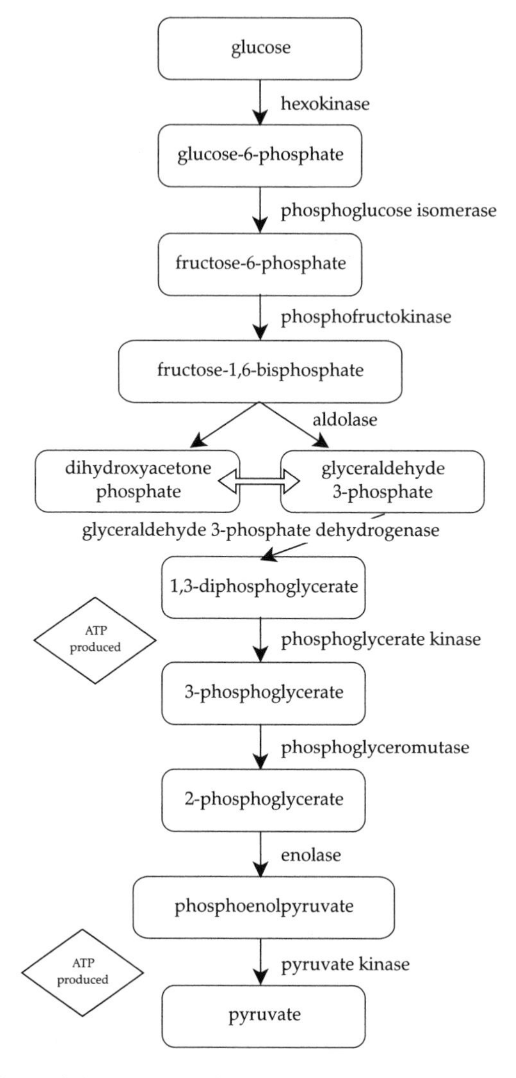

Figure 3.6 The Embden–Meyerhof–Parnass (EMP) glycolytic pathway.

way (HMP) (Fig. 3.7). In strictly chemical terms such a description is true; glucose 6-phosphate is diverted out, undergoes a range of chemical conversions, and then fructose 6-phosphate and glyceraldehyde 3-phosphate feed back into the EMP pathway. The EMP provides the cell with its major intermediate for energy generation (pyruvate), the PPP provides pentose sugars for nucleotide synthesis (which includes coenzymes and energy carriers as well as RNA and DNA), erythrose phosphate for the synthesis of aromatic amino acids through the shikimic acid pathway,

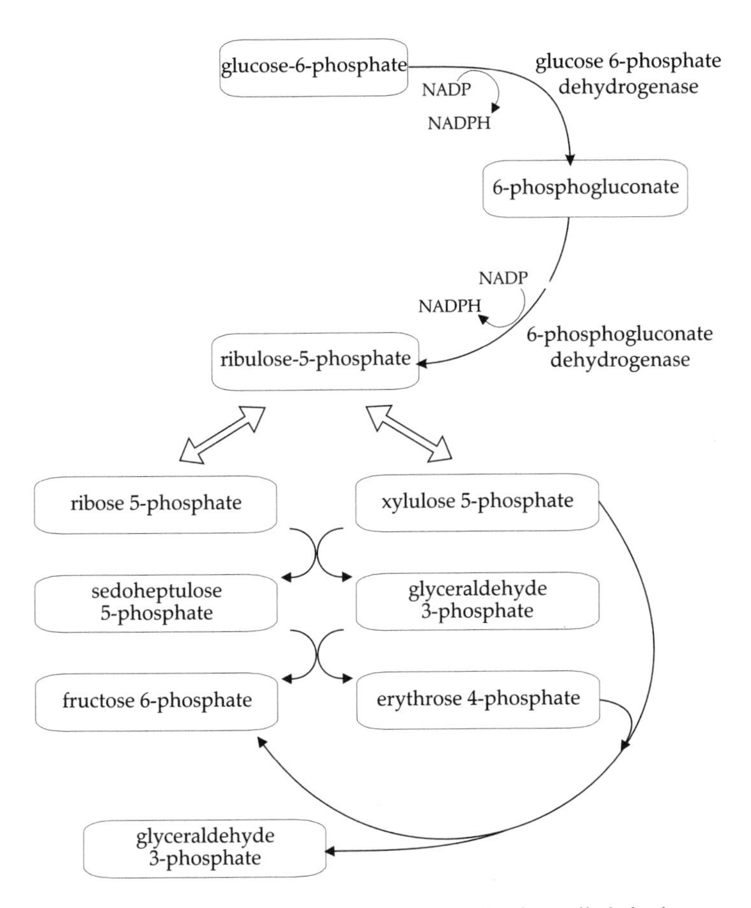

Figure 3.7 The pentose phosphate pathway (PPP), also called the hexose monophosphate pathway (HMP).

and $NADPH_2$ – the coenzyme which is most often used in biosynthetic reactions that require reducing power, especially fat and oil synthesis. So although the PPP can theoretically achieve complete glycolysis (six cycles through the reaction sequence would completely oxidise a molecule of glucose to CO_2), it is more likely to be involved in furnishing biosynthetic intermediates. The PPP does, of course, also provide a route for utilisation of pentose sugars which become available as carbon sources, and for interconverting hexose and pentose phosphates.

The *Entner–Doudoroff (ED)* pathway proceeds via 6-phosphogluconate to 2-keto-3-deoxy-6-phosphogluconate, which gives rise to pyruvate and glyceraldehyde 3-phosphate (Fig. 3.8). It is a common glycolytic pathway in bacteria, but has been demonstrated in only a few fungi though where it does occur it seems to be the major glycolytic route.

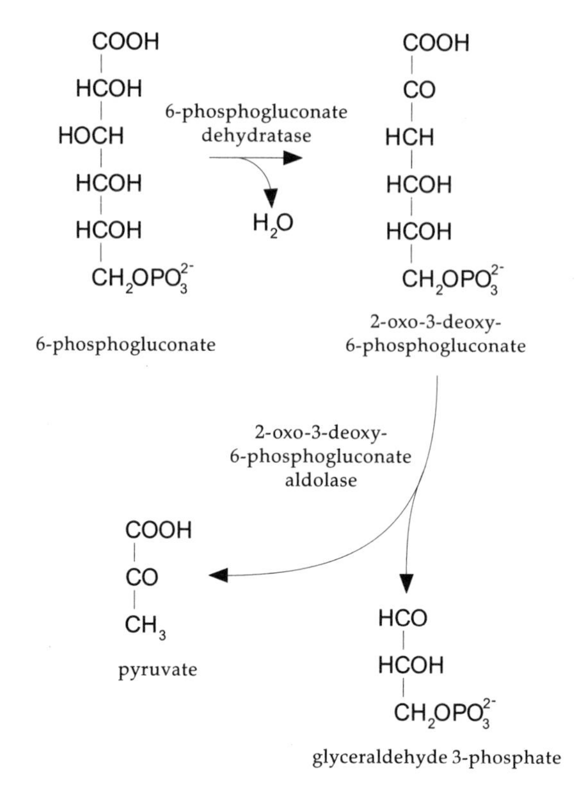

Figure 3.8 The Entner–Doudoroff (ED) pathway.

The use of different glycolytic pathways in any cell will reflect the relative contribution their intermediates are required to make to the functions of the cell; they will change with age, activity and nutrition. In general, since the PPP provides intermediates for biosynthesis, use of this pathway increases in rapidly growing and in differentiating cells, and is minimised in those which are resting or quiescent. Stimulation of PPP activity can often be demonstrated by imposing nutritional conditions requiring expenditure of $NADPH_2$, as by the provision of ammonium or nitrate as sole nitrogen sources.

3.10.2 The tricarboxylic acid (TCA) cycle: oxidation of pyruvate

Whichever glycolytic pathway produces pyruvate, this latter molecule is formed in the cytoplasm and must then be transported into the mitochondrion where it is converted to acetyl coenzyme A (acetyl-CoA). This step is achieved by the pyruvate dehydrogenase complex – a combination of enzymes which first decarboxylate pyruvate and then transfer the resulting acetyl group to coenzyme A.

The TCA cycle is cyclic because its 'end-product', oxaloacetate, reacts with acetyl-CoA to introduce what remain of the pyruvate carbon atoms into a reaction sequence (Fig. 3.9) the primary function of which is to convert pyruvate formed in glycolysis entirely to CO_2, the released energy being captured primarily in $NADH_2$. The overall stoichiometry is that one molecule of pyruvate with three molecules of water forms three molecules of CO_2 and releases 10 protons, the latter appearing in the form of three molecules of $NADH_2$, and one each of $FADH_2$ and the 'high energy' compound GTP. The succinate dehydrogenase enzyme is bound to the inner mitochondrial membrane (and, because of this, is often used as a marker for the presence of mitochondria in fractionated cell extracts), the other enzymes of the TCA cycle occur in the mitochondrial matrix.

A common variant of the TCA cycle is the *glutamate decarboxylation loop* in which 2-oxoglutarate is aminated to glutamate rather than being oxidatively decarboxylated to succinate. The glutamate is decarboxylated to 4-aminobutyrate; transamination between the latter and 2-oxoglutarate yielding succinate semialdehyde which, on oxidation, feeds back into the TCA cycle as succinate (Fig. 3.9). The enzymes of this loop have been found to be at high activity in fruit bodies, especially caps, of the basidiomycete *Coprinus cinereus* (Moore and Ewaze, 1976; Moore, 1984a). The glutamate decarboxylation loop is the normal route of TCA metabolism in this organism and also operates in *Agaricus bisporus* (see section 4.2.2; for review see Kumar and Punekar, 1997).

3.10.3 Oxidative phosphorylation

Through glycolysis and the TCA cycle, all of the carbon contained in the substrate glucose is released as CO_2. However, very little energy is released, most of it being trapped in $NADH_2$ (with a small amount in GTP and $FADH_2$). The energy represented in the form of the reduced coenzymes is recovered as ATP through the electron transport chain, located on the inner mitochondrial membrane, in the process known as *oxidative phosphorylation*.

The electron transport chain transfers electrons from the reduced coenzymes through a series of reactions until the electrons are finally passed to oxygen, reducing it to water. Stepwise transfer of electrons

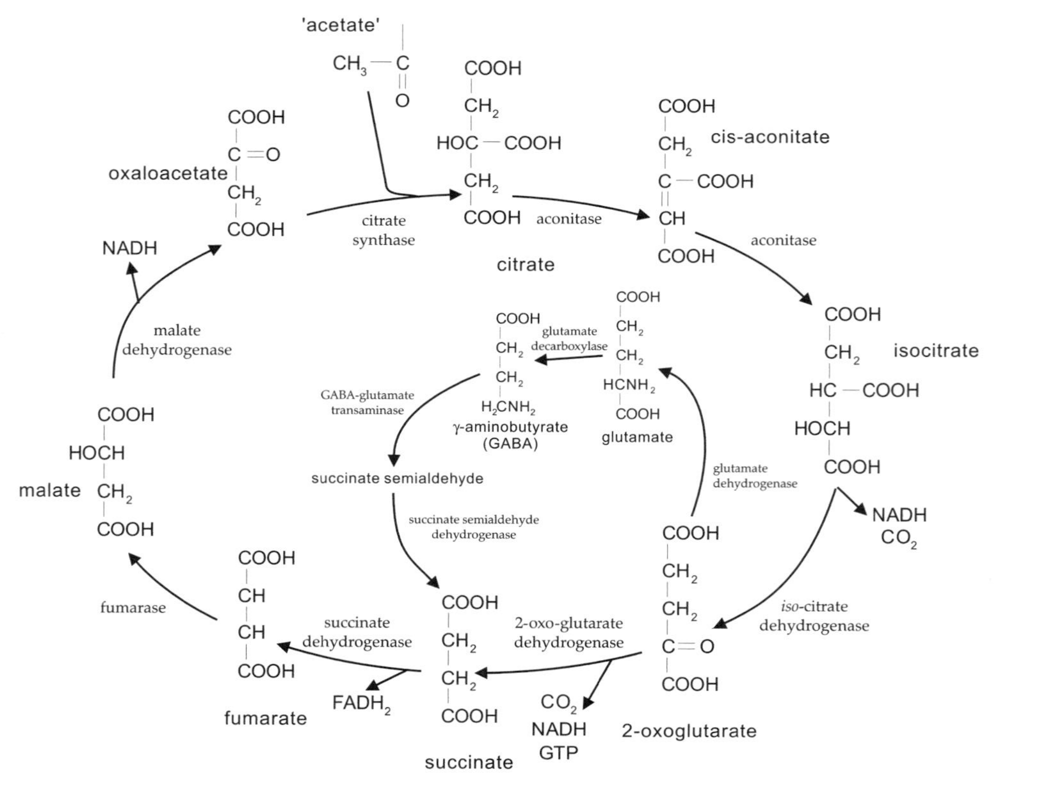

Figure 3.9 Oxidation of pyruvate: the tricarboxylic acid (TCA) cycle. The glutamate decarboxylation loop is shown in the centre of the main cycle.

between components of the electron transport chain leads to the pumping of protons from the mitochondrial matrix into the intermembrane space (Bowman and Bowman, 1996). The resulting proton gradient (the pH in the intermembrane space is about 1.4 units lower than that of the matrix) is used to generate ATP.

Transfer of a pair of electrons from one molecule of $NADH_2$ to oxygen leads to proton pumping at three sites in the chain, at each of which the consequent proton gradient can be used to synthesise one molecule of ATP. The ATP is synthesised by an enzyme complex located on the matrix side of the inner mitochondrial membrane. As protons move down a channel in this complex (the channel, known as the F_0 sector, is composed of at least four hydrophobic subunits forming the proton channel located in the membrane) the associated F_1 sector (containing five different subunits) projects into the matrix and is responsible for ATP synthesis (Bowman and Bowman, 1996).

3.10.4 Gluconeogenesis and the synthesis of carbohydrates

Although I am primarily concerned here with catabolism, the pathways described permit, with a few modifications, sugar synthesis. It has been stressed above that glycolysis and the TCA cycle provide opportunities for the fungus to make use of a very wide range of potential carbon and energy sources; but one which is successfully growing on acetate, for example, is clearly required to synthesise all of those compounds which have more than two carbon atoms chained together. In such circumstances glycolysis cannot simply be reversed because the steps governed by kinases (hexokinase, phosphofructokinase and pyruvate kinase) are irreversible, so for these steps in particular additional enzymes are required for gluconeogenesis (Fig. 3.10). The first steps in conversion of pyruvate to carbohydrate are carried out by pyruvate carboxylase, which synthesises oxaloacetate which is then decarboxylated and phosphorylated to phosphoenolpyruvate by phosphoenolpyruvate carboxykinase. The phosphoenolpyruvate can then be converted to fructose 1,6-bisphosphate by reversal of the EMP pathway, but an additional enzyme, fructose bisphosphatase, is required to generate fructose 6-phosphate. As the sugar phosphates are readily interconvertible, once this compound is formed oligosaccharide and polysaccharide synthesis can proceed. The structures of many of the polysaccharides formed have been shown earlier in this chapter.

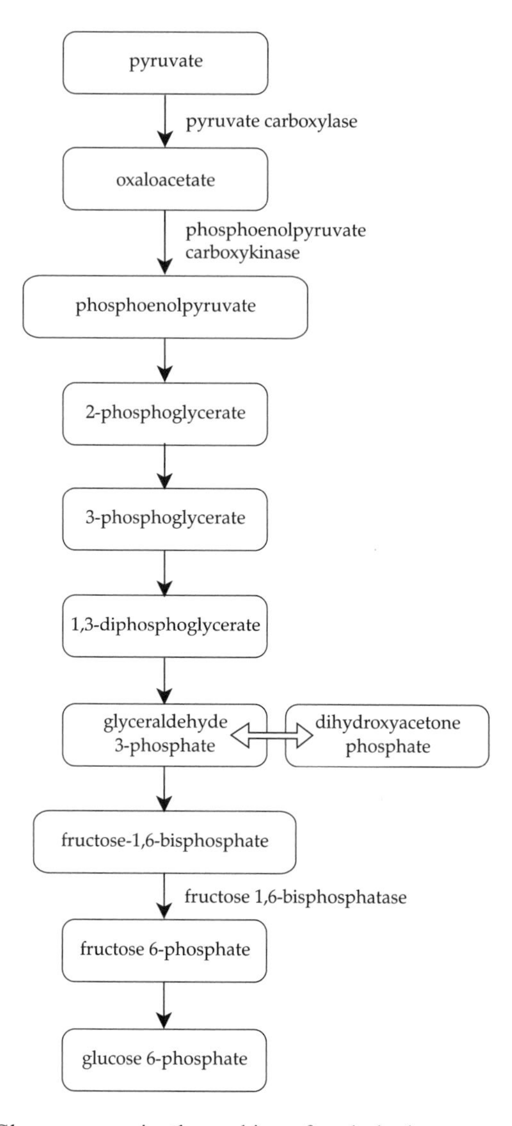

Figure 3.10 Gluconeogenesis: the making of carbohydrates.

Glycolysis and gluconeogenesis are obviously alternatives which demand close control to assure metabolic balance. I have already noted the sorts of controls which are exercised over phosphofructokinase the key glycolytic control point, it is interesting to note that fructose bisphosphatase responds inversely to the same molecules, being allosterically activated by citrate but inhibited by AMP.

Before leaving carbohydrate metabolism, it is worth mentioning here that the sugar alcohol mannitol and the disaccharide trehalose are almost always found among the water-soluble cytoplasmic carbohydrates in fungi; trehalose is the most widely-distributed sugar in fungi. Mannitol and trehalose seem to serve as transient storage compounds (i.e. molecules capable of immediate mobilisation when required). For example, both have been identified as substrates used for the metabolism associated with spore germination (discussion in Blumenthal, 1976). However, mannitol can certainly serve an osmoregulatory function in the marine fungus *Dendryphiella* (Jennings and Austin, 1973) and may serve the same purpose in fruit bodies of the cultivated mushroom, *Agaricus bisporus*, in which it can be accumulated to concentrations of up to 50% of the dry weight (Hammond and Nichols, 1976); similar concentrations have been encountered in fruit bodies of *Lentinula edodes* (Tan and Moore, 1994; see section 4.2.5). In *A. bisporus*, mannitol is synthesised by reduction of fructose by an NADP-linked mannitol dehydrogenase.

Trehalose synthesis/accumulation/degradation cycles occur at a number of stages in development of a fungus so it is quite clear that this sugar is the 'common currency' of the fungal carbohydrate economy. It is synthesised, as trehalose 6-phosphate, by the enzyme trehalose phosphate synthase from glucose 6-phosphate and the sugar nucleotide UDP-glucose.

3.11 Fat catabolism

Fats are molecules of glycerol in which the three hydroxyl groups are replaced with three fatty acid molecules. In degradation the first step is carried out by lipase which removes the fatty acids from the glycerol (see Fig. 3.8). The latter can be converted to glyceraldehyde 3-phosphate and thereby enter glycolysis, but it represents only about 10% by weight of a fat molecule, the bulk being represented by the fatty acids which consist of long carbon chains (e.g. palmitic acid, C_{16}; stearic acid, C_{18}).

These chains are degraded by sequential removal of the terminal two carbon atoms in the form of an acetyl group attached to CoA (Fig. 3.11). Because the cleavage occurs at the second (β) carbon atom, this process is called β-oxidation and it takes place in the mitochondrial matrix. Each such cleavage is oxidative, enzymes passing the H atoms to the coenzymes NAD and FAD. Thus, degradation of palmitic acid requires seven cleavages and yields eight molecules of acetyl-CoA (which enter

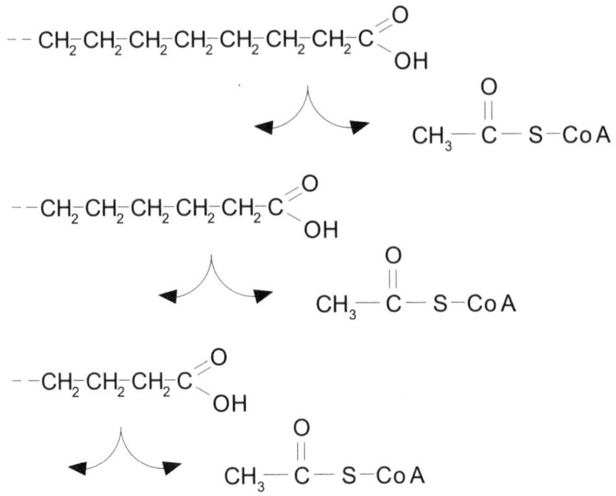

$$--CH_2-CH_2-CH_2-CH_2-CH_2-CH_2-CH_2-COOH \rightarrow CH_3-CO-S-CoA$$

$$--CH_2-CH_2-CH_2-CH_2-CH_2-COOH \rightarrow CH_3-CO-S-CoA$$

$$--CH_2-CH_2-CH_2-COOH \rightarrow CH_3-CO-S-CoA$$

Figure 3.11 β-oxidation of fatty acids.

the TCA cycle), seven $NADH_2$ and seven $FADH_2$ (both of which enter the electron transport chain for oxidative phosphorylation). Oxidation of fatty acids releases considerable amounts of energy; for example, one molecule of palmitic acid will give rise to about 100 molecules of ATP. This is why fats are such effective energy storage compounds.

3.12 Nitrogen metabolism

Ultimately, all nitrogen in living organisms is derived from the native element in the atmosphere. Each year an amount between 100 and 200 million tonnes of atmospheric nitrogen is reduced to ammonium by the nitrogenase enzyme system of *nitrogen-fixing* bacteria and bluegreen algae.

3.12.1 Formation of amino groups

If amino groups are not available directly by absorption of amino acids, they have to be formed and the most immediate source is by the assimilation of ammonium. The only route of ammonium assimilation which can be considered pretty well universal in fungi is the synthesis of glutamate from ammonium and 2-oxoglutarate by the enzyme glutamate dehydrogenase. Many filamentous fungi and yeasts have been shown to produce two glutamate dehydrogenase enzymes; one linked to the coenzyme NAD and the other linked to NADP.

As can be appreciated from Figure 3.12, the interconversion of 2-oxoglutarate and glutamate is a reaction which occupies a central position in metabolism and is one at which important pathways in both carbon metabolism and nitrogen metabolism come together. The reaction is readily reversible and it is often considered that NAD-linked glutamate dehydrogenase (NAD-GDH) has a deaminating or catabolic role (glutamate $\longrightarrow$ 2-oxoglutarate + ammonium), while the NADP-linked enzyme provides the aminating or anabolic function (2-oxoglutarate + ammonium $\longrightarrow$ glutamate). Stewart and Moore (1974) showed that the NAD-GDH predominated in mycelia of *Coprinus cinereus* whether they were grown with glutamate as sole nitrogen source (which would demand deamination) or with ammonium (which requires amination). The key seems to be that this reaction is so important in intermediary metabolism that different organisms have evolved different patterns of *endogenous* regulation, especially in relation to morphogenetic processes. Certainly, in *Coprinus* the NADP-GDH normally appears at high activity only in the cap tissue of the mushroom fruit body where it is located in the basidia apparently protecting meiosis and sporulation from inhibition by ammonium, i.e. acting as an ammonium detoxifier rather than ammonium assimilator (see sections 4.2.3 and 6.5.3).

This example aside, NADP-GDH is generally the most important enzyme involved in ammonium assimilation in mycelia and its activity is often increased when ammonium is provided as a growth-limiting, sole nitrogen source. However, in some fungi an alternative enzyme system appears to scavenge for ammonium when it becomes limiting, this is the glutamine synthetase/glutamate synthase system. Glutamine synthetase is widely, perhaps universally, distributed and is responsible for synthesis of glutamine. However, glutamine synthetase can have a high affinity for ammonium and, in combination with glutamate synthase (which converts glutamine + 2-oxoglutarate to two molecules of glutamate), forms an ammonium assimilation system which can recover ammonium even when this molecule is present at low concentrations. The net result (2-oxoglutarate + NH_4^+ $\longrightarrow$ glutamate) is the same as the reaction promoted by NADP-GDH, but the cost is higher as glutamine synthetase uses ATP to make glutamine. The glutamate synthase mechanism is common in bacteria, but is encountered in fewer fungi, but has been demonstrated in *Neurospora crassa*, *Aspergillus nidulans* and several yeasts and mycorrhizas.

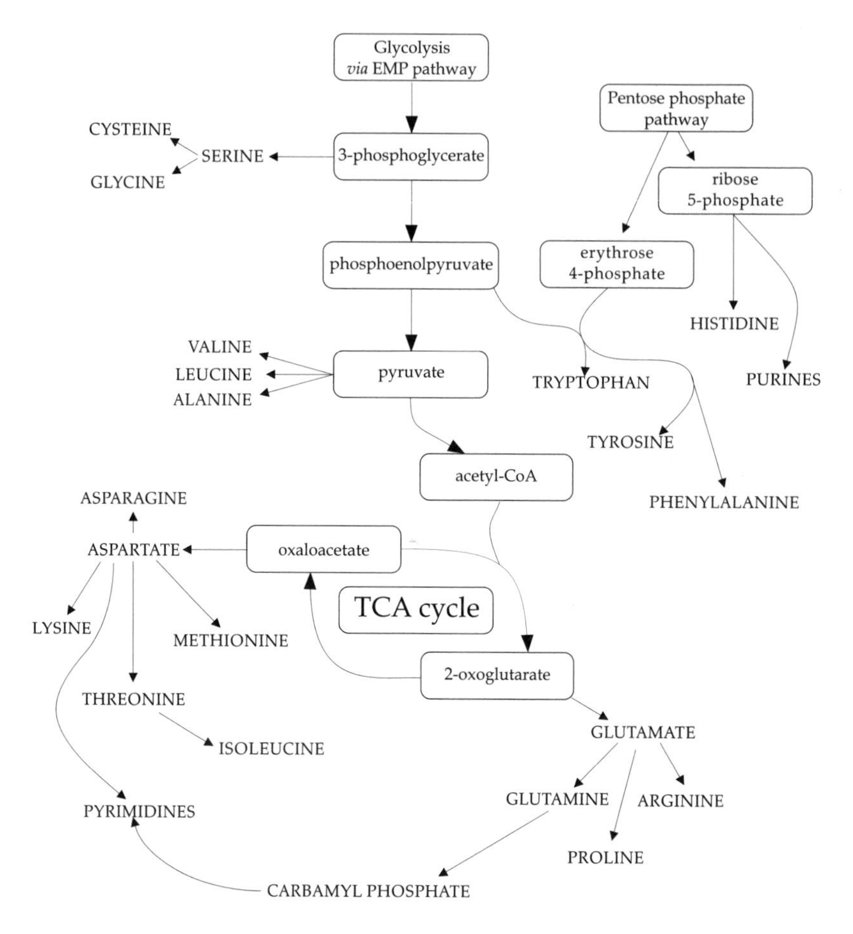

Figure 3.12 A flow chart illustrating pathways of nitrogen redistribution by showing the metabolic origins of amino acids, purines and pyrimidines.

Some yeasts and a larger number of filamentous fungi can utilise nitrate as sole source of nitrogen. Chemically, nitrate is first converted to nitrite which is then converted to ammonium, but the enzymic steps are quite complicated and not yet fully understood. The complexity of the reaction is reflected in the large number of mutant genes, in both *Neurospora* and *Aspergillus*, which have been found to affect nitrate assimilation (Marzluf, 1996). The first stage is performed by nitrate reductase which, generally in fungi, has a molybdenum-containing co-factor and requires NADPH (NADH in at least some yeasts). Nitrate is thought to bind to the molybdo-cofactor of the enzyme prior to being reduced by removal of an oxygen atom. Removal of this allows the nitrate formed to be bound to nitrite reductase, through its nitrogen atom, for the reduction to ammonium. The ammonium formed from nitrate is immediately used for the reductive amination of 2-oxoglutarate to glutamate.

Conversion of NO_3^- to NH_3 is a chemical reduction requiring considerable energy expenditure. In fact the equivalent of four $NADPH_2$ molecules (880 kJ of energy) are used to reduce one NO_3^- ion to NH_3, which is additional to the energy demand for assimilation of the ammonium (one $NADPH_2$ is used for assimilation via NADP-GDH, 1 $NADPH_2$ + 1 ATP for assimilation through glutamine synthetase and glutamate synthase). Given these additional energy demands, it is not surprising that the nitrate reduction machinery is produced only when nitrate is the sole available source of nitrogen, being induced by nitrate and rapidly repressed by the presence in the medium of ammonium or alternative sources of reduced nitrogen.

3.12.2 *Disposal of excess nitrogen*

The constituents of living cells are in a continual state of flux; all components being subjected to turnover as old materials are catabolised and new ones synthesised. When proteins and other nitrogen-containing compounds are broken down, either as part of this turnover process or as externally supplied nutrients, the carbon can be disposed of as CO_2, hydrogen as water and nitrogen either as ammonium or as urea. The use of protein as a *carbon* source has been discussed above. In these circumstances the organism (animal, plant or fungus) suffers an excess of nitrogen and must excrete it. Experiments with the basidiomycetes *Agaricus bisporus*, *Coprinus cinereus* and *Volvariella volvacea* have shown that one third to one half of the nitrogen contained in the protein given as substrate is excreted as ammonium into the medium (Kalisz *et al.*, 1986). Obviously this resulted in very considerable increases in the ammonium concentration in the medium, about 40 mM for each organism in the 25 ml culture volumes used in these *in vitro* experiments. Since ammonium is toxic this type of nutritional strategy must rely on rapid dissipation/dilution of the excreted ammonium in nature. In terrestrial mammals metabolising protein the toxicity of ammonium is avoided by excretion of urea formed through the urea cycle (Fig. 3.13). However, the enzyme urease seems generally to be constitutive in fungal mycelia so any urea formed is likely to be dissimilated to NH_3 and CO_2. Nevertheless there are circumstances in which fungi accumulate urea. Especially large

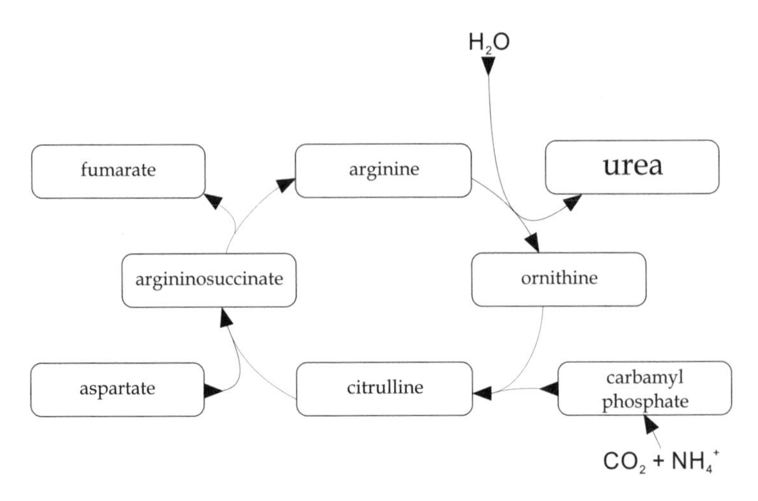

Figure 3.13 The urea cycle for disposal of excess nitrogen. The ammonium molecule which contributes to carbamyl phosphate and the amino group of aspartate are both 'discarded' into the urea molecule.

accumulations have been found in basidiomycete fruit bodies (Reinbothe and Tschiersch, 1962; Reinbothe *et al.*, 1967), where it seems likely that it acts as an 'osmotic metabolite' controlling water entry into cells during expansion. Urea also accumulates in the cap of the agaric *Coprinus cinereus*, probably for the same purpose, and in this case urease activity has been shown to be lacking in cap cells, though present in other fruit body tissues and constitutive in mycelia (Ewaze *et al.*, 1978). Thus, the capacity to synthesise, and even accumulate, urea is well developed in fungi but it seems that it is ammonia which is excreted to dispose of excess nitrogen.

3.13 Secondary metabolism

When all nutrients are available to a fungus its rate of nutrient utilisation is maximised and the growth rate is exponential. As a nutrient becomes depleted the rate of growth slows and eventually stops. The progress of metabolism is correspondingly altered, a number of special biochemical mechanisms appear (or become amplified) and a range of novel secondary metabolites are produced. The subject is huge, and this section is intended only to give an indication of the chemical versatility of fungi; it is not intended to be a comprehensive treatment.

3.13.1 Definition

The term secondary metabolite seems to have been coined by J. D. Bu'Lock (1961) in preference to continued use of the phrase 'natural products', on the grounds that some of the materials so described were a good deal less natural than others. He contrasted the metabolism leading to secondary metabolites with general metabolism (which we might more tidily call primary metabolism), observing that primary metabolism exhibits basic biochemical patterns which are common to widely differing organisms, and he defined secondary metabolites as being of restricted distribution (exhibiting even species-specific distribution) and having no obvious function in general metabolism.

This final phrase should be read: 'no *obvious* function'. Formation of secondary metabolites is common in plants and microorganisms (conventionally, animals are not widely considered to produce secondary metabolites) and many organisms expend considerable effort and energy in secondary metabolism so some selective advantage must accrue. Possible roles and functions will be discussed in section 3.13.4, but it is important here to appreciate that secondary metabolism *is* functional, though (i) we may be too ignorant of the biochemistry and physiology involved to appreciate the function and/or (ii) most of our interest (and many an example among the fungi) is likely to be concentrated on abnormal cultures grown in abnormal conditions to yield a product of commercial value. There are many potential differences between primary and secondary metabolites and there are examples of both which have commercial value. Consequently, definitions tend to become blurred. The most reliable one, though, remains that secondary metabolites are '. . . restricted in their distribution, being found in less than every species in a single family.' (Campbell, 1984). Thus, secondary metabolism is a common feature. It consists of a relatively small number of enzymological processes (often of relatively low substrate specificity) which convert a few important intermediates of primary metabolism into a wide range of products (Bu'Lock, 1967). These later stages of secondary metabolism are so varied that individual secondary metabolites tend to have the narrow species-distribution which is their essential character.

3.13.2 Main fungal pathways and products

Primary and secondary metabolism are coextensive – they can occur at the same time in the same cell and draw carbon-containing intermediates from the same sources. In many respects, however, secondary metabolism has a greater biochemical complexity than is evident in primary metabolism.

A few secondary metabolites are derived directly from glucose without cleavage of the glucose carbon chain, such as the kojic acid produced by *Aspergillus* species (Turner, 1971; Turner and Aldridge, 1983), but the majority have their origins in a small number of the intermediates in pathways dealt with above. Acetyl CoA is, perhaps, the most important, being a precursor for terpenes, steroids, fatty acids and polyketides; phosphoenolpyruvate and erythrose 4-phosphate initiate synthesis of aromatic secondary metabolites through the shikimic acid pathway by which aromatic amino acids are synthesised, whilst other secondary metabolites are derived from other (non-aromatic) amino acids (Fig. 3.14).

3.13.2.1 The mevalonic acid pathway: terpenes, carotenoids and sterols

Terpenes, carotenoids and steroids are widely distributed in nature and though the fungal products are not unusual in that sense, many of the end-products have chemical structures which are unique to fungal metabolism. All of these compounds are related by the occurrence within their structures of five carbon atoms arranged as in the hydrocarbon isoprene. The terpenes are the simplest of the naturally occurring isoprenoid compounds and are derived by condensation of a precursor isoprene unit to form an isoprenoid chain (Fig. 3.15).

Strictly, terpene is the name of the hydrocarbon with the formula C_9H_{16}, but the term is used generally to refer to the ten-carbon isoprenoids which may be open-chain (i.e. acyclic) or cyclic compounds. Synthesis of acyclic terpenes starts when dimethylallylpyrophosphate condenses with a molecule of isopentenylpyrophosphate to form the 9-carbon monoterpene geranylpyrophosphate (Fig. 3.16); addition of another isopentenylpyrophosphate gives the 15-carbon sesqui- (one-and-a-half) terpene farnesylpyrophosphate; and of another, the 20-carbon diterpene geranylgeranylpyrophosphate, and so on (Turner, 1971; McCorkindale, 1976). As many as 24 isoprene units may condense in this way, though secondary metabolites usually have between two and five. Not all such molecules are secondary metabolites; the ubiquinones of the respiratory chain are polyprenoid quinones (Stone and Hemming, 1967). Sesquiterpenes are the largest group of terpenes isolated from fungi. Most of the fungal sesquiterpenes are based on carbon skeletons which can be derived by cyclisation of farnesylpyrophosphate as

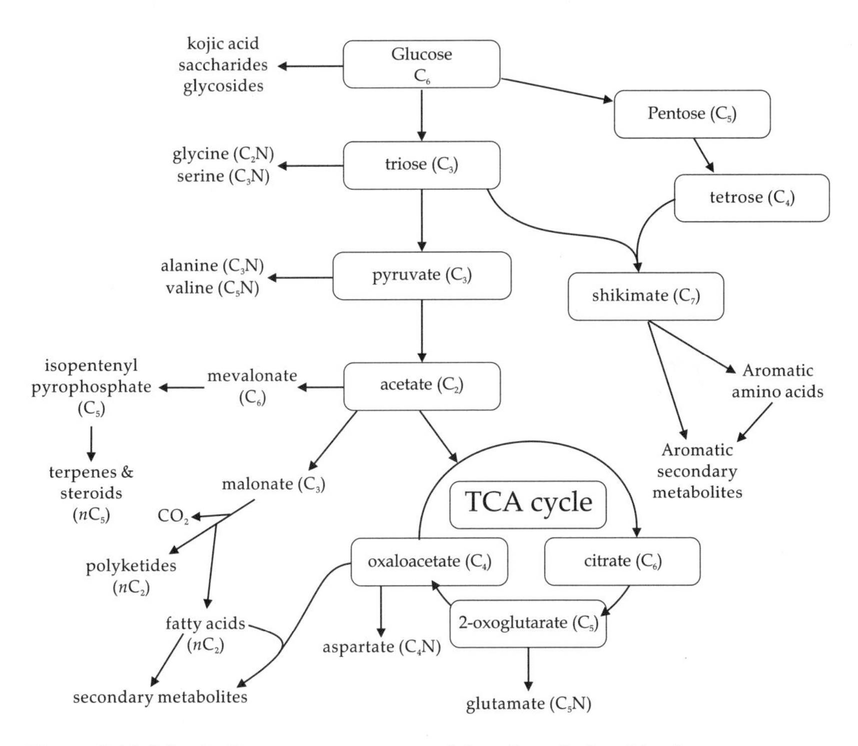

Figure 3.14 Metabolic route map summarising the relationships between primary metabolism and the major pathways for synthesis of secondary metabolites. Compare with Fig. 3.12.

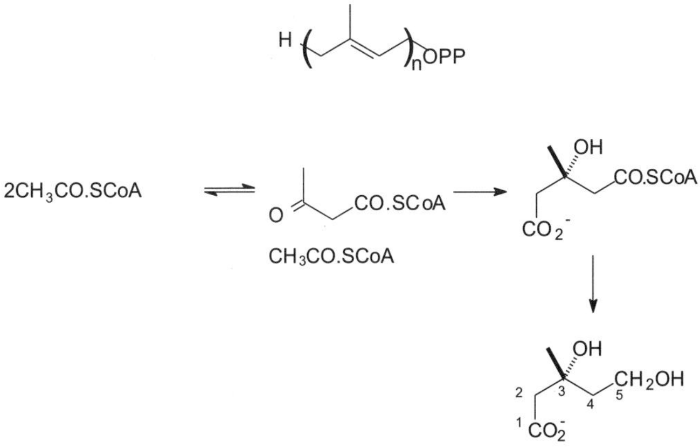

H ( )n OPP

2CH3CO.SCoA ⇌ CH3CO.SCoA (O, CO.SCoA) → (OH, CO.SCoA, CO2-) → (OH, CH2OH, CO2-; 1 2 3 4 5)

Figure 3.15 Derivation of terpenes from isoprene units. The basic isoprenoid repeat structure is shown at the top; the lower panel shows the synthesis of mevalonate from two molecules of acetyl-coenzyme A.

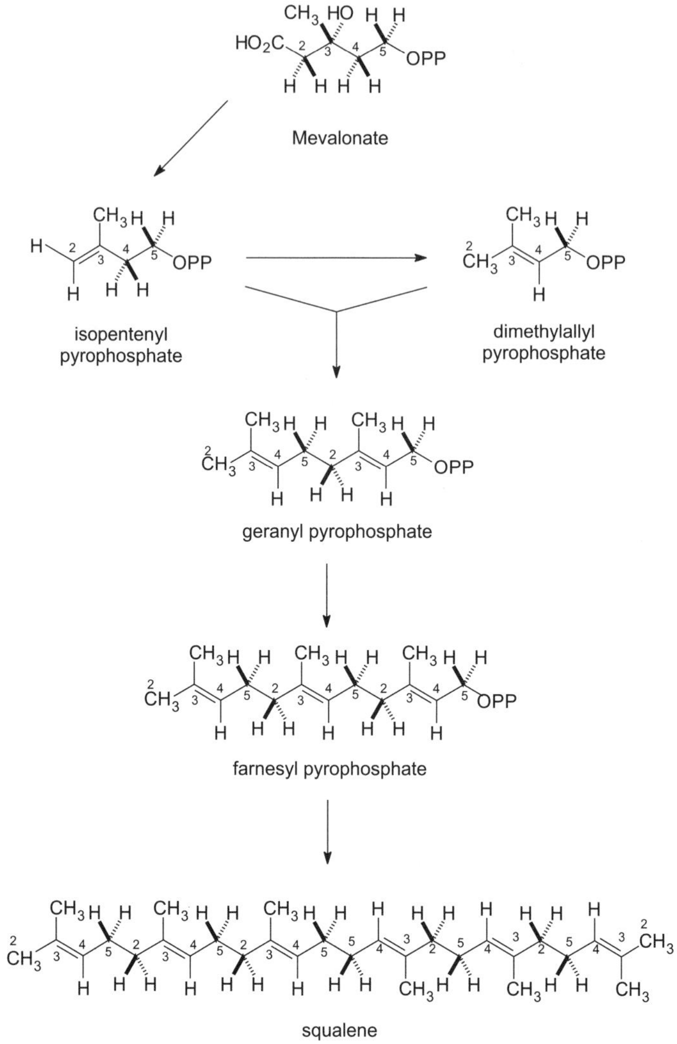

Figure 3.16 Synthesis of acyclic terpenes by successive condensations. Dimethylallylpyrophosphate condenses with isopentenylpyrophosphate to form geranylpyrophosphate. Redrawn after Turner (1971).

suggested in the schemes shown in Fig. 3.17 (most of which are plausible possibilities rather than defined metabolic pathways).

Diterpenes are derived by cyclisation of geranylgeranyl pyrophosphate (Fig. 3.18) and include the gibberellins, a mixture of plant growth promoters which were first isolated from culture filtrates of the plant

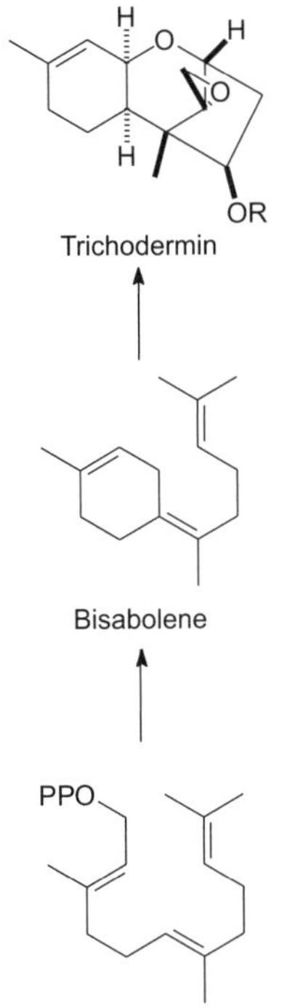

Trichodermin

Bisabolene

Farnesyl pyrophosphate

Figure 3.17 Sesquiterpenes: possible cyclisations of farnesylpyrophosphate to produce the trichothecane nucleus.

pathogen *Gibberella fujikuroi* (the sexual form of *Fusarium moniliforme*) and only later shown to be endogenous plant hormones.

The major significance of the triterpenes (C_{30}) is that the acyclic triterpene squalene is the precursor of sterols. Squalene is formed by a head-to-tail condensation of two sesquiterpene (farnesylpyrophosphate) units; it is so called because it is found in high concentration in shark liver oil, but in fungi as well as in other organisms sterols (and cyclic triterpenes) are derived by cyclisation of squalene oxide (Fig. 3.19). Cholesterol is the quantitatively predominant sterol in animals where it serves to control membrane fluidity, and ergosterol, so named because it was first isolated from the ergot fungus (*Claviceps purpurea*), probably fulfils a similar role in fungi, influencing permeability characteristics of the membrane (Hemmi *et al.*, 1995; Parks and Casey, 1995) and as it is unique to fungi, ergosterol synthesis is a potential target for antifungal agents (Barrettbee and Dixon, 1995). However, other sterols, even cholesterol,

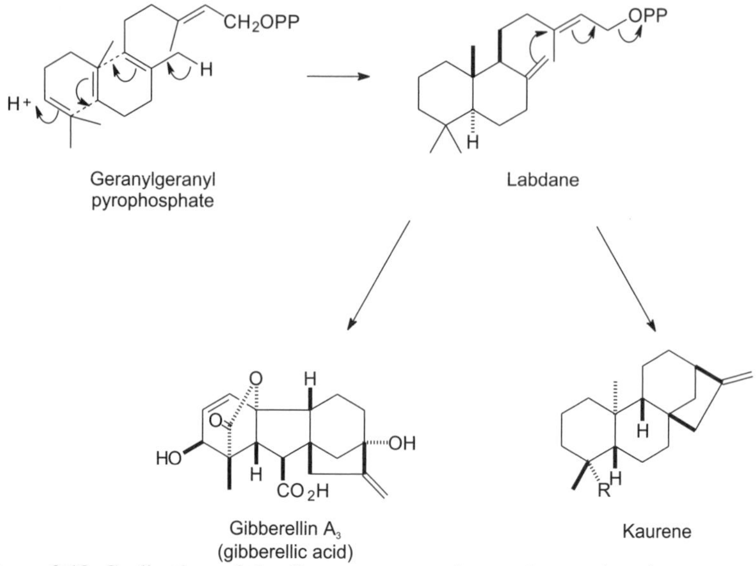

Geranylgeranyl pyrophosphate

Labdane

Gibberellin A_3 (gibberellic acid)

Kaurene

Figure 3.18 Cyclisation of the diterpene geranylgeranyl pyrophosphate producing gibbane and kaurane structures. Gibberellic acid is a plant growth hormone.

occur commonly and a very wide range of sterols has been detected in fungi (Goodwin, 1973; Weete and Gandhi, 1996) which have scope for characterising strains (Muller *et al.*, 1994). Lipid, sterol and phospholipid contents differed between yeast and mycelial forms of *Candida albicans* and *Mucor lusitanicus*, so there is a morphogenetic connection, too (Goyal and Khuller, 1994; Funtikova *et al.*, 1995). The sequential order of steps in sterol synthesis is well established as mutations in nearly every step of the yeast sterol biosynthetic pathway have been induced and selected (Parks *et al.*, 1995).

In much the same manner as squalene is formed by head-to-tail condensation of sesquiterpenes, head-to-tail condensation of two diterpenes generates the C_{40} carotenoid pigments. These include acyclic compounds like lycopene, the pigment responsible for the red colour of tomatoes, and monocyclic and dicyclic compounds like γ-carotene and β-carotene (Fig. 3.20). The carotenes and some keto-derivatives are widely synthesised (Goodwin, 1976; Armstrong and Hearst, 1996) and carotenoids are good taxonomic markers for some fungi (Valadon, 1976), though not all fungi produce them. Mutants of *Mucor circinelloides* with defects in all enzymatic steps from farnesyl pyrophosphate to β-carotene, together with some regulatory mutants, have been described (Navarro *et al.*, 1995) so the pathway is well established and open to detailed analysis.

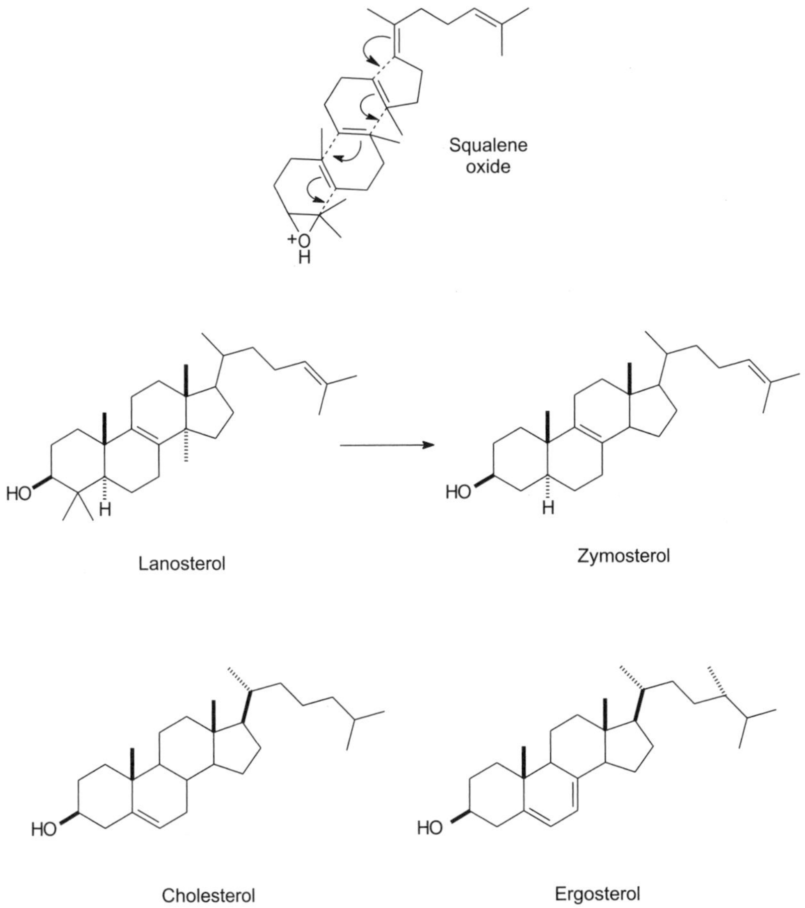

Squalene oxide

Lanosterol

Zymosterol

Cholesterol

Ergosterol

Figure 3.19 Sterols (and cyclic triterpenes) are derived by cyclisation of squalene oxide.

Continued condensation of isoprene units can produce terpenes with very long carbon chains. Natural rubber is a highly polymerised isoprene compound, containing of the order of 5000 isoprene units. The latex of the major commercial source, the rubber palm *Hevea brasiliensis*, is an aqueous colloidal suspension of particles of the hydrocarbon. The milk-like juice which is exuded from broken fruit bodies of *Lactarius*, stems of *Mycena* and gill edges of *Lacrymaria* is sometimes also called 'latex' but is chemically very different from rubber palm latex, being only superficially similar in appearance. The fungal product probably differs in structure between genera, but the 'latex' or 'milk' of *Lactarius rufus* has been shown to contain mannitol, glucose and lactarinic acid

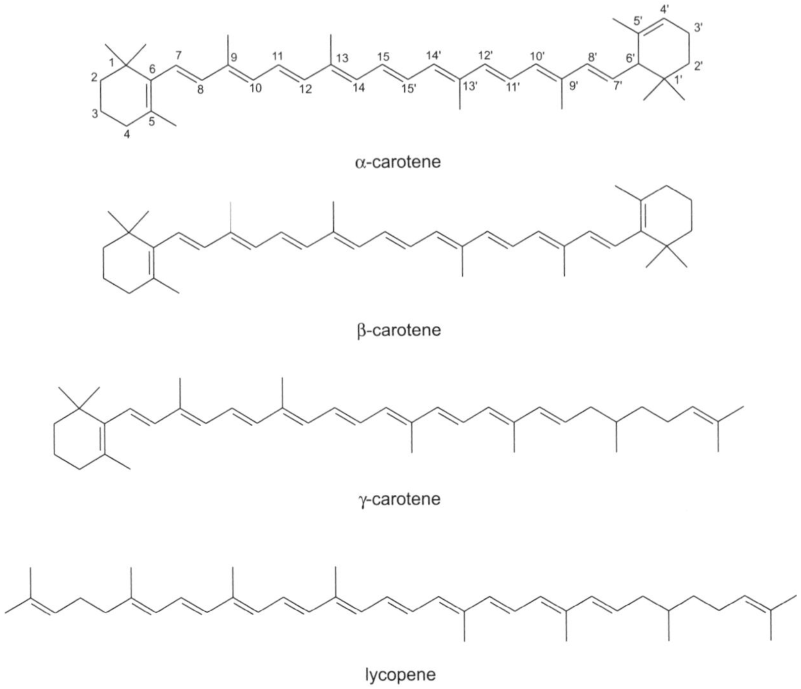

α-carotene

β-carotene

γ-carotene

lycopene

Figure 3.20 Fungal carotenoid pigments.

($CH_3[CH_2]_{11}CO[CH_2]_4COOH$; structural formula shown in Fig. 3.25). The latter is a modified fatty acid (6-oxo-octadecanoic acid, also known as 6-ketostearic acid).

3.13.2.2 The malonate pathway: polyketides

More secondary metabolites are synthesised through the polyketide pathway in fungi than by any other pathway (Turner, 1976). Polyketides are characteristically found as secondary metabolites among ascomycetes, and especially the imperfect deuteromycetes; they are rarely encountered in basidiomycetes and are produced by only a few organisms other than fungi. Polyketides are more correctly described as poly-β-ketomethylenes, the fundamental acyclic chain from which they are derived being comprised of -CH_2CO- units. Just as linear poly-isoprenoid chains can 'fold' and cyclise, so too the polyketides can cyclise to produce a wide range of different molecules (Fig. 3.21). As might be expected from the chemical nature of the repeating unit, synthesis of polyketides involves transfer of acetyl groups which are ultimately derived from acetyl-CoA. In fact the

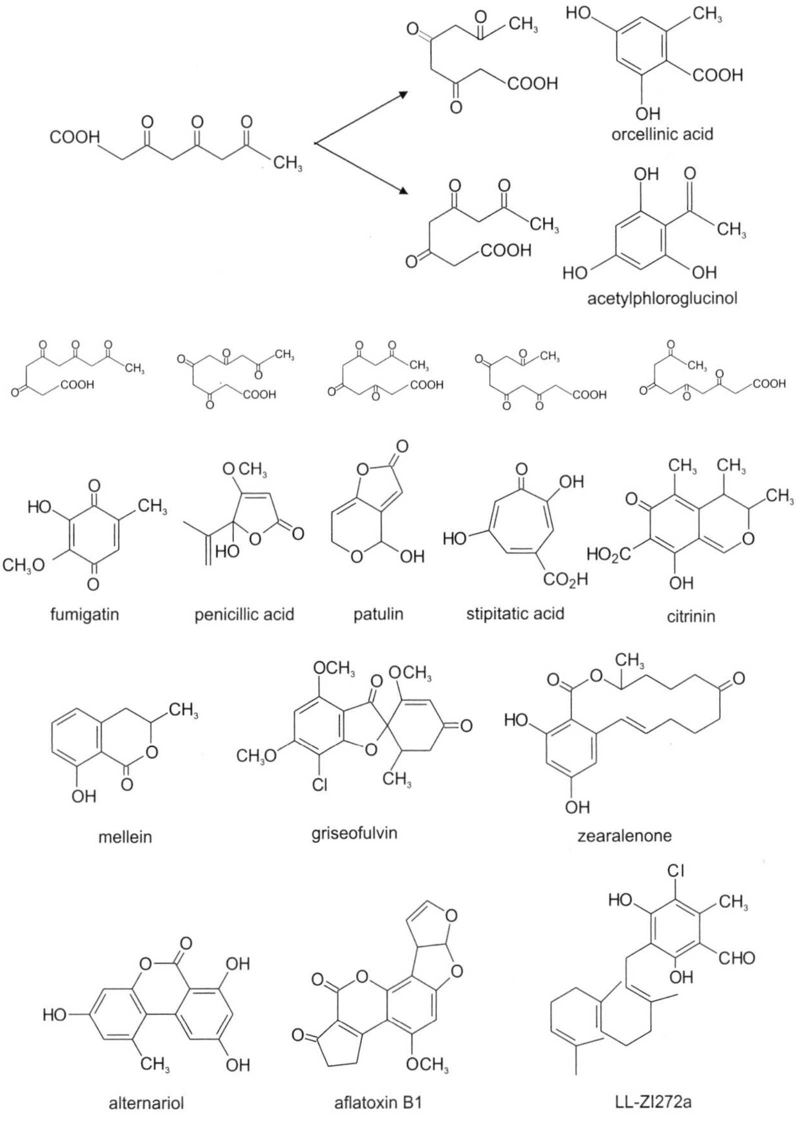

Figure 3.21 Polyketide chains can fold in a variety of ways and internal aldol condensations form closed aromatic rings. The alternative cyclisations of a tetraketide are shown at the top, together with the potential products orsellinic acid and acetylphloroglucinol. A pentaketide can cyclise in five ways, and these are shown schematically across the centre of the figure. The bottom section shows the structures of a small selection of polyketides discussed in the text in section 3.15.2.2. LL-Z1272a is an antibiotic isolated from *Fusarium* species in which the polyketide-derived aromatic rings have a sesquiterpene substituent.

synthesis of polyketides has a lot in common with fatty acid biosynthesis. Fatty acids are catabolised by stepwise removal of 'acetyl-units' (β-oxidation, see Fig. 3.11) but their synthesis requires malonyl-CoA which is formed by carboxylation of acetyl-CoA by the enzyme acetyl-CoA carboxylase: $CH_3COSCoA + HCO_3^- + ATP \longrightarrow COOH.CH_2COSCoA + ADP + Pi$.

Synthesis of fatty acids is carried out by a complex of enzymes called the fatty acid synthetase system (Walker and Woodbine, 1976; Kohlwein *et al.*, 1996). The reactions occur between substrate molecules which are chemically bound to the enzyme proteins; precursors are not found free in the cytoplasm. In the first reaction 'acetate' from acetyl-CoA is transferred to a peripheral sulfydryl group on the enzyme complex (releasing CoA), then a malonyl group (from malonyl-CoA) is transferred to an adjacent (central) sulfydryl group on the protein and a condensation reaction occurs between the enzyme-bound substrates to form acetoacetate which is still bound to the enzyme at the central sulfydryl group [CH_3COCH_2CO-S-enzyme]. This is then reduced, dehydrated and reduced again to form a butyryl-enzyme complex [$CH_3CH_2CH_2CO$-S-enzyme]. For chain lengthening, the butyryl residue is transferred to the peripheral sulfydryl so that a malonyl group can be brought in to the central one, permitting repetition of the condensation, reduction, dehydration, reduction cycle. Thus, malonyl-CoA provides all the carbon of long chain fatty acids except for the two terminal atoms, which derive from the 'acetate' initially introduced to the peripheral sulfydryl binding site.

This brief description of fatty acid synthesis can be echoed by a description of polyketide synthesis which is initiated by condensation of an acetyl unit with malonyl units, requires the respective CoA derivatives, and seems to occur on the enzymes involved. Again, free precursors are not found. Precise details of the mechanism of polyketide synthesis are still uncertain, however, but the similarities observed imply that the chain-building mechanism must be much the same between fatty acids and polyketides. However, for polyketides the reduction process does not occur and successive rounds of condensation generate polymers with the -CH_2CO- ('ketide') repeating unit, the number of which can be used to designate the chain formed as a triketide, pentaketide, octaketide, etc. The methylene (-CH_2-) and carbonyl (=C:O) groups of these chains can interact in internal aldol condensations to form closed aromatic rings (Fig. 3.21). A great variety of cyclisations thus become possible and this, in part, accounts for the wide range of polyketides which are encountered. There are limits, however; it seems that compounds in which the uncyclised residue of the CH_3-end of the chain is shorter than that from the -COOH end are not found (Turner, 1976).

The variety of cyclisations only partly accounts for the range of observed polyketides because other chemical modifications, especially dehydration, reduction, oxygenation by hydroxylation, decarboxylation, substitution and oxidation can generate further derivatives (Turner, 1976), and differences in the order in which such reactions occur can result in an enormous variety of actual and potential biosynthetic pathways.

Polyketide secondary metabolites are too numerous to be documented here (see Turner, 1971; Turner and Aldridge, 1983) but they include (Fig. 3.21) the tetraketides orsellinic acid, fumigatin (from *Aspergillus fumigatus*), penicillic acid (from *Penicillium* species, and *not* to be confused with penicillins which are modified dipeptides, patulin and tropolones like stipitatic acid (from *Penicillium stipitatum*). Pentaketides include the antibiotic citrinin (originally isolated from *P. citrinum*, but commonly found in ascomycetous fungi) and mellein, derivatives of which can be useful taxonomic and phylogenetic characters in Xylariaceae (Whalley and Edwards, 1987). Heptaketides include the antifungal agent griseofulvin (originally isolated from *Penicillium griseofulvum*) and alternariol (from *Alternaria* species). Nonaketides include zearalenone, a toxin produced in stored grain contaminated by *Fusarium*, the aflatoxins which are produced in mouldy groundnut meal by *Aspergillus* species (Moss, 1996), and the tetracycline antibiotics produced by actinomycetes. Other, more complex compounds also arise and there are some compounds in which polyketide-derived aromatic rings are attached to sesquiterpenes, such as the antibiotic LL-Z1272a isolated from *Fusarium* species.

Sterigmatocystin and the aflatoxins are among the most toxic, mutagenic, and carcinogenic natural products known. The sterigmatocystin biosynthetic pathway in *Aspergillus nidulans* is estimated to involve at least 15 enzymatic activities, while certain *A. parasiticus*, *A. flavus*, and *A. nomius* strains contain additional activities that convert sterigmatocystin to aflatoxin. Brown *et al.* (1996) characterised a 60 kb region of the *A. nidulans* genome and found it contained 25 co-ordinately regulated transcripts representing most, if not all, of the functions needed for sterigmatocystin biosynthesis.

3.13.2.3 Other secondary metabolic pathways

The previous two sections deal with special aspects of metabolism which give rise to particular classes of secondary metabolite, but there are many opportunities in primary metabolism for secondary metabolites to arise by greater or lesser modification of primary intermediates. Thus, important secondary metabolites can be formed as derivatives of amino acids, and fatty acids and are briefly discussed below.

The primary role of the shikimate–chorismate pathway is the synthesis of the aromatic amino acids phenylalanine, tyrosine and tryptophan (Fig. 3.14), but the pathway provides intermediates for the synthesis of other aromatic compounds as secondary metabolites (Turner, 1971; Towers, 1976). Plants synthesise a particularly large variety of compounds by this route, though it is not so widely used in fungi, where the polyketide pathway is used more to make aromatic ring compounds. Nevertheless, numerous shikimate–chorismate derivatives of quite common occurrence in plants have been isolated from fungi, such as gallic acid, pyrogallol, and methyl *p*-methoxycinnamate (Fig. 3.22). Among compounds more particularly associated with fungi are the ergot alkaloids (from *Claviceps purpurea*) which include ergocristine and lysergic acid amide and the hallucinogenic principle of the original 'magic' mushroom (*Psilocybe*), psilocybin which are all essentially tryptophan derivatives. Also, basidiomycete pigments like gyrocyanin, which oxidises to a blue pigment in injured fruit bodies of *Gyroporus cyanescens*, and hispidin, a polymer of which may be responsible for toughening of the fruit bodies of *Polyporus hispidus* (Bu'Lock, 1967) are derivatives of the shikimate–chorismate pathway.

Non-aromatic amino acids may also be modified (Wright and Vining, 1976). The muscarines and muscaridines (Fig. 3.23), which are the main toxic constituents of *Amanita muscaria* (but are also found in *Inocybe* and *Clitocybe* species), are synthesised from glutamate (Turner and Aldridge, 1983). Agaritine and related molecules like glutaminyl hydroxybenzene (or GHB) from *Agaricus* spp. (originally *A. bisporus*, the cultivated mushroom) are substituted (strictly, *N*-acylated) glutamate molecules, while the antifungal agent variotin (from *Paecilomyces varioti*) is γ-aminobutyric acid (GABA, shown in Fig. 3.9) *N*-acylated with a hexaketide (Fig. 3.23). Volatile amines formed by decarboxylation of neutral amino acids form part of the distinctive odours of some fungi; the fungus causing stinking smut of wheat (*Tilletia tritici*) produces large quantities of trimethylamine, and the aroma of Camembert cheese depends on the formation of *N*-dimethyl methioninol by *Penicillium camemberti*.

A variety of fungal secondary metabolites are derived from peptides; two or more amino acids linked through a peptide bond. Among those derived from dipeptides are the penicillins and cephalosporins (Weil *et al.*, 1995a,b). These could rate as the most important compounds ever to be isolated from fungi as they gave rise to a new era in medicine and a new branch of biotechnology. The basic structure of both antibiotics is

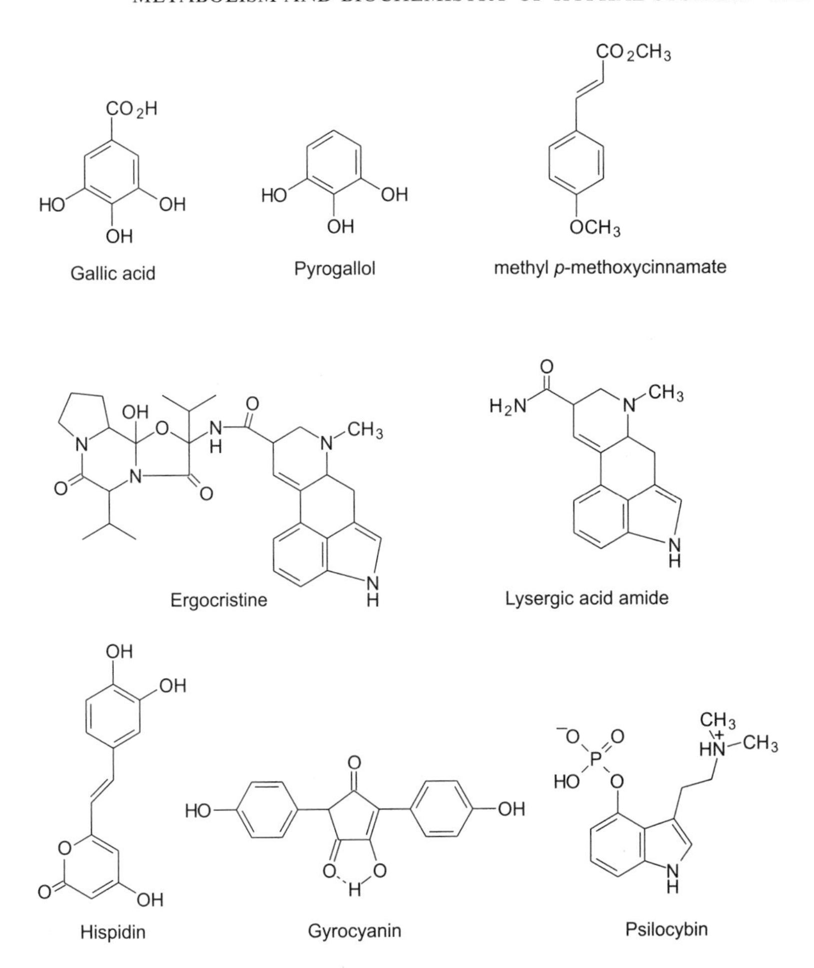

Figure 3.22 The shikimate–chorismate pathway produces a variety of aromatic compounds as well as the amino acids phenylalanine, tyrosine and tryptophan. Gallic acid, pyrogallol, and methyl *p*-methoxycinnamate are relatively simple compounds produced by many plants as well as fungi. Ergocristine, which is one of the ergot alkaloids (from *Claviceps purpurea*), lysergic acid amide and psilocybin, the hallucinogenic principle of the original 'magic' mushroom (*Psilocybe*), are all essentially tryptophan derivatives. Gyrocyanin is a product of *Gyroporus cyanescens* which oxidises to a blue pigment in injured fruit bodies, and hispidin is the precursor of a polymer which seems to be responsible for toughening the fruit bodies of *Polyporus hispidus*.

derived from cysteine and valine (Fig. 3.24) though the variable acyl group may come from another amino acid as in penicillin G where the acyl group is phenylalanine (illustrated in Fig. 3.24). Biosynthesis involves the tripeptide β-(L-α-aminoadipyl)-L-cysteinyl-D-valine, which is also illustrated in Fig. 3.24, but does not involve ribosomes. Instead,

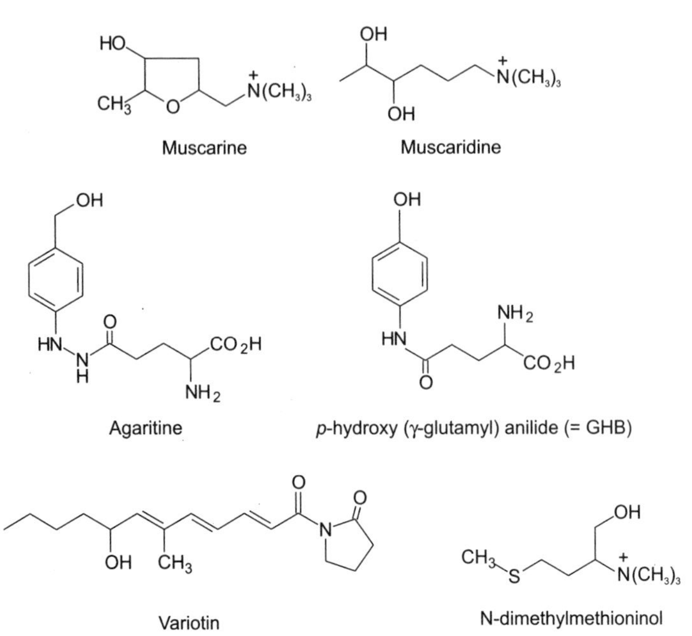

Figure 3.23 Non-aromatic amino acids may also be modified and accumulated as secondary metabolites. Muscarine and muscaridine, the main toxins of *Amanita muscaria*, are synthesised from glutamate, and agaritine and *p*-hydroxy (γ-glutamyl) anilide (also known as glutaminyl hydroxybenzene, or GHB) from *Agaricus* spp., are *N*-acylated glutamate molecules. Agaritine may account for up to 0.3% of the dry weight in A. bisporus fruit bodies and GHB may be involved in controlling basidiospore dormancy and is the most likely precursor for *Agaricus* melanin in the spore walls (Hammond and Wood, 1985). Variotin is an antifungal agent isolated from *Paecilomyces varioti* and is γ-aminobutyric acid *N*-acylated with a hexaketide. *N*-dimethylmethioninol is a volatile amine formed by decarboxylation of methionine by *Penicillium camemberti*. It is responsible for the aroma of Camembert cheese.

amino-acid activating domains of peptide synthetases determine the number and order of the amino acid constituents of the peptide secondary metabolites. *In vitro* reconstruction of the gene sequences of these multifunctional enzymes produces hybrid genes that encode peptide synthetases with altered amino acid specificities able to synthesise peptides with modified amino acid sequences (Stachelhaus *et al.*, 1995). Also derived from peptides are the *Amanita* toxins of which there are many, but which can be represented by α-amanitin and phalloidin. Because of the resemblance between *Amanita phalloides* and edible field mushrooms, these toxins are involved in the majority of cases of mushroom poisoning (e.g. O'Brien and Khuu, 1996). Many fungi produce siderophores for

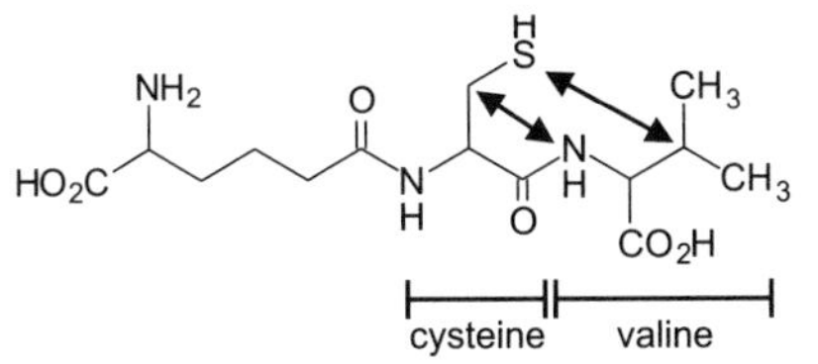

δ-(α-amino-adipoyl)cysteinylvaline

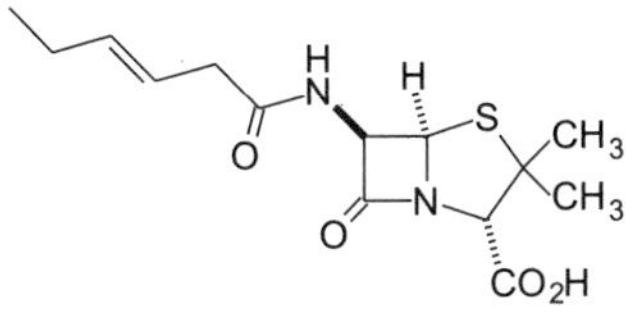

Penicillin F

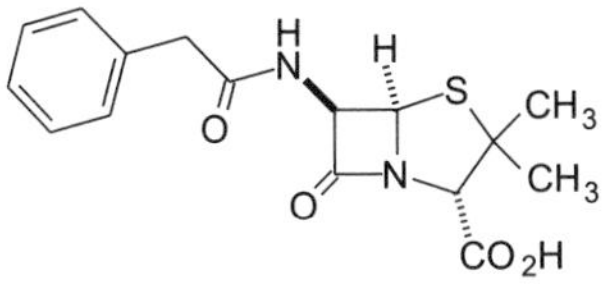

Penicillin G

Figure 3.24 The penicillin antibiotics are representative of secondary metabolites which are derivatives of peptides. This figure shows the structural formulae of two penicillins. At the top is the precursor δ-(α-amino-adipoyl)cysteinylvaline. The cysteine and valine residues of this compound are identified and the double-headed arrows show the bonds which have to be made to create the penicillin nucleus.

acquiring iron which, though an essential nutrient, is not readily available in aquatic or terrestrial environments or in animal hosts. These iron-binding compounds also originate as peptides formed from modified amino acids (Turner, 1971; Turner and Aldridge, 1983; Wright and Vining, 1976; Winkelmann, 1986; Guerinot, 1994).

Chemical modification of fatty acids produces a variety of secondary metabolites, most particularly the polyacetylenes, many of which have been obtained from basidiomycetes. These compounds have straight carbon chains varying between C_6 and C_{18}, though C_9 and C_{10} are most common in fungi. They are conjugated acetylenes (i.e. with triple bonds between adjacent carbons, e.g. hexatriyene from *Fomes annosus*, Fig. 3.25), or systems containing both ethylenic (i.e. with double bonds between carbon atoms) and acetylenic structures (e.g. nemotinic acid, Fig. 3.25). Polyacetylenes are derived from fatty acids by a series of dehydrogenation reactions (Turner, 1971; Turner and Aldridge, 1983).

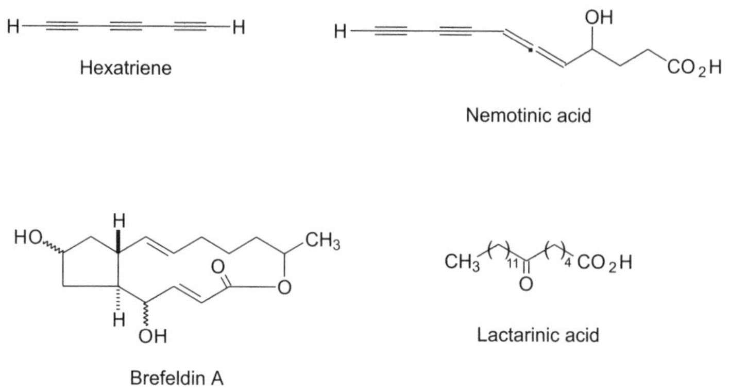

Hexatriene

Nemotinic acid

Brefeldin A

Lactarinic acid

Figure 3.25 Chemical modification of fatty acids produces a variety of secondary metabolites including polyacetylenes and cyclopentanes. Lactarinic acid is a major component of the 'milk' or 'latex' which exudes from injured fruit bodies of *Lactarius* spp. Lactarinic acid is 6-ketostearic acid and is very different from true latex which is an isoprene compound.

Also formed from fatty acids are cyclopentanes like brefeldin, which has been extracted from *Penicillium*, *Nectria* and *Curvularia* species (Fig. 3.25). Lactarinic acid was referred to above (section 3.13.2.1).

3.13.3 Regulation

Secondary metabolism characterises the periods occurring after one or more nutrients become growth-limiting. Indeed, Bu'Lock (1961) describes cell growth and secondary biosynthesis as incompatible, competing processes; the synthesis of secondary metabolites being suppressed while the cells in a culture are multiplying. A further implication is that a fungal culture passes through a sequence of distinct phases: initially it is not nutrient-limited and primary metabolism and cell growth and proliferation predominate, but those activities use nutrients and eventually one becomes limiting. When that happens, secondary metabolic activities are triggered.

It is important to appreciate that this fungal culture will be, for the most part, a diverse population of cells of different age and metabolic state so the behaviour observed for the culture as a whole is the average of all the members of this population. Surface cultures, whether the substratum is solid or liquid, are particularly diverse and inhomogeneous; submerged cultures a little less so as, at least, they lack surface and aerial components. Synchronised cultures, if they can be obtained, provide by definition a population of cells which are progressing through their cell cycles in unison, so the observed behaviour of the culture most accurately reflects individual cell behaviour. Synchronised cultures are, of course, very artificial; the natural state of fungal cultures being the unsynchronised, inhomogeneous one. This means that the descriptions of the phases in the growth of a culture which follow can be interpreted at a number of levels: they can be seen as averaged descriptions of the 'life experiences' of individual cells, or as averaged accounts of happenings in the whole culture. The latter is the usual interpretation, but then the ability to observe the expected phases will depend on the degree of homogeneity in the population. Surface cultures are usually too inhomogeneous to permit identification of the phases; submerged batch cultures (i.e. one batch of liquid medium inoculated with a single starter culture and cultivated to the desired end-point) are usually sufficiently uniform to demonstrate the major characteristics.

The phases through which such a batch culture progresses have been defined and named as the balanced phase, storage phase and maintenance phase by Borrow *et al.* (1961); or the trophophase (balanced phase) and idiophase (storage phase + maintenance phase) by Bu'Lock (1965). Descriptions like this are convenient for describing the course of a fermentation but further than this, they do describe coordinated changes in physiology which are quite readily measurable. The balanced phase is the period of rapid growth extending from the onset of growth to the time of exhaustion of the first nutrient. This is the period of rapid proliferation and both the chemical composition and the microscopic appearance of the mycelium remain constant throughout the balanced phase of growth. Both features are remarkably independent of the initial nutrient concentrations of the medium. The storage phase is one in which reserves of carbon and other elements are accumulated in mycelium by redistribution of the components of cells which have ceased rapid proliferation. The dry weight and fat and carbohydrate content of the mycelium increased in nitrogen-limited cultures given sufficient glucose though at about the time that the dry weight reaches its maximum, the uptake of nutrients other than glucose ceases. Timing of individual events within this phase differs, implying that nutrient exhaustion initiates a number of independent metabolic changes which proceed during the storage phase. Among these changes are those processes which are called secondary metabolism. There is, of course, a transition between the balanced and storage phases; the events occurring during the transition depend on the nature of the limiting nutrient. The onset of storage phase is signalled by cessation of cell proliferation; but during storage phase the biomass continues to increase as reserves are redistributed and accumulated. Eventually, however, the mycelial dry weight (biomass) reaches a maximum and this defines the start of maintenance phase. During this phase gross features like dry weight, nucleic acid content and total carbohydrate remain approximately constant but there is considerable turnover as carbon sources, either from the medium or mobilised from internal reserves, are metabolised. During the course of this metabolism secondary metabolites can be accumulated to their maximum levels.

Thus, the fundamental control factor initiating secondary metabolism seems to be nutrient limitation, but any of a wide range of nutrients can be effective in this and, as we have seen, there is a wide range of secondary metabolic pathways which might be initiated.

3.13.4 Role

Secondary metabolites may be obtained by chemical extraction of material collected in the field, the traditional way in which organic chemists obtained and catalogued 'natural products', but the special attraction of fungi is that so many of them will produce secondary metabolites when grown in the laboratory in liquid media. This ability has proved a great convenience for academic study of secondary metabolism and has been the essential prerequisite of its exploitation for commercial purposes. It must be stressed, though, that except for yeasts and the water moulds growth in liquids and, especially, growth in submerged culture, is physiologically abnormal for fungi. Furthermore, many of the fungal strains most favoured for commercial production of secondary metabolites are biochemically defective in some way; a strain which 'wastes' nutritional resources in order to overproduce a commercially useful product being the organism of choice (Aharonowitz, 1980). In attempting to assign a role (or roles) in nature to secondary metabolism and secondary metabolites it must be recognised that our basic information is fragmentary, is concentrated in areas which have relevance to commercial exploitation and is, therefore, inevitably at its weakest in areas relating to the natural environment and natural behaviour of the fungus.

Nevertheless, a consistent argument can be developed along the lines that, rather than being some sort of luxurious biochemical extravagance, secondary metabolism does serve important physiological roles. Bu'Lock (1961) clearly stated that we must suppose that formation of secondary metabolites confers selective advantage on the producing organism. He further observed that secondary metabolites are known in such variety that no single intrinsic property accounting for production of secondary metabolites in general can be found that is common to all of them. Bu'Lock suggested that the selective advantage of secondary metabolism might be that the synthetic activities that characterise it may serve to '... maintain mechanisms essential to cell multiplication in operative order when that cell multiplication is no longer possible.'

In other words it is cell multiplication which drives balanced growth by providing a continuously expanding sink for the products of primary metabolism. When that growth process stops, secondary metabolism, as a differentiation process, provides an alternative sink permitting a number of general synthetic and nutrient uptake mechanisms to continue operating, especially during the maintenance phase, but without requiring close integration of processes because the end-product (the secondary metabolite) is of no special significance. This fundamentally economic explanation could well provide a general, very unspecific, role to secondary metabolite production and thereby account for its common occurrence. However, an unsatisfactory aspect of this notion is that the other characteristic of secondary metabolism, namely that secondary metabolites are individually of such restricted distribution as often to represent species-specific markers, must then be left to chance. That is, it may be that all fungi need to embark upon secondary metabolism as their growth becomes nutrient-limited but the particular metabolite that an individual species synthesises is an arbitrary 'choice'. Bu'Lock (1961) argued that advantages other than the economic one (such as ecological advantages conferred by antibiotics; structural and physiological advantages due to synthesis of pigments and polyphenols in cell walls, etc.) should be viewed as incidental to the main economic role.

An alternative (and not necessarily exclusive) argument has been developed. Campbell (1984) points out that in several ways secondary metabolism is more sophisticated in its biochemistry, especially in the stereospecificity of terpene and polyketide biosyntheses. In this respect secondary metabolism represents a considerable evolutionary advance over primary metabolism and must therefore be of very considerable selective advantage. He also emphasises the distributional definition of secondary metabolites, that they are metabolites which are (individually) of limited or restricted distribution, and points out the Darwinian prediction that competition would be most severe between varieties of the same species or species of the same genus. This being the case, it might be expected that variety- or species-specific processes providing competitive advantage would arise as a consequence, and some would result in variety- or species-specific metabolites.

Campbell (1984) lists a number of physiological processes in which limited distribution of the metabolites is a necessary prerequisite of the function. These include sex hormones, pheromones and tropic agents; hormones and growth factors involved in morphogenesis; agents affecting spore germination and outgrowth; chelating agents; structural and extracellular protective agents; and host-specific toxins. The majority of the processes in this list contribute in some way to fungal morphogenesis. In addition, protection or defence is especially important and is viewed in its widest interpretation as taking in processes as diverse as cell wall reinforcement, toxin production (e.g. to deter grazing animals) and cell-cell communication (Gloer, 1995; Shearer, 1995). Thus, there is no shortage of physiological functions which depend upon uniqueness of the effective molecule. As our knowledge of the biological nature of these processes increases, so our understanding of the part played by secondary metabolites will increase. Indeed, many of the generalised interpretations of the function of 'secondary metabolism' may well be overtaken by events as knowledge accumulates of the molecular architecture of the cell. Several polypeptides, including the nuclear lamins, several vesicular transport proteins, the oncogene product Ras and fungal peptide pheromones require the post-translational attachment of a farnesyl group, an isoprenoid lipid moiety derived from mevalonate, to the carboxyl-terminus of the protein (Caldwell *et al.*, 1995; Dimsterdenk *et al.*, 1995). Should this protein prenylation be labelled as an aspect of secondary metabolism, or is the isoprenoid pathway just another way of making function-specific post-translational modifications to peptides rather than a secondary metabolic route? Which is more significant, the biochemistry or the language used to describe it?

The study of secondary metabolism was dominated in its early years by the desire of organic chemists to document the occurrence of natural products. It is an area of biological chemistry which has a particular history and we are still left with conventions and distinctions which are purely matters of definition or semantics. For example, it was stated above that, conventionally, animals are not considered to produce many secondary metabolites. This is a matter of the convention in the use of the term, rather than an indication of a profound biochemical difference between the different groups of organisms. For the sorts of reasons discussed immediately above, chemicals employed in defence, offence, communication and control in fungi, bacteria and plants are expected to have limited distribution and therefore to be classified as secondary metabolites. Animals produce toxins, antitoxins, hormones and pheromones too. They are not called secondary metabolites because their function has often been discovered before isolation of the chemical itself, so there is no history of long lists of natural animal products which lack obvious function, 'secondary metabolites'. It's a matter of semantics, but it seems that such instances as the spider which excretes a moth pheromone (Stowe *et al.*, 1987) to attract prey to a last supper are genuine examples of what in a plant or fungus would be called secondary metabolism rather than chemical mimicry – the phrase used in zoological vocabulary. A more extreme example is the presence of the hallucinogen bufotenin in the toad *Bufo* and in a species of the toadstool *Amanita* (Metzenberg, 1991). This has been mentioned before as another possible example of horizontal transmission at some stage in evolution (section 1.7), but why should it be described by different terms in the two organisms?

4

Physiological factors favouring morphogenesis

The biochemical capabilities summarised in Chapter 3 are used in a variety of ways during morphogenetic processes in fungi. In this chapter I will describe and discuss some of these. Throughout this account from now on my emphasis will be on finding general principles to weave into an overall model of fungal morphogenesis. Experimental observations, of course, deal with specific cases under particular conditions. I will describe some of these specific cases but, as I cannot describe them all, I have selected examples which seem to me to illustrate particular points which need to be included in any overall picture of morphogenesis. Other examples can be found in the literature and I will make reference to some of them. Furthermore, by attempting to generalise from the few cases I can illustrate, I will be relying on making inductive inferences, some of which will, inevitably, be more reliable than others. It is necessary to take the risk of being wrong in order to get some view of the overall picture.

The variety of data in the literature indicates that the major events in the growth and differentiation of a mycelium are essentially as follows. During growth of the vegetative hyphae a large proportion of the nutrients adsorbed from the substratum is stored. Different organisms use the different storage compounds (polysaccharide, protein, oils and fats and polyphosphates) to different extents. The nature of the stored materials may differ, too. Polysaccharide may be stored in the form of glucans in secondary wall layers and/or as glycogen granules scattered in the cytoplasm. However, the essential point is that the hyphae absorb sufficient nutrients to support their active vegetative growth *and* to allow accumulation of reserve materials, often in particular, frequently swollen, cells in various parts of the mycelium. While this is proceeding, neighbouring hyphae, often those of the surface or more aerial parts of the mycelium, may interact to form centres of rapid but self-restricting growth and branching which become the hyphal aggregates or mycelial tufts, perhaps 100–200 μm in diameter, that are the 'initials' of the reproductive structure the organism can produce. Frequently, and especially in culture, these aggregates are formed in great number over the whole surface of the colony. As supplies of nutrients in the medium approach exhaustion repression of the morphogenesis of these hyphal aggregates is lifted and they proceed to develop further. Illumination may be required, either to promote further morphogenesis, or to direct development into one of a small number of morphogenetic pathways. Particular temperatures may also be required for particular pathways of development. Usually, only a small number of the first-formed hyphal aggregates undergo further development and these become the focus for translocation of nutrients, mobilized from the stores in other parts of the colony and transported through the hyphal network to the developing reproductive structures.

Madelin (1956b) provided the clearest demonstration of this. He noted that although primordia of fruit bodies of *C. lagopus* were usually more or less uniformly dispersed within an annular zone of the colony, enlarging and mature fruit bodies were usually very uneven in their distribution. Arguing that this implied that certain favourably placed young fruit bodies initiate a flow of nutrients in their direction which is disadvantageous to the less favoured ones which thus fail to mature, Madelin (1956b) investigated the effect of dividing the colony into two separate halves. Physically separated halves of cultures gave almost equal yields of fruit body dry weight, whereas the two sides of an intact culture could differ in fruit body dry weight yield by as much as 10:1. In the latter case it is concluded that the 'minority' half is exporting its nutrients to the 'majority' half. When the two halves are separated early in growth, this redistribution of nutrient cannot occur and the two half-cultures each produce about half the usual yield of fruit bodies.

The interpretation that fruit body distribution is governed by a flow of nutrients towards particular developing fruit bodies rather than, say, localised nutrient depletion or inhibition of development is based on two main observations. First, only a comparatively small number of primordia develop into mature fruit bodies. Yet experiments in which the amount of substrate was varied (discussed at the end of section 4.3.1) suggest that fruit body size can be adjusted to match available substrate. If fruit body size is related to local nutrient supply, one would expect that all of the primordia on a colony would develop into mature but small fruit bodies, using those quantities of materials which are locally available. Second, a crop consisting of several fruit bodies will often develop as a group, so that any inhibitory action is unlikely. The concept that nutrients flow towards a favoured centre would permit several neighbouring primordia to mature in a clump, whilst still withholding nutrients from unfavourably situated primordia. It is important to add that yields of vegetative mycelium from culture halves in these experiments were not significantly different. Consequently, the postulated flow of materials is spread throughout the colony. The whole mycelium contributes more or less uniformly to development of its fruit bodies.

The mechanism which relates fruit body size to available substrate, especially across an entire culture, is unknown. However, in section 4.2.4 experiments are described in which cAMP was administered to cultures of *Coprinus cinereus* (Kuhad *et al.*, 1987) in an asymmetric manner. As a homogeneous response was obtained in a time much less than either the rates of growth or diffusion would allow, rapid transmission of some sort of signal must have occurred.

Development usually proceeds in a series of steps which may be coordinated by particular environmental cues (illumination, temperature, atmosphere) and often involve sweeping re-allocation of cellular components. Within the young fruit body, therefore, new accumulations of 'stored' nutrients arise, and there may be a number of these accumulation-mobilisation-translocation-accumulation cycles during the development of the reproductive structure. Some of the metabolic changes which can be observed to occur endogenously in developing reproductive structures can be mimicked in experimental mycelial cultures, but whether the control events which these experiments reveal actually participate in the developmental sequence is open to question.

Nevertheless, there does seem to be a general correlation between nutrient exhaustion of the medium and the onset of multicellular morphogenesis (see Fig. 4.2 and examples below). This sort of correlation emerged at the end of the 19th century, particularly in the work of Klebs (1898, 1899, 1900), who stated categorically that reproduction is induced by nutritional deficiency. According to Klebs, initiation of reproduction is not a property of the mycelium but is due to the stimulus of external conditions. However, his interpretations are too simplistic and as Klebs

Table 4.1. *Dry weights of mycelium and fruit bodies during development by* Coprinus lagopus *(Madelin, 1956).*

	Dry weight (mg)			
Age (days)	Mycelium	Fruit body	Total	Developmental stage
9	26	0	26	Initiation of primordia
10	30	0	30	
11	32	4	36	Enlargement
12	37	5	42	
13	22	19	41	Maturation and
14	31	18	49	deliquescence
15	30	30	60	
16	39	14	53	Deliquescence and
17	43	20	63	start of second crop

experimented with *Saprolegnia*, it is unlikely that his principles apply to fungi. In most fungi, reproduction is not an alternative to vegetative hyphal growth but an aspect of the differentiation of vegetative hyphae. Inevitably, therefore, most fungi must conflict with the view of Klebs (1900) that conditions favouring reproduction are always more or less unfavourable for vegetative growth. Continued growth of the vegetative mycelium over the period when fruit bodies are being produced is clearly evident in data for *Schizophyllum commune* (see Fig. 4.2) and *Coprinus lagopus* (Table 4.1). It is, of course, essential that the mycelium continues to thrive in order to provide support and sustenance to the developing fruit bodies.

Thus, even though fruiting does indeed often coincide with nutrient exhaustion of the medium, this does *not* mean that development is prompted by starvation conditions. Only preconditioned mycelium is capable of undergoing morphogenesis. The preconditioned mycelium must be beyond a particular minimum size, perhaps be of a particular minimum age, and the underlying nature of both these preconditions is that the mycelium has been able to accumulate sufficient supplies of reserve materials to support development of the minimum reproductive structure. Exhaustion of a particular metabolite from the medium or substrate is a signal which prompts morphogenesis in a mycelium which is *not* starving, but is healthy and well-provisioned. Exhaustion of one or more constituents of the medium changes the balance of nutrient flow. If the medium is no longer fully supportive, the requirements of active hyphal growth can no longer be met by import from outside the hypha and the balance must shift from 'reserve material accumulation' to 'reserve material mobilisation'. That change from balanced growth to growth under limitation in external nutrient supply is what signals the onset of morphogenesis. This is not to say that metabolic support of morphogenesis is fully internalised. Cellular differentiation *is* an expression of unbalanced growth, precipitated by one or more changes in the balance of metabolism, and itself causing further cycles in which cellular components are re-allocated, but nutritional dependence on the external substrate can still be demonstrated for early stages in development at least.

4.1 Nutrition

The nutrients required for reproduction are much the same as those that are needed for vegetative growth, though the different phases may exhibit different 'preferences' for the form in which nutrients are made available. Formation of reproductive structures involves synthesis of materials of a nature or in a concentration not normally encountered in the vegetative mycelium, so distinctive nutritional requirements are not unexpected and much of the early literature concentrates on these features as 'requirements' for reproduction (Lilly and Barnett, 1951; Cochrane, 1958; Hawker, 1966; Turian, 1966, 1978; Takemaru and Kamada, 1969). Some of the more unusual stimulatory additives will be discussed below. In this section attention will be given to major carbon and nitrogen source nutrients. Other descriptive studies of the physiological responses to specific environmental conditions have dealt with numerous variations in external conditions and some attempt will be made to interpret how these regulate morphogenetic sequences resulting in fruit bodies of a particular sort.

4.1.1 Carbohydrates in the substratum

Simple sugars tend to favour asexual spore production while disaccharides and polysaccharides are especially good carbon sources for production of fruit bodies (Hawker, 1939). Glucose often represses fruit body production, even in very low concentrations (Weste, 1970a,b; Moore-Landecker, 1975, 1987), but Hawker (1939) showed that whilst perithecia of *Sordaria* were not formed on a medium containing 5% glucose, they did form if the same amount of sugar was supplied in small increments. Most physiological studies deal with nutrient concentration as the experimental variable; this being technically very easy to control. Hawker's work, on the other hand, stresses the importance of rate of supply and ease of use of substrates as determinants of their value in promoting fruit body formation. For example, Hawker and Chaudhuri (1946) showed that it was the rate with which a fungus could hydrolyse a carbohydrate which determined its ability to promote fruit body formation. Carbohydrates which were hydrolysed at an intermediate rate, providing adequate but not excessive concentrations of hexose over a long period allowed production of the largest number of fruit bodies (Fig. 4.1). In addition to concentration, Hawker (1947) found that the ease with which a carbohydrate was phosphorylated is an important

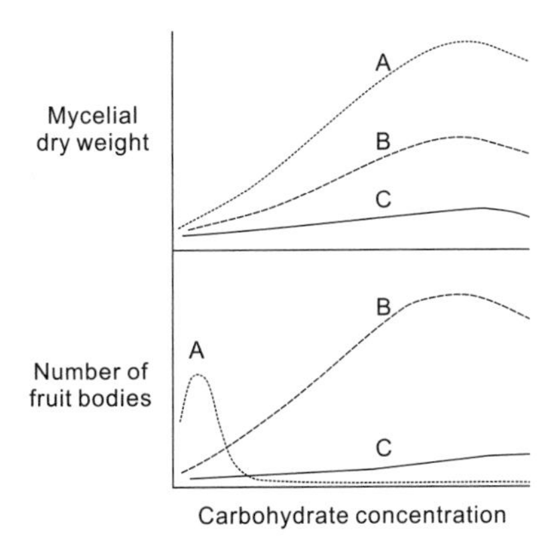

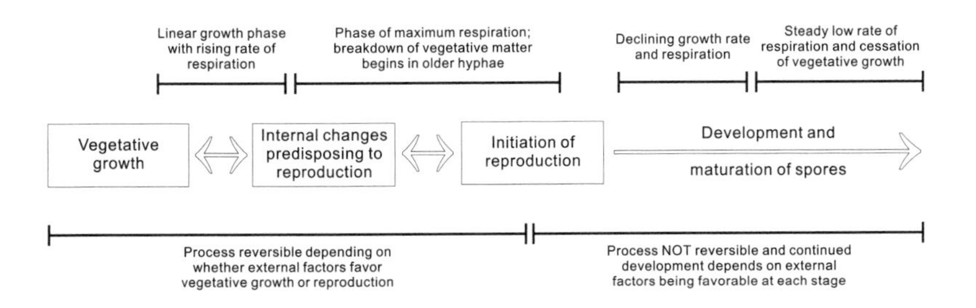

Figure 4.1 Top panel shows typical graphs relating mycelium and fruit body biomass production to carbohydrate concentration for (A) a readily used simple carbohydrate like glucose, (B) a more complex oligomeric or polymeric carbohydrate which the fungus can efficiently degrade and utilise, and (C) a poorly-used carbohydrate. Note that substrate category B is generally more supportive of fruit body production. Revised after Hawker and Chaudhuri (1946). The lower panel is a flow chart of the overall relation between growth, respiration and sporulation. Revised from Hawker (1966).

factor in determining the effect of that carbohydrate on fruit body formation. Despite the enormous volume of research which has been done since Hawker's experiments (for reviews see Moore-Landecker, 1993; Jennings, 1995), the greatest insight has come from her work. I think the key understandings are, first, that the conditions under which experiments are carried out in the laboratory are totally abnormal as far as the organism is concerned, and, second, that nutrients which are inferred to be 'favourable' for fruiting are those which allow the organism to exert its own intrinsic controls over the progress of its metabolism.

4.1.2 Nitrogen sources in the substratum

Similar conclusions are reached when attention turns to 'the best nitrogen source', which usually proves to be one amino acid or a mixture of amino acids. For example, potassium nitrate supports the production of more protoperithecia by *Cochliobolus miyabeanus* than other nitrogen sources, but L-arginine or DL-valine are required to produce the most mature perithecia (Kaur and Despande, 1980). This type of observation suggests that the formation of fruit body initials may be an activity of the vegetative mycelium. It is their further development which constitutes the fundamental 'mode switch' into the fruit body morphogenetic pathway. In most cases ammonium salts fail to support fruit body development (Singh and Shankar, 1972; Ghora and Chaudhuri, 1975a; Asina *et al.*, 1977; Kaur and Despande, 1980; Dehorter and Perrin, 1983). *Venturia inaequalis* forms ascomata on medium containing ammonium salts only when calcium carbonate is added to control the pH (Ross and Bremner, 1971), so at least some of the deleterious effects of ammonium salts may be due to their influence on the pH of the medium (Chapter 3). However, the metabolite repression caused by ammonium ions (Chapter 3) may be another cause. Fruit body formation in some fungi is favoured by provision of protein as source of nitrogen. The entomogenous fungus *Cordyceps militaris* is an example, and Basith and Madelin (1968) suggest an interpretation similar to Hawker's: that the favourable effect is due to slow extracellular digestion providing a low level of amino acids over a long period. Several basidiomycetes (*Agaricus bisporus*, *Coprinus cinereus* and *Volvariella volvacea*) have been shown to be able to use protein as a carbon source as efficiently as they use glucose as a sole source of carbon (Kalisz *et al.*, 1986), so an advantage of protein is that it serves as a source of carbon, nitrogen and sulfur; the utilisation of these elements being under the control of the organism. In more natural cultivation conditions, *A. bisporus* and a wide range of other filamentous fungi have been shown to be able to degrade dead bacteria and to utilise them as sole source of carbon, nitrogen, sulfur and phosphorus (Fermor and Wood, 1981; Grant *et al.*, 1986).

Higher carbon than nitrogen concentrations are usually required for fruit body production but the optimum C:N ratio varies from around 30:1 to about 5:1 (references in Moore-Landecker, 1993). High concentrations of amino acids tend to delay and/or depress maturation of fruit bodies even in organisms in which fruit body formation is optimal on media containing lower concentrations of amino acids (Ross and Hamlin, 1965; Moore-Landecker, 1975). The adverse effects of high concentrations of amino acid may result from the production of large quantities of ammonium as a nitrogen-excretion product (Chapter 3). When grown on protein as sole carbon source, nitrogen needs to be excreted from the mycelium; when this happens *in vitro* the ammonium concentration of the medium *increases* during mycelial growth quite drastically (Table 4.2). One third to one half of the supplied protein-nitrogen was assimilated and metabolised to ammonia by batch cultures of three basidiomycetes during use of protein as a carbon source (Kalisz *et al.*, 1986).

4.2 Adaptations of metabolism

I have stressed above my belief that, whilst there is a continuing need for nutrient supplies from the substrate, the most important aspect of the mode switch between a vegetative hypha and a hypha undergoing a morphogenetic process is that the regulation of metabolism devolves to the inner environment of the hypha. Under natural circumstances the regulatory influence of the external environment would then be minimised to influencing choice of pathways, etc. In the unnatural circumstances of Petri dish or fermenter the experimenter can still, of course, blunder in with his bottle full of purified chemical and thereby disturb the entire course of events.

4.2.1 Turnover of cellular polymers

Normal morphogenesis involves major change in the biochemical composition of many parts of the cell and, as indicated above, considerable turnover of existing cellular components. Histochemical studies of developing fruit bodies revealed something about the scale of this process.

Table 4.2. *Concentration of ammonium in the media of cultures of* Agaricus, Coprinus *and* Volvariella.

Organism	Days growth	Supplements to the medium				
		Protein	Protein + glucose	Protein + ammonium	Protein + glucose + ammonium	Glucose + ammonium + sulfate
	0	15 (2)	19 ± 3 (2)	857 ± 8 (2)	827 ± 21 (2)	815 (2)
	9	72 ± 4 (3)	12 ± 2 (3)	982 ± 27 (3)	862 ± 26 (2)	927 ± 42 (3)
	17	243 ± 28 (3)	96 ± 16 (2)	1165 ± 17 (2)	815 ± 39 (3)	675 ± 16 (3)
Agaricus	25	287 ± 1 (2)	278 ± 31 (3)	1381 ± 15 (3)	969 ± 6 (2)	685 ± 42 (3)
	33	399 ± 12 (3)	708 ± 77 (3)	1381 ± 88 (3)	1279 ± 33 (3)	661 ± 23 (3)
	41	533 ± 9 (3)	883 ± 19 (3)	1388 ± 36 (3)	1424 ± 62 (2)	640 ± 32 (3)
	49	561 ± 16 (2)	1029	1471 ± 43 (3)	1522 ± 36 (3)	718 ± 16 (3)
	0	32 ± 18 (6)	7 ± 3 (4)	773 ± 64 (6)	627 ± 38 (6)	560 ± 80 (4)
	3	34 ± 5 (8)	4 ± 3 (6)	686 ± 102 (8)	785 ± 130 (6)	561 ± 133 (6)
	6	179 ± 28 (8)	154 ± 144 (5)	841 ± 72 (8)	484 ± 130 (8)	352 ± 61 (6)
Coprinus	9	344 ± 82 (7)	455 ± 61 (6)	867 ± 185 (8)	534 ± 292 (7)	393 ± 38 (3)
	12	569 ± 60 (8)	821 ± 39 (6)	1121 ± 199 (8)	486 ± 94 (6)	301 ± 100 (4)
	15	782 ± 100 (8)	957 ± 121 (6)	1177 ± 197 (8)	1123 ± 286 (7)	424 ± 47 (4)
	0	14 ± 1 (4)	12 ± 3 (4)	1088 ± 36 (4)	1081 ± 22 (4)	1078
	4	76 ± 3 (6)	62 ± 31 (6)	1041 ± 17 (6)	1126 ± 244 (6)	1016
	9	107 ± 7 (6)	4 ± 2 (6)	1109 ± 21 (6)	978 ± 96 (6)	890
Volvariella	14	358 ± 24 (6)	71 ± 38 (6)	1228 ± 75 (6)	809 ± 64 (6)	789
	19	1043 ± 211 (6)	346 ± 88 (6)	1860 ± 94 (6)	1007 ± 280 (6)	750
	24	1475 ± 148 (6)	984 ± 64 (6)	2005 ± 186 (6)	1395 ± 202 (6)	721

Entries in the table are the means ± standard error for assays of the number of replicate cultures shown in brackets. The initial concentration of ammonium is shown by the 0 days growth samples which are assays of uninoculated medium. The flasks contained 25 ml of medium supplemented with protein alone, protein + 1% glucose, protein + 30 mM NH_4Cl, protein + 1% glucose + 30 mM NH_4Cl, or a protein-free medium containing 1% glucose + 30 mM NH_4Cl + 0.05% $MgSO_4$.

Kosasih and Willetts (1975) studied formation of apothecia from sclerotia of *Sclerotinia sclerotiorum*, showing that accumulations of lipids, glycogen, and proteins in sclerotial hyphae were mobilised to support apothecial development. Metabolism in the apothecial initials featured enhanced activities of several synthetic and hydrolytic enzymes, together with increased concentrations of protein and glycogen, implying degradation of the sclerotial reserves to produce (presumably temporary) stores of new polymers as well as metabolic energy release. As the apothecial initial elongated protein and glycogen were accumulated at the base of the elongating stem, but as the disc differentiated, accumulations of these materials became evident in the subhymenium and hymenium.

Histochemical studies of *Pyronema domesticum* apothecia showed that gametangia and other parts of young apothecia had high levels of protein, DNA, RNA, phospholipids, and carbohydrates other than glycogen (Moore-Landecker, 1981a,b). The author claims this to indicate a high metabolic activity which accompanies the rapid growth which occurred in these structures. Glycogen accumulations (together with accumulations of protein, phospholipids, mucilage, and other carbohydrates) were localised outside the apothecia in sterile cells of the mycelia. These accumulations decreased in all cells as the apothecium matured, suggesting translocation of their breakdown products from mycelium to apothecium and then, within the apothecium, from the gametangia into the ascogenous hyphae.

The distribution of nitrogen and carbohydrate between mycelium and fruit bodies of the basidiomycete *Schizophyllum commune* is shown in Fig. 4.2. Wessels (1965) used a medium-replacement technique to demonstrate that external carbon and nitrogen sources were required for formation of fruit body initials, but only external carbon source was required for further development of the initials (endogenous nitrogen supplies were sufficient). Nutritional support of the final stages of maturation of the fruit body caps bearing hymenia was completely endogenous, requiring no external sources of nitrogen or carbon. Indeed, this final stage, formation of the fruit body cap, normally occurred only after glucose was exhausted from the medium (Fig. 4.2). Decline in cellular carbohydrate, particularly breakdown of cell wall polysaccharides, was associated with this stage of development. An alkali-insoluble cell wall component, which was called R-glucan, was the main fraction of the wall to be broken down (Niederpruem and Wessels, 1969; Wessels, 1969). R-glucan contained both β-(1⟶6) and β-(1⟶3) linkages and was distinct from another major wall fraction (S-glucan) which was alkali-soluble. S-glucan constituted the bulk of the cell wall material left after mobilisation

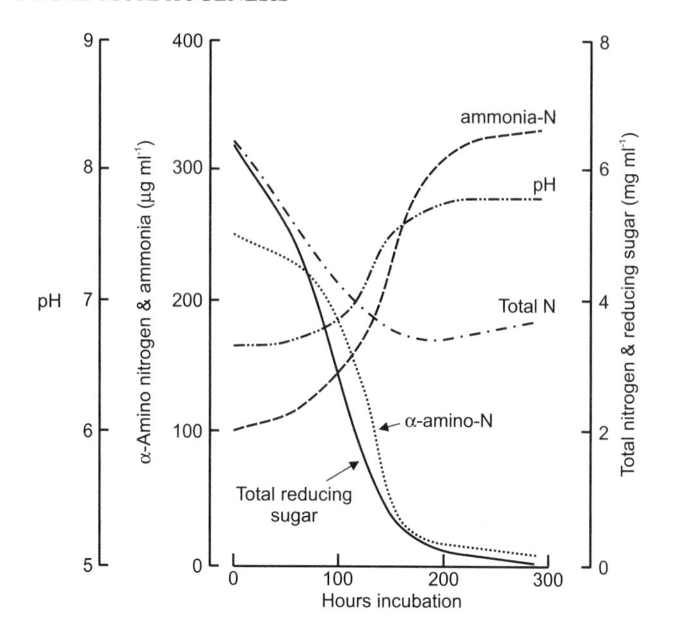

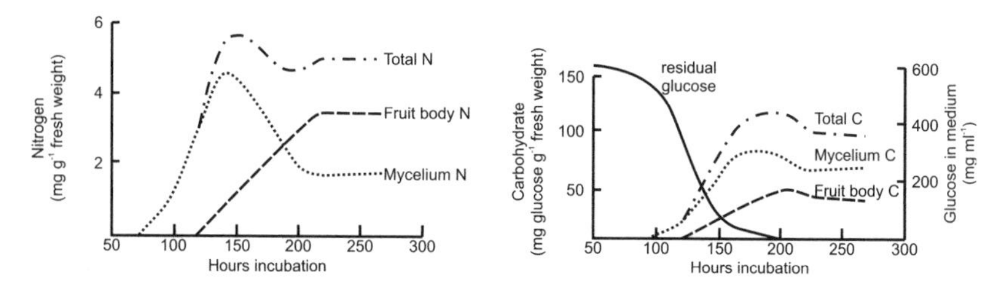

Figure 4.2 The nutrient connection with fruit body formation. The top panel records the changes which occur in the medium during growth of a mycelium of *Coprinus cinereus* as it produces fruit bodies (Stewart and Moore, 1974). The first fruit body initials were observed after about 100 h, meiosis occurred at about 200 h and spore discharge occurred after about 250 h. Note that the medium became exhausted of sugars and α-amino nitrogen as fruiting was initiated. However, total nitrogen in the medium remained high because large amounts of ammonium were produced (following use of amino acids as carbon sources). The ammonium accumulated in these *in vitro* cultures. In nature the alkaline pH (itself a result of proton pumping by the hyphae associated with nutrient uptake) would release ammonia gas for dissipation by wind and rain. The bottom panel shows changes in nitrogen and carbohydrate in fruiting cultures of *Schizophyllum commune* (Wessels, 1965). In these cultures, also, fruit bodies appeared after about 100 h and steadily matured. Note that, again, the medium becomes exhausted of glucose shortly after fruiting commences. However, these observations demonstrate that accumulation of carbohydrate into the mycelium occurs for most of the next 100 h and the mycelium always has more carbohydrate than the fruit bodies. On the other hand, nitrogen is redistributed from mycelium to fruit body tissues.

of the R-glucan. Studies of the mobilisation process indicated that cell wall degradation correlated with cap development and was controlled by changes in the level of a specific R-glucanase enzyme as well as by differences in the susceptibility of cell wall glucans to the enzymes. R-glucanase activity was repressed by glucose in the medium (Wessels, 1966, 1969, 1994a,b). Thus, the model which emerged from these studies is that while glucose remains available in the medium carbohydrate is temporarily stored in the form of R-glucan in the walls of mycelial and fruit body primordium hyphae. During this phase of net R-glucan synthesis the R-glucanase is repressed by glucose in the medium, but when this is exhausted the repression is lifted, R-glucanase is synthesised and by breaking down the R-glucan it provides substrate(s) which are specifically required for cap development.

The R-glucan of *Schizophyllum commune* is chemically very different from glycogen, which is an α-1$\longrightarrow$4 /α-1$\longrightarrow$6 linked glucan (see Chapter 3) and is alkali-soluble, but glycogen seems to serve a similar sort of function in *Coprinus cinereus*. In early stages of vegetative growth of the dikaryotic mycelium of *C. cinereus,* glycogen is accumulated in bulbous cells of the submerged mycelium (Madelin, 1960; Waters *et al.*, 1975b). Glycogen accumulated in vegetative cells of the dikaryotic mycelium is lost when fruit bodies are formed, and losses of mycelial mass are correlated very closely with gain in mass of fruit bodies produced by the mycelium (Madelin, 1960). Glycogen is involved in various aspects of vegetative morphogenesis (Waters *et al.*, 1975b; Jirjis and Moore, 1976), but has also been implicated repeatedly in fruit body development. The material has been identified as glycogen by enzyme digestion methods (Waters *et al.*, 1975b) and by the spectroscopic properties of the iodine-complex of the extracted polysaccharide (Jirjis and Moore, 1976).

C. cinereus produces fruit bodies after exposure to light and a temperature downshift. When temperature and light regimes prevent fruiting, the mycelium produces sclerotia only, and Jirjis and Moore (1976) found that decline in the quantity of glycogen in the dikaryotic mycelia coincided with the appearance of mature sclerotia under such conditions. Sclerotia are resistant survival structures which pass through a period of dormancy before utilising accumulated reserves to 'germinate', often by producing fruiting structures. Dormant sclerotia may survive for several years (Sussman, 1968; Coley-Smith and Cooke, 1971; Willetts, 1971), owing their resistance to a rind composed of tightly-packed hyphal tips which become thick-walled and pigmented (melanised) to form an impervious surface layer. The medulla forms the bulk of the sclerotium, and its cells (and those of the cortex where present) may accumulate reserves of glycogen, polyphosphate, protein and lipid. *C. cinereus* produces sclerotia on both the aerial and submerged parts of the mycelium, but in some strains those formed by the two regions are quite different in origin and structure (Waters *et al.*, 1975a). Both types of sclerotia serve as perennating organs, germinating when adverse conditions improve. Submerged sclerotia are regions of the submerged mycelium separated from the rest by a unicellular rind layer of thick-walled cells. Aerial sclerotia are compact globose bodies (about 250 μm diameter) consisting of two quite distinct and specialised multicellular tissues (Waters *et al.*, 1975a,b). Aerial sclerotia were produced most abundantly in conditions which strongly favoured mycelial growth. With 25 mM ammonium tartrate as nitrogen source the number of sclerotia formed increased as the initial maltose concentration of the medium was increased, reaching a maximum at 25 mM maltose. The length of time required for sclerotia to mature was also shortest on media which initially contained 25 mM maltose. Changing the concentration of ammonium tartrate (in media containing 25 mM maltose) did not affect the total number of sclerotium initials which were formed but affected their maturation. Sclerotia matured most rapidly with 5 mM ammonium tartrate; higher ammonium concentrations first delayed and then prevented maturation (Moore and Jirjis, 1976). The high C:N ratio required for maturation is a reflection of the importance of carbohydrate turnover in the process. Glycogen is synthesised and accumulated in young sclerotia, but is not the end storage-product. For long term storage, much of the carbohydrate is converted into a form of secondary wall material. Many of the inner cells of the mature sclerotium have walls thickened on their inner surfaces by very large, loosely-woven, fibrils (Fig. 4.3; Waters *et al.*, 1972) the development of which coincides with the disappearance of glycogen from the cells (Fig. 4.4; Waters *et al.*, 1975b).

Another important point about the requirement for a high C:N ratio for sclerotium maturation is that it may provide a mechanism for allocating carbohydrate in favour of fruit bodies. Under normal environmental conditions dikaryons produce both sclerotia and fruit bodies but the latter have the ability to excrete ammonia into the medium and this may delay sclerotium maturation (see Fig. 4.2). The result might simply be the avoidance of competition for metabolites between the two morphogenetic sequences; by delaying completion of the vegetative resistant organs the sexual reproductive structure will be favoured. In the extreme, delaying maturation of sclerotia would allow their accumulations of glycogen to be re-exported to developing fruit bodies prior to its being converted to secondary wall materials, which may be less readily reallocated.

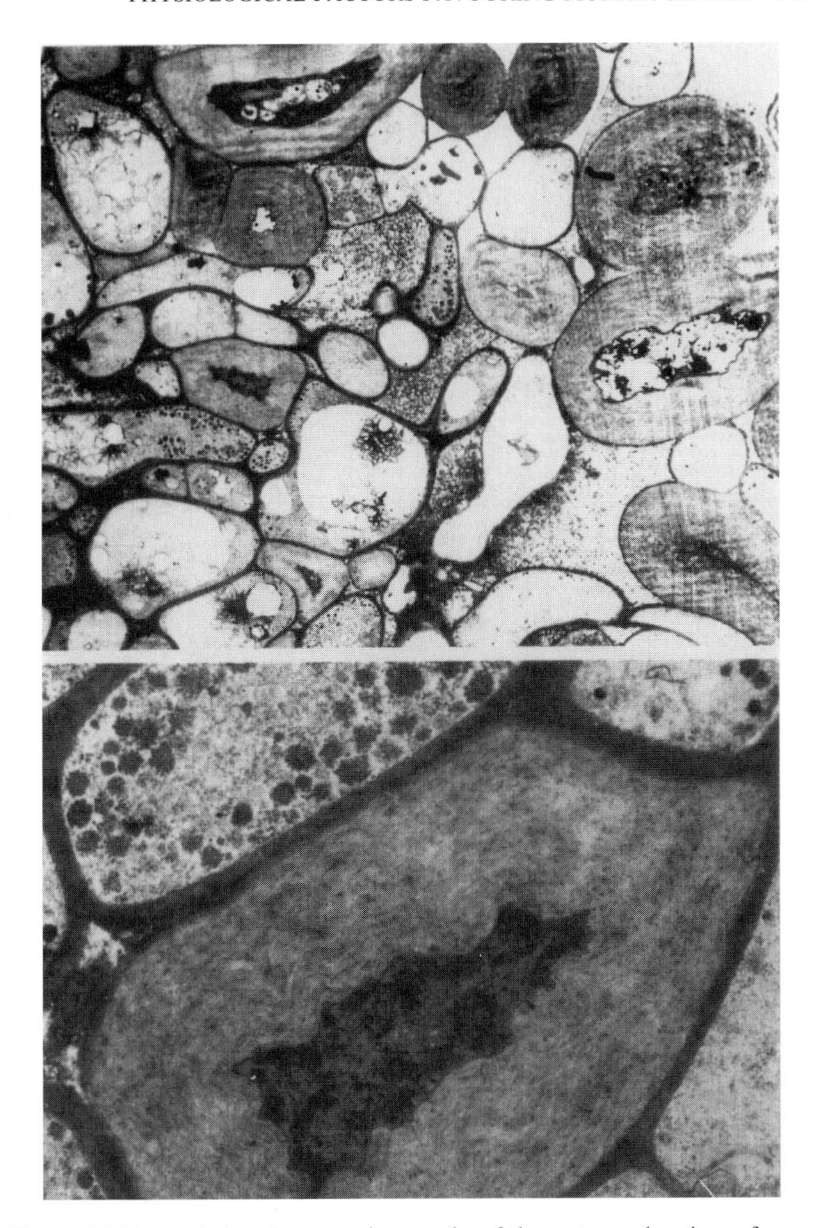

Figure 4.3 Transmission electron micrographs of the mature sclerotium of *Coprinus cinereus* to illustrate the prevalence of cells with secondary wall thickening. The bottom image contrasts such a cell with a neighbour which has an apparently normal wall but heavy accumulation of glycogen in the cytoplasm in the form of electron-dense clusters of glycogen granules. Top image, ×2700; bottom image, ×18,000.

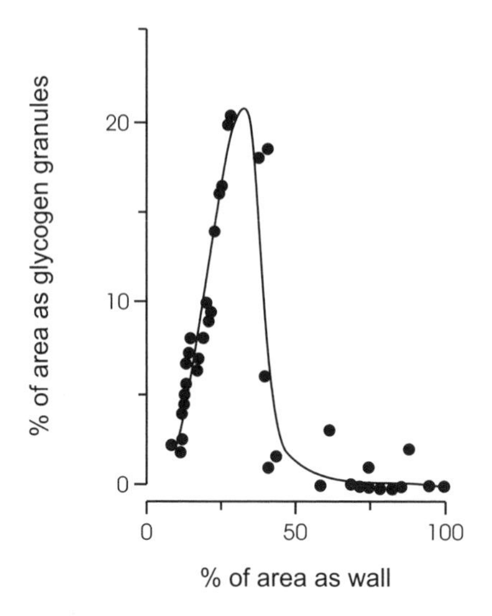

Figure 4.4 Development of secondary wall in medullary cells of sclerotia of *Coprinus cinereus* coincides with the disappearance of glycogen from the cells. Waters *et al.* (1975b) obtained these data from electron micrographs; the area of the section occupied by glycogen granules (illustrated in Fig. 4.3) on the one hand and by the cell wall on the other hand were each expressed as a percentage of the total area of the cell. In some sections the lumen of the cell was totally occluded by secondary wall.

Accumulations of glycogen in fruit body tissue are evident at very early stages of development (Matthews and Niederpruem, 1973), deposited initially in the base of the stem, glycogen subsequently disappears from this location as accumulations grow in the cap (Moore *et al.*, 1979) where it becomes concentrated in subhymenial tissues (Bonner *et al.*, 1957; McLaughlin, 1974). As much as 2 mg glycogen is eventually accumulated in each fruit body (Moore *et al.*, 1979; Kuhad *et al.*, 1987); it is utilised almost totally by the time mature spores are shed. The amount of glycogen utilised by an average fruit body would yield about 25 joules of metabolic energy – if expressed instantaneously as heat energy, this would be sufficient to raise the temperature of an average fruit body cap by 17°C!

In very young primordium initials polysaccharide deposits are evident in the stem base and along the edge of the first recognisable cap tissues. As development proceeds the deposits in the stem decline as glycogen accumulation in the gill tissues becomes emphasised. It appears that accumulation in the gill coincides with depletion of the stem base deposit and is probably initiated at about the time that karyogamy starts the meiotic division (Table 4.3; Moore *et al.*, 1979). The gill accumulations

Table 4.3. *Comparison of specific activities of glutamate dehydrogenases and mycelial growth of the BC9/6,6 monokaryon of* Coprinus cinereus *on contrasting media.*

Medium composition	Enzyme activity*		Mycelial dry weight (mg l^{-1})
	NAD-GDH	NADP-GDH	
25 mM acetate + 25 mM urea	177	5	244
25 mM acetate + 25 mM glutamate	207	7	259
Dung extract	219	0	432

*Enzyme activity is shown as nmol substrate used min^{-1} mg^{-1} protein. Data from Stewart & Moore (1974).

are clearly localised in the subhymenial tissue. The major increase of glycogen concentration in this location being initiated as karyogamy became evident. The onset of large-scale utilisation of accumulated glycogen was a post-meiotic event and staining intensity lessened as spore formation was initiated. No polysaccharide deposits can be detected in specimens bearing mature spores. These observations suggest that glycogen formed by the vegetative mycelium is translocated into fruit bodies to provide for their development, and is shifted in stages through the fruit body, being finally accumulated in the spore-bearing tissue and used for sporulation. However, the true relationship between glycogen and sporulation seems to be less direct than this.

In *Saccharomyces cerevisiae* glycogen synthesis takes place in both sporulating and non-sporulating strains but only sporulating strains degrade the polysaccharide (Pontefract and Miller, 1962; Roth and Lusnak, 1970; Kane and Roth, 1974; Clancy *et al.*, 1982; Katohda *et al.*, 1988). Such a close linkage between glycogen degradation and sporulation in *Coprinus* is also feasible. Meiosis is highly synchronised in *C. cinereus*. Lu (1982) estimated that in a single fruit body of *C. cinereus* there were approximately 1 to 3 $\times$ 10^7 basidia, of which 70 to 75% were at the same stage of meiosis. However, there are many other developmental processes, such as basidiospore formation, stem elongation and cap expansion, that occur during fruit body maturation which have not been distinguished. Another problem is that most studies in the literature have relied on subjective descriptions of fruit body development, mostly based on morphology, to describe the developmental timescale of the specimens studied (Madelin, 1956a,b; Takemaru and Kamada, 1972;

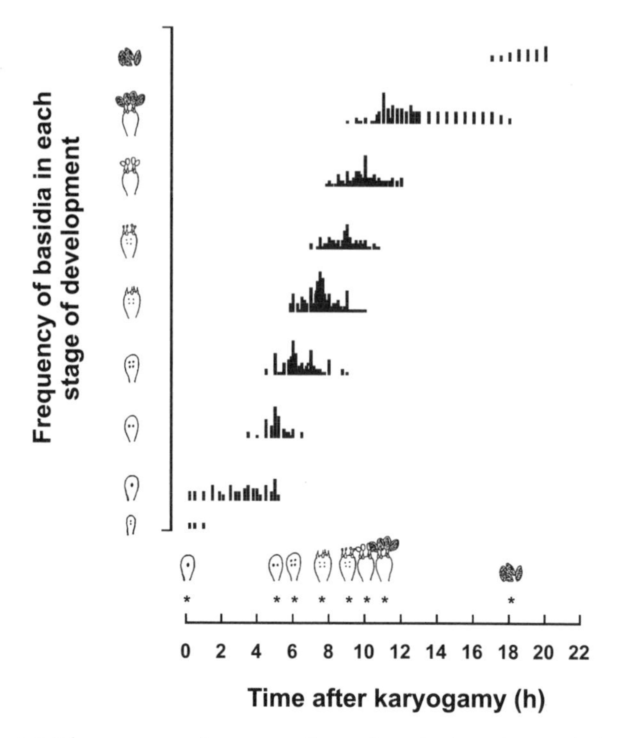

Figure 4.5 Time course of progress through meiosis and sporulation in *C. cinereus*. The abscissa shows elapsed time. Samples were removed from fruit bodies at regular time intervals and sample populations of hymenial cells observed in squashes were quantitatively categorised into the developmental stages represented in the cartoons on the ordinate. Time 0 h is arbitrarily set to the start of karyogamy. Other timings emerged from the observations (as the medians of the distributions shown): meiosis I occurred at 5 h, meiosis II at 6 h, sterigmata appeared at 7.5 h, basidiospores started to appear at 9 h, began pigmentation at 11 h and mature basidiospores were being discharged from 18 h. The asterisks on the abscissa emphasise these stages. Redrawn after Hammad *et al.* (1993a).

Matthews and Niederpruem, 1973; Morimoto and Oda, 1973; Moore *et al.*, 1979). Pukkila *et al.* (1984) introduced nuclear staining to define the stage of development of the whole fruit body in terms of progress through meiosis and basidiospore formation in the hymenium, and Hammad *et al.* (1993a) developed the technique to provide an objective and reliable time base (Fig. 4.5).

Use of this timescale in the careful analysis of individual fruit bodies showed that the temporal pattern of glycogen accumulation/degradation was very variable among 23 different fruit body caps when judged on a common developmental timescale (Ji and Moore, 1993). This indicates that glycogen content was not closely associated with meiosis or basidiospore formation even though glycogen degradation usually started from about the onset of the latter when data averaged from a number of fruit bodies was considered. Glycogen phosphorylase was the only glycogen degradation enzyme found in cap tissues. No α-amylase activity was detectable in cell-free extracts of *C. cinereus* fruit body caps. Glycogen phosphorylase activity first increased when sterigmata appeared (7.5 h after karyogamy); activity increased more than ten-fold between then and basidiospore maturation. Glycogen synthase activity was measured in the same fruit bodies and the results showed that activity was low during sporulation but there was a four-fold increase during basidiospore maturation, indicating that glycogen synthesis and degradation could occur simultaneously and that the ability to synthesise glycogen was maintained throughout the period when glycogen degradation was most rapid (Ji and Moore, 1993).

Glycogen synthase activity was detectable (and increased with time) even during the phase of most rapid glycogen degradation. Analysis of glycogen contents during *in vitro* gill development and in sporeless, stem elongationless and cap expansionless mutants showed that glycogen degradation was not specific for any one of these processes. Experiment showed that nutrient translocation occurred mainly in the stem to cap direction and glycogen influx to the cap continued into the phase of glycogen degradation in the cap (Ji and Moore, 1993). The overall conclusion must be that glycogen accumulated in the cap was required for all the processes of later cap development, i.e. basidiospore formation, cap expansion, stem elongation, etc., but was not specifically or exclusively utilised for any one of these processes. It is also clear that an organised intracellular nutrient translocation circuit operates from mycelium to stem to cap and the major supply route to the gills runs radially from the cap apex through the cap flesh. However, the stem receives some nutrients from the cap and in young fruit bodies gills can be supplied by translocation of nutrients through the connections which exist between the stem and the edges of primary gills. Clearly, as *C. cinereus* fruit bodies develop to maturity their glycogen content goes through a distinct series of synthesis–degradation cycles, as it does during mycelial development. Such is the efficiency of storage of energy in glycogen that only about 6% of the energy obtainable by complete oxidation of each glucose residue (i.e. a little more than two out of 37 ATP molecules) would be used in the cycle glucose $\longrightarrow$ glycogen $\longrightarrow$ glucose $\longrightarrow$ glycogen. Advantages resulting from translocation of mono- or disaccharides rather than a polysaccharide could well compensate for this expenditure of energy. Glycogen synthesis–degradation cycles are a reasonably efficient way of distributing carbohydrate. Furthermore, the fact that enzymes for both synthesis and degradation can be demonstrated in crude extracts of the same tissue (Ji and Moore, 1993) suggests that the processes may be taking place in different cells of that tissue. A futile cycle (synthesise–degrade–synthesise) could occur if both enzymes occurred in the same cell. Compartmentation into different cells would avoid this.

Protein does not seem to play a significant role in these large scale nutrient distributions in the *C. cinereus* fruit body. Cytochemical tests reveal protein to be concentrated in the upper regions of the stem and, particularly, in the presumptive cap in young primordia containing stem base localised glycogen. Also, protein in gill tissue was clearly localised in the hymenium at a time when glycogen was concentrated in the subhymenium (Moore *et al*., 1979). Measurements of protein content in the caps and stems of fruit bodies at different stages of development (Table 4.3) showed no very conspicuous changes in the stem. Both soluble protein (extracted by grinding in buffer at 0°C and filtering off the debris prior to precipitation with trichloroacetic acid) and total protein (estimated after the ethanol-soaked sample had been homogenised and dissolved in strong NaOH at 100°C) declined in proportion of the fresh weight in parallel with the decline of the dry weight as the tissue developed. Protein estimated in this way was thus maintained at a virtually constant proportion of the stem dry matter throughout fruit body development. Similar results have been reported by Gooday (1977) who used slightly different methods of analysis, and by Kamada *et al.* (1976) who examined the final phase of rapid stem elongation in *C. cinereus.* In contrast to the stem the total protein content of the cap was maintained at a constant proportion of the fresh weight. Thus, as development proceeds the fraction of cap dry matter represented by protein must increase quite significantly. This, too, has been observed to be the case in the final phase of development in the Japanese strains of *C. cinereus* (Kamada *et al*., 1976).

The glycogen content of the stem does decline during elongation in the later stages (Table 4.3), but only a small quantity of the material is involved so it is unlikely that this compound plays a key role in providing for elongation. Thus neither glycogen nor protein, the two most readily identified reserve materials have a central function in stem elongation. Gooday (1974a) found that stem elongation in *C. cinereus* had no requirement for connection either with the cap or the parental mycelium and elongation can occur in isolated stems in the absence of added water or nutrients, thus implying that the process can be an autonomous endotrophic one. Hammad *et al.* (1993a) demonstrated another aspect of this cap/stem interplay by showing that whilst the stem may not be *dependent* on the presence of the cap, its elongation certainly benefits considerably from the presence of the cap. Intact fruit bodies elongated about 25% more than decapitated ones (amounting to 2–3 cm greater length). Elongation of the stem in *C. cinereus* is dependent on an enormous increase in the volumes of the constituent cells which is accompanied by very active metabolism of wall polysaccharides which show rapid fluxes during elongation (Kamada and Takemaru, 1977b), the content of chitin increasing particularly during development (Gooday, 1972a, 1975b). The importance of rapid wall synthesis is emphasised by the facts that the specific activity of chitin synthase increases during stem elongation (Gooday, 1973a) and that the elongation is inhibited by Polyoxin D (Gooday, 1972b, 1975b), a nucleoside antibiotic which is known to be a powerful competitive inhibitor of chitin synthase in *C. cinereus* (Gooday *et al.*, 1976). However, the wall remains unthickened and most of the cell interior is occupied by vacuoles, implying that the stem is supported by a hydrostatic skeleton. The ingress of water into the stem during its development is quite dramatic. In extending from about 15 mm to about 75 mm the stem of an 'average' fruit body doubles its fresh weight with hardly any change in dry weight (Moore *et al.*, 1979), absorbing nearly 200 mg of water. The turgor pressure of stem cells remains almost constant throughout the period of stem elongation (Kamada and Takemaru, 1977a) so an appropriate osmotically active solute must be formed and accumulated in the stem cells in parallel with the absorption of water and synthesis of wall.

A mixture of simple carbohydrates appears to be the best candidate for the stem osmoticum. Rao and Niederpruem (1969) fractionated the alcohol-soluble carbohydrate pools and were able to show a three-fold increase in the quantity of trehalose as the stem increased in length from 40–90 mm, this sugar finally accounting for almost 18% of the dry weight. These results were confirmed by Gooday (1977). Trehalose is a non-reducing sugar, and even though glucose levels decline (Rao and Niederpruem, 1969; Gooday, 1977) reducing sugars have been found to remain constant at 5–6 mg g^{-1} fresh weight as the primordium develops into the mature fruit body (Ewaze *et al.*, 1978). The change in the ratio of fresh weight to dry weight in the stem as it elongates is such that while reducing sugars will account for about 6% of the dry weight of the 15 mm tall primordial stem, they will account for at least 12% of the dry weight of the mature 75 mm tall stem. Together, therefore, trehalose and alcohol-soluble reducing sugars represent about 30% of the dry weight of the mature stem.

In *Agaricus*, mannitol can amount to as much as 50% of the fruit body dry weight (Hammond and Nichols, 1976; see section 4.2.5), but it does not seem to have any role in *C. cinereus*. Total polyols never exceed 6% of the dry weight and decline in quantity as the fruit body develops (Darbyshire, 1974). Thus, although there is good reason to suppose that simple carbohydrates serve an osmoregulatory function in the developing stem of *C. cinereus*, no one sugar can be identified as being of prime importance; it seems more likely that a diverse metabolism contributes several components to an osmoregulatory cocktail. It must be emphasised that I am referring here to what might be called the adjustable osmoticum. Because of the low dry weight to fresh weight ratio (1:11 in 15 mm stems, 1:22 in 75 mm stems (Moore *et al.*, 1979)) the bulk of the osmotic pressure of the stem cell must be contributed by compounds of low molecular mass and it is probable that inorganic ions make a very important contribution, as they do in plants (Cram, 1976). However, although the sugar content accounts for only a small part of the overall osmolarity of the cell, it represents a portion which is readily adjusted by metabolism of cell constituents; a property which is not shared by the inorganic constituents of the cell. Sufficient enzyme activities have been detected in extracts of stem tissues to show wide-ranging activity in intermediary metabolism (Stewart and Moore, 1971; Moore and Ewaze, 1976; Taj-Aldeen and Moore, 1983), and the adverse effects of the variety of metabolic inhibitors tested by Gooday (1974a) emphasises the view that stem elongation is dependent upon large scale and diverse metabolic activity.

4.2.2 Intermediary carbon metabolism

Moore and Ewaze (1976) measured the specific activities of representative enzymes of the pentose phosphate cycle, Embden–Meyerhof–Parnas (EMP) pathway and the tricarboxylic acid (TCA) cycle in *Coprinus cinereus* fruit bodies at different stages of development. Care must be exercised in interpretation of these enzyme specific activities. Although usually considered to be synchronised, development of the cap in *Coprinus* is not fully synchronous. The oldest hyphae are at the free edge of the cap and it is in this region that morphogenetic changes are initiated. Each successive change then passes as a developmental wave which, over about 5 h, migrates from the free edge towards the cap to stem junction. An extract of a single cap thus provides an estimate of enzyme activity which is an average for a 5 h segment of the developmental sequence, not an estimate for a single point in time and consequently small-scale fluctuations are obscured. Separate analysis of slices cut from individual caps (as used for the glycogen analyses of Ji and Moore (1993) described above), facilitate more detailed description but were not used by Moore and Ewaze (1976). Another caveat is that conclusions must be based solely on enzyme activities measured under optimal conditions *in vitro*. Better understanding would require knowledge of *in vivo* concentrations of substrates and other metabolites as a complement to determinations of enzyme activities.

The enzymes assayed as representative of the pentose phosphate cycle (glucose-6-phosphate dehydrogenase and 6-phosphogluconate dehydrogenase) always showed lower activities than the two enzymes of the EMP glycolytic pathway (glucose-6-phosphate isomerase and aldolase). More significantly, perhaps, the first enzyme of the pentose phosphate cycle, glucose-6-phosphate dehydrogenase, declined considerably in activity, particularly in the stem but also in the cap, from about 15 mm tall fruit bodies onwards. These results suggest that the EMP pathway represents the major route of sugar catabolism in the fruit body (and mycelium) of *Coprinus*. The pentose phosphate cycle plays a minor role and seems to be completely dispensed with early in stem development but to be maintained to the completion of meiosis in the cap. This contrasts with the *Agaricus bisporus* fruit body in which the pentose phosphate cycle appears to be the major pathway of glucose metabolism (Le Roux, 1967) and is part of a very drastic difference between the mannitol metabolism of these two fungi (Dütsch and Rast, 1972; Hammond and Nichols, 1976). On the other hand, changes in the specific activities of glucose-6-phosphate dehydrogenase and isomerase recorded for *Coprinus* were similar to those reported by Schwalb (1974) to occur during fruiting of *Schizophyllum commune*. In both *Coprinus* and *Schizophyllum*, an initial slight increase in the glucose-6-phosphate dehydrogenase activity over the basal mycelial level was followed by a noticeable decline as the fruit body developed, whereas the specific activity of the isomerase changed relatively much less.

Assay of enzymes of the TCA cycle in *Coprinus* revealed high activities of both isocitrate dehydrogenases, and of the succinate and malate dehydrogenases. However, activity of 2-oxoglutarate dehydrogenase was not detected although a variety of extraction methods were tried. In a search for possible bypass reactions to an inoperative 2-oxoglutarate dehydrogenase, enzymes of the glyoxylate shunt were assayed. This pathway has been shown to be related to asexual morphogenesis in *Blastocladiella* (Cantino and Lovett, 1964; Cantino, 1966), *Neurospora* (Turian and Combépine, 1963) and *Aspergillus* (Galbraith and Smith, 1969) but in *C. cinereus* isocitrate lyase activity did not exceed the repressed levels characteristic of glucose-grown mycelium (Casselton *et al.*, 1969) and crude extracts of fruit body caps (but not their stems) contained a low-molecular weight inhibitor of this enzyme. Presence of the inhibitor and the very low isocitrate lyase activity detected show that the glyoxylate shunt plays no part in fruit body metabolism in *C. cinereus*.

The only other alternative link which can restore the TCA cycle by acting as a bypass to the 2-oxoglutarate dehydrogenase step is the glutamate decarboxylation loop (see Fig. 3.9). This is the sequence followed by the TCA cycle in the central nervous system (Baxter and Roberts, 1960) and it is evident that it is also operative in *Coprinus* since the first two enzymes of the loop, glutamate decarboxylase and 4-aminobutyrate aminotransferase, had patterns and levels of activity similar to those of isocitrate dehydrogenase (Moore and Ewaze, 1976). The glutamate decarboxylation pathway also operates in *Agaricus bisporus*; the radioactivity of [^{14}C]-glutamate appeared in 4-aminobutyrate, succinate, malate and aspartate after only 2 h exposure of *A. bisporus* fruit body tissues to the isotopically labelled material (Piquemal *et al.*, 1972) and the enzymes of the glutamate decarboxylation shunt occur in *Agaricus* spores (Rast *et al.*, 1976). Thus, in both *Coprinus* and *Agaricus* the glutamate decarboxylation pathway is probably a normal part of the TCA cycle (see Fig. 3.9).

4.2.3 Intermediary nitrogen metabolism

An active TCA cycle in *Coprinus* presumably requires glutamate dehydrogenase as a component of the glutamate decarboxylation sequence. There are two glutamate dehydrogenase enzymes in *Coprinus cinereus* which show distinctive patterns of regulation during development of the fruit body. The NAD-linked enzyme (NAD-GDH) is derepressed about threefold in both cap and stem, while the NADP-linked enzyme (NADP-GDH) is derepressed in cap tissues only, remaining at a very low level of activity in mycelium, stem and basidiospores (Stewart and Moore, 1974). This difference in NADP-glutamate dehydrogenase activity between cap and stem is a striking example of developmentally-related, tissue-specific gene regulation and the function of the enzyme is now known.

Coprinus is not alone in having two GDH enzymes linked specifically to NAD or NADP. Many filamentous fungi and yeasts have been shown to be similarly equipped. It is often considered that NAD-GDH has a catabolic role (glutamate $\longrightarrow$ 2-oxoglutarate + ammonium) while the NADP-linked enzyme is anabolic (2-oxoglutarate + ammonium $\longrightarrow$ glutamate). This suggestion was first made for *Neurospora crassa* by Sanwal and Lata (1961) and for *Saccharomyces cerevisiae* by Holzer *et al.* (1965). Sanwal and Lata (1962) further developed the idea by suggesting that these two enzymes are reciprocally (or concurrently) regulated. These concepts have almost reached the stage of being accepted as truisms, so it is worth emphasising that not all the data supports this view. Stine (1968) concluded that reciprocal (concurrent) regulation of these two enzymes did not occur in germinating conidia of *N. crassa*; while Dennen and Niederpruem (1967) showed that NADP-GDH activity of *Schizophyllum commune* monokaryotic mycelium was depressed by transfer from glutamate medium to ammonium medium but increased by the reciprocal transfer; the NAD-linked enzyme being largely unaffected by these treatments (which is not expected if NADP-GDH is devotedly anabolic and NAD-GDH catabolic). The position was reviewed by Casselton (1976) and the identification of a glutamate synthase/glutamine synthetase ammonium assimilation system in many of these fungi further complicates the issue. For *Coprinus*, no evidence has been obtained for a glutamate synthase activity. Furthermore, the NAD-GDH of the mycelium shows high activity whether the growth conditions demand amination or deamination (Table 4.3) and the enzymes are not reciprocally regulated, so we can interpret the *Coprinus* system as one in which the NAD-GDH serves the mycelium for amination or deamination as nutritional conditions require, whilst NADP-GDH is reserved for metabolic purposes related specifically to morphogenesis.

As indicated above, the ubiquitous occurrence of NAD-GDH is probably related to the requirement for 2-oxoglutarate amination for completion of the tricarboxylic acid (TCA) cycle in *Coprinus* (Moore and Ewaze, 1976). If this is so, then what is the function of NADP-GDH? The two enzymes are kinetically quite different (Al-Gharawi and Moore, 1977). The NADP-linked GDH exhibited positively cooperative interactions with the substrates 2-oxoglutarate and NADPH, negatively cooperative kinetics with $NADP^+$ and was extremely sensitive to inhibition of deamination activity by ammonium. On the other hand, NAD-GDH showed positive cooperativity with NADH but Michaelis–Menten kinetics with all other substrates and was only mildly inhibited by reaction products. The substrate cooperativity observed in the NADP-GDH amination reaction is interpreted as a switch mechanism which allows substrate (2-oxoglutarate) to accumulate to a threshold level of about 4 mM before maximum enzyme activity is released (Al-Gharawi and Moore, 1977).

The origin of $NADPH^+$ required by the NADP-GDH reaction is a problem. The pentose phosphate pathway plays only a minor role in carbohydrate metabolism in the developing fruit body cap (see above; Moore and Ewaze, 1976). Since the TCA cycle activity is considered to be greatly amplified as the cap develops (see below), $NADH^+$ could be used to regenerate $NADPH^+$. Enzymes responsible for this cycling may well be found among the numerous NAD-dehydrogenase and NADP-dehydrogenase isozyme activities (some of which seem to show development-related changes in level) detected by specific enzyme staining methods following electrophoresis in polyacrylamide (Jirjis and Moore, 1979; Moore and Jirjis, 1981).

Definition of the metabolic context within which NADP-GDH is derepressed in the developing fruit body have concentrated on enzyme surveys and metabolite measurements (Ewaze *et al.*, 1978). These studies found three enzymes which, like NADP-GDH, showed great increases in activity in the developing cap while remaining at low levels (or declining in activity) in the stem: these three enzymes are glutamine synthetase (GS), ornithine acetyltransferase (OAT) and ornithine carbamyltransferase (OCT). A fourth enzyme, urease, showed the reverse behaviour, being absent from the cap though present in the stem (Table 4.4) and constitutive in mycelium (Table 4.5). Greater activity of succinate dehydrogenase in the cap than in the stem, and increased isocitrate dehydrogenase activity in the fruit body as development proceeds indicates that TCA cycle activity is considerably amplified in the cap of *Coprinus cinereus*.

The effects of these changes in enzyme activity are evident from the concentrations of metabolites in developing fruit body tissues (Table 4.6) which show that arginine and urea accumulate in the cap as development proceeds. The level of arginine increased whether quantified in terms of tissue fresh weight (Table 4.6a) or dry weight (Table 4.6b); but while urea content on a dry weight basis increased by a factor of 2.5, the urea concentration on a fresh weight basis was essentially unchanged during cap development. Urea was the only compound to behave like this, leading to the conclusion that during cap development synthesis of urea is amplified, urea accumulates and water is driven osmotically into the cells in which urea accumulation is taking place. There is certainly a need for considerable water uptake during the later stages of cap development, for cell expansion is absolutely central to the whole morphogenesis of the developing cap (Moore *et al.* 1979). The changes in cap morphology

Table 4.4. *Some enzyme activities in fruit bodies of* Coprinus cinereus.

Enzyme	Tissue	Enzyme activity* at developmental stage†: 3	4	5
NADP-linked glutamate	Cap	150	267	830
dehydrogenase	Stem	12	15	103
Glutamine	Cap	300	950	2067
synthetase	Stem	85	290	550
Ornithine acetyl	Cap	16	37	34
transferase	Stem	8	7	5
Ornithine carbamyl	Cap	353	693	1240
transferase	Stem	153	436	310
NAD-linked glutamate	Cap	270	553	1183
dehydrogenase	Stem	363	483	1287
Urease	Cap	55	70	38
	Stem	143	570	4300

*Except for urease, enzyme activities are shown as nmol substrate used min^{-1} mg^{-1} protein. Urease activity is shown as pmol substrate used min^{-1} mg^{-1} protein.
†A stage 3 primordium is beginning to make spores, the stage 4 fruit body has a full complement of mostly unpigmented spores, and the stage 5 fruit body is releasing mature spores. Data from Ewaze *et al.* (1978).

Table 4.5. *Specific activity of the enzyme urease in monokaryotic and dikaryotic mycelium of* Coprinus cinereus.

Nitrogen source in the medium	Enzyme activity: Monokaryon	Dikaryon
50 mM urea	3.1	1.5
50 mM ammonium tartrate	2.1	3.2
25 mM urea + 25 mM ammonium tartrate	4.8	3.6

Urease activity is expressed as nmol $^{14}CO_2$ liberated from [^{14}C]-urea min^{-1} mg^{-1} protein. The media contained 222 mM glucose as carbon source and cultures were incubated for 4 days at 37°. Data from Ewaze *et al.* (1978).

which characterise the maturation process can be accounted for by cell inflation and this depends on osmotic influx of water driven by the substrate accumulations which are a consequence of the metabolic shifts discussed here. The indication is, then, that the changes in enzyme activity described above occur specifically in the cap in order to provide for amplification of the urea cycle leading to accumulation of urea as an osmotic metabolite. Feeding of live tissue slices with (U-^{14}C)-citrulline (Table 4.7) confirmed that urea synthesis occurs *in vivo* and further illustrated its accumulation in the cap but not in the stem (the latter, of course, having a high urease activity). Urea synthesis has also been demonstrated in *Agaricus bisporus* and the puffball, *Lycoperdon*. In a series of papers in the 1960s, Reinbothe and colleagues used a variety of techniques to investigate the urea biosynthetic pathways in both these basidiomycetes. The primary concern of these studies was to demonstrate the presence of the pathway (Reinbothe and Tschiersch, 1962; Reinbothe, 1964; Wasternack and Reinbothe, 1967; Reinbothe *et al.*, 1969). Unfortunately, as the main conclusion was that "... rapid formation and extensive accumulation of urea by fruit bodies of some pileated fungi seems to be the plant analogue of mammalian and amphibian liver" (Reinbothe and Tschiersch, 1962) the truly *fungal* significance of the observations were rather lost in inappropriate comparisons with plants. Nevertheless, it is clear that the ornithine cycle reactions leading to urea occur in both *Agaricus* and *Lycoperdon*. Feeding with radiolabels shows preferential direction of urea to the hymenium which was the location of the greatest urea accumulation and also of the highest urease activity. Feeding with ^{14}C-urea showed that released CO_2 was not re-assimilated, but the authors claim that urea-N was used in chitin and protein. Urea accumulated in developing puffballs up to their second stage of development; thereafter the content decreased to zero in ripe sporulating fruits. The authors suggest that urea serves as a special form of nitrogen reserve in *Lycoperdon* (Reinbothe *et al.*, 1967).

Urea is probably the most significant, but is not the only compound to accumulate in developing cap tissue of *Coprinus cinereus*. Arginine content increases by a factor of four (concentration by a factor of two) as the primordial cap develops to maturity, though both content and concentration decline in the stem during this time. This situation can be interpreted as a means by which the activity of arginase is regulated. In *Coprinus* this enzyme has a K_m of 100 mM and a V_{max} of 1.6 μmol substrate used min^{-1} (mg protein)$^{-1}$. If the accumulated arginine occupies the same metabolic compartment as the enzyme, it can be calculated that the flux through the arginase reaction is likely to increase at least by a factor of two to three in the cap while declining in the stem as development proceeds from immature fruit body stage (post-meiotic but spore formation only just starting) to mature fruit body discharging spores. Indeed the arginine accumulation will lead to at least a 6-fold greater flux through the arginase reaction in the cap than in the stem even though

Table 4.6. *Concentrations of some metabolites in developing fruit body tissues of* Coprinus cinereus.

	Fruit body stage 3*		Fruit body stage 5	
Metabolite	Cap	Stem	Cap	Stem
(a) Quantified in terms of μmol g^{-1} fresh weight				
Alanine	0.8	0.5	3.9	0.8
Arginine	1.9	1.8	4.3	0.8
Glutamate	1.9	1.4	3.8	0.9
Ornithine	0.5	0.5	1.1	0.2
Urea	2.1	1.2	2.8	1.1
(b) Quantified in terms of μmol per average fruit body (dry weight basis)				
Alanine	0.3	0.2	3.2	0.4
Arginine	0.9	0.4	3.5	0.3
Glutamate	0.8	0.3	3.1	0.4
Ornithine	0.2	0.1	0.9	0.1
Urea	0.9	0.2	2.3	0.5

*A stage 3 primordium is beginning to make spores and the stage 5 fruit body is releasing mature spores. Data from Ewaze *et al.* (1978).

Table 4.7. *Metabolism of [U-^{14}C]-citrulline by intact fruit body tissues of* Coprinus cinereus.

		Radioactivity (cpm mg^{-1} dry weight) recovered in		
Incubation time (min)	Tissue	Citrulline	Arginine	Urea
30	Cap	17	425	44
60	Cap	707	1222	351
120	Cap	172	731	526
30	Stem	33	219	19
60	Stem	565	964	101
120	Stem	395	860	95

there is little difference between the *in vitro* measurements of arginase activity in the two tissues. This further strengthens the view that the urea cycle is very specifically amplified in developing cap tissues (see Fig. 4.6). These processes do operate *in vivo*: after 1 h incubation with (U-^{14}C)-glutamate, 22% of the radioactivity recovered from cap tissues appeared

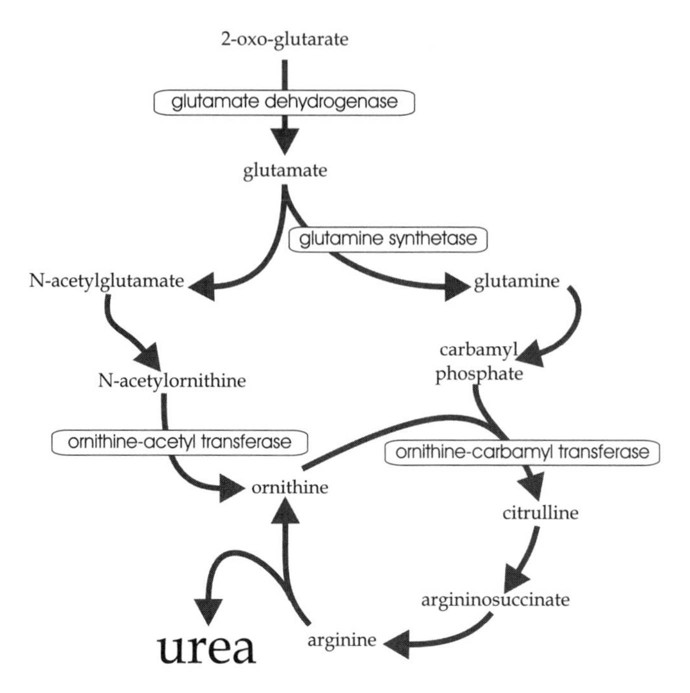

Figure 4.6 Four enzymes which are derepressed together in the developing cap of *Coprinus cinereus*, contributing to urea cycle amplification and accumulation of urea.

as 4-aminobutyrate, 54% as malate + succinate + citrate, and 6% as arginine. The corresponding figures for the stem were, respectively, 38%, 35% and 2%; i.e. the glutamate decarboxylation loop is used efficiently in both cap and stem, but arginine accumulation only occurs in the cap (Ewaze *et al.*, 1978).

All of the results discussed so far derive from analyses of normally developing fruit bodies. Extensive use has been made of a medium transfer experiment to study regulation of these enzymes in mycelium using a monokaryotic strain in which the same enzymic events can be induced (Stewart and Moore, 1974; Al-Gharawi and Moore, 1977). The routine technique was to grow mycelium of the monokaryotic strain number BC9/6,6 (ATTC 42725) in a rich medium for 4–5 days to obtain a good mycelial yield and then harvest, wash and resuspend it in a salts medium containing 100 mM pyruvate and no nitrogen source for further incubation. The result of this treatment is that BC9/6,6 produces high activities of NADP-GDH, GS, OAT and OCT and shows diminished activities of NAD-GDH and urease (Table 4.8). Significantly, increases in activity of NADP-GDH and GS are highly correlated (correlation coefficient 0.94) implying some form of coordinate control. The response to the medium transfer is sensitive to inhibition by cycloheximide and experiments using differential isotopic labelling before and after transfer showed that the NADP-GDH protein is synthesised *de novo* (Jabor and Moore, 1984). The time course of this derepression showed no lag period (Fig. 4.7) but during the first 5 h after transfer NAD-GDH also increased in activity (by about 40%); only when incubation was continued overnight did the recorded activity of NAD-GDH prove to be less than that measured for NADP-GDH. Categorically, therefore, the NAD- and NADP-linked glutamate dehydrogenases in *Coprinus* are not reciprocally regulated.

Varying the composition of the medium used in this transfer technique provided the opportunity to establish which metabolites are involved in the regulation of NADP-GDH. Very few compounds share with pyruvate the ability to cause significant increases in NADP-GDH activity only glucose, fructose, dihydroxyacetone, acetate and propan-1-ol were effective as inducers. Particularly noteworthy is that alanine (2-aminopyruvate) was ineffective, as were TCA cycle intermediates. A mutant mycelium known to lack the enzyme acetyl-CoA synthetase failed to show induction of NADP-GDH activity on acetate medium, although normal induction occurred on medium containing either glucose or pyruvate. This indicates that NADP-GDH induction requires synthesis of acetyl-CoA and that this latter compound is the probable intracellular regulator (Moore, 1981c).

Among compounds likely to prevent the induction caused by pyruvate only ammonium, urea, arginine and citrulline (and possibly ornithine) were effective. Urea and urea-cycle intermediates are most probably rapidly metabolised to ammonium which is the active repressor and is effective at very low concentrations (Fig. 4.8). However, external concentrations of ammonium which are able to prevent induction are not accompanied by elevated internal ammonium levels, so it must be concluded that though the external level of ammonium determines whether induction will occur, the intracellular signal must be another compound. Similarly, neither methylamine nor hydroxylamine had any inhibiting effect on the pyruvate-promoted induction of NADP-GDH. Since both these compounds are useful structural analogues of the ammonium molecule the implication, again, is that some metabolic derivative of ammonium is the effective molecule rather than ammonium itself. Ammonium repression in *Aspergillus* and *Neurospora* (Arst and Cove, 1973; Hynes, 1975; Grove and Marzluf, 1981) uses glutamine as the effective intracellular molecule, and in view of the close correlation between induction of NADP-GDH and GS in *Coprinus* (Table 4.8) this is also likely to be

Table 4.8. *Specific activities of selected enzymes in mycelia of the monokaryon BC9/6,6 before and after transfer to medium containing 100 mM pyruvate with no nitrogen source.*

	Enzyme activity			
Treatment	Urease	Glutamine synthetase	NADP-linked GDH	NAD-linked GDH
Initial control	8	16	36	550
3 h after transfer	7.7	160	140	524
18 h after transfer	2.1	330	260	350
24 h after transfer	1.7	440	290	260

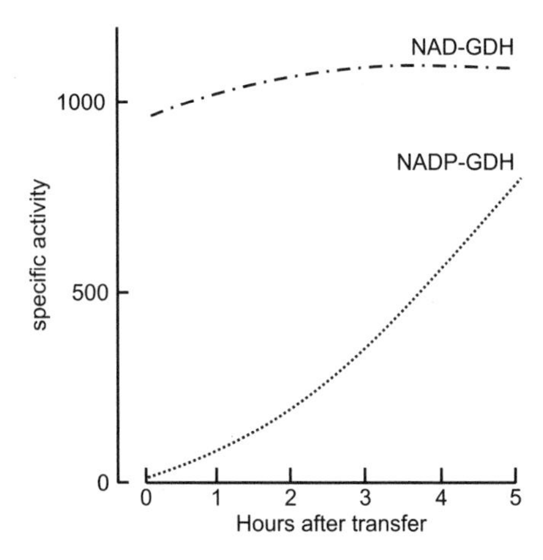

Figure 4.7 Time course of induction of NADP-linked glutamate dehydrogenase activity following transfer of mycelium of strain BC9/6,6 to a nitrogen-free medium supplemented with 100 mM pyruvate. The NAD-linked enzyme activity was largely unchanged during this period.

true for *Coprinus*. The regulation mechanism is therefore interpreted as one in which the enzyme is induced by elevated intracellular levels of acetyl-CoA providing external levels of ammonium are low.

Correlation of these mycelial experimental results with endogenous events in the fruit body rests on the large quantities of glycogen which are metabolised in the fruit body cap (see above), and elevated activities of both isocitrate dehydrogenase and succinate dehydrogenase in the cap

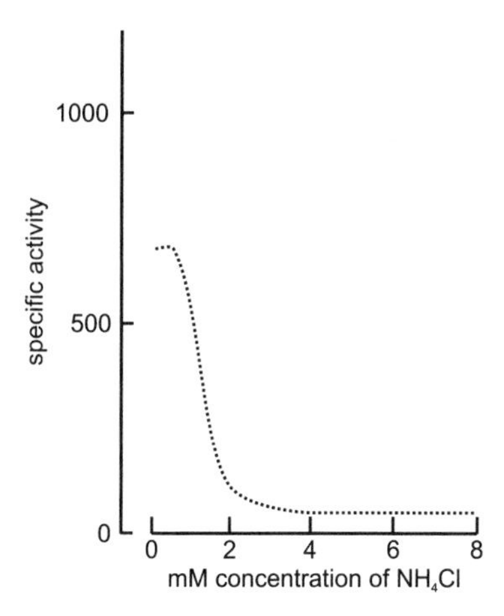

Figure 4.8 Effect of the inclusion of ammonium chloride in the pyruvate transfer medium on activity of NADP-GDH in mycelia of strain BC9/6,6. Mycelia were harvested for assay 20–22 h after transfer to the pyruvate/ammonium chloride medium.

(relative to the stem), implying enhanced metabolism through the TCA cycle (Moore and Ewaze, 1976). Furthermore, the cap always contains less free ammonium than does the stem and the concentration of this metabolite drastically declines as the primordium develops into the mature fruit body (Ewaze *et al.*, 1978; Table 4.9). Metabolism of large amounts of glycogen with so little available ammonium could obviously lead to accumulation of acetyl-CoA to levels sufficient to induce NADP-GDH and associated enzymes.

A fundamental question is why NADP-GDH needs to be induced at all in the developing cap. NAD-GDH is always present in the tissues which acquire the NADP-linked enzyme; so NADP-GDH is introduced as an *additional* component of nitrogen metabolism. Some of the reactions discussed above, particularly the glutamate decarboxylation loop, can clearly cycle ammonium in the sense that successive reactions in the same pathway involve both amination and deamination. However, the evidence points to a considerable amplification of TCA cycle activity and, moreover, nitrogen is being removed from availability by being accumulated as urea and other nitrogenous compounds. Thus there will be a need for enhanced ammonium assimilation and as the K_m for ammonium of NADP-GDH (2 mM) is some ten times lower than that of NAD-GDH (18.8 mM) the former enzyme is a better candidate for the task of ammonium scavenging than the latter. Since induction of NADP-GDH is very closely correlated with induction of GS (Ewaze *et al.*, 1978), in *Coprinus*

Table 4.9. *Some enzyme and metabolite levels in developing fruit bodies of* Coprinus cinereus.

		Developmental stage		
Enzyme	Tissue	3	4	5
NADP-GDH	Cap	47	420	780
	Stem	29	11	20
NAD-isocitrate	Cap	50	150	200
dehydrogenase	Stem	90	200	220
Succinate dehydrogenase	Cap	12	27	62
	Stem	28	16	10
Glycogen content transferase	μg per fruit body	1857	1361	97
	% in cap	97	96	86
Ammonium content	Cap	32	11	9
(μmol g^{-1} fresh weight)	Stem	90	40	22

*Enzyme activities are shown as nmol substrate used min^{-1} mg^{-1} protein. Mycelia and fruit body primordia contain approx. 125 μmol ammonium g^{-1} fresh weight. A stage 3 primordium is beginning to make spores, the stage 4 fruit body has a full complement of mostly unpigmented spores, and the stage 5 fruit body is releasing mature spores.

NADP-GDH and GS together form an ammonium scavenging system. *Coprinus* is not the only basidiomycete to have this novel ammonium scavenging process. Cultures of *Sporotrichum pulverulentum* grown under nitrogen-limiting conditions also showed simultaneously increased levels of both NADP-linked glutamate dehydrogenase and glutamine synthetase (Buswell *et al.*, 1982). Under normal circumstances, however, these ammonium scavenging enzymes of *Coprinus* appear only in a fruit body cap which is well provided with nutrients and well separated from the substrate so ammonium *assimilation* is an unlikely reason for the derepression of these enzymes.

Cytochemical analysis of frozen sections showed that from the earliest stages of development NAD-GDH was uniformly distributed in the basidia, and as the fruit body developed an increasing proportion of cells of the subhymenium came to express enzyme activity (Elhiti *et al.*, 1979). The NADP-GDH was initially limited to basidia in isolated patches of the hymenium. As development proceeded the proportion of basidia exhibiting NADP-GDH activity increased, as did the proportion of subhymenial cells showing this enzyme activity. Islands of tissue expressing NADP-GDH were formed first, then these expanded and united (Fig. 4.9). No NADP-GDH activity was detectable in sections of stem tissue.

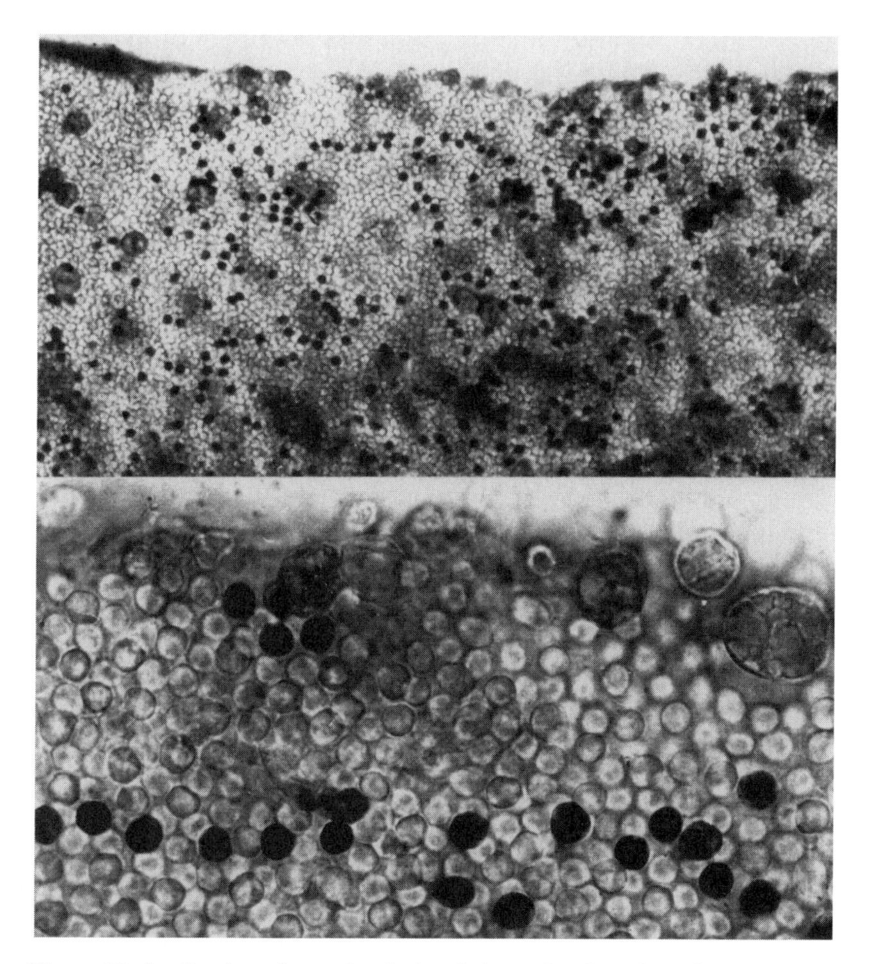

Figure 4.9 Application of cytochemical techniques for detection of NADP-GDH activity *in situ* in live tissues shows that at the earliest stages of development NADP-GDH is found in a few scattered basidia. These photographs are surface views of young hymenium treated so that cells containing NADP-GDH will stain darkly. Apart from the few large cystidia, all the cells visible in these tissues are young basidia. Clearly, only a minority contain the enzyme at this time and those that do stain may emerge as branches from the same tramal hyphae, because they seem often to fall into lines in the hymenium (bottom image).

Electron microscopic cytochemistry showed that the NADP-GDH activity is localised in cytoplasmic vesicles, mainly at the peripheral region of basidia and very young spores, sometimes being found adjacent to the cell walls, suggesting that these enzymes are transported to the vicinity of the plasma membrane (Elhiti *et al.*, 1986, 1987). Such a location would be consistent with protection of the meiotic apparatus against ammonium inhibition. Increase in NADP-GDH activity was initiated as karyogamy became evident in normally-developing fruit bodies. Enzyme activity stabilised for about 4 h during meiosis, but resumed on the conclusion of the second meiotic division, and continued to increase until spore maturation (Moore *et al.*, 1987).

Basidia of *C. cinereus* are committed to their developmental pathway, continuing through meiosis and sporulation even when excised from their parental fruit body (Chiu and Moore, 1988a; see section 6.5.1). Of a range of compounds tested, only ammonium and its immediate structural analogues, and glutamine together with some of its structural analogues, were able to inhibit basidium differentiation. Growth was not inhibited; instead the differentiation inhibitors caused vegetative hyphal tips to grow out from regions of the basidium expected to be in active growth during sporulation (Chiu and Moore, 1988b). Moore *et al.* (1994) showed that the stage in sporulation between the end of meiosis and emergence of sterigmata is most sensitive to ammonium inhibition. Thus, the purpose of the derepression of NADP-GDH during the development of the *C. cinereus* fruit body cap is to protect the meiotic apparatus against inhibition by ammonia, though the exact identity of the ammonia-sensitive function remains to be established.

A number of enzyme and metabolite transport systems are regulated by ammonium concentration in *Aspergillus nidulans* and there is evidence that NADP-GDH also has a role in this. The regulated systems include: nitrate reductase (Cove, 1966; Pateman and Cove, 1967); xanthine dehydrogenase (Scazzocchio and Darlington, 1968); acetamidase and formamidase (Hynes and Pateman, 1970); glutamate uptake (Pateman *et al.*, 1974); urea uptake (Dunn and Pateman, 1972); and extracellular proteinases (Cohen, 1972). The intracellular concentration of ammonium ions is the important regulatory factor for some of these, but for others the fungus monitors the extracellular ammonium, possibly with a membrane bound system (Pateman *et al.*, 1973). Mutants of *Aspergillus* which lack NADP-GDH are also defective in both internal and external ammonium concentration monitoring functions, suggesting that this enzyme has a key role in the regulatory system (Pateman *et al.*, 1973).

4.2.4 Adenosine 3′:5′-cyclic monophosphate (cAMP)

Because of its involvement in the concerted migration and growth leading to fruit body formation in the slime mould *Dictyostelium discoideum*

(Newell, 1978) a great deal of interest has been focused on the likely role of adenosine 3′:5′-cyclic monophosphate (cAMP) in controlling differentiation in other organisms. Addition of cAMP to a dark-grown culture of the ascomycete *Saccobolus platensis* (Galvagno *et al.*, 1984) induces apothecium development and an increased level of adenylate cyclase and decreased cyclic AMP phosphodiesterase. Inhibitors of this latter enzyme also induce apothecia in the dark, perhaps by causing cyclic AMP to accumulate, although an increase in the level of cAMP in response to light could not be demonstrated.

Interest was heightened by the analogies which could be drawn between the metabolism of glycogen in animals (Soderling and Park, 1974; Roach and Larner, 1976) and the polysaccharide transformations in *C. cinereus,* discussed above, which prompted the suspicion that cAMP could be involved in fruit body morphogenesis. Indeed, a very close parallel between glycogen metabolism in mammals and that in *C. cinereus* emerged from the researches of I. Uno and T. Ishikawa in the 1970s (Uno and Ishikawa, 1982). Most of the work was concentrated on certain mutant monokaryons able to produce monokaryotic fruit bodies (Uno and Ishikawa, 1971), but the dikaryon (the normal mycelial origin of fruit bodies) was included as a control to relate the events to normal circumstances. It was shown that the A and B mating type factors are involved in the coordinate regulation of adenylate cyclase, phosphodiesterase and a cAMP-dependent protein kinase (Swamy *et al.*, 1985). The activities of adenylate cyclase and phosphodiesterase were shown to increase in dikaryotic mycelia exposed to illumination (Uno *et al.*, 1974) and fertile mycelia (the dikaryon normally, or fruiting-competent monokaryotic mutants) accumulated cAMP at the onset of fruiting. The accumulation continued to a late primordial stage, but then the amount of cAMP declined as the primordium matured (Uno *et al.*, 1974). The nucleotide was shown to activate glycogen phosphorylase and inhibit glycogen synthetase *in vitro* (Uno and Ishikawa, 1973a,b, 1974, 1976, 1978), which is significant, since translocation of glycogen from the mycelium to the developing fruit bodies, and then between different tissues of the fruit body (see above) is such an important feature of fruit body development (Moore *et al.* 1979). Although this work implied some sort of connection between cAMP metabolism, glycogen metabolism and formation of the fruit body (perhaps mediated directly by modulation of glycogen synthetase/ phosphorylase activities, or through the agency of the cAMP-dependent protein kinase), throughout their analyses Uno and Ishikawa made extractions of mycelia together with any fruit bodies they might have produced. No attempt was made to separate the two structures.

Kuhad *et al.* (1987) did partition the cAMP between the component parts of the fruit body of *Coprinus cinereus*; fruit bodies being sampled at different stages of development and assayed both for cAMP and glycogen contents (Table 4.10). The greatest content of cAMP, about 90 pmol mg^{-1} dry wt, was observed in fruit body initials. The amount of cAMP per fruit body increased during development to reach a maximum at stage 3 (as defined in Table 4.10), followed by a rapid decline. Tissue content (dry weight basis) varied with the nature of the tissue analysed and its stage of development but in no case did the content in any part of the developing fruit body significantly exceed 50% of the level observed in fruit body initials (Table 4.10).

Administration of cAMP to mature mycelial cultures enhanced net accumulation of glycogen by delaying utilisation of the polysaccharide. The cAMP was administered within the body of a preformed colony by application of 20 μl of a 40 pM solution of dibutyryl-cAMP (0.8 fmol) to small wells cut into the agar medium. Accumulation of glycogen promoted by cAMP administration was detectable within 3 h on some media and occurred uniformly over the whole culture despite the non-uniform method of administration of cAMP (Kuhad *et al.*, 1987). Thus, rather than cAMP polarising glycogen transport, the experiments imply rapid transmission of some other sort of signal to orchestrate a homogeneous response. Lateral and radial transmission of this (unknown) signal occurred at a rate much faster than either the rates of growth or diffusion.

Dikaryotic mycelium contained cAMP levels varying between 0.5 and 2.9 pmol mg^{-1} during 9 days growth at 37°C. In fruit body initials cAMP was accumulated to a level some 30 times greater than that of the parent mycelium (Table 4.10). This very high concentration of cAMP was not maintained, though it remained 510 times higher than the mycelial level for most of the life of the fruit body, and as the fruit body matured, total content of cAMP increased in parallel with increasing content of glycogen. There was a broad parallel in the proportional distribution of cAMP and glycogen; i.e. at stage 2 the largest proportion of both was in the stem, at stage 3 the largest proportion of both was in the cap. However, cAMP content reached a maximum and started to decline much earlier than did the glycogen. At the stage when the majority of the glycogen was being utilised, the concentration (dry weight basis) of cAMP in fruit body tissues (particularly the cap) was in rapid decline, approaching the range found in mycelium.

Accumulation of cAMP generally signifies a net deficit of glucose. The involvement of cAMP in regulation of glycogen metabolism in animals

Table 4.10. *Cyclic-AMP and glycogen contents of different parts of the fruit body of* Coprinus cinereus *at various stages of development.*

		Glycogen		cAMP	
Stage	Tissue	μg per fruit body	mg g^{-1} dry wt	pmol per fruit body	pmol mg^{-1} dry wt
0	Initials	70	74	53	90
1	Primordium	271	80	105	14
2	Cap	47	21	104	37
	Stem	506	68	127	19
	Cap	407	40	410	39
	Stem top	28	16	96	30
3	Stem middle	64	26	169	49
	Stem base	289	56	123	35
	Cap	1420	41	274	6
	Stem top	27	5	211	31
4	Stem middle	78	10	168	20
	Stem base	209	17	175	11
	Cap	10	4	20	5
	Stem top	10	4	20	5
5	Stem middle	22	6	20	3
	Stem base	38	8	18	2

Initials are spherical to subglobose aggregates up to 2 mm in diameter which appear about 88 h after inoculation; the larger initials have gill lamellae and dense polysaccharide deposits in the basal bulb of the stem. Stage 1 primordia are ovoid structures, 2–6 mm in height, appearing about 110 h after inoculation, they have well developed gill tissues and dense polysaccharide deposits in the subhymenium as well as in the base of the stem, karyogamy occurs towards the end of this stage. Stage 2 primordia are 6–9 mm tall fruit bodies found about 132 h after inoculation, the partial veil is intact initially but becomes ±free, histochemical staining of polysaccharide continues to intensify in the cap but that in the stem base lessens, meiosis occurs during this stage; stage 3 fruit bodies occur about 144 h after inoculation, they are 10–20 mm tall, partial veil free, stem begins to elongate slowly, meiosis is completed in this stage and basidia begin to form sterigmata; stage 4 fruit bodies occur 150–156 h after inoculation, they are 15–45 mm tall, basidiospore pigmentation begins to appear and polysaccharide deposits in subhymenium begin to disperse, stem continues to elongate slowly; stage 5 fruit bodies are mature, they are found about 160 h after inoculation, the stem elongates rapidly, the cap opens and spores are discharged. These developmental descriptions are taken from Moore *et al.* (1979) where illustrations can be found.

consists in elevated cAMP levels enhancing glycogen utilisation (by coordinated inhibition of synthesis and activation of breakdown) so as to relieve the glucose deficit and restore the energy charge. The negative correlation between glucose content of the medium and cAMP content of the *Coprinus* mycelium (Uno and Ishikawa, 1974); and the observed effects of cAMP on *Coprinus* glycogen synthetase (Uno and Ishikawa, 1978) and phosphorylase (Uno and Ishikawa, 1976), phosphodiesterase (Uno and Ishikawa, 1973b), and location of cyclase (Uno and Ishikawa, 1975) imply close parallels between *Coprinus* and higher eukaryotes. The enzymic and other *in vitro* data seem to be consistent with the view that cAMP synthesis regulates the amount of glycogen in *Coprinus* mycelia which are producing fruit bodies by inhibiting glycogen synthetase and activating glycogen phosphorylase (Uno and Ishikawa, 1982).

However, during development of the fruit body, elevated levels of glycogen were mirrored in elevated levels of cAMP and the stage in fruit body development during which the bulk of the glycogen was finally used was the stage which had the lowest levels of cAMP (Kuhad *et al.*, 1987). The essential questions are whether the observed correlations between glycogen and cAMP in *Coprinus*, *in vivo*, always reflect utilisation of glycogen to relieve a metabolic glucose and energy deficit, and what relationship they have to fruit body development. For a vegetative mycelial hyphal cell suffering a glucose deficit the most immediate remedy lies in control of glucose uptake rather than polysaccharide degradation. It is thus likely that cAMP stimulation of glycogen accumulation observed in plate cultures reflects an additional level of control of the allosteric glucose transport system (Moore and Devadatham, 1979; Taj-Aldeen and Moore, 1982) which, by furnishing additional supplies of glucose 6-phosphate consequentially over-rides any inhibitory effect the cAMP may have on activity of glycogen synthetase. A regulatory circuit of this sort, involving the relative abilities of glucose 1-phosphate and glucose 6-phosphate to interact with cAMP in the control of enzymes important to fruit body development could well be the basis of the apparent ability of sugar phosphate supplementation of the medium to induce fruit body formation in certain ascomycetes (Hawker, 1948; Buston and King, 1951; and see discussion below).

In the fruit body, accumulation of cAMP occurs in locations which are involved in mobilising glycogen for translocation rather than for immediate metabolism. It is unlikely that glycogen is translocated as a polysaccharide. Thus, if translocation of glycogen requires its breakdown; one could readily understand how, while the polysaccharide is in transit, high tissue contents of both cAMP and glycogen could be maintained (since the cell in which the glycogen is broken down is exporting the product, it could still experience a glucose deficit). Eventually, in the cap of the mature fruit body the glycogen is finally utilised and the consequent increase in energy charge results in a major reduction of cAMP level. Consequently, a specific role for cAMP in glycogen breakdown as part of a carbohydrate translocation process is likely, but any other function, as a morphogen for example, is unlikely. This is not a great surprise as cAMP does not control differentiation in all slime moulds; in another cellular slime mould, *Polysphondylium violaceum*, the aggregation signal is a small peptide (Wurster *et al.*, 1976; Bonner, 1977).

4.2.5 Mannitol accumulation

Mannitol is one of a group of compounds which are known as compatible solutes, being defined as low molecular weight, neutral compounds which can be accumulated to high concentrations within the cell without causing enzyme inhibition or other metabolic disturbances (Brown and Simpson, 1972; Brown, 1978). One of the roles ascribed to mannitol is that of osmoregulation, that is, service as an osmoticum for the maintenance of a high osmotic or suction pressure inside the cell to establish turgor and create an inflow of water, especially when the substrate is of high osmolality (Lewis and Smith, 1967; Ellis *et al.*, 1991). Rast (1965) and Hammond and Nichols (1976) reported concentrations of mannitol as high as 50% of the dry weight of the fruit body in *Agaricus bisporus*. As the mannitol turnover rate was low and the apparent cellular concentration remained at about 150 mM (Hammond and Nichols, 1976), mannitol was assigned an osmotic function, being thought to provide for support and expansion of the fruit body. In contrast, in *Coprinus cinereus*, a similar function as an osmoticum is ascribed to urea on the grounds that its accumulation is paralleled by influx of water (content of urea increases during fruit body development when expressed on a dry weight basis but not on a fresh weight basis, see above). The total polyols of the *Coprinus* fruit body do not exceed 6% of the dry weight and decline in concentration as the fruit body develops (Darbyshire, 1974). Further, neither mannitol nor NADP-linked mannitol dehydrogenase (the enzyme responsible for mannitol synthesis) have been detected in *Coprinus* (Rao and Niederpruem, 1969) and the NAD-linked mannitol dehydrogenase activity shows no correlation with glucose-6-phosphate dehydrogenase (Moore and Ewaze, 1976).

A quantitative GLC analysis for mannitol confirmed these findings and extended the analysis to *Lentinula edodes*. The occurrence of mannitol in *L. edodes* has not previously been noted. Mycelia of *L. edodes* had a low level of mannitol (about 1% on a dry weight basis) compared to the fruit body stem and cap (20–30%; Table 4.11). The highest level of mannitol

Table 4.11. *Mannitol content (dry weight basis) of tissues of* Lentinula edodes, Coprinus cinereus *and* Agaricus bisporus *as determined by gas chromatography.*

Tissue	RRT*	μmol mannitol mg^{-1} (w/w)	% mannitol (w/w)
L. edodes mycelium	1.2	0.07 ± 0.02	1.29 ± 0.28
L. edodes young cap	1.23	1.16 ± 0.07	21.16 ± 1.21
L. edodes young stem	1.22	1.68 ± 0.07	30.57 ± 1.22
C. cinereus young cap	0	0	0
A. bisporus young cap	1.22	2.54 ± 0.41	46.33 ± 7.46

was observed in the cap of *A. bisporus* (close to 50%), whilst no mannitol was detectable in the cap of *C. cinereus* (Tan and Moore, 1994).

Synthesis of mannitol in *A. bisporus* is mediated by an NADPH-dependent mannitol dehydrogenase using fructose as substrate (Edmundowicz and Wriston, 1963), the NADPH being obtained through the pentose phosphate pathway (PPP)(Dütsch and Rast, 1972); a greater proportion of glucose oxidation occurs via the PPP in the fruit body than in the mycelium in *A. bisporus* (Hammond and Nichols, 1976, 1977; Hammond, 1977). This contrasts with *C. cinereus*, in which the Embden–Meyerhof–Parnas pathway represents the major route of sugar catabolism in both fruit body and mycelium, with the PPP playing only a minor role and being completely dispensed with in early fruit body development (Moore and Ewaze, 1976; Moore, 1984a; see above).

Considering the large amount of mannitol which is accumulated by *L. edodes*, it is not surprising to find that glucose catabolism in this species resembles that of *A. bisporus*, with the PPP being the primary catabolic route in particular tissues of the fruit body. The study used radiorespirometry of ^{14}C-1/^{14}C-6 ratios in CO_2 respired from specifically-labelled glucose fed to *L. edodes* tissues. The pentose phosphate pathway was very high in activity in the fruit body but low in mycelium, the highest activity being recorded in tissues which were biosynthetically most active (young gills), requiring the reducing power of NADPH generated through the PPP. Extensive conversion of ^{14}C-3,4-labelled glucose to $^{14}CO_2$ in the mycelium underlined the important role of the Embden–Meyerhof–Parnas pathway (EMP) in that tissue. Comparative studies with *C. cinereus* gills confirmed the comparatively lesser importance of the PPP in this organism.

Enzymic determination corroborated the metabolic pattern deduced from radiolabelling. Activity of PPP enzymes (glucose-6-phosphate dehydrogenase and 6-phosphogluconate dehydrogenase) were three times as high in the fruit body compared to the mycelium, highest activity being in the young cap. EMP enzymes (fructose-1,6-bisphosphate aldolase and glucose-6-phosphate isomerase) were more active in mycelium than in fruit body, and also more active than PPP enzymes within the mycelium itself. Within the fruit body, specific activities of EMP and PPP enzymes were about the same. Enzymic activity in mature cap of *C. cinereus* resembled the pattern in the mycelium rather than the fruit body of *L. edodes*; EMP enzymes were very much more active than PPP.

Mannitol is, therefore, assumed to serve an osmoregulatory function in both *A. bisporus* and *L. edodes*. If the intracellular volume is similar to the 2.1 $\mu l\ mg^{-1}$ dry weight which has been measured for *Coprinus cinereus* (see Chapter 3), and the mannitol is assumed to be uniformly distributed within that volume, then the concentration of mannitol in the *L. edodes* stem would be of the order of 500 mM, and in the cap 750 mM. These values can be considered only as gross approximations, but concentrations like this would provide a formidable driving force for water through the fruit body and protect the most exposed tissues against desiccation.

It is particularly significant that the two species differ in the way in which the fruit body expands. In *A. bisporus* fruit body expansion occurs by hyphal *inflation*, whilst in *L. edodes* hyphal *proliferation* is responsible. Thus, the two very different basidiomycete strategies for tissue expansion can be facilitated by the same metabolite. Interestingly, *C. cinereus* (which expands by inflation) and *Schizophyllum commune* (which expands by proliferation) also show many metabolic similarities despite the fundamental difference between their strategies of expansion. Remembering that within the fruit body of *C. cinereus* different metabolic mechanisms are employed to provide adjustable osmotica for cell inflation in cap (osmoticum: urea with some related nitrogen compounds) and stem (osmoticum: a mixture of simple sugars)(see discussion above), then the lesson we learn here is that totally different tactics can be used to achieve the same strategic end. The evolutionary 'choice' between different metabolic mechanisms which enable a morphogenetic process to be put into effect, have been made independently of the 'choice' between different cell biological processes which contribute to that morphogenesis.

4.3 Environmental variables

4.3.1 Nature of the substratum

Experimenters can easily overlook the fact that most aspects of *in vitro* culture are alien to most fungi. In particular, the nature and volume of the culture vessel and the viscosity of the medium can have considerable impact on behaviour of the fungi. In their study of spiral growth of hyphae, Trinci *et al.* (1979) showed that the branching frequency as well as the direction of hyphal growth was altered by the concentration of agar used in the medium (Fig. 4.10). Madelin *et al.* (1978) suggested that spiral growth results from mechanical rolling caused by the axial rotation of the growing apex interacting with the substratum. The differential effect of the agar concentration demonstrated by Trinci *et al.* (1979) supports this interpretation and demonstrates that the physical nature of the substratum can influence morphogenesis at the hyphal level directly, without intervening physiological interactions. As natural substrata are likely to vary in physical form on a very small scale, this sort of mechanical influence on hyphal morphogenesis could have impact on initiation of fruit body structures. Interestingly, *Magnaporthe salvinii* forms more perithecia on medium gelled with 1.2–2% agar than with higher or lower concentrations (Tsuda *et al.*, 1982).

Liquid media are abnormal substrata for most fungi and many are unable to form fruit bodies when cultivated on or in liquids (Weste, 1970a; Asina *et al.*, 1977), or are less productive than when cultured on solid medium (Ghora and Chaudhuri, 1975b), or have an altered reaction to other environmental factors; for example, a higher light intensity is required to induce fruiting in *Nectria ditissima* when grown on a liquid rather than a solid medium (Dehorter and Perrin, 1983).

The quantity of substratum is also important. Buller (1931; p. 165) discussed the requirement for a minimum amount of mycelium to support a minimum fruit body in *Coprinus*. In considering the ‘social organisation’ of the fungus, he argued that one of the functions of hyphal fusions between germlings produced by different spores is to ensure the rapid formation of that minimal size mycelium encompassing a corresponding minimum quantity of substrate: ‘For the production of so large a structure as a fruit-body of *Coprinus sterquilinus* a very considerable amount of mycelial contents is required. To obtain this amount of mycelial contents a single simple monosporous colony must become master of a certain minimum mass of the nutrient substratum’. This consideration may be reasonably obvious for large mushroom fruit bodies, but it seems

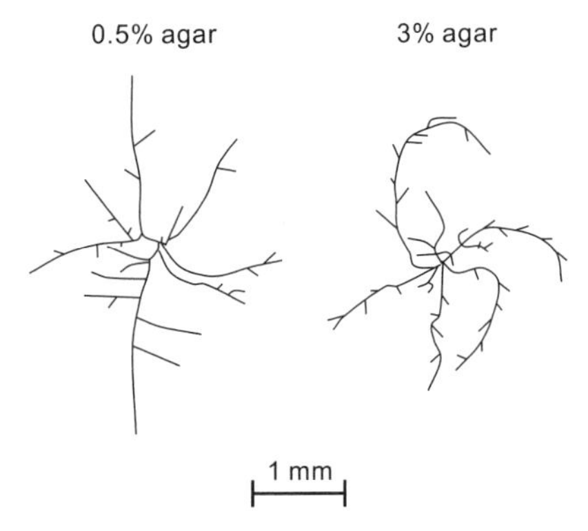

Figure 4.10 Effect of the substratum (as determined by agar concentration) on hyphal morphogenesis of mycelia of *Mucor hiemalis* grown for 24 h. The higher gel-strength agar causes spiral growth. Redrawn after Trinci (1984).

to apply even to species which produce smaller fruiting structures. For example, Ghora and Chaudhuri (1975a,b) showed that abundance of perithecia of *Chaetomium aureum* differed in tube, Petri dish, or flask cultures, and *Magnaporthe salvinii* forms more perithecia in Petri dishes containing 15 ml of medium than with either 10 or 20 ml (Tsuda *et al.*, 1982).

Madelin (1956b) tested the proposition (Voderberg, 1949) that fruiting in small species of *Coprinus* (called *C. lagopus* but possibly *C. radiatus* or *C. cinereus*) was independent of the quantity of available nutrients by measuring mycelium and fruit body yields of disks of medium. He found that disks of all sizes bore fruit bodies and, with exceptions in the smallest sizes, the dry weight of fruit body tissue was approximately directly proportional to the volume of substrate. On most disk sizes (7.5–30 mm diameter) similar numbers of fruit bodies matured, suggesting that fruit body size was adjusted to match available substrate. On larger disks (40 and 50 mm diameter), the average number of mature fruit bodies was slightly increased. Fruiting began earlier on the smaller discs and although fruit bodies were considerably larger on the bigger disks, the time required for maturation was much the same in all cases. Madelin (1956b) found no evidence for the existence of a minimal quantity of substrate for fruiting of *C. lagopus* to take place, but argued that this did not conflict with Buller's statements relating to *C. sterquilinus* because the fruit body of *C. lagopus* is so much smaller than that of *C. sterquilinus* that the size of the minimum is likely to be below the range Madelin had tested.

4.3.2 Constraints edges and injuries

Probably related to volume of substratum is the commonly encountered feature that sexual reproduction is initiated when the growing mycelium reaches an obstacle like the edge of the dish or barriers placed on to the surface of the medium (Hawker, 1966). In the laboratory this has been demonstrated by Buston *et al.* (1953) and Buston and Rickard (1956) with *Chaetomium globosum*, by Bahn and Hock (1973, 1974) and Molowitz *et al.* (1976) with *Sordaria macrospora*, by Pollock (1975) with *Sordaria fimicola* and by Long and Jacobs (1974) with *Agaricus bisporus*. This is the so-called 'edge effect', for which the most commonly believed (but probably erroneous, as is explained below) explanation is that the walls of the culture container (or other obstruction) act as diffusion barriers, resulting in the build up of (unknown) 'staling' substances at the colony margin which either directly induce fruiting (Buston and Rickard, 1956) or inactivate inhibitors of fruiting (Chet *et al.*, 1966; Chet and Henis, 1968). Increased formation of perithecia occurs in *Sordaria macrospora* in response to addition of biotin and arginine (Bahn and Hock 1974; Molowitz *et al.* 1976); arginine being called a 'perithecia inducing substance' by Bahn and Hock (1974). Sugar phosphates were suggested to be the inducers in *Chaetomium globosum* (Buston and Rickard, 1956) because Hawker (1948) and Buston and King (1951) showed that addition of sugar phosphates to the medium induced fruiting in *Chaetomium globosum* and *Sordaria*. However, the reliability of this conclusion must be in serious doubt because the prevalence of intracellular and extracellular phosphatases would probably mean that added sugar phosphates would be hydrolysed, making the effective concentrations employed (0.05% glucose 1-phosphate, which is approximately 2 mM, added to wells cut into the medium) very uncertain indeed. Whether such an addition of sugar phosphate (specifically) has a significant physiological effect or whether impurities, or even the edge of the well cut in the medium, could have been significant contributors to the observed effects remain open questions. What is most important about these experiments is that they demonstrated that relatively simple metabolites *might* be involved in determining fruiting behaviour.

Robinson (1926) investigated the edge effect on the formation of apothecia by *Pyronema confluens*, and its relation to incident light and nitrogen source. He concluded that reproductive structures only arose on solid medium when growth had been checked, either by physical or chemical means. Interestingly, a correlation between an arrest to hyphal elongation and the induction of fruit bodies had previously been deduced by Klebs (1900). The most critical experiments have been those of Lysek (1976) using *Podospora anserina*, and, especially, MacDonald and Bond (1976) who studied the edge effect in *Sordaria brevicollis* cultured on cornmeal agar.

Lysek (1976) studied the number and distribution of perithecia of *Podospora anserina* using different means to achieve 'staling' of hyphal elongation. Higher numbers of perithecia were found on both sides of lines of contact between two colonies as they met, after the mycelium grew into a poor medium, and when growth of the mycelium was partially inhibited by addition of the sugar L-sorbose (10 mM sorbose in a medium containing 100 mM fructose). The clear conclusion which Lysek (1976) reached was that fruit body formation "mainly occurs at those parts of the mycelium which have terminated their vegetative growth . . . the cessation of hyphal elongation . . . initiate(s) reproduction". Reduced nutrient supply may cause this check to growth, or it may be a response to physical barriers, injuries or added substrates. Hock *et al.* (1978) observed a correlation between the ability of a colony (of *Sordaria macrospora* on chemically defined medium) to form a large number of side branches and fruit body formation. They suggested that a check to growth of leading hyphae results in branch formation on the older mycelium and the resulting increase in hyphal density allows fruit body formation. Inoue and Furuya (1970) also found that perithecia of *Gelasinospora reticulospora* were induced preferentially on the older mycelium (in this case after an inductive light treatment) and gradually spread to younger parts of the colony.

MacDonald and Bond (1976) established that protoperithecia of *Sordaria brevicollis* were formed only after the mycelium reached the edge of the vessel whatever the volume of medium used and irrespective of the size and shape of the Petri dish. Experiments in which cornmeal agar was re-used, mycelium being inoculated onto cellophane discs to enable removal from the dish, demonstrated that protoperithecial initiation was not stimulated by exhaustion of a metabolite or by the build up of a metabolite/inhibitor in the medium. Other experiments using a growth chamber which allowed chemicals to diffuse between two mycelia without allowing them to come into physical contact, showed that a colony which had already formed protoperithecia did not cause a second colony in the other half of the growth chamber to produce protoperithecia any earlier than normal. These experiments exclude fruiting inducers, counter-inhibitors and staling substances as being instrumental in causing protoperithecium development.

When *Sordaria brevicollis* was grown from a cornmeal agar central zone onto a ring of unsupplemented water agar the colony formed protoperithecia before it had reached the edge of the Petri dish and shortly after the growing mycelial margin had crossed the transition in the medium (MacDonald and Bond, 1976). By varying the size of the central zone it was established in every case that protoperithecium formation occurred shortly after the mycelial front crossed the boundary between rich and poor medium. These results are in contrast to those obtained on single medium plates. Controls comprising two physically different zones of cornmeal agar formed protoperithecia, as usual, only after the mycelium reached the edge of the plate. When the colony was grown from a central zone of double strength cornmeal agar onto a peripheral zone of normal cornmeal agar there was a small stimulation of protoperithecial formation shortly after the colony grew past the medium transition point but before the plate edge was reached. Taken together, these experiments show that a physical barrier is not absolutely necessary for the 'edge effect'. The first-described experiments showed there was no dependence on accumulation of a particular chemical, and the subsequent ones revealed the stimulus for protoperithecial production to be the perturbation of growth as the colony crossed the transitions between different substrata. MacDonald and Bond (1976) concluded that the important determining factor is the disturbance in metabolism which results either from encountering the edge of the dish or a major change in nutritional value of the substrate. Thus, different sorts of barrier and different sorts of medium transition are able to disturb the progress of metabolism sufficiently to initiate fruit body formation.

Pollock (1975) showed that physical injury could also stimulate perithecium formation in *Sordaria fimicola* grown on cornmeal agar. Diffusible inhibitors of perithecium development were postulated to be responsible for the pattern of perithecial development in a colony following injury. Formation of fruit bodies and other multicellular organs (the latter including 'zone lines' of sclerotial tissue in wood; Hubert, 1924) in response to mechanical injury has been reported in a range of species, including *Alternaria solani* (Rand, 1917), *Collybia velutipes* (Bevan and Kemp, 1958; and see the 'regeneration' phenomena discussed below), *Phytophthora* (Reeves and Jackson, 1974; Reeves, 1975) and *Armillaria* and *Stereum* in decaying wood (Lopez-Real, 1975; Lopez-Real and Swift, 1975, 1977). Most information about response to mechanical injury has come from studies of monokaryotic fruiting in the basidiomycete *Schizophyllum commune* (Leonard and Dick, 1973; Leslie and Leonard, 1979a,b). In this organism, fruit bodies are normally produced on the dikaryotic mycelium (see Fig. 5.4), but under certain conditions monokaryotic mycelia can fruit, though less vigorously than the dikaryon. However, in both mycelia the production of fruit bodies depends on the interaction of morphogenetic genes, mating type factors, and environmental conditions and monokaryotic fruiting is a valuable model system for studies of development and differentiation (Stahl and Esser, 1976). A word of warning here is that the morphology of these monokaryotic fruit bodies is highly variable, and their categorisation as 'fruit bodies' can be rather subjective in some cases. Structures are produced which range from barely detectable localised aggregations of hyphae, through elongated, antler-like growths devoid of hymenial surface, on to fully mature, expanded fruit bodies which are fully comparable to those produced by the dikaryon (see discussion in sections 5.4.2 and 5.4.3). Nevertheless, Leslie and Leonard (1979a,b) were able to show that the fruiting response to mechanical injury was genetically determined and involved at least four genes. They assigned the genes known to be involved in determining haploid fruiting into four sets, two of which involved response to injury, one response to a 'fruit inducing substance', with the fourth being made up of two genes determining spontaneous formation of haploid fruit bodies. Again, therefore, the indications are that there exist a number of different parallel routes leading to fruit body initiation.

4.3.3 Light

Light has a different effect on formation of reproductive structures in different organisms, either increasing or decreasing their number, affecting their development, or determining whether or not they are produced (Carlile, 1970; Tan, 1978; Elliott, 1994). In general, the most effective parts of the spectrum are the near ultraviolet and blue wavelengths and there are indications that the photoreceptor involved in fruit body morphogenesis may be membrane bound (Gressel and Rau, 1983; Durand, 1985) and it has been called the mycochrome system (Kihara and Kumagai, 1994). In some fungi levels of cAMP and activities of dehydrogenase, enzymes have been found to respond very rapidly to changes in illumination (Chebotarev and Zemlyanukhin, 1973; Cohen, 1974; Daniel, 1977) and coenzyme levels and the pathway of carbohydrate metabolism can also be drastically affected by light (Graafmans, 1977; Kritsky, 1977). In *C. cinereus*, development of the activity of chitinase and of NADP-linked glutamate dehydrogenase in the cap was dependent upon the light signal that triggered maturation of the fruit body primordium (Ishikawa and Serizawa, 1978). Clearly, the pattern of illumination received by the mycelium and fruit body primordium may well control the metabolic changes which characterise fruit body development.

Many ascomycete vegetative mycelia require exposure to light before they will produce fruit bodies (ascomata) and show specificity not only for particular wavelengths but also for a particular dosage of light radiation. For example, perithecium production by *Neurospora crassa* requires 12 s exposure to blue light with an intensity characterised by an irradiance of 1.05 W m^{-2}, but 240 s exposure to light of irradiance 5.25×10^{-2} W m^{-2} (Innocenti *et al.*, 1983). This 'reciprocity' between light intensity and time of exposure has been demonstrated in *Pleospora herbarum* (Leach and Trione, 1966), *Nectria haematococca* (Curtis, 1972), and *Gelasinospora reticulospora* (Inoue and Furuya, 1975) as well as *N. crassa*. In all these fungi the requirement is for a single relatively short exposure (seconds or minutes) to induce perithecium formation. In other species more complex exposure sequences are required. As an example, perithecium formation by *Nectria galligena* requires exposure to light (irradiance 7.515 W m^{-2}) for 12–16 h on each day from the 7th to the 21st day of culture (Dehorter and Lacoste, 1980). This is, presumably, an expression of the sort of light exposure experienced by the organism in nature. In species of the basidiomycete *Coprinus*, sequential light exposures are responsible for initiating and programming fruit body morphogenesis. In *C. congregatus* (Manachère, 1971, 1985, 1988; Durand, 1983), *C. domesticus* (Chapman and Fergus, 1973) and *C. cinereus* (Morimoto and Oda, 1973; Lu, 1974), light stimuli are required at two stages during mushroom fruit body production. The initiation of primordium formation has a requirement for brief exposure to low intensity light, and further development of the fruit body, particularly differentiation of the cap, depends on additional periods of illumination. Again, blue (400–520 nm) to near-ultraviolet (320–400 nm) light is the most effective (Madelin, 1956b; Tsusue, 1969; Lu, 1974; Manachère and Bastouill-Descolonges, 1983) in *Coprinus* spp. and in *Schizophyllum commune* (Perkins, 1969; Perkins and Gordon, 1969). Detailed action spectra have been established for *S. commune* (Perkins and Gordon, 1969; Yli-Mattila, 1985), *Psilocybe cubensis* (Badham, 1980), *C. congregatus* (Durand and Furuya, 1985) and *Pleurotus ostreatus* (Richartz and MacLellan, 1987), and in the polypore *Favolus arcularius* the two photosensitive stages exhibit different action spectra (Kitamoto *et al.*, 1972, 1974a).

Another complicating factor is that dark periods are also important in the 'light response'. Growing hyphae of *Gelasinospora reticulospora* need at least 30 h of uninterrupted darkness to become light-competent. If this inductive dark phase is interrupted by brief exposure to near-ultraviolet or blue light, subsequent perithecium formation is prevented (Inoue and Furuya, 1974). After dark induction, exposure to blue light (wavelengths between 350–520 nm) is required for perithecium formation (Inoue and Furaya, 1975), but a further exposure to near-ultraviolet light (wavelengths shorter than 350 nm) prevents perithecium formation (Inoue and Watanabe, 1984). In *Coprinus*, primordia do not mature unless further exposed to illumination and four or five phases have been recognised in the light–dark exposure cycle which is required for complete development (Manachère, 1970; Kamada *et al.*, 1978), including requirements for dark periods and an interplay with temperature (Lu, 1972, 1974; Morimoto and Oda, 1974; Durand, 1983). If the mushroom primordia are kept in darkness the base of the stem elongates to produce etiolated fruit bodies, though these have been variously termed 'dark stipes' (Tsusue, 1969) 'oversized stipes' (Lu, 1974) and 'long slender stalks' (Morimoto and Oda, 1973). Buller (1931, pp. 112–116) used the term 'pseudorhiza' and showed that it is the solid stem base which is elongated in the dark. He argued that during normal development 'light inhibits the elongation of the stipe-base and so permits the growth-energy of the fruit-body to be concentrated earlier than would otherwise be the case upon the enlargement of the pileus and the elongation of the stipe-shaft'. Lu (1974) has also suggested that the light signals control the disposition of resources. In view of the translocation of glycogen between the different tissues in the developing primordium (see above), it may be that light signals activate polysaccharide translocation during maturation.

A differential effect of light on different fruit body tissues has also been observed in some ascomata. *Sclerotinia sclerotiorum* forms apothecial initials from sclerotia more quickly in the light. Subsequent formation of apothecial stems takes place independently of illumination, but disk formation at the stem apex requires light (Letham, 1975). *S. trifoliorum* forms elongated immature apothecia in the dark, which may be analogous to the etiolated stems of dark-grown fruit bodies in *Coprinus*, but exposure to near-ultraviolet light is required for formation of asci in the disk (Honda and Yunoki, 1975). Light also stimulates development of ascogenous hyphae, asci and ascospores during apothecial maturation in *Pyronema domesticum* although excessive exposure inhibits ascospore maturation. Sterile tissues and paraphyses of the apothecium require less light than the ascogenous hyphae and are also repressed by excessive exposure (Moore-Landecker, 1979a,b). In *Ascobolus immersus* a higher light intensity is required for completion of meiosis than for apothecial initiation. Inadequate light exposure whilst the asci are in mid-pachytene causes them to vacuolate and fail to complete meiosis (Lewis, 1975). *Sordaria fimicola* (Ingold and Dring, 1957) and *S. macrospora* (Walkey and Harvey, 1967a,b) respond to blue light with an increase in ascospore discharge but in many cases periodicity in ascospore discharge persists in continuous light or dark (Walkey and Harvey, 1967a,b; Pady and Kramer, 1969; Kramer and Pady, 1970).

Multiple photoresponses (and especially different action spectra) suggest multiple photoreceptors or that very different metabolic pathways are invoked after photoreception. At least two different photosensitive systems appear to operate in fungi, one stimulated by near ultraviolet and the other by blue light. Conidiation of many ascomycetes occurs only after illumination with near ultraviolet (or 'black light'). Trione and Leach (1969) extracted a substance which they called P310 from its maximum absorption at 310 nm, which developed in irradiated cultures of many species. Compounds, called mycosporines, with P310 characteristics have been found in several fungi. However, although they are closely correlated with sporulation, mycosporines are found quite widely in organisms as diverse as cyanobacteria and echinoderms and are now generally thought of as UV-absorbing photoprotectants.

Because their absorption spectra parallel the action spectra of the blue light photoresponses, carotenes and flavins appear to be the best candidates for photoreceptors (Carlile, 1970; Tan, 1978; Song, 1980). Inhibition of photoresponses by flavin inhibitors (Tschabold, 1967; Schmidt, 1980) and positive response to treatment with oxidising agents (Moore-Landecker, 1983) focuses attention on flavins. Of particular interest is the correlation between flavin-mediated photoreduction of cytochrome-*b* by nitrate reductase and photoresponsiveness of a mutant of *Neurospora crassa* (Muñoz and Butler, 1975; Klemm and Ninnemann, 1978, 1979). It appears that the primary photoreceptors are free flavins and redox reactions in nitrate reductase constitute the earliest reactions in the photoresponse (Fritz and Ninnemann, 1985). This enzyme, acting as a cytochrome, is thought to initiate a sequence of oxidation-reduction cycles which, presumably, amplify the light stimulus and convey it into metabolism (Gressel and Rau, 1983). As it is likely that quinones and phenols are involved in these 'downstream' redox reactions, this may account for the frequent claims for the involvement of phenoloxidases in the initiation of various aspects of fungal reproduction (Ross, 1982a,b).

This brings the nutritional status of the medium into consideration of photosensitivity. Indeed, the dependence of photomorphogenesis (and the edge effect) on the cultivation medium has been evident for many years. Robinson (1926) concluded that, though also light-dependent, exhaustion of nitrate ions appeared to be a prerequisite for apothecium formation in *Pyronema confluens*. Excess nitrogen also inhibits photoinduction of protoperithecia in *Neurospora*. Blue light accelerates protoperithecium formation (Degli-Innocenti *et al.*, 1983) but a colony grown on the usual Vogel's medium does not produce them. Reduction of the concentration of NH_4NO_3 by a factor of 100 is necessary to achieve photoinduction (Sommer *et al.*, 1987). Carotenoid synthesis was unaffected by these changes, suggesting that the photoreception may not have involved carotenes.

In *Coprinus cinereus* the requirement for the first light stimulus can exhibit a very low threshold and is medium-dependent. Primordia can be formed during incubation in continuous darkness on a potato extract + 2% sucrose medium (Tsusue, 1969; Morimoto and Oda, 1973), but the same strain exhibits a minimum requirement of 6 h exposure to white light of 500 1x for formation of primordia when grown on a medium containing glucose (2%), peptone (0.2%) and yeast extract (0.2%) (Morimoto and Oda, 1973). Similarly, aeration can substitute for light requirement in *Pyronema domesticum* which forms apothecia in the dark when cultures are placed in a strong air current, though exposure to near-ultraviolet light is required in sealed cultures (Moore-Landecker and Shropshire, 1982, 1984). These observations imply that there may be an interaction between the photosensitive apparatus and the nutritional status of the mycelium. Since sucrose is a very poor carbon source for *C. cinereus* (Moore, 1969b) it may be that the requirement for the initial light exposure is reduced in threshold as the medium approaches exhaustion.

An important point emphasised by Ross (1982a, 1985) is that the mycelium is not uniform and different parts may vary in their competence to respond to stimuli. This is particularly relevant to the effects of light because it is so easy to find variations in shading in most natural substrates. This consideration applies equally to other factors but can be more easily modelled with radiation than with most other factors because light can be applied locally using microbeams (Galun, 1971; Inoue and Furuya, 1978).

4.3.4 Temperature

The production of fruit bodies typically occurs over a much more restricted range of temperature than that which will support mycelial growth (Moore-Landecker, 1993). Optimum temperatures have been determined for fruit body production in many species. Different isolates exhibit different optima and maxima, and in many cases the response to temperature changes with developmental stage. For example, temperatures between 21 and 24°C favour perithecium and ascospore development in *Nectria haematococca*, but prior to fertilization the optimum is in the range 18 to 24°C (Dietert *et al.*, 1983). Similarly, early stages of development in *Cochliobolus miyabeanus* are favoured by lower temperatures than the later stages (Kaur and Despande, 1980). Several reports demonstrate that suboptimal temperatures are more likely to impair development of the ascogenous hyphae than of the sterile hyphae, resulting in infertile ascomycete fruit bodies (Weste, 1970b; Singh and Shankar, 1972; Moore-Landecker, 1975; Fayret and Parguey-Leduc, 1976; Tsuda *et al.*, 1982).

In basidiomycetes, there is little information about the impact of temperature except in species adopted as laboratory subjects or for commercial cultivation. Descriptions of conditions required by cultivated species frequently mention the need for a temperature shiftdown (and lower CO_2 concentrations). Chang and Hayes (1978) record that lowered temperature and lowered CO_2 are required by *Agaricus bisporus*, *Coprinus cinereus*, *Flammulina velutipes*, *Kuehneromyces mutabilis*, *Lentinula edodes*, *Pholiota nameko*, *Pleurotus ostreatus*, *Stropharia rugosa-annulata* and *Volvariella volvacea*. This list includes compost-grown fungi as well as some log-grown wood degraders, and is not unrepresentative of the wider community of saprotrophic mushrooms so it may be that most basidiomycetes require a temperature downshift. Flegg (1972, 1978a,b) found that the optimum temperatures for mycelial growth and for production of fruit body initials by *Agaricus bisporus* was 24°C in each case. The temperature downshift was required for further development of those initials. If the temperature was maintained at 22 to 24°C the initials did not develop beyond a cap diameter of about 2 mm. The fruit bodies developed normally when the temperature was lowered to 16°C. The functional significance is unknown, although it is a reasonable presumption that key developmental processes depend on temperature-sensitive proteins. It is clear that this is more than a generalised chemical response to temperature level; particular steps in the developmental sequence have distinct temperature requirements, and if those requirements are not met then developmental abnormalities result. The research has mostly been done with very broad developmental definitions ('production of fruit bodies', 'maturation of fruit bodies', etc.) and very few specific functions have been identified. One example is that suboptimal temperatures increase the frequency of three- and four-spored basidia in the (normally two-spored) cultivated mushroom, *Agaricus bisporus* (Elliott and Challen, 1984). Another example is basidial differentiation of *Rhizoctonia solani* being induced and increased by a cold temperature shock (Noel *et al.*, 1995). Unfortunately, there is too little information available to allow any real understanding of the effect of temperature.

4.3.5 Aeration

The atmosphere is a complex mixture of gases and volatile metabolites. There is some evidence that volatiles other than the normal components of the atmosphere can affect fungal growth and development (Stoller, 1952) and/or are associated with specific stages of development (Turner *et al.*, 1975). However, only oxygen and carbon dioxide will be considered here. As the above discussions of metabolism imply, fruit body development requires oxidative metabolism (glycolysis and TCA cycle activity are often amplified) and, consequently, considerable supplies of oxygen. Good aeration is, not surprisingly therefore, associated with successful fruiting. Something which has not been investigated is how oxygen is supplied to the inner tissues of large fruit bodies. That this could be a problem worthy of investigation is suggested by the observations which have been made of oxygen tension in the pellets which are formed in liquid fermenter cultures. Even under optimal growth conditions, oxygen tension has been shown to fall to zero within 150 μm of the surface of pellets of *Penicillium chrysogenum* in such cultures (Huang and Bungay, 1973; Wittler *et al.*, 1986). Glucose levels also decline rapidly beneath the surface of such pellets (Cronenberg *et al.*, 1994). Thus, although every effort may be made to ensure that growth conditions are optimised in fermenter cultures, pelleted cultures are oxygen-limited because of insufficient oxygen transfer from the surface of the pellet (Yano *et al.*, 1961; Kobayashi *et al.*, 1973). Yet, even fairly 'ordinary' mushrooms may be composed of masses of tissue which have regions which may be several millimeters away from the nearest atmosphere-exposed surface, and there are other instances (bracket fungi, larger mushrooms, puffballs and large sclerotia like those of *Polyporus mylittae*) in which perfectly healthy tissue may be situated several centimetres from the surface. The sort of detailed analysis of nutrient gradients which has been carried out with pellets of *Penicillium* has not been done with more normal hyphal aggregates. The mechanisms which safeguard gas exchange in such deep-seated tissues are completely unknown.

It is not just a matter of supply of oxygen, carbon dioxide is extremely important also. Elevated carbon dioxide concentrations are required for perithecium formation by *Chaetomium globosum* which produces the maximum number in 10% carbon dioxide (Buston *et al.*, 1966), and air containing 5% carbon dioxide accelerates production of perithecia and ascospores and affects ascospore discharge in some pyrenomycetes (Hodgkiss and Harvey, 1972). In *Neurospora tetrasperma*, oxygen and carbon dioxide are required for differentiation; ascogonial coils form in low oxygen conditions (0.5% oxygen) or in the absence of carbon dioxide but perithecium maturation requires an oxygen concentration of 5% and a carbon dioxide concentration of 1% (Viswanath-Reddy and Turian, 1975). In *Agaricus,* high carbon dioxide concentration promote mushroom stem elongation whereas cap enlargement is enhanced by lowered carbon dioxide concentrations (Lambert, 1933; Turner, 1977).

Turner (1977) showed that increase in size of *Agaricus bisporus* fruit bodies after removal from the substrate was less than half that for those left on the compost, though excised fruit bodies did complete their development, with cap opening and ripening occurring at a smaller final size. Much the same is also true for *Coprinus cinereus* (Hammad *et al.*, 1993a). However the effect of CO_2 was the same in excised and intact *A. bisporus* fruit bodies. In both cases increased elongation of the stem occurred in the presence of CO_2, whereas cap and gills expanded and spores ripened more rapidly when CO_2 was removed. This effect was evident in the ratio of cap: stem weight which, for a representative fruit body, was 2.2:1 (fresh weight) or 2.4:1 (dry weight) in the presence of CO_2, but 2.6:1 (fresh) or 3:1 (dry) when CO_2 was removed with KOH. Thus, CO_2 affects the quantity of tissue in the two organs as well as their shape and rate of development. Tschierpe (1959) also found cap:stem weight ratios to be reduced by elevated CO_2 levels, and Le Roux (1968) recorded that, in fruit bodies treated with CO_2, the ratio between weight of gills and weight of cap flesh was reduced, indicating delayed development of the gills. Naturally-accumulated respiratory CO_2 was sufficient to cause these effects. Turner (1977) pointed out that Long and Jacobs (1974) had found that exogenous CO_2 is required for growth of mycelium and mycelial strands into casing soil, rate of strand growth into sterile casing soil being proportional to CO_2 concentration between 22 and 370 ppm. CO_2 concentration also controlled initiation of fruiting in non-sterile casing soil (initiation does not occur in sterile casing), the optimal range being 340–1000 ppm. San Antonio and Thomas (1971) also demonstrated a requirement for CO_2 by hyphae of *A. bisporus* growing on agar. There are reasons to believe, then, that CO_2 concentration has a controlling role in several stages of fruit body development in *A. bisporus*. Turner (1977) suggests that its morphogenetic effect on maturation of the fruit body may have ecological advantage. She argues that CO_2-enhanced elongation of the stem would raise the gills away from the surface of the substrate where the concentration of CO_2 might be expected to be higher than in the wider atmosphere because of the respiratory activity of microorganisms in the casing soil.

Initiation of fruit body formation by dikaryotic cultures of *Schizophyllum commune* was promoted by decreased growth of apical and increased growth of subapical hyphae at the colony margin. Rapid differentiation required short cells in the mycelium so that branches were closer together and able to aggregate. The short cells were induced by aeration and appropriate temperatures (Raudaskoski and Viitanen, 1982; Raudaskoski and Salonen, 1984). High CO_2 levels promote formation of long hyphal compartments which do not provide a good basis for rapid differentiation. Raudaskoski and Salonen (1984) suggest an ecological correlation here, arguing that a wood degrader like *S. commune* is likely to experience elevated CO_2 levels within the wood as respiratory CO_2 accumulates. Mycelium which reaches the surface of the wood, however, will be exposed to good aeration and, specifically, CO_2 levels reduced to the atmospheric normal. Such mycelium will be able to form the shortened cells and more compact branching habit and be predisposed to fruit body formation, providing temperature and illumination regimes are suitable to induction of fruiting.

4.3.6 Physiological generalisations

Despite the wide range of factors which promote fruiting, I believe that three generalisations can be extracted from these observations on fruiting physiology. First, the organism internalises nutrients rapidly in order to gain regulatory control over nutrient access and distribution. By so doing the vegetative mycelium becomes competent to produce multicellular structures like fruit bodies. Second, factors which promote fruiting, whether physical or chemical, seem to work by disturbing the normal progress of cellular metabolism. It is the disturbance itself which is the effective factor. Consequently, there appear to be a number of parallel pathways covering some stages of fruit body development so, for these stages, different factors seem to be interchangeable (e.g. a particular nutritional state may replace a particular illumination requirement). Third, even relatively simple developmental pathways (e.g. sclerotium formation) can be subdivided into stages (at least, initiation, development and maturation) and there seems to be a need for successive signals (successive metabolic disturbances) to keep the developmental process rolling. Without the required sequence of disturbances the developmental pathway stops and stagnates. More complex structures, for example, the mushroom fruit body, emphasise these features and even physiological studies suggest that such complex developments are made up of parallel pathways. The physiology of the hyphal system influences both progress through the branch points at which the parallel pathways originate and progress from step to step within individual pathways.

5

The genetic component of hyphal differentiation

The basic genetic architecture of fungi is fairly typically eukaryotic, and all the major principles of genetics apply – Mendelian segregations, recombination, chromosomal structure, gene structure, etc. (Clutterbuck, 1995a). Nevertheless, there are some differences between most fungi and most of the rest of the eukaryotes. These will be summarised here; detailed information can be obtained from Carlile and Watkinson (1994), Elliott (1994) and Chiu (1996).

Fungi have a generally smaller genome size than other eukaryotes (Clutterbuck, 1995b). The yeast, *Saccharomyces cerevisiae*, for example, has a haploid genome of about 15×10^6 base pairs, which is less than four times the size of the genome of the bacterium *Escherichia coli* (Table 5.1). Fungal nuclei are consequently difficult to study by conventional cytological procedures because both nuclei and chromosomes are small, variable in shape and indistinct by conventional microscopy. Progress in understanding nuclear changes during the fungal life cycle has been slow, though in recent years the study has benefited greatly from application of electron microscopy and molecular techniques for analysing the karyotype.

Fungal mitotic divisions are intranuclear, unlike most animals and plants, which means that their division spindle is formed within an intact

Table 5.1. *Karyotypes of some fungi*

Fungus	Chromosome number by microscopy	Chromosome number by PFGE	Genome (Mb)	Chromosome size (Mb)
Agaricus bisporus	8, 9, 12	13	26.8; 34	1.2–3.5
Aspergillus nidulans	8	4 + 2 doublets	31	2.9–5.0
Coprinus cinereus	4, 12, 13	13	ND	1.0–5.0
Letinula edodes	8	8	33	2.2–7.0
Neurospora crassa	7	3 + 2 doublets	47	4–12.6
Phytophthora megasperma	13, 14; 17–23; 30–40	13; 9–10 ND	46.5; ND; ND	1.4–6.8; 2.5–6.1; 3.0–6.2
Pleurotus ostreatus	8–10	6; 9; 10	20.8; 31.3; > 39.5	2.1–5.2; 1.1–5.7; 2.3– > 6
Podospora anserina	7	5 + 1 doublet	34	3.8–6.0
Saccharomyces cerevisiae	17	15 + 1 doublet	15	0.225–2.21
Schizophyllum commune	8	4 + 2 doublets	35–36	1.2–3.5
Volvariella volvacea	9	5 + 4 doublets	34	1.3–5.4

ND, not determined; PFGE, pulsed field gel electrophoresis. Data from Chiu (1996); see also Clutterbuck (1995b).

nuclear membrane. This makes the progress of the division even more difficult to see and study, but does not appear to affect the biological consequences of the mitotic division.

The majority of fungi are haploid for most of their life cycles, diploidy being limited to a short period immediately prior to meiosis. A major biological impact of this arrangement has been the evolution of processes to bring together two haploids so that genetically different nuclei can co-exist in the same cytoplasm; these processes are the incompatibility mechanisms which regulate cytoplasmic and nuclear compatibility. Through their action hyphae from different haploid parental mycelia can safely approach each other, undergo hyphal fusion to create a channel between themselves and then exchange cytoplasm and nuclei, so that a heterokaryotic mycelium is formed. The heterokaryon then grows normally as a vegetative mycelium until conditions are right for sexual reproduction to take place. At this stage, cells of the heterokaryon go through karyogamy followed by the meiotic division, and then generate and distribute progeny spores.

5.1 Nuclear divisions

5.1.1 Mitosis

Intranuclear mitotic nuclear division with the nuclear membrane remaining intact or nearly so is usual in fungi (Aist and Wilson, 1968; Taylor, 1985; O'Donnell, 1994; Lü and McLaughlin, 1995). Interphase resting nuclei appear as spherical structures with diffuse chromatin in the light microscope, while migrating interphase nuclei are elongated and have a distinct nucleolus. The spindle pole body replicates in S-phase (Taylor, 1985; Heath, 1994), nuclear volume decreases and chromatin condenses. The dividing nuclei have a nucleolus, which is lost at the anaphase stage of mitosis, and steadily enlarge until daughter nuclei are formed.

Anaphase movements were found to be randomised in relation to the long axis of the hypha (Aist and Wilson, 1968). This means that the mitotic division spindle does not have a preferred orientation. In the first stage of anaphase, chromatids move to the poles of the spindle, and in the second stage of anaphase the two poles of the division spindle move further apart. Daughter chromatids do not separate from one another synchronously and lagging chromosomes have often been reported (Kameswar *et al.*, 1985; Lü and McLaughlin, 1995). The final stage of mitosis, telophase, follows one of three patterns: (i) median constriction, separating the entire nucleoplasm into the two daughter nuclei; (ii) a double constriction which incorporates only a portion of the parental nucleoplasms into each daughter nucleus, the rest being discarded and degraded; and (iii) formation of new daughter nuclear membranes, separate from the parental one, enclosing the chromosomes and a small portion of the nucleoplasm into the daughter nuclei while the bulk of the parental nucleoplasm and its membrane are discarded and degraded (Heath, 1994). In multinucleate hyphae, mitotic divisions are usually asynchronous, but in some fungal tissues, such as mushroom stems, nuclear division is rapid and the degree of synchrony appears to be increased (Doonan, 1992; Chiu, 1993). However, the higher fungi do seem to have a looser connection between cell differentiation and nuclear number and ploidy than is usual in plants and animals. In the cultivated

Table 5.2. *Numbers of nuclei in the cells of the stems of some mushrooms. Data from Chiu (1996).*

Fungus	Range
Agaricus bisporus	4–32
Armillaria mellea	1
Coprinus cinereus	2–156
Flammulina velutipes	2–32
Lentinula edodes	2–8
Pholiota nameko	2–4
Volvariella bombycina	5–30

mushroom, *Agaricus bisporus*, cells of vegetative mycelial hyphae have 6–20 nuclei (Colson, 1935; Kligman, 1943) and those in the mushroom fruit body have an average of six nuclei (Evans, 1959), though stem cells can range up to 32 (Murakami and Takemaru, 1980). In *Coprinus cinereus*, which has a strictly regular vegetative dikaryon, cells of the fruit body stem can become multinucleate, with up to 156 nuclei, by a series of consecutive conjugate divisions (Stephenson and Gooday, 1984; Gooday, 1985), a peculiarity exhibited by other agarics (Table 5.2). *Armillaria* species have diploid tissues in the fruit body (see below), so ploidy level and nuclear number are both variable, and evidently controlled by factors other than those imposed by the need to assemble multicellular structures.

5.1.2 Meiosis

Meiosis in heterothallic fungi progresses through stages fairly typical for eukaryotic haploids. The major round of DNA replication is premeiotic, occurring before karyogamy. Indeed, it was research with the ascomycete *Neotiella* which first demonstrated this aspect of meiosis (Westergaard and von Wettstein, 1970). Meiosis takes place in the meiocytes (basidia or asci) and chromosome behaviour in meiosis follows the Mendelian laws of segregation, independent assortment, linkage and crossing over. Detailed description of meiotic nuclear behaviour can be found in the literature (Raju and Lu, 1973; Wells, 1977; Lu, 1993, 1996; Pukkila, 1994).

Particular landmarks in the meiocyte development pathway are: commitment to recombination, commitment to the first and second meiotic divisions (haploidisation), commitment to spore formation and commitment to spore maturation (Lu, 1982; McLaughlin, 1982; Ross and Margalith, 1987; Chiu and Moore, 1990c; Padmore *et al.*, 1991; Hawley and Arbel, 1993; Klein, 1994; Pukkila, 1994).

Meiosis is traditionally understood to comprise synapsis, recombination and segregation (Hawley and Arbel, 1993; Moens, 1994). Premeiotic DNA replication and formation of the synaptonemal complex are independent events (Kanda *et al.*, 1989a, 1990; Pukkila, 1994). Importantly, chromosome pairing and synapsis are distinct from one another in terms of both mechanism and timing (Padmore *et al.*, 1991). DNA homology is not required for synapsis (Hawley and Arbel, 1993; Pukkila, 1994). At prophase I, homologous chromosomes align prior to appearance of synaptonemal complex, and at this time DNA strand breaks (providing sites for meiotic recombination) usually occur. In meiotic recombination, heteroduplex DNA (hDNA), which refers to the hybrid DNA formed following strand exchange and with one or more mismatched base pairs, is an essential intermediate. DNA–DNA recombination events are not sufficient to ensure disjunction (Hawley and Arbel, 1993; Pukkila, 1994), but chiasmata serve the vital function of balancing the spindle forces on the kinetochores of homologous chromosomes to ensure their proper co-orientation at metaphase I (Hawley and Arbel, 1993; Moens, 1994). The frequency of homologous recombination is 100–1000-fold higher during meiosis than during mitosis. As in other eukaryotes, meiosis I is a reductional division and meiosis II an equational division. The meiotic II division shares a number of functions with mitosis. Unlike other eukaryotes, the nuclear membrane remains intact through prophase I in fungi.

Fungal chromosomes are usually very small but karyotypes can be studied cytologically using special staining (Lu, 1993) and *in situ* hybridisation techniques (Scherthan *et al.*, 1992). Techniques are now available for resolving most fungal karyotypes by pulsed field gel electrophoresis (PFGE). Subsequent southern hybridisation with homologous probes can establish the ploidy levels of different isolates (Barton and Gull, 1992; Büttner *et al.*, 1994); gene amplification associated with differentiation (Pukkila and Skrzynia, 1993); loss of supernumerary chromosomes (usually less than 1 Mb in size and dispensable); and generation of chromosome-length polymorphisms (Zolan *et al.*, 1994; Carvalho *et al.*, 1995). Unlike higher plants, polyploidy is not common in higher fungi. Instead, fungi seem to be diverse in terms of number of nuclei per cell (Table 5.2) and chromosome length polymorphisms are widespread in both sexual and asexual species. This implies a general genome plasticity which is tolerant of wide variations in DNA contents per nucleus even to the extent that within the same species, different strains may have very different genome sizes (Tooley and Carras, 1992; Martin, 1995). The fungal genome also tolerates variation in the lengths of tandemly repeated sequences, for example, repeats of rRNA genes, and dispensable chromosomes and dispensable chromosome regions also occur. Many of these karyotype variations seem to be genetically neutral, but this sort of genomic variation may be advantageous by providing additional sequences in which mutations which might allow adaptation to new environments can occur with minimum risk to the basic genotype (Zolan, 1995).

In contrast to this natural tolerance of genome plasticity, artificially synthesised polyploids seem to suffer defects. Triploid and tetraploid fruit bodies of *Coprinus cinereus* produced various abnormalities, including prolonged meiotic division or aborted meiosis and formation of many abnormal and inviable spores (Murakami and Takemaru, 1984). In triploid *Saccharomyces cerevisiae*, triple-synapsis was common whereas synapsed bivalents instead of multiple synapsis was the rule in tetraploid yeast strains (Loidl, 1995).

5.1.3 Post-meiotic events

Many ascomycetes carry out one postmeiotic mitosis leading to the production of eight ascospores per ascus. Ascospores are endospores, being formed within the ascus which is the remains of the meiocyte (Fig. 5.1). Ascospore formation occurs by infolding of membranes around the daughter nuclei to delimit eight spores, each of which then forms a spore wall. Ascus cytoplasm left outside the spores, called the epiplasm, may provide nutrients to the maturing spores, contribute to the outer layers of the spore wall, or contribute to the osmotic potential of the ascus to aid subsequent spore discharge. Ascus and ascospore morphogenesis has been reviewed by Read and Beckett (1996). Ascomycetes such as *Ascobolus immersus*, *Neurospora crassa* and *Sordaria fimicola*, produce ordered tetrads (in which the position of the meiotic products reveals their origin as first or second division segregation) whereas the remaining ascomycetes and the basidiomycetes produce unordered tetrads.

Basidiospores are exospores, being formed as outgrowths of the meiocyte (basidium). There are usually four basidiospores, produced externally on outgrowths of the basidial wall which are called sterigmata (Fig. 5.2). The walls of the sterigma and of the early spore initial are contin-

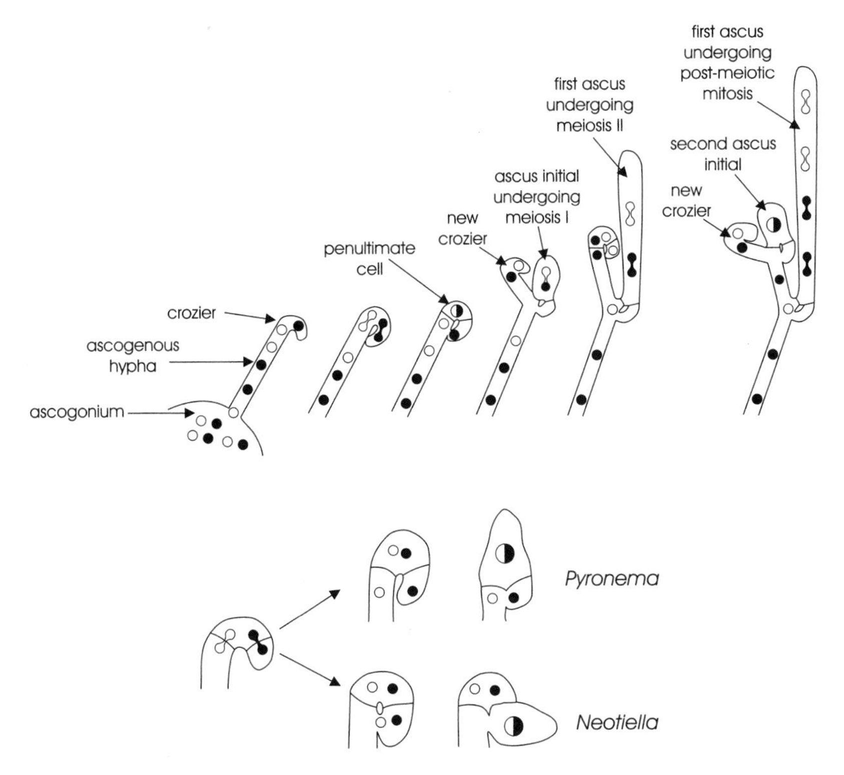

Figure 5.1 Diagram of ascus formation (adapted and redrawn from Webster, 1980). Hyphal fusion or similar mating between male and female structures results in nuclei moving from the male into the female to form an ascogonium in which male and female nuclei may pair but do not fuse (dikaryon). Ascogenous hyphae grow from the ascogonium. Most cells in these hyphae are dikaryotic, containing one maternal and one paternal nucleus, the pairs of nuclei undergoing conjugate divisions as the hypha extends. In typical development, the ascogenous hypha bends over to form a crozier. The two nuclei in the hooked cell undergo conjugate mitosis and then two septa are formed, creating three cells. The cell at the bend of the crozier is binucleate but the other two cells are uninucleate. The binucleate cell becomes the ascus mother cell, in which karyogamy takes place. In the young ascus meiosis results in four haploid daughter nuclei, each of which divides by mitosis to form the eight ascospore nuclei. Karyogamy may occur in the penultimate cell of the crozier (top diagram and *Pyronema* in the lower diagram) or the terminal or stalk cell (*Neotiella* in the lower diagram, which is adapted and redrawn from Read and Beckett, 1996).

uous and homologous. Starting with the spherical growth of the apex of the sterigma the spore initial grows further to attain the species-specific form and dimension. Nuclei migrate into the maturing spore and protection for the spore is increased by formation of further wall layers which might be pigmented and ornamented. There are various patterns of post-

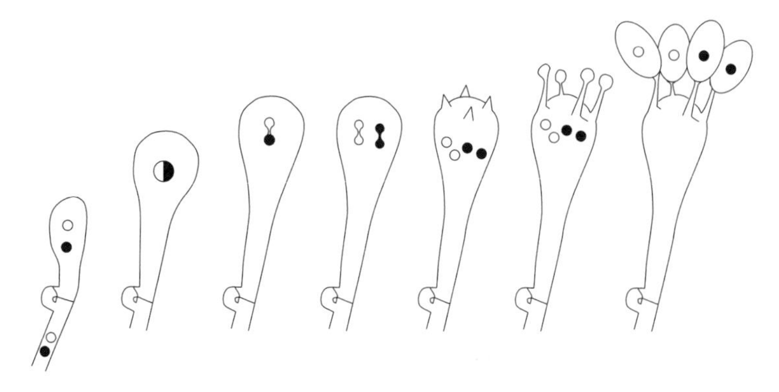

Figure 5.2 Diagram of basidium formation. In a 'classic' homobasidiomycete the basidium arises as the terminal cell of a dikaryotic hyphal branch which inflates and undergoes karyogamy and meiosis. At the conclusion of the meiotic division four outgrowths (sterigmata) emerge from the basidial apex and inflation of each sterigma tip produces the basidiospore (an exospore, produced outside the meiocyte in contrast to the endospores of ascomycetes). Nuclei then migrate from the basidium into the newly formed basidiospores. Mitosis may take place within the basidiospores before they are discharged. Comparison of this diagram with Fig. 5.1 will indicate readily how tempting it is to suggest some evolutionary relationship between crozier formation and the early stages of basidium and clamp connection.

meiotic nuclear behaviour in basidiomycetes (Chiu, 1996). The four meiotic products may migrate into spores without further division, or a post-meiotic mitosis may take place, either within the basidium prior to nuclear migration into the spores, or in the spores after migration of the meiotic daughter nuclei (Evans, 1959; Kühner, 1977; Chiu and Moore, 1993; Hibbett *et al.*, 1994a; Jacobson and Miller, 1994; Petersen, 1995).

Spore formation processes require precise nuclear positioning and/or active nuclear movement. Various cytoskeletal structures have been associated with nuclear division, nuclear migration and cytokinesis (McKerracher and Heath, 1986; Osmani *et al.*, 1990; Clutterbuck, 1994). Kamada and colleagues (1989a,b and 1993) induced and isolated benomyl-resistant (*ben*) mutants which were defective in structural genes for α- or β-tubulin in *Coprinus cinereus*. The *ben* mutants were blocked in migration of nuclei during formation of dikaryotic hyphae with clamp cells, but migration of nuclei into developing spores was unaffected. This may indicate that there are at least two nuclear movement systems, only one being dependent on tubulin. An important functional difference may be that microtubules are involved in the pairing of the two conjugate nuclei in the dikaryon (Kamada *et al.*, 1993).

The spindle pole body (the fungal equivalent of the centrosome of other eukaryotes; SPB in abbreviation) seems to be a key player in nuclear positioning and migration. The SPB goes through a characteristic duplication cycle in meiosis and mitosis. Analysis of mutants indicates that structural modification of the SPB in meiosis II is required for the endospores of *Saccharomyces cerevisiae* and *Schizosaccharomyces pombe* to be nucleated properly (Esposito and Klapholz, 1981; Hirata and Schimoda, 1994). In homobasidiomycetes, the SPB always leads the daughter nucleus migrating into the maturing basidiospore.

Nuclear migration after meiosis influences progeny genotypes and phenotypes when fewer spores are formed than there are nuclei to accommodate. Details of such events in several ascomycetes have been illustrated by Raju and Perkins (1994), but the most important instance of this phenomenon, in commercial terms, is in *Agaricus bisporus* which usually forms only two spores on the basidium, each of which receives two of the meiotic daughter nuclei. As the diploid nucleus prior to meiosis was heterozygous at the mating type factor, there are two progeny nuclei of each mating type in the basidium prior to spore formation. Consequently, three different genotypes are possible in the spores according to which nuclei migrate together: ($A + A$) and ($a + a$), both of which would germinate to produce sterile homokaryons, and ($A + a$) which would germinate into a fertile heterokaryon. This phenomenon is called secondary homothallism, because a single progeny spore gives rise to a fertile mycelium (see below). Elliott and Challen (1983) used a mathematical approach which assumed random segregation of nuclei to deduce a segregation ratio of 2:1 for heterokaryotic to homokaryotic progeny in any such secondarily homothallic species which packs two compatible nuclei into the same spore. Yet in isolates of *Agaricus bisporus* collected from nature which actually bore 2-, 3- and 4-spored basidia (but at different frequencies), a significant deviation from the predicted ratio was observed (Kerrigan *et al.*, 1993).

Among the possible explanations for this observation is that nuclear segregation may not be random; the migration mechanism may be able to sort nuclei as well as transport them and may be influenced by environmental conditions. This possibility is important for morphogenesis. Nuclear migration into the basidiospore is just a specialised instance of nuclear migration in general. If the process can be selective for nuclear genotype then it raises the further possibility that exercise of this sort of discrimination during formation of tissues by heterokaryons could give rise to segregation of nuclei of specific genotype into particular regions of the tissue.

5.2 Sexuality in fungi

Unlike most animals and plants, after mating a persistent and independent heterokaryotic or diploid phase exists in fungi. The sexual cycle proper is initiated only under particular environmental circumstances. The heterokaryon is capable of indeterminate growth and, with very few exceptions, is capable of producing asexual spores which are extremely effective in dispersing the organism. These are usually produced in such very large numbers that even small quantities of substrate might be expected to produce a sufficient number for mutation alone to provide the variation on which selection might operate. If this is (or can be) the case, then there most be some specific reason(s) why so many fungi invest more resources into a more complex sexual reproduction phase. Although there are fungi (the Deuteromycotina) which reproduce asexually only, the majority of fungi do have a sexual cycle, yet sex must have selective advantage, otherwise sexual stages will be replaced by asexual ones entirely (Maynard Smith, 1978).

The main point of contrast between asexual and sexual reproduction is that the latter brings together nuclei derived from different individuals, whereas in asexual reproduction, only the genetic constitution of the one individual is multiplied by mitotic nuclear divisions. When the individuals taking part in sexual reproduction differ in genotype the fusion nucleus is heterozygous and the products of the meiotic division (i.e. the progeny if the organism is haploid, or the gametes if the organism is diploid) can have recombinant genotypes. New combinations of characters are created in a single sexual cycle and appear in the next generation to be exposed to selection. Promotion of genetic variability through out-crossing is the advantage which is most usually associated with sex, in the expectation that variability is needed for the species to evolve to deal with competitors and environmental changes. A great deal of evidence shows that out-crossing promotes variability, and that asexual lineages change little with time. This sort of evidence seems to support the view that variability in the population enables the organism to survive evolutionary challenges. Sex may be an important means to enhance the overall rate of adaptation, therefore (Hurst and Peck, 1996).

This, though, is a group selectionist interpretation (the argument maintains that it is the group or population to which the individual belongs which benefits from the variation generated in an individual meiosis). Current theory emphasises that selection acts on individuals, so any feature supposed to confer a selective advantage must do so because it benefits either the individual itself or its immediate progeny

(Dawkins, 1976; Carlile, 1987). Bernstein *et al.* (1985) argued that repair of DNA damaged by mutation or faulty replication could be the crucial advantage of the meiotic sexual cycle. By bringing unrelated haploid nuclei together to form a diploid (out-crossing ensuring heterozygosis), the sexual cycle enables damage in one chromosome to be repaired by recombination with its homologue. Mutations can be recessive and damaging, and different ones will be found in different mitotically-generated cell lines. If recessive adverse mutations are masked by non-mutant alleles in the nuclei of the other parent, then the formation of the diploid (or heterokaryon or dikaryon) by out-crossing will itself benefit the mated individual. Out-crossing can also give rise to heterozygous advantage, where the heterozygote has selective advantage over either of its homozygous parents. This has been demonstrated in plants and animals and has also been demonstrated in the yeast *Saccharomyces cerevisiae*.

A more radical view is that the sexual cycle of eukaryotes arose by infection of early cells by genome parasites. The argument (Bell, 1993) draws parallels with bacterial conjugative plasmids which ensure their own spread by directing gene transfer between different cells. Though this is primarily a horizontal transfer aimed at spreading the agent through a population belonging to the current generation, the originally infectious elements involved are thought to have become 'domesticated' into vertical transmission when sex of the host cell became associated with reproduction. This argument seeks to explain the distant origin of the sexual cycle, but once the supposed infective sexual element is domesticated the maintenance of the sexual cycle in its original host (which is now, of course, a symbiotic partnership) will depend on the selective advantages to the host (as outlined above) becoming evident.

5.2.1 The mycelium as an individual

The essential first step in the sexual cycle is to bring together nuclei from different individuals. This occurs by conjugation (cell fusion) in the unicellular fungi, yeasts, and hyphal anastomosis (fusion between hyphae) in filamentous fungi. To achieve nuclear transfer the cytoplasms of the fusing hyphae must be brought into continuity with each other, a process which involves breakdown of two hyphal (cell) walls and union between two separate plasma membranes.

During normal mycelial growth vegetative hyphae usually avoid each other (negative autotropism) to promote exploration and exploitation of the available substrate. Anastomosis requires that hyphae grow towards each other (positive autotropism). How and why the usual avoidance reactions between hyphae are reversed are unknown. Hyphal fusions are common even within the same thallus system because, as the mycelium matures, fusions convert the initially radiating system of hyphae into a fully interconnected (and three-dimensional) network. This enables transport of nutrient and signalling molecules anywhere in the colony. Thus, as part of normal mycelial development, most fungi are fully equipped with the machinery necessary for hyphal tips to target onto other hyphae (Fig. 5.3) and to produce and externalise the enzymes needed for anastomosis. The problem, therefore, is not how to promote hyphal fusion, the mechanism is there already. Rather, it is how to regulate hyphal anastomosis so that its genetic advantages can be realised without hazard. If genetically different hyphae are to interact to take advantage of the sexual cycle, for example, nuclear and cytoplasmic control requirements are very different.

Maximum selective advantage is gained from sexual reproduction when the nuclei of the two parents are genetically as different as possible. In contrast, the security of the cell requires that if cytoplasms are to mingle then they must be as similar as possible. The hazards at the cytoplasmic level are that hyphal anastomosis carries the risk of exposure to infection with alien genetic information arising from defective or harmful cell organelles, viruses or plasmids (Nauta and Hoekstra, 1994). Protection against alien DNA is provided by one to several nuclear genes which limit completion of hyphal anastomosis between colonies to those which carry the same alleles of those genes and therefore belong to the same 'genotypic family'. This is vegetative compatibility. It is a self/non-self recognition system. Mycelia which possess the same vegetative compatibility genes (or alleles) are said to belong to the same vegetative compatibility group (v-c group). When the colony margins of two fungal isolates from nature grow to meet each other the leading hyphae may mingle without interacting, or hyphal anastomoses may occur between their branches. If the colonies involved are not compatible the cells immediately involved in anastomosis are killed. Vegetative compatibility (also called vegetative, somatic or heterokaryon incompatibility) prevents formation of heterokaryons except between strains which are sufficiently closely related to belong to the same v-c group. Other compatibility factors (the mating type factors, see below) will then determine the ability of nuclei which are brought together to undergo karyogamy and meiosis to complete the sexual cycle.

Slime moulds have a fusion incompatibility system which determines the ability to fuse. But the type of vegetative compatibility most com-

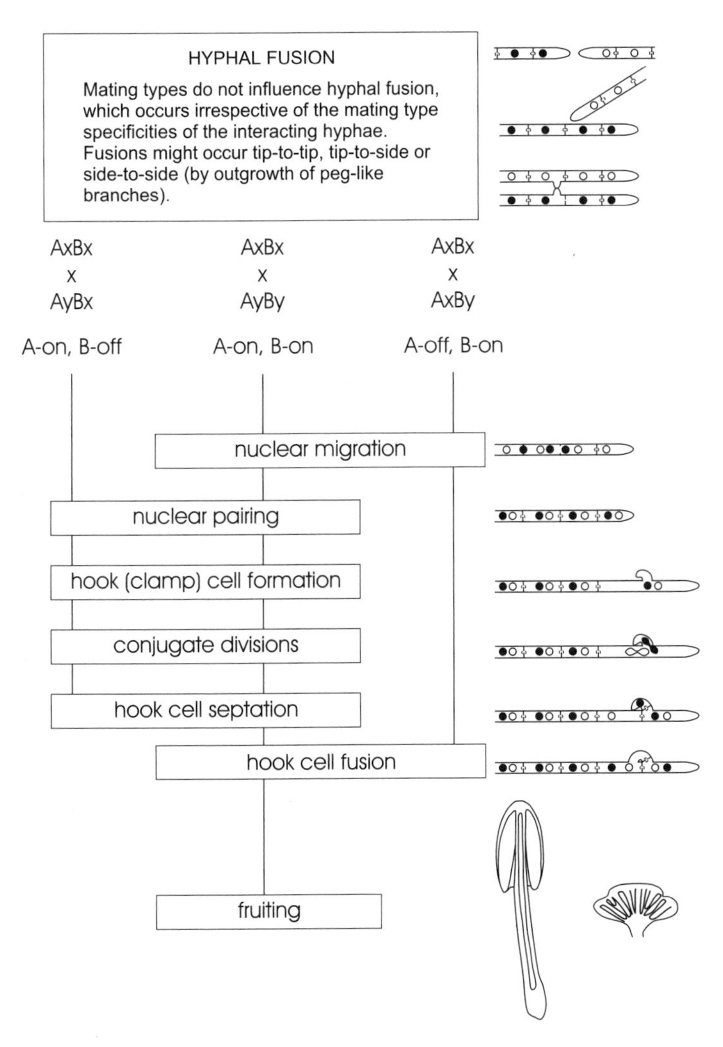

Figure 5.3 Flow chart diagram of *A* and *B* mating type factor activity in the homobasidiomycetes *Coprinus cinereus* and *Schizophyllum commune*. See text for details. From Chiu and Moore (1998).

monly encountered in fungi is a post-fusion mechanism. Consequently, hyphal anastomosis (fusion of hyphae or branches) can be described as being promiscuous, it occurs frequently and without preference with regard to partner. However, compatibility of the cytoplasms which fusion brings together then determines whether cytoplasmic exchange will progress beyond the initial interaction. The v-c system kills hyphal compartments involved in incompatible anastomoses. This strategy prevents transfer of cytoplasmic components and organelles between more than a few hyphal compartments if incompatible strains meet, and those parts of the hyphae are then sacrificed. If the incompatibility reaction is slow, a virus or plasmid may be communicated to adjacent undamaged cells before the anastomosing compartments are killed.

Vegetative compatibility is a genetically defined differentiation process which leads to all the hyphae of a mycelium being genetically and physiologically distinct from the hyphae of another mycelium which belongs to a different v-c group. This genetic differentiation is not usually accompanied by any difference in morphology. Membership of a v-c group cannot usually be established from the hyphal structure. Like immune reactions in animals, v-c group typing can only be done by challenging one strain with another and assessing their joint reaction.

Compatibility reactions (vegetative and sexual together) define the fungal individual. In genetical and evolutionary terms a population consists of individuals which are able to interbreed. Individuals are important in evolution because selection operates on individuals. Populations are important in systematics because the species, which is the fundamental unit of biological classification, is conventionally defined in terms of mating success and production of viable offspring. This is a genospecies, or biological species, which is delimited by mating tests between fungal isolates collected over a wide geographical area. This obviously applies to fungi which have a sexual process but in Deuteromycotina, which lack sexual reproduction, the concept of a genospecies is meaningless. Rather, individuals can be delimited as a 'taxospecies' on the basis of their similarity in as many phenotypic characteristics as can be examined. Molecular methods of analysis and computer aided numerical analysis have become increasingly important in recent years.

Successful delimitation of a new species leads to provision of a Latin binomial as its specific name in accordance with the International Code of Botanical Nomenclature (the nomenspecies). Genospecies, taxospecies and nomenspecies do not always coincide because selection in different geographical areas may cause local populations to adapt into distinct taxospecies which, despite their differences, can interbreed when combined artificially (i.e. different taxospecies can be one genospecies). In contrast, mating barriers may arise, separating populations which are still sufficiently similar to remain in one taxospecies into different genospecies.

When the mycelium is considered to be the fungal individual it is important to recognise that interactions are not limited to those between homokaryons. Heterokaryons interact and their relationships are also governed by their compatibility systems. Mating in the Basidiomycotina can occur between a dikaryon and a monokaryon, the former contributing nuclei to the latter to form a new dikaryon (Buller, 1930, 1931, 1941). Natural populations of heterothallic basidiomycetes are composed of vegetatively incompatible, genetically and physiologically distinct individual heterokaryotic (secondary) mycelia (Rayner and Todd, 1979; Todd and Rayner, 1980) called genets (Rayner and Boddy, 1988). The nature of the mycelial growth form enables mycelia to alter their organisational pattern by varying the distribution of their populations of nuclei and mitochondria and, perhaps, other organelles (Rayner *et al.*, 1995b). Ramsdale and Rayner (1994) found that ratios of nuclear genotypes in conidia from heterokaryotic strains of *Heterobasidion annosum* depended on the closeness of the relationship between the homokaryons used to make the heterokaryons and on the growth rates of the parental homokaryons. Ramsdale and Rayner (1996) compared strains of *H. annosum* from Britain and Northern European populations and found considerable differences between allopatric matings (allopatric describes a group of similar organisms that could interbreed but do not because they are geographically separated) and sympatric matings (sympatric describes a group that, although in close proximity and theoretically capable of interbreeding, do not interbreed because of differences in behaviour). Asymmetry in nuclear ratios as high as 9:1 in favour of invasive nuclei were observed, though the heterokaryons had phenotypes characteristic of resident nuclei. Such findings indicate that phenotype-genotype relationships diverge greatly on geographical separation. Maintenance of a common phenotype as genotypes diverge is presumably another aspect of the genome plasticity referred to in section 5.1.2.

Under a stable environment, haploidy and diploidy show equal selective advantage (Jenkins, 1993). A prolonged dikaryotic or heterokaryotic stage, however, may show hybrid vigour and can segregate into homokaryons if adverse selection pressure is imposed. Vegetative compatibility systems maintain the individuality of a mycelium and enable it to compete with other mycelia of the same species for territory and resources. The territory conquered by an individual mycelium may be minute, perhaps a small fraction of a rotting twig, or as enormous as an entire forest. A mycelium of *Armillaria bulbosa* has been growing in a Canadian forest for an estimated 1500 years, ranking as probably the largest and oldest living thing on Earth (Smith *et al.*, 1992).

When individuals do exchange nuclei it is the mating systems which then regulate sexual exchange between the vegetatively compatible mycelia.

5.2.2 *Mating control systems*

Most fungi have polyphasic life cycles, passing through successive phases as haploid (N), dikaryotic (N + N) (or heterokaryotic with the two nuclear types in random ratio) and diploid (2N), and exhibit diverse forms of sexuality to achieve the transition from (N) to (N + N). Higher fungi are commonly heterothallic: haploids are self-sterile but cross-fertile with a compatible haploid. True (primary) homothallism is also encountered and is exemplified by *Aspergillus nidulans* in which a haploid conidium can give rise to hyphae which are able to enter the sexual reproduction pathway. Secondary homothallism, resulting from formation of heterokaryotic spores as in *Agaricus bisporus*, was referred to above. Homothallism is not a concern at the moment. The purpose of this section is to describe the mating type genes which are involved in regulating outcrossing and sexual development. I do not intend to attempt a detailed review of mating type genes; rather, I will concentrate on features which seem to have direct relevance to morphogenetic processes, which are the structures of the mating type DNA sequences and their ability to produce specific growth factors and regulate morphogenesis. For further details see Chiu and Moore (1998) and other references in this section, and for discussion of the evolution of different reproductive systems in filamentous ascomycetes see Nauta and Hoekstra (1992a,b).

Many fungi have two mating types; examples are *Neurospora crassa*, the brewer's yeast *Saccharomyces cerevisiae*, and the grass rust *Puccinia graminis*. In these cases the 'mating type' of a culture depends on which allele it possesses at a single mating type locus (this is unifactorial incompatibility) and successful mating can only take place between (yeast) cells or mycelia that have different alleles at the mating type locus (Nelson, 1996). This simple statement immediately brings out the first peculiarity in mating types, for although the different mating type loci occupy equivalent positions on homologous chromosomes, they are *not* alleles. For one thing the chromosomal regions which control mating type contain several coding sequences so should not be called 'genes'; the term mating type factor is preferred. More importantly, Metzenberg (1990) pointed out that the term allele implies that different forms of a gene are very similar and encode equivalent proteins; this is not true for mating type genes of the ascomycetes where the alternative mating type loci contain dissimilar DNA sequences and encode unrelated proteins; the term idiomorph was introduced to denote dissimilar mating type loci by analogy with the major histocompatibility complexes in humans (Metzenberg and Glass, 1990).

5.2.2.1 *Mating type factors of S. cerevisiae*

The yeast life cycle features an alternation of a haploid phase with a true diploid phase, in this respect differing from filamentous ascomycetes in which the growth phase after anastomosis is a heterokaryon. Haploid yeast cells have one of two mating types which are symbolised a and α. Karyogamy (nuclear fusion) follows fusion of cells of opposite mating type and the first bud after these events contains a diploid nucleus. Because haploid meiotic products mate soon after meiosis while they are in close proximity, most natural cultures of yeast are diploid. The diploid state is maintained by mitosis and budding until specific environmental conditions induce sporulation. These requirements are: a deficiency in nitrogen and carbohydrate; good aeration; and acetate or other carbon sources which favour the glyoxylate shunt. The entire cell is then converted into an ascus in which meiosis occurs and haploid ascospores are produced. Ascospore germination re-establishes the haploid phase, which is itself maintained by mitosis and budding.

Mating phenotypes in *S. cerevisiae* are controlled by a complex genetic locus called *MAT* where two linked genes are harboured (a1, a2 for mating type a and α1, α2 for mating type α). The *MAT*α locus encodes the divergently transcribed α1 and α2 polypeptides (Fig. 5.4), and *MAT*α encodes polypeptides α1 and α2. The a2 polypeptide is a repressor of transcription of α-factor (in α-cells), whilst a1 represses a-specific genes in a-cells. The α1 protein activates transcription of genes coding for α-pheromone and α-factor surface receptor. In a/α diploids, a1 and α2 polypeptides form a heterodimer which represses genes specific for the haploid phases. Heterozygosity at *MAT* signals diploidy and eligibility to sporulate (i.e. even partial diploids carrying *MAT*a/*MAT*α will attempt to sporulate).

Although *S. cerevisiae* is heterothallic, a clone of haploid cells of the same mating type frequently sporulates, producing both a and α progeny. This results from mating type switching controlled by the gene *HO* (HOmothallic). Rare occurrence of mating type switching (about once in 10^5 divisions) is the phenotype of allele *ho*, whereas in strains carrying *HO* the switch occurs at every cell division. On either side of the *MAT* locus (and on the same chromosome) there are silent ('storage') loci, one for each mating type, *HML* and *HMR* (Fig. 5.5; see Schmidt and Gutz,

Saccharomyces cerevisiae

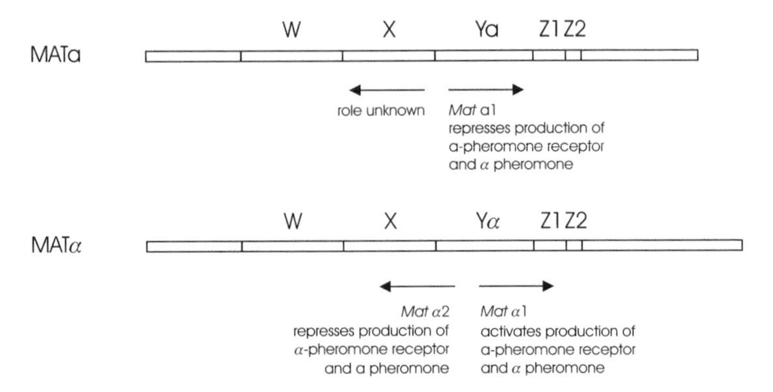

In a/α diploids, the MATa1/MATα2 heterodimer protein activates meiotic and sporulation functions, and represses haploid functions (turning off α-specific functions by repressing MATα1, a-specific functions being repressed by MATα2 alone

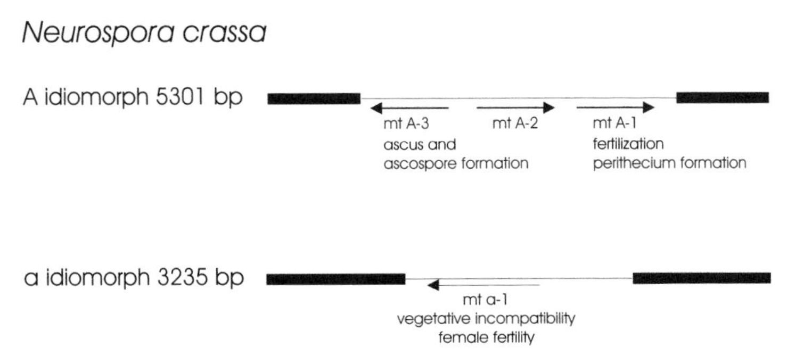

Figure 5.4 Functional regions of mating type factors in *Saccharomyces cerevisiae* (top) and *Neurospora crassa* (bottom). In both panels the arrows indicate direction of transcription and the legends beneath the arrows indicate functions of the gene products. In *S. cerevisiae* of mating type a, a general transcription activator is responsible for production of a-pheromone and the membrane-bound α-pheromone receptor. In the *N. crassa* illustration the black bars represent the conserved DNA sequences which flank the idiomorphs, the latter shown as lines. These diagrams are oriented so that the centromere is on the left, consequently the centromere-distal sequence is on the right. From Chiu and Moore (1998).

1994). *HO* encodes an endonuclease which creates a double-strand break at locus *MAT*, and switching involves replacement of information at the *MAT* locus by that at either *HML* or *HMR* by an intrachromosomal recombination event.

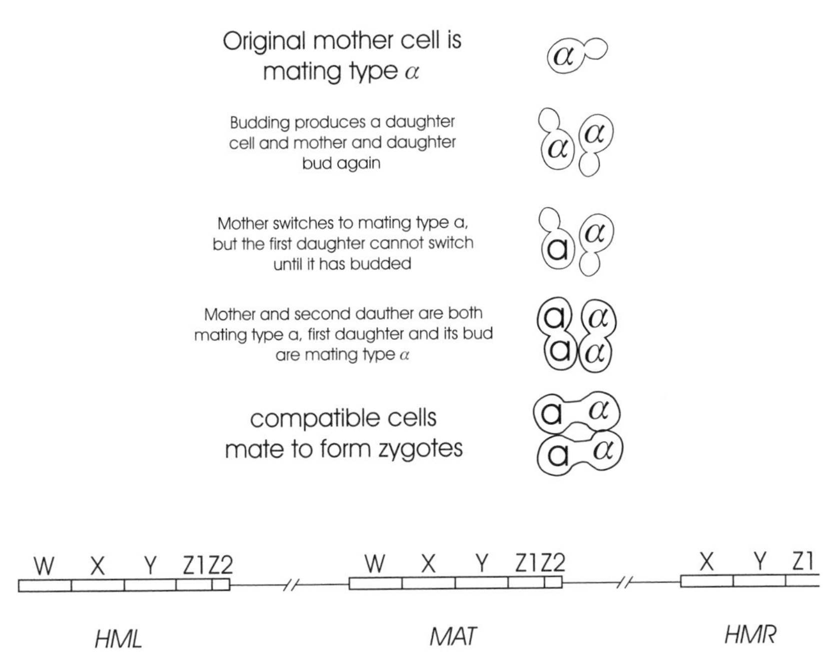

Figure 5.5 Top: pattern of mating type switching in *Saccharomyces cerevisiae* showing the consequences of a mating type switch in one mother cell. Bottom: the three loci involved in mating type switching, HML, MAT and HMR, are located on the same chromosome. From Chiu and Moore (1998).

5.2.2.2 Mating type factors in Neurospora crassa

Species of *Neurospora* exhibit at least four markedly contrasting mating strategies but I only wish to consider the bipolar heterothallism (with mating types *A* and *a*) of *N. crassa*. Unlike *S. cerevisiae*, the mating type genes are present in a single copy per genome, and *A* and *a* do not share homology (hence the name idiomorph). Unusually, in *N. crassa* the mating type locus is one of ten heterokaryon incompatibility (*het*) loci which are active during vegetative growth, preventing the formation of viable *a* + *A* heterokaryons. Yet this mating type-controlled heterokaryon incompatibility is suppressible by an unlinked wild type suppressor gene called tol^{+} ('tolerant'). Under nitrogen starvation, strains of either mating type develop female structures (protoperithecia and trichogynes). Mycelia and asexual spores (macroconidia or microconidia) of the strain of the opposite mating type can serve as the male in a sexual cross and secrete mating type specific pheromones which orient trichogynes to cells of the opposite mating type.

The *A* idiomorph is 5301 bp long and generates at least three transcripts (A-1, A-2 and A-3), with the former two transcribed in the same direction (Fig. 5.4). Saupe *et al.* (1996) found that 85 amino acids at the N-terminal end of the *A* product were sufficient for female fertility. A region from position 1 to 111 conferred a vegetative incompatibility function, while amino acids from position 1 to 227 were required for male-mating activity. The mating type-specific mRNA was expressed constitutively in vegetative cultures, and both before and after fertilisation. Transcript *A*-1, which shows a high degree of similarity to *MAT*α1 of *S. cerevisiae,* is essential for fertilisation and fruiting body formation and the other two transcripts are involved in post-fertilisation functions, ascus and ascospore formation. The *A*-3 transcript has DNA binding ability, indicating its potential function as a transcriptional factor and shares 50% similarity to the mating type polypeptide (*Mc*) of the fission yeast *Schizosaccharomyces pombe.*

Bases 2923–4596 of the *a* idiomorph give rise to a single *a*-1 transcript which encodes a single polypeptide (288 amino acids) belonging to high-mobility group proteins with DNA-binding activity. This is the domain for mating function whereas amino acids 216–220 of mt *a*-1 act in vegetative incompatibility (Philley and Staben, 1994). The separation of vegetative incompatibility from both mating and DNA binding indicates that vegetative incompatibility functions by a biochemically distinct mechanism. All *a* idiomorphic DNA sequences between 1409 and 2923 are non-essential (Chang and Staben, 1994). Although products of both A and a idiomorphs have DNA binding features characteristic of transcription factors, the genes which they might regulate are not known. However, Staben (1995) has described and illustrated a model which assumes that during vegetative heterokaryon growth gene products induced by *a*-1 and *A*-1 interact in a pathway dependent on the *tol* gene to provide a vegetative incompatibility response. Under nitrogen limitation a new set of 'female-specific genes' is expressed which represses vegetative incompatibility and promotes formation of protoperithecia. Subsequent interaction of *a*-1 and *A*-1 (in a manner dependent on the *A-2* region of the *A* idiomorph) induces perithecium differentiation and ascosporogenesis.

5.2.2.3 Mating type factors in some basidiomycetes

Many basidiomycetes have two unlinked mating type factors (designated A and B), which is described as a bifactorial incompatibility system. In this case also, compatibility requires that two mycelia have different alleles, but in this instance both mating type factors must differ (Kothe, 1996).

The tetrapolar system of *Ustilago maydis* (cause of smut disease of maize) has one multiallelic mating type factor and one with only two alleles. Cell fusion (between yeast-like budding cells called sporidia) is controlled by the biallelic '*a*' mating-type locus which induces structures similar to conjugation tubes and initiates the transition between the yeast and filamentous growth forms. The *a*1 allele comprises 4.5 kb of DNA, and the *a*2 allele 8 kb. Two mating type specific genes have been identified in these regions (Fig. 5.6): *mfa*1 (in *a*1) and *mfa*2 (in *a*2) code for precursors of the farnesylated pheromones and *pra*1/*pra*2 which encode pheromone receptors.

The multiallelic '*b*' locus determines true hyphal growth form, pathogenicity and prevents diploid cells fusing. The *b* mating type factor in *U. maydis* contains two genes separated by a 260 bp spacer region. The genes are called *bE* and *bW* (East and West) and they are transcribed in opposite directions. They code for polypeptides of 473 and 629 amino acids, respectively, with a high degree of variation at the amino terminal end and highly conserved regions with similarities to the homeodomain or DNA-binding regions of known transcription factors at the carboxy-terminal ends. The bW and bE proteins are thought to form a heterodimer which might act as an activator of genes required for the sexual cycle and/or repressor of haploid-specific genes. Inherent in this model is the assumption that the complex comprised of bE and bW from the same allele is always inactive (perhaps the homeodomains are not properly exposed). Function requires that the proteins come from different alleles.

Mating type factors of the homobasidiomycetes *Coprinus cinereus* and *Schizophyllum commune* govern a major change in mycelial morphology and growth pattern as they convert the sterile parental homokaryons (also called monokaryons) into a fertile heterokaryon (specifically a dikaryon) (Fig. 5.3). In *C. cinereus*, for example, the homokaryon has uninucleate cells and produces abundant uninucleate arthrospores (oidia). The dikaryon has binucleate cells with characteristic clamp connections (see Fig. 5.3) at each septum. It does not produce oidia but under appropriate nutritional, temperature and illumination conditions it does produce the mushroom fruit bodies. Because of the clear phenotypic differences between homokaryons, heterokaryons and dikaryons these mating type factors have been subjects for classical genetic studies for many years.

C. cinereus and *S. commune* exhibit tetrapolar heterothallism determined by two mating type factors, called *A* and *B*, which are located on different chromosomes. The natural population contains many different mating types which behave in crosses as though they are multiple alleles at the two mating type loci. However, classical genetic analysis showed that

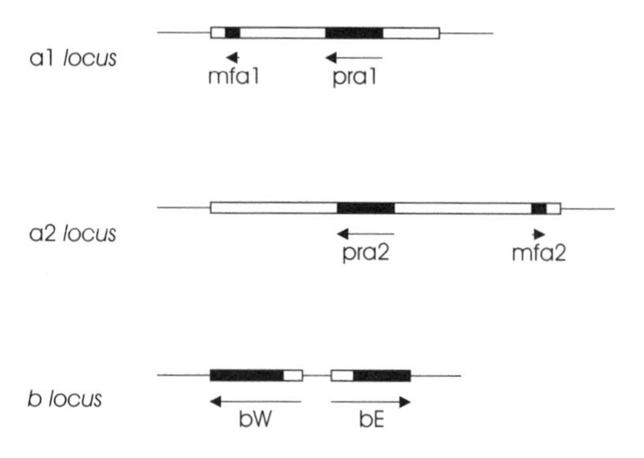

Figure 5.6 Schematic representations of the structures of the a and b mating type loci of *Ustilago maydis*. Alleles of the a locus consist of mating-type specific (i.e. variable) DNA sequences (4500 base pairs in a1, 8000 base pairs in a2), here shown as open boxes, within which are the genes for mating (*mfa* and *pra*). The b locus has two reading frames, bW and bE, which produce polypeptides containing domains of more than 90% sequence identity (shown as black boxes) and variable domains (open boxes) which show 60 to 90% identity. Arrows indicate the direction of transcription. From Chiu and Moore (1998).

each mating type factor is a complex genetic region comprising Aα, Aβ, Bα and Bβ subloci. The subloci also exhibit multiple allelism and recombination between subloci can generate new mating type specificities.

Compatible mating with formation of clamp connections and conjugate nuclei in the mated hyphae requires heterozygosity at both *A* and *B* (A-on, B-on). Mating type factor *A* controls nuclear pairing, clamp connection formation and synchronised (conjugate) mitosis whereas mating type locus *B* controls nuclear migration and clamp cell fusion (Fig. 5.3). Nuclei migrate through the existing mycelium after breakdown of the septa between adjacent cells. Heterokaryons may be formed in matings which are homozygous for one of the mating type factors. When *A* factors are the same (A-off, B-on; common-A heterokaryon), nuclear migration occurs but no clamp connections form. Mating of strains with the same *B* factor (A-on, B-off; common-B heterokaryon) forms a heterokaryon only where the mated monokaryons meet because nuclear migration is blocked. Terminal cells of common-B heterokaryotic hyphae initiate clamp connections and nuclei divide but the hook cell fails to fuse with the subterminal cell and its nucleus remains trapped.

Molecular analyses revealed that *A* mating type factors are composed of many more mating type genes/gene pairs than classical genetic analysis revealed (Fig. 5.7; Kües and Casselton, 1992; Casselton and Kües, 1994).

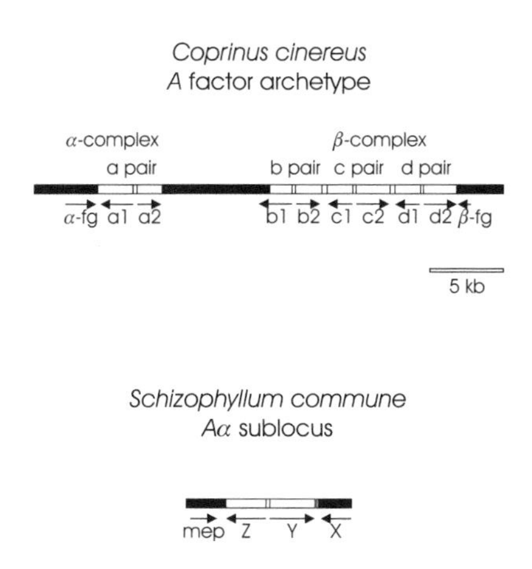

Figure 5.7 Schematic representations of the structures of parts of the A mating type factors in *Coprinus cinereus* and *Schizophyllum commune*. Arrows show the direction of transcription. The (predicted) archetypal A factor from *Coprinus cinereus* has four pairs of functionally redundant genes (a, b, c and d) which feature the homeodomain 1 (HD1 in a1, b1, c1 and d1) and homeodomain 2 (HD2 in a2, b2, c2 and d2) sequences. Interaction between HD1 and HD2 proteins is the basis of the compatible reaction. A factors examined in different strains of *C. cinereus* isolated from nature contain different combinations, and different numbers, of these genes. In *Schizophyllum commune* the mating type genes are called X and Y and carry HD1 and HD2 respectively. Again, different alleles are found in different natural mating types; indeed, the Z gene is absent in the Aα1 mating type. The sequences shown as mep and α-fg are homologous, encoding a metalloendopeptidase; X is a flanking gene of unknown function. From Chiu and Moore (1998).

The gene pairs encode two families of proteins which have homeodomain regions (HD1 and HD2). Homeodomain regions were first identified in relation to DNA regions of sequence similarity (the homeobox) in genes of the fruit fly *Drosophila melanogaster* which control position-specific differentiation of body regions. The proteins with homeodomains are transcription factors which bind to regulatory regions of eukaryotic genes and interact with RNA polymerase to modulate transcription of the gene or gene family which possess the regulatory sequence to which the transcription factor binds. Thus, individual transcription factors can integrate transcriptional control of numerous genes. Another aspect of the control, however, is provided by the presence or absence of particular transcription factor binding sites in the regulatory sequence of the gene. Genes which bind transcription factors active in all cell types will be expressed in all cells. On the other hand, if a gene binds factors produced only under certain conditions or only in certain cells, then that gene will be expressed in a condition-specific or cell-specific manner.

The homeodomain sequence does vary, but there is a highly conserved 12-base sequence which is characteristic of proven transcription factors from animals, fungi and plants. The presence of that sequence in the genes which make up the *A* mating type factors of *C. cinereus* strongly suggests that they may encode transcription factors. The N-terminal regions of these proteins are essential for choosing a compatible partner but not for regulating gene transcription. The compatible reaction required for sexual development is triggered by dimerisation between HD1 and HD2 proteins from the different *A* mating type factors of the compatible mating (Fig. 5.8; Asante-Owuss *et al.*, 1996). As was stressed above in relation to the similar model of heterodimerisation of bW and bE proteins produced by the *b* mating type factor in *U. maydis,* this interpretation demands that any complex comprised of HD1 and HD2 from the same mating type factor is always inactive. Function requires that the proteins come from different mating type factors. In the context of the regular dikaryon of these homobasidiomycetes, this can be restated as: function requires that the HD1 and HD2 proteins come from different nuclei.

Analysis of the *B* mating type factor is under way. The a subunit of the B factor of *Schizophyllum commune* encodes several pheromone and receptor genes (Wendland *et al.*, 1995) which might be involved in controlling the growth of the clamp connection. Ásgeirsdóttir *et al.* (1995) have demonstrated that *S. commune* hydrophobin genes show differential expression under the control of the mating-type factors. Hydrophobin SC3 is expressed in monokaryons and dikaryons, whilst hydrophobins SC1 and SC4 (and a gene, SC7, encoding a hydrophilic wall protein) are expressed in dikaryons but not monokaryons. None of these four genes was expressed in common-A heterokaryons, indicating that the *B*-on phenotype represses SC3 and that compatibility at both *A* and *B* mating type factors is required for activation of SC1, SC4 and SC7. Immunolabelling revealed that SC3p was produced by aerial dikaryotic hyphae but not by hyphae making up the fruit-body tissue. In contrast, SC7p was secreted into the extracellular matrix of the fruit body tissue, but aerial dikaryotic hyphae did not produce dikaryon-specific transcripts. The authors argue that this suggests that aerial dikaryotic hyphae do not express the *B*-on pathway. Noting that in aerial hyphae the two nuclei were further apart than in a typical dikaryon, they tentatively suggested that effective interaction of different mating type *B* genes in *S. commune* requires proximity of the two nuclei containing them. Disruption of the binucleate state may be a novel mechanism of position-dependent gene control during cell differentiation. Since

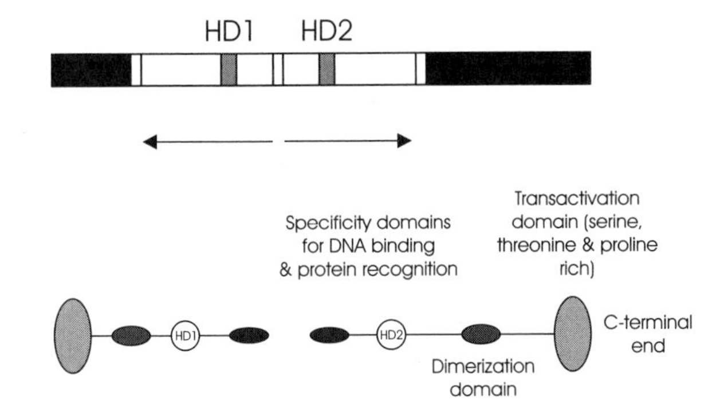

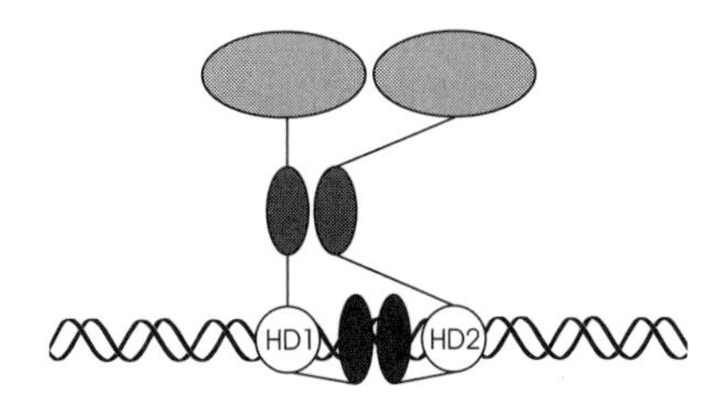

Figure 5.8 Schematic diagram showing a model of homeodomain protein production, structure and interactions involved in A mating type factor activity in *Coprinus*. From Chiu and Moore (1998), redrawn and adapted after Casselton and Kües (1994).

microtubules participate in the pairing and positioning of the two conjugate nuclei in dikaryotic cells in *Coprinus cinereus* (Kamada *et al.*, 1993; see section 2.3) there may be an organised cytoskeletal mechanism to regulate nuclear proximity for this purpose.

5.2.2.4 Overview

Mating type genes act as master regulator genes producing transcription factors which control, by activation and repression, the expression of several to many other genes (downstream regulation) which then impact on the sexual development pathway. They immediately affect the morphology, specific cell–cell interactions, morphogenesis and behaviour of the cells, opening new pathways of development. Features affected by mating type factors are summarised (and generalised from the different organisms studied) in Table 5.3.

Table 5.3. *Generalised summary of the main features of mating type gene function.*

Mating type genes are complex genetic regions with several reading frames
Mating type loci have definable positions on the chromomes but 'alleles' are actually idiomorphs with different sequences encoding different products
Mating type genes in yeast can switch position to change the mating phenotype of the cell
Mating type gene products are transcription factors
Mating type genes activate transcription
Mating type genes repress transcription
Mating type genes regulate production of pheromones and pheromone receptors which orchestrate highly specific cell-cell interactions
Mating type genes regulate cell morphology and growth pattern (including cell shape, extension growth rate, branching frequency and angle)

The mating type factors are sophisticated control elements in the sexual reproduction pathway in fungi, but they are more than that. Mating type factors, especially those of basidiomycetes, are the most complex genetic control devices of which we have any knowledge. They point the way towards possible strategies for genetic control of other morphogenetic processes. Suggesting how expression of families of genes contributing to a specific function might be regulated in a coordinated way by control sequences which generate transcription factors under a specific set of circumstances. Turning back to Table 5.3, the possibility exists that the phrase 'mating type' might be replaced by some other phrase representing another morphogenetic function. If the mating type factors are realistic paradigms for master control factors, then 'mating type' in Table 5.3 might be replaced by 'fruit body initiation', or by 'hymenium formation' or any of a range of other phrases referring to discrete developmental stages in fungal morphogenetic pathways.

5.3 Shape and form in yeasts and hyphae

Despite a generally smaller genome size in fungi (Table 5.1), only a small proportion of the genome is associated with any particular morphogenetic process. This is a general feature of development in eukaryotes. The emphasis in morphogenetic gene regulation is on differential expression of activity rather than on large scale replacement of one set of gene products by another. Fungal examples have been revealed in relation to the comparison between homokaryotic and dikaryotic phenotypes in *Schizophyllum* (Zantinge *et al.*, 1979; de Vries *et al.*, 1980; de Vries and Wessels, 1984; Wessels *et al.*, 1987), the transition from vegetative state to fruit body formation in *Coprinus* (Yashar and Pukkila, 1985; Pukkila and Casselton, 1991), perithecium formation in *Neurospora* (Nasrallah and Srb, 1973, 1977, 1978) and *Sordaria* (Broxholme *et al.*, 1991), sclerotium development in *Sclerotinia* (Russo *et al.*, 1982), and sporulation in *Saccharomyces*. In the yeast example, only 21–75 of the estimated 12,000 genes were found to be specific to meiosis and ascospore formation (Esposito *et al.*, 1972).

In contrast, Timberlake (1980) studied poly(A) RNA sequences accumulated in conidiating cultures of *Aspergillus nidulans*, comparing spores germinated for only 16 h (vegetative hyphae) and conidiating cultures grown for 40 h. He found that 11–18% of sequences occurring in sporulating cultures were not detectable in vegetative hyphae (Timberlake, 1980) and that 6% of the unique sequences were expressed during conidiation (Timberlake and Marshall, 1988). This type of comparison shows the sum total of differences between cultures grown for 16 and 40 h together with differences due to the differentiation associated with conidiation together, possibly, with other age-related differences (e.g. secondary metabolism) which may have no relation to conidiation. Comparisons between fruiting and non-fruiting cultures referred to in the previous paragraph involved cultures of similar age which, for environmental or genetic reasons, differed in their ability to undergo a morphogenetic change. The two types of comparison represent opposite ends of the spectrum. One is all-inclusive, the other is as exclusive as technically possible. The latter approach seeks to identify genes on which the morphogenetic change is causally dependent, and seems to show that there are relatively few of these. The all-inclusive approach identifies the subset of the genotype, which turns out to be a large minority, which contribute to the morphogenetic change. The gene subsets involved in different morphogenetic events differ (this is what is meant by 'differential gene expression'), but do have shared components. Even very different pathways of morphogenesis may share aspects of cell differentiation, such as particular parts of primary metabolism, cell inflation, wall thickening, accumulation of metabolites, etc. The problem is to distinguish between the causal and the merely contributory features.

5.3.1 Yeast-mycelial dimorphism

The life cycle of many pathogenic fungi features two characteristic growth forms: a yeast-like form and a filamentous mycelial form. This dimorphism is observed in plant pathogens like *Ustilago maydis* and human pathogens like *Candida albicans*. Non-pathogens also show dimorphism, including species of *Mucor* and, to the extent that the mating projection is pseudohyphal, *Saccharomyces cerevisiae* itself (Figs 5.9 and 5.10; Gow, 1995b). Dimorphism of *Ustilago maydis* is governed by the *a* and *b* mating type loci (see above), the *a* factor is necessary for conjugation tube formation and the *b* locus produces true hyphal filamentous growth (Banuett and Herskowitz, 1994). Filamentous growth is dependent on stimulation by the mating pheromone pathway and on a panel of genes whose expression is strictly limited to the filamentous phase and are directly or indirectly regulated by the mating type factors (Bolker *et al.*, 1995). This model of mating type regulation has encouraged a search for molecular switches of similar sorts in organisms in which the dimorphism is not part of the sexual cycle. These being seen as model systems to study cellular differentiation (Datta, 1994) and in which to identify the sort of 'master control factors' to which reference was made at the end of the previous section.

However, the search is not straightforward because of the wide range of metabolic and environmental factors which influence or govern dimorphism. Even in *Ustilago maydis* the dimorphic switch responds to environmental and metabolic conditions. Acid pH of the medium is sufficient to induce development of the mycelial form, suggesting that growth at low pH overcomes the control processes governed by the *b* mating type factor (Ruiz-Herrera *et al.*, 1995b). Metabolically, the cAMP-dependent protein kinase signalling pathway is involved in controlling morphogenesis in *U. maydis*. Disruption of the gene encoding adenylate cyclase results in a constitutively filamentous phenotype; budding being restored by growth in the presence of cAMP (Gold *et al.*, 1994).

Since such relatively unspecific environmental/metabolic effects can be seen in an organism in which the dimorphism is known to be regulated by master control genes (the mating type factors), it is not surprising that there is vigorous debate about causality in those organisms which lack an obvious master control genetic element. Unspecific effects are readily demonstrable: dimorphic alterations during temperature shifts in *Paracoccidioides brasiliensis* were preceded by significant changes in protein synthesis during the yeast to mycelium transition (temperature downshift 36–26°C), though there were few changes during the mycelium to

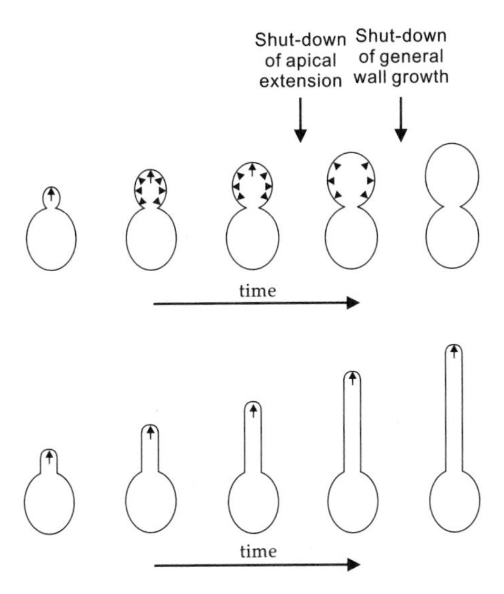

Figure 5.9 The sites and pattern of wall deposition during bud and germ tube development in yeasts like *Candida albicans* emphasise differences between the yeast and hyphal growth form. During yeast budding (top panel) wall synthesis is initially localised at the point of emergence (vertical arrow) but after a brief period of apical extension wall growth becomes more generalised around the cell (arrow heads), which may be called spherical growth in spores, and the cell takes on a rounded morphology. In hyphal growth (bottom panel) apical extension growth persists. Apical extension growth *without* general wall growth creates the hyphal growth form. Redrawn after Orlowski (1994).

yeast differentiation (26–36°C; Da Silva *et al.*, 1994). More specific differences can also be found, such as promotion of differentiation of the dimorphic yeast *Yarrowia lipolytica* from yeast to mycelia by a gene product which stimulates Golgi secretory function (Lopez *et al.*, 1994). But these are all associations rather than causes. Even in more extensively-studied fungi, there is considerable debate over the relative importance of differential gene expression and the pattern of metabolism. For example, in *Mucor* dimorphism the chemistry of the wall is similar in both phases, what distinguishes them is the way in which the wall is synthesised: isodiametric in the yeast form, apical and vectorial in the hyphae. Various enzyme activities and physiological processes alter during the dimorphic change, but none seem to be strictly causal (Orlowski, 1994, 1995). Cyclic AMP and other signalling molecules, and enzymes governing their intracellular concentrations, show consistent dimorphism-related patterns of change. Similarly, cytoskeletal components and their protein kinase regulators are involved in apical growth, but again a

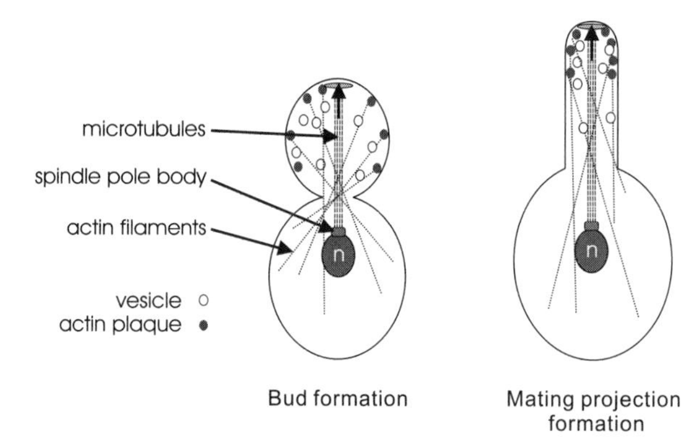

Figure 5.10 Polarity and distribution of cytoplasmic components in *Saccharomyces cerevisiae* during bud formation (left) and formation of the mating projection (right), the latter interpreted as analogous to hyphal growth. A bundle of parallel microtubules extends from the spindle pole body to the point of emergence (as signified by an arrow in Fig. 5.9). Disruption of microtubules does not affect bud formation so the microtubules are thought to position the nucleus at the bud neck and orient the spindle pole body. Actin microfilaments are thought to direct secretory vesicles to the sites of active wall growth. Actin granules (plaques) are distributed evenly around the periphery of expanding yeasts, but are localised near the apex of elongating hyphal projections. Thus, the same cellular structures are involved in apical and generalised wall growth; it is the distribution of those structures which distinguishes the two growth modes. Redrawn after Orlowski (1994).

causal link is lacking. A similar story can be told for dimorphism of *Candida albicans* (Cannon *et al.*, 1994; Gow *et al.*, 1995) in which gene expression has been extensively studied. However, the studies reveal complex alterations in gene expression during the dimorphic transition, with most genes examined showing transient or persistent increases or decreases in mRNA levels. Further complication is added by strain- or medium-dependence of morphogenesis-specific gene expression of two chitin synthase genes (CHS2 and CHS3) and three aspartyl proteinase genes which (in the affected strain and/or effective medium) were transcribed preferentially in the hyphal form.

Even *Saccharomyces cerevisiae* strains, if they are starved for nitrogen, undergo a developmental transition to a filamentous pseudohyphal form. This dimorphism requires elements of the MAP kinase signal transduction pathway. MAP kinase is activated by treatment with agents promoting proliferation and/or differentiation. In yeast the genes concerned (STE20, STE11, STE7, and STE12) are essential for mating pheromone response (Roberts and Fink, 1994) emphasising, again, involvement of signalling pathways in the dimorphic transition.

Egorova and Lakhchev (1994) point out that yeast cell morphology results from interaction of the two processes: polarised growth and cytokinesis (with subsequent division of the cell wall). Imbalance of the equilibrium of these two central processes lead to formation of morphological variants. This idea of imbalance causing morphological change(s) could be a valuable generalisation.

When osmotic stress is applied to filamentous water moulds their growth can become disorganised with weak and malformed hyphal walls, the filaments taking on shapes like budding yeasts (Kaminskyj *et al.*, 1992; Money and Harold, 1992, 1993). In discussing this, Johnson *et al.* (1996) also raise the question of the role played by imbalance in biosynthetic processes in driving differentiation. Some chemostat-grown fission yeasts cultured at low dilution rate (i.e. under nutrient limitation) divide asymmetrically, yielding daughter cells of unequal volume (Vraná, 1983a,b) and fission yeast shaped like round-bottom flasks were induced by treatment with aculeacin A which inhibits β-glucan synthesis, though α-glucan synthesis continued apparently normally (Miyata *et al*, 1980, 1985, 1986a,b). Johnson *et al.* (1996) concluded that "imbalanced glucan synthesis led to aberrant cell shapes, and that all else was a consequence of the cells following their usual rules for extensile growth and division while coping with the deformed cell shapes that acted as unusual templates for further morphogenesis. The immediate cause . . . might well have been an unbalanced synthesis or wall assembly that was only casually, *not causally,* correlated with reduced turgor pressure [or nutrition]."

It is not clear to me whether the contrast between balanced growth and unbalanced growth is being used here strictly in accord with use of these terms in describing kinetics of growth of microorganisms in *in vitro* culture. However, this usage is implied, and would certainly be appropriate. In terms of microbial growth kinetics the term 'balanced growth' describes the growth which occurs when all nutrients are available in sufficient quantity and the microbial cells can synthesise all of their components in balance (Campbell, 1957). Unbalanced growth occurs when some limitation, nutritional or environmental, adversely affects synthesis of one or more of the cellular components, yet growth persists and the cells which are formed are 'abnormal' (i.e. they differ from those produced by balanced growth). The term unbalanced growth was introduced by Cohen and Barner (1954) to describe the continued growth of bacteria in conditions in which there was no DNA synthesis. The relevance to the present discussion is that the growth pattern of a differentiated cell is 'unbalanced' in comparison with the growth pattern of an undifferentiated vegetative cell. The direction, progress and extent of the imbalance is precisely what defines the state of differentiation. Yet there are numerous ways in which unbalanced growth can be precipitated. By analogy, therefore, perhaps there are numerous ways in which a state of differentiation can be initiated. The master genetic control elements may be involved in defining and providing for the events which *could* take place, rather than being causally involved in what *will* take place. Causality may rest with altered temperature, pH, nutrition, etc., which expedite change. The manner of the change may depend on the past history of the cell and the future avenues for change which that history has made possible. In a crude analogy, bricks may be necessary to build a house, but the manufacture of bricks does not determine the shape of the house, nor even that it should be built. And, in the absence of bricks, a house could be built of timber.

5.3.2 Conidiation

Much further progress has been made in establishing the details of the genetic control of conidiospore formation, especially in *Aspergillus nidulans*. Conidia are asexual spores of ascomycetes which can survive in a non-growing state for longer periods than vegetative hyphae. They arise on specialised hyphae called conidiophores and the spores themselves are essentially rounded off hyphal segments, often more or less spherical in shape, which detach for dispersal (Adams, 1995; Ebbole, 1996).

Formation of conidia by surface cultures of *A. nidulans* occurs after about 16 h hyphal growth, this period of vegetative growth being required to make the cells competent to respond to the induction process. Induction requires exposure to air and is probably a reaction to cell-surface changes at air-water interfaces (Navarro-Bordonaba and Adams, 1994). After induction, some mycelial hyphae produce aerial branches which become conidiophore stalks (Fig. 5.11). The cell from which the branch emerges is the conidiophore foot cell, which is distinguishable from other vegetative cells by having a brown pigmented secondary wall thickening on the inside of its original wall (Smith *et al.*, 1977). The stalk grows apically until it reaches a length of about 100 μm when the apex swells to form the conidiophore vesicle which has a diameter of about 10 μm. A single tier of numerous primary sterigmata, called metulae, then bud from the vesicle and secondary sterigmata, the phialides, bud from the exposed apices of the metulae. The phialides are the stem cells which then undergo repeated asymmetric divisions to form

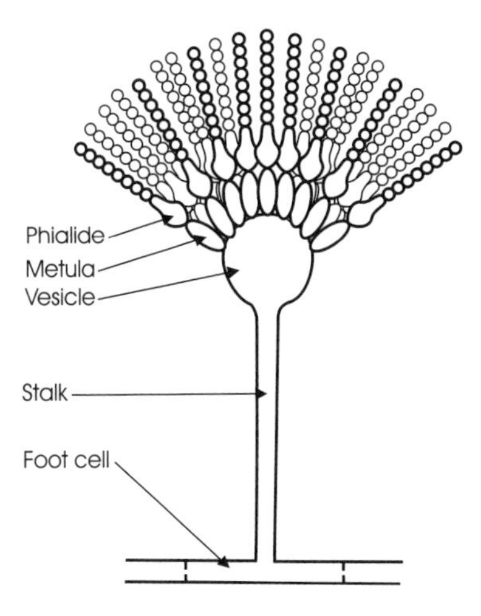

Figure 5.11 Diagrammatic structure of the conidiophore of *Aspergillus nidulans*.

the long chains of conidia which are approximately 3 μm in diameter (Cole, 1986; Mims *et al.*, 1988; Miller, 1990; Sewall *et al.*, 1990; Timberlake, 1990).

5.3.2.1 Genetics of conidiation in Aspergillus nidulans

Classical genetic analysis, by isolation and analysis of mutants, has been used to establish the basic genetic outline of the process (Clutterbuck, 1977, 1978; Champe *et al.*, 1981; Timberlake *et al.*, 1983; Gwynne and Timberlake, 1984; Timberlake and Hamer, 1986; Timberlake, 1987). Martinelli and Clutterbuck (1971) estimated that between 300 and 1000 loci were concerned with conidiation, by comparing mutation frequencies at loci affecting conidiation with those for other functions. Analysis of mRNA species indicated that approximately 6000 were expressed in vegetative mycelium, and an additional 1200 were found in cultures which included conidiophores and conidia; 200 of these additional mRNAs being found in the conidia themselves (Timberlake, 1980, 1986). As mentioned above, this method does not distinguish conidiation-specific mRNAs from those coincidentally associated with conidiation.

Only about 2% of aconidial mutants of *A. nidulans* had defects in stages concerned with conidiophore growth and development. By far the majority (83%) were defective in the preconidiophore stage, and 15% were affected in conidium germination or pigmentation

(Martinelli and Clutterbuck, 1971). 85% of conidiation mutants were also defective for vegetative hyphal growth. Attaining competence therefore seems to involve the largest number of gene functions. Of the few genes which seem to determine conspicuous developmental events in conidiophore morphogenesis, two in particular play a key role. These are the 'bristle' (*brlA*) gene which has defects in vesicle and metula formation, and 'abacus' (*abaA*) in which conidia are replaced by beaded lengths of hypha (Clutterbuck, 1969). These two loci are each represented by thirty or more mutant alleles, and no other mutations have been identified which affect these stages of conidiophore morphogenesis. Temperature-sensitive mutants, study of epistatic interactions and RNA transcript detection studies have indicated that *brlA* is required during vesicle, metula and phialide stages, and *abaA* acts during conidial budding from the phialide (Martinelli, 1979; Johnstone *et al.*, 1985; Timberlake *et al.*, 1985; Boylan *et al.*, 1987). A third gene with regulatory properties is *wetA* which is defective in an early stage of spore maturation. Conidia of *wetA* lack pigment, hydrophobicity and autolyse after a few hours (Clutterbuck, 1969) and fail to express a range of spore-specific mRNAs (Gwynne *et al.*, 1984). The *wetA* gene transcript is lacking in *brlA* and *abaA* mutants (Boylan *et al.*, 1987). Gene expression patterns and epistasis between the genes in double mutants suggests that these three genes act in the order *brlA* ⟶ *abaA* ⟶ *wetA* (Martinelli, 1979; Zimmerman *et al.*, 1980; Boylan *et al.*, 1987). There are many other *A. nidulans* mutations which affect a variety of specific functions in sporulation, but these three genes *brlA*, *abaA* and *wetA* seem to be the key control elements.

5.3.2.2 Molecular aspects of the control of conidiation in Aspergillus nidulans

A striking feature of the mutational analysis of conidiophore development in *A. nidulans* is that each phenotype is represented by mutation in just one locus (Clutterbuck, 1969, 1972). This may suggest that the genes which are isolated in mutation analysis are regulators which integrate the expression of other genes which are not themselves specific to conidiation. Molecular analyses support this interpretation (Timberlake, 1990, 1993; Timberlake and Clutterbuck, 1993). The model which has emerged (Figs 5.12 and 5.13) is as follows. The amino acid sequence of the *brlA* product contains zinc fingers near the carboxy-terminus. Zinc fingers are secondary structures found in many transcription factors in which a Zn atom is bound at the bend in a loop of the polypeptide chain. These are

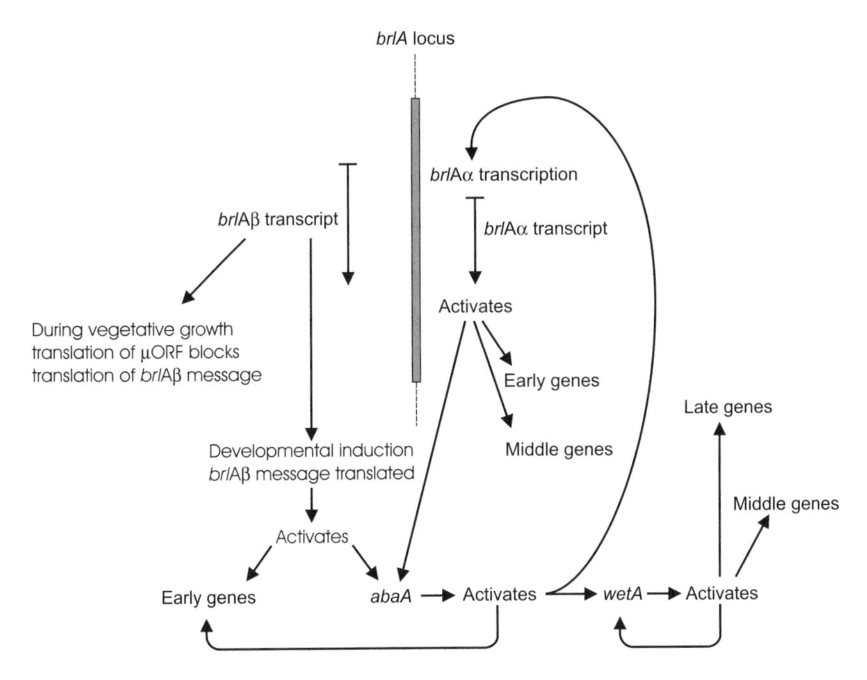

Figure 5.12 Summary of the genetic regulatory circuit for conidiophore development in *Aspergillus nidulans*.

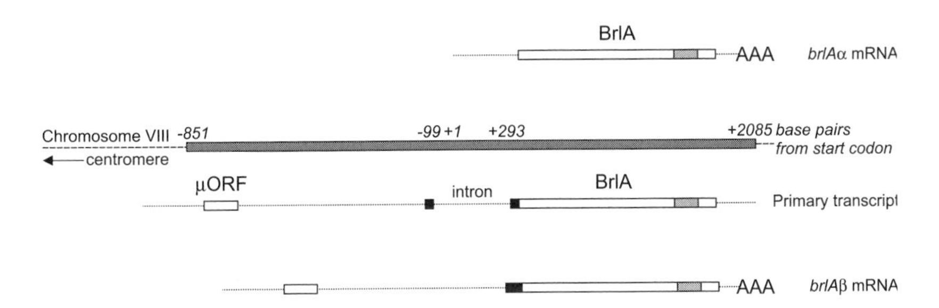

Figure 5.13 Structure of the *brlA*α locus of *Aspergillus nidulans*. The *brlA*α mRNA is shown at the top, above a shaded box which represents the BrlA segment of chromosome VIII. The *brlA*β primary transcript and mRNA are shown in the lower half of the figure. The *brlA*α sequence is a single exon encoding a Cys⟶2-His⟶2 zinc finger polypeptide (location of the zinc fingers shown as a shaded box within BrlA). The *brlA*β sequence contains one intron. The polypeptide encoded by *brlA*β contains an additional 23 amino-terminal residues (corresponding sequence shown as a black box) and the transcript has a short upstream Open Reading Frame (ORF) which regulates translation of *brlA*β. Redrawn after Timberlake, 1993.

responsible for directing sequence-specific DNA-binding. Presence of such sequences in the *brlA* product suggest that it is a nucleic acid binding protein. The *brlA* sequence is therefore thought to encode a positively acting transcription regulator which is required for transcriptional activation of developmentally regulated target genes.

However, phenotypes of some *brlA* mutants which have only partially lost function (in which target genes show varied effects out of proportion to the loss of *brlA* function) suggest the *brlA* product has different affinities for different target genes. The *brlA* locus consists of overlapping transcription units (Fig. 5.13), the downstream unit being designated *brlA*α and the upstream unit *brlA*β. The two share the same reading frame for most of their length but *brlA*β has an additional 23 amino acid residues at the amino-terminal end of that reading frame, and its transcript also possesses an ATG-initiated reading frame of 41 amino acid residues (called μORF) near its 5′ terminus. The two transcription units are needed for normal conidiophore development but the two BrlA peptides they encode can substitute for each other. Their functional difference seems to be in the very earliest stages of the initiation of development. The *brlA*β transcript can be detected in vegetative hyphae but the BrlA peptide is not translated from the transcript because translation initiation at μORF represses translation from the downstream (BrlA) reading frame (Han *et al.*, 1993). Thus, the competent hypha is primed to undertake conidiophore development, only this translational repression maintaining vegetative growth and preventing irreversible activation of the conidiation pathway. Timberlake (1993) has described the activation of the conidiation pathway as 'translational triggering' because if the repression caused by μORF can be overcome the *brlA*β transcript will be translated and BrlA will activate conidiation. Further, Timberlake (1993) suggests that the translational trigger may be a way of making development sensitive to the nutritional status of the hypha, as nitrogen limitation (a common environmental signal for initiation of sporulation) could reduce aminoacyl-tRNA pools and disturb translational regulation by μORF. Activation of *brlA* depends on a gene called *flbA* which encodes a mRNA that is expressed throughout the *A. nidulans* asexual life cycle (Lee and Adams, 1994). The sequence encodes a polypeptide with some similarity to a *Saccharomyces cerevisiae* protein which is required by yeast cells to resume growth following prolonged exposure to yeast mating pheromone **a**. The *flbA* protein is thought to contribute to a signalling pathway in *Aspergillus* which distinguishes between continued vegetative growth and conidiophore development.

Activation of *brlA* is, therefore, seen as the first step in conidiophore development, and its product in turn activates a panel of conidiation-specific genes among which are *rodA* (encodes a hydrophobic component of the conidium wall), *yA* (encodes a *p*-diphenol oxidase (laccase) responsible for conversion of yellow spore pigment to green), and, directly or jointly with *medA*, the next regulator, *abaA*. The *abaA* product is also a DNA-binding protein with characteristics which suggest that it is a transcriptional regulator which enhances expression of *brlA*-induced structural genes. The *brlA* and *abaA* genes are reciprocal activators, because *abaA* also activates *brlA*. Of course *brlA* expression must occur before *abaA* can be expressed, but the consequential *abaA*-activation of *brlA* reinforces the latter's expression and effectively makes progress of the pathway independent of outside events. The *abaA* product also activates additional structural genes and the final regulatory gene, *wetA*, which activates spore-specific structural genes. Since *brlA* and *abaA* are not expressed in differentiating conidia, *wetA* is probably involved in inactivating their expression in the spores. Expression of *wetA* is initially activated in the phialide by sequential action of *brlA* and *abaA*. There is, however, evidence that *wetA* is autoregulatory. Positive autoregulation of *wetA* maintains its expression after the conidium has been separated (physically or cytologically) from the phialide. Spatial organisation of gene expression of this sort is also imposed upon the core regulators by the genes *stuA* and *medA*. Mutants in *stuA* form diminutive (*stu*nted) conidiophores with unthickened walls. This locus is classified as an auxiliary regulator as a number of conidiophore-expressed transcripts are missing in *stuA* mutant. Medusa (*medA*) mutants form conidia on top of multiple layers of metulae; these mutants are also sterile and unable to form cleistothecia. The *stuA* gene is complex, two transcripts being produced from distinct transcription start signals, both having short open reading frames (mini-ORFs) in their leader sequences and there is some evidence for translational regulation of *stuA* expression. Both *stuA* transcripts increase in concentration by a factor of about 50 when cells become developmentally competent, and there is an additional 15-fold increase in *stuA* expression (which requires *brlA* activity) following developmental induction (Miller *et al.*, 1992). The *medA* gene interacts with *brlA* but it is not yet clear how. However, the *medA* transcript level declines following developmental induction.

Non-regulatory development-specific genes have been categorised into four classes on the basis of transcript accumulation in strains carrying mutations of the regulators (Mirabito *et al.*, 1989). Class A genes are involved in early development and are activated by *brlA* or *abaA* or both, but independently of *wetA*. Class B genes are involved in late (spore-related) functions and are activated by *wetA* independently of *brlA* or *abaA*. Genes put into classes C and D are thought to encode phialide-specific functions and their activation requires the combined activity of all three regulators.

The genetic structure revealed in this analysis is significant because it demonstrates that the conidiophore developmental process is naturally divided into sequential steps. Translational triggering exposes a mechanism which can relate a developmental pathway to the development of competence on the one hand, and to initiation in response to environmental cue(s) on the other hand. Further, the reciprocal activation, feedback activation and autoregulation seen in the core regulatory sequence reinforces expression of the whole pathway, making it independent of the external environmental cues which initiated it. This has been called feedback fixation by Timberlake (1993) and it results in developmental determination in the classic embryological sense. Many of the *Aspergillus* conidiation mutants are also defective in sexual reproduction (Clutterbuck, 1969; Kurtz and Champe, 1981; Butnick *et al.*, 1984a,b). Thus, another conclusion to be drawn from these *A. nidulans* mutants is that there is some economy of usage of morphogenetic genes in different developmental processes. Presumably, different developmental modes employ structural genes that are not uniquely developmental, but function in numerous pathways, having their developmental-specificity bestowed upon them by the regulators to which they respond. This is epitomised in the idea that the key to eukaryote development is in the ability to use relatively few regulatory genes to integrate the activities of many others.

5.3.2.3 Conidiation in Neurospora crassa

N. crassa forms two types of conidium, microconidia and macroconidia. Microconidia are small uninucleate spores which are essentially fragmented hyphae. They are not well adapted to dispersal and are thought to serve primarily as 'male gametes' in sexual reproduction. Macroconidia are more common and more abundant, they are large multinucleate, multicellular spores produced from aerial conidiophores. Conidiation (and sexual reproduction, too) in *N. crassa* seems to respond more to environmental signals than to complex genetic controls like those operating in *Aspergillus* (Springer and Yanofsky, 1989). Macroconidia are formed in response to nutritional limitation, desiccation, change in atmospheric CO_2, and light exposure (blue light is most effective, and though light exposure is not essential, conidia develop faster and in greater numbers in illuminated cultures). In addition, a circadian rhythm provides a burst of sporulation each morning. When induced to form conidia, the *Neurospora* mycelium forms aerial branches which grow away from the substratum, form many lateral branches which become conidiophores and undergo apical budding to produce conidial chains.

The genetics of conidiation has been studied by means of mutation and molecular analysis (Matsuyama *et al.*, 1974; Selitrennikoff, 1974; Berlin and Yanofsky, 1985a,b; Davis, 1995; and further references below). There are some parallels in terms of types of mutants obtained with *N. crassa* and *A. nidulans*, and a particular example would be the hydrophobic outer rodlet layer which is missing in the *N. crassa* 'easily-wettable' (*eas*) and *A. nidulans rodA* mutants (Beever and Dempsey, 1978; Claverie-Martin *et al.*, 1986; Stringer *et al.*, 1991; Bell-Pederson *et al.*, 1992; Lauter *et al.*, 1992). Despite such functional analogies, there is no underlying similarity between the genetic architectures used by these two organisms to control conidiation. Importantly, there is no evidence for regulatory genes in *N. crassa* which are similar to the *brlA–abaA–wetA* regulators of *A. nidulans*. Nevertheless, a large number of mutants have been isolated which have defects in particular stages of conidiation though there is a general absence of analysis at the molecular level (discussed in Navarro-Bordonaba and Adams, 1994). A number of conidiation (*con*) genes are known which encode transcripts which become more abundant at specific stages during conidiation (Berlin and Yanofsky, 1985a; Roberts *et al.*, 1988; Roberts and Yanofsky, 1989; Sachs and Yanofsky, 1991). At least four of these genes are expressed in all three sporulation pathways in *Neurospora* (macroconidia, microconidia and ascospores) but others have specific localisation to macroconidia (Springer and Yanofsky, 1992; Springer *et al.*, 1992). However, many of the *con* genes can be disrupted without affecting sporulation (Springer *et al.*, 1992); so, despite being highly expressed during sporulation, they presumably encode redundant or non-essential functions.

5.4 Sexual reproductive structures

The conidiation mutants of *Aspergillus* and *Neurospora* make it clear that mycelium has a number of alternative developmental pathways open to it: continuation of hyphal growth, production of asexual spores, and progress into the sexual cycle. Sexual reproduction predominates over conidiation in many strains of *A. nidulans* collected from the wild

(Croft and Jinks, 1977) when grown on normal media; laboratory strains carry a mutation (*veA;* velvet) which shifts the balance towards conidiation. However, mutation to increased sexual reproduction at the expense of conidiation were frequent amongst those reported by Martinelli and Clutterbuck (1971). In contrast, some of the *Aspergillus* conidiation mutants (including *medA*, *stuA*, *yB*, and *acoA*) also exhibit defects in sexual reproduction (Clutterbuck, 1969; Kurtz and Champe, 1981; Butnick *et al.*, 1984a,b) suggesting shared functions in the different morphogenetic pathways. Unfortunately, far less attention has been given to sexual reproduction in these ascomycetes than to conidiation.

5.4.1 Reproductive structures in ascomycetes

In Ascomycotina, the sexually produced *ascospores* are contained in *asci* (singular: *ascus*) enclosed in an aggregation of hyphae termed an *ascoma.* Ascomata are not formed from hyphae that have taken part in the meiotic cycle, instead they arise from non-dikaryotic sterile hyphae that surround the ascogonial hyphae of the centrum. A variety of ascomata exist, including the open cup-like 'discocarps' of *Peziza*, the flask-like perithecium (found, for example in *Neurospora* and *Sordaria*) and the completely closed cleistothecium formed by, for example, *Aspergillus* (Turian, 1978; Chadefaud, 1982a,b,c; Read, 1994).

Almost all the recent research on sexual development in *Neurospora* has been aimed at understanding mating type structure and function. Johnson (1976) used heterokaryons in which one nucleus carried a recessive colour mutant as genetic mosaics to show that perithecia of *Neurospora* arise from an initiating population of 100 to 300 nuclei, and that the perithecium wall is composed of three developmentally distinct layers. Later, he identified 29 complementation groups (equivalent to functional genes) which were involved in perithecium development (Johnson, 1978), but this research has not been followed up. Ashby and Johnstone (1993) used constructs including the *Escherichia coli* -glucuronidase as a reporter gene to study development of ascomata in *Pyrenopeziza brassicae* and also revealed three tissue layers showing differential expression of the mating types. Both mating types were expressed in one of the layers, but the two mating types were expressed separately in each of the other two layers. The significance of extensive tissue layers in which only one mating type is expressed is unknown but there may be analogies with the differential expression of genes in dikaryotic hyphae of *Schizophyllum commune* which is thought to depend on change in proximity of nuclei carrying the mating-type factors (Ásgeirsdóttir *et al.*, 1995; see section 5.2.2.3).

Cleistothecium development in *Aspergillus nidulans* has been fully described (Champe and Simon, 1992; Yager, 1992) but the developmental observations have not been accompanied by extensive genetic analysis as yet. Apart from the involvement of conidiation mutants coincidentally noted above, a β-tubulin gene has been shown to be essential for sexual reproduction (Kirk and Morris, 1991), and laccase activity is specifically located in cleistothecium primordia (Hermann *et al.*, 1983). Laccase enzymes (see section 3.2.2) have been associated with several asexual and sexual reproductive processes in ascomycetes and basidiomycetes. Phenoloxidases (e.g. tyrosinase and laccase) function in lignin degradation by white-rot fungi but they are also involved in pigment production, and polymerisation of phenols may have a role in strengthening cell walls during development. Bu'lock (1967) suggested that phenoloxidases may crosslink hyphal walls into coherent aggregates during morphogenesis. Laccase is required for pigmentation of conidia in *A. nidulans* (Clutterbuck, 1972; see section 5.3.2.2) and a second laccase enzyme, electrophoretically and antigenically distinct from the conidial laccase, is localised to the hyphae of the young cleistothecia and the Hülle cells which surround them (Hermann *et al.*, 1983). This cleistothecial laccase was produced prematurely in an arginine mutant which also formed cleistothecia precociously (Serlupi-Crescenzi *et al.*, 1983) so it appears that laccase activity is associated with morphogenesis of the ascomycete fruit body (Read, 1994). Esser (1956, 1968) established a correlation between phenoloxidase activity and fertility in *Podospora anserina*, though there was no correlation between pigmentation and protoperithecial formation. Esser and Minuth (1970) detected 390 kDa, 80 kDa and 70 kDa forms of laccase in *P. anserina* and Prillinger and Esser (1977) analysed eight mutants deficient in laccase, identifying five laccase structural gene loci. A correlation between tyrosinase activity, melanin production and protoperithecial formation in *N. crassa* was established by Hirsch (1954). Mycelia grown under optimal conditions for protoperithecial production had abundant tyrosinase activity and this appeared at the same time or just before protoperithecia were formed. Several mutants which did not produce protoperithecia lacked tyrosinase activity (Barbesgaard and Wagner, 1959). However, mutations in the structural gene for tyrosinase are not affected in protoperithecial production. Rather, other mutants which influence tyrosinase synthesis indirectly were female sterile and manipulation of growth conditions could restore tyrosinase activity to these mutants but not fertility (Horowitz *et al.*,

1960). Evidently, tyrosinase is on the same regulatory circuit as protoperithecium production and tyrosinase activity might normally be necessary but was not sufficient in itself to safeguard protoperithecial formation.

There is an analogous increase in phenoloxidase activity accompanying the formation of fruit body initials in various basidiomycete species (Leonard, 1971; Leonard and Phillips, 1973; Turner, 1974; Wood, 1980c; Leatham and Stahmann, 1981). The *cohesiveless* mutant of *Schizophyllum commune* has lost the ability to form hyphal aggregates and has no phenoloxidase activity in monokaryons (Leonard, 1972). Dikaryons of *cohesiveless* could not form hyphal aggregates or fruiting bodies but regained phenoloxidase activity. Leonard (1972) concluded, as did Horowitz *et al.* (1960) from their studies of tyrosinaseless mutants of *Neurospora crassa,* that phenoloxidase is necessary but not sufficient for the formation of fruit body initials.

Although there is a simple correlation between the levels of laccase activity and the development of fruit body primordia in *Coprinus congregatus* when enzyme activity was measured in extracts of whole mycelium, localisation studies refuted the simple correlation (Ross, 1982b). Instead, the correlation seems to be with sensitivity to photo-induction. Light was able to induce primordia only in those regions of the mycelium which had high laccase activity. Ross (1982b) suggested that laccase had a role in attaining developmental competence.

A number of mutations have been described which can suppress or modify incompatibility reactions in *Podospora anserina* (Belcour and Bernet, 1969; Bernet, 1971; Delettre and Bernet, 1976; Labarère and Bernet, 1977). Nearly all these modifying (*mod*) mutants affect protoperithecial formation and fertility. Several *mod* mutants have altered proteinase enzymes, others are thought to be affected in ribosomal structure and translation (Bernet *et al.*, 1973; Boucherie *et al.*, 1976) or to suffer plasma membrane defects (Asselineau *et al.*, 1981; Bonneu and Labarère, 1983). Some of the mutants are defective in protoperithecial production and/or ascospore germination, but in others protoperithecial production is increased and occurs earlier than in the wild type (Durrens, 1983). Durrens and Bernet (1985) blended the observations into a model in which the key feature of fungal development was seen as the achievement of a quiescent state (competence?) which is prerequisite for protoperithecial development. Durrens and Bernet (1985) suggest the quiescent state evolved as a survival mechanism under conditions of nutrient limitation. Their interpretation equates production of the protoperithecium with production of a vegetative survival structure and is interesting in

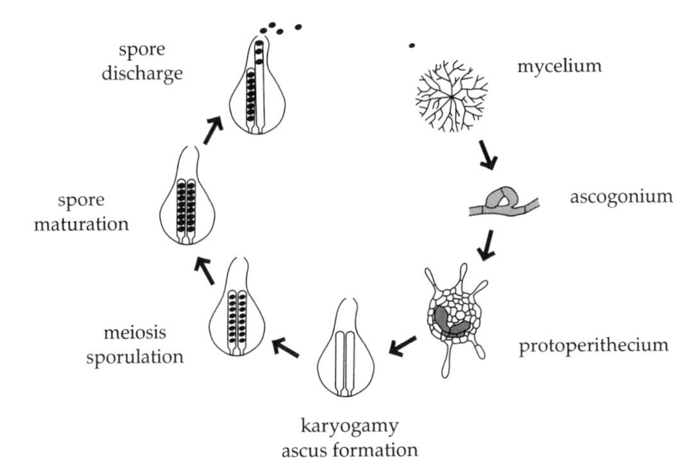

Figure 5.14 Life cycle diagram and perithecium developmental pathway of *Sordaria macrospora* (from Moore, 1995; after Esser and Straub, 1958). A variety of mutants are known which block the pathway at each of the stages represented by arrows, so the whole pathway is interpreted as being essentially a single sequence. Contrast this with the multiple parallel 'subroutines' which seem to characterise basidiomycete fruit body development (Figs 5.16 and 5.17).

view of the non-perithecial multicellular structures observed in *Sordaria* (Broxholme *et al.*, 1991; Read, 1994) and the close relationship which has been demonstrated between sclerotium production and the fruit body initiation pathway in the basidiomycete *Coprinus cinereus* (Moore, 1981a; see section 5.4.4).

Esser and Straub (1958) used classical genetic approaches, namely identification of variant strains, application of complementation tests to establish functional cistrons, construction of heterokaryons to determine dominance/recessive and epistatic relationships (to indicate the sequence of gene expression) to establish a 'developmental pathway' for perithecium formation in *Sordaria* (Fig. 5.14). The fruit-bodies of basidiomycetes, which include mushrooms, toadstools, bracket fungi, puffballs, stinkhorns and bird's nest fungi, are all examples of *basidiomata* which bear the sexually produced *basidiospores* on *basidia*. For these reproductive structures the picture revealed by classical genetic approaches is less clear. One reason for this is that fruit bodies in basidiomycetes are normally formed by secondary mycelia which are heterokaryotic. The co-existence of two (or more) nuclei (and, therefore, two or more genotypes) makes it difficult to study the genetics of development by conventional means. On the other hand, fruiting by monokaryotic mycelia has been recorded in a number of basidiomycetes (Stahl and Esser, 1976; Elliott, 1985) and these strains have allowed a start to be made on the genetic control of fruit body development.

5.4.2 *Monokaryotic fruiting in basidiomycetes*

Stahl and Esser (1976) used conventional genetic crosses between monokaryotic fruiting strains of *Polyporus ciliatus*, finding three unlinked genes involved in monokaryotic fruiting: fi^+, which was thought to initiate monokaryotic fruiting, fb^+, which is seen as being responsible for 'moulding' the structure of the fruit initiated by fi^+ into a fruit body. The third gene, mod^+, appeared to direct development into a futile pathway leading to formation of non-fruiting mycelial masses called stromata. In the dikaryon mod^+ inhibited fruiting, but neither fi^+ nor fb^+showed any expression even when homozygous. The way in which these genes function is unknown. A very similar genetic system was found in analogous experiments with the agaric *Agrocybe aegerita* (Esser and Meinhardt, 1977). Again, one gene, fi^+, was identified as being responsible for initiation of monokaryotic fruiting, and a second, fb^+, was considered to be responsible for modelling the initiated structures into fruit bodies. A contrast with the genes found in the polypore, *Polyporus*, was that the *Agrocybe* genes were found to influence fruiting in the dikaryon as well. Fertile fruit bodies were produced only by dikaryons carrying at least one allele of both fi^+ and fb^+.

Raper and Krongelb (1958) studied some monokaryotic fruiting strains (called *hap*) of *Schizophyllum commune*. They found there was no correlation between monokaryotic and dikaryotic fruiting, and argued that monokaryotic fruiting was probably under polygenic control. The polygene complex involved may have been identified by Esser *et al.* (1979), who identified four genes in *S. commune* which controlled monokaryotic fruiting. In this species they found two 'fruiting initiation genes' (fi-1^+ and fi-2^+, either of which alone allowed differentiation into fruit body 'initials' of about 2–3 mm in size; when both were present, fruit body stems 6–8 mm long were formed. A third gene (fb^+) determined formation of complete monokaryotic fruit bodies. The fourth gene (st^+) prevented expression of the others. A monokaryon carrying st^+ produced only stromata and a homozygous st^+/st^+ dikaryon was also unable to fruit. The other three genes had no affect on differentiation of fruit bodies in the dikaryon but did influence how quickly the dikaryon fruited. Dikaryons which were homozygous for all three monokaryotic fruiter genes fruited most rapidly.

Dikaryons which did not carry any of the monokaryotic fruiter alleles were the slowest to fruit, but they did form fruiting bodies and that clearly implies a major difference in the impact which these genes have on the fruit body development pathway in the two types of mycelium. Barnett and Lilly (1949) also reported increased frequency of fruiting in dikaryons made from monokaryotic fruiters in *Lenzites trabea*.

Leslie and Leonard (1979a,b) examined genetic factors enabling monokaryons of *Schizophyllum commune* to initiate fruiting bodies in response to mechanical and chemical treatments and identified eight genes involved in four distinct pathways (Fig. 5.15). The authors placed the operation of these genes at a stage prior to formation of aggregations of cells without defined shape (Leslie and Leonard, 1979b) so they may be distinct from those identified by Esser *et al.* (1979), which produce structures with a recognisable (stem-like) shape. However, morphological descriptions are the only basis for judgement about relationships between these systems.

The wide range of genetic factors involved in monokaryotic fruiting mirrors the range of physiological conditions which are able to promote such fruiting. The frequency of monokaryons able to fruit differs drastically between genera: 27% of *Sistotrema* isolates formed monokaryotic fruit bodies (Ullrich, 1973), 7% of *Schizophyllum* strains did so (Raper and Krongelb, 1958), but only one of 16 monokaryons of *Coprinus cinereus* tested by Uno and Ishikawa (1971). Most 'monokaryotic fruits' are abnormal structures; usually incomplete, sterile or both. This raises the question of whether genes which influence fruiting in monokaryons are relevant to the normal process of dikaryotic fruiting. Some of the genes identified in monokaryons do show expression in the dikaryon (Stahl and Esser, 1976; Esser and Meinhardt, 1977; Esser *et al.*, 1979) but the role which they might play is obscure. The only molecular observation made on this phenomenon is that Horton and Raper (1991) identified a DNA sequence which induced monokaryotic fruiting in strains of *Schizophyllum commune* into which it was introduced by transformation. Unfortunately, there is no indication as to how it operates. Another peculiarity is the induction of dikaryotic fruiting bodies on originally monokaryotic cultures of *Coprinus cinereus* when the latter were subjected to nutritional stress for several weeks to several months (Verrinder Gibbins and Lu, 1984). The authors suggest that nutritional stress may trigger a mating type switch which results in a conventional dikaryon being established. However, they did not carry out any mating type analysis on the progeny spores formed by the fruit bodies formed on initially monokaryotic cultures.

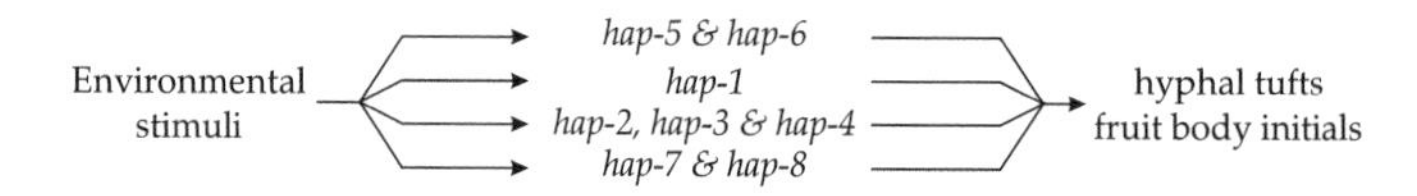

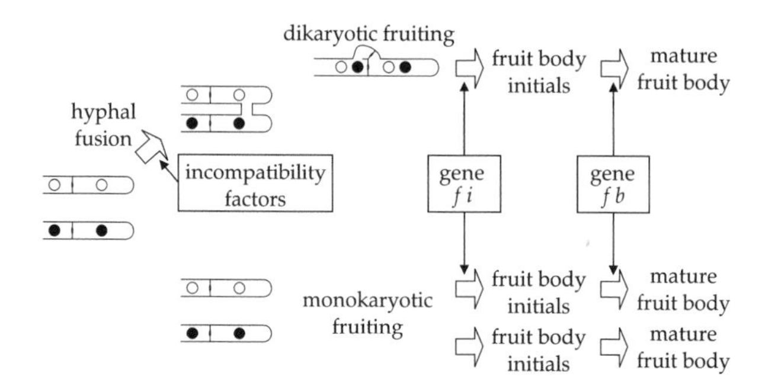

Figure 5.15 Models for the genetic control of fruit body development in basidiomycetes. The top panel shows the genes involved in monokaryotic fruiting in *Schizophyllum commune* (after Leslie and Leonard, 1979b) in which *hap*-5 and *hap*-6 control the spontaneous fruit body initiation, *hap*-1 alone or *hap*-2, -3 and -4 acting together control fruiting as a response to injury, and *hap*-7 and *hap*-8 determine fruiting in response to applied chemicals. The bottom panel is a proposed model for the action of major genes controlling mushroom formation in *Agrocybe aegerita* (after Esser and Meinhardt, 1977).

5.4.3 Dikaryotic fruiting in basidiomycetes

The only organism in which any concerted attempt has been made to study the genetic control of fruit body formation by the dikaryon is *C. cinereus*. Dikaryons of *C. cinereus* can form sclerotia and basidiomata, monokaryons may also form sclerotia but normally do not form basidiomata. Initial steps in the development of both sclerotia and fruit bodies have been described separately and the descriptions are remarkably similar (Matthews and Niederpruem, 1972; Waters *et al.* 1975a). In the formation of both structures, development from the mycelium involves similar patterns of hyphal aggregation so the likeness observed may indicate a shared initial pathway of development or coincidentally analogous separate, but parallel, pathways. These possibilities were distinguished with the aid of monokaryons unable to form sclerotia, a phenotype which segregated in crosses as though controlled by a single major gene. Four *scl* (sclerotium-negative) genes were found; one, *scl*-4, caused abortion of developing fruit body primordia even when paired in the dikaryon with a wild type nucleus but the other *scl* genes behaved as recessive alleles in such heteroallelic dikaryons and were mapped to existing linkage groups (Waters *et al.*, 1975a,b). Later, Moore (1981a) showed that homoallelic dikaryons (dikaryons in which both nuclei carried the same *scl* allele) were unable to form either sclerotia or fruit bodies. Since these single genetic defects blocked development of both dikaryon structures it was concluded that in the initial stages sclerotia and basidiomata share a common developmental pathway governed by the *scl* genes (Fig. 5.16). When they mutate they are usually recessive so the pathway can proceed only in the heteroallelic dikaryon where the missing *scl* function is provided by the nucleus from the other parent.

Takemaru and Kamada (1971, 1972) attempted to characterise the basic genetic control of dikaryon fruit body development in *Coprinus cinereus* (under the name *C. macrorhizus*) by searching for developmental abnormalities among the survivors of mutagen-treated fragments of dikaryotic mycelium. Including spontaneous mutations, a total of 1594 were identified out of 10,641 dikaryotic survivors tested, and were classified into categories on the basis of the phenotype of the fruit body produced. The categories were:

1. *knotless,* no hyphal aggregations are formed;
2. *primordiumless,* aggregations are formed but they do not develop further;
3. *maturationless,* primordia are produced which fail to mature;
4. *elongationless,* stem fails to elongate but cap development is normal;
5. *expansionless,* stem elongation normal but cap fails to open;
6. *sporeless,* few or no spores are formed in what may otherwise be a normal fruit body.

Since dikaryotic mutagen survivors were isolated, the genetic defects identified are all dominant. Elongationless mutants have been used to study stem elongation (Kamada and Takemaru, 1977a,b, 1983), and sporeless mutants have been used to study sporulation (Miyake *et al.*, 1980a,b). These mutants suggest that fruit body development is organised into different pathways which are genetically separate. Prevention of meiosis still permits the fruit body to develop normally, demonstrating, as do monokaryotic fruit bodies, that meiosis and spore formation are entirely separate from construction of the spore-bearing structure. It is also very significant that mutants were obtained with defects in either cap expansion or stem elongation. Both processes depend on enormous cell

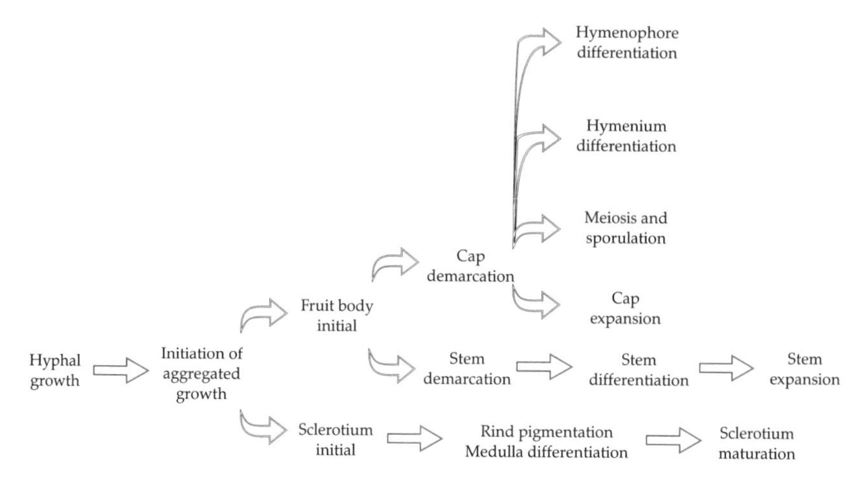

Figure 5.16 Genetically distinct pathways involved in sclerotium and fruit body development in *Coprinus cinereus* . From Bourne *et al.* (1996), revised after Moore (1981).

inflation, and the fact that they can be separated by mutation indicates that the same result (increase in cell volume) is achieved by different means (Moore *et al.*, 1979).

There is a problem in accounting for the induction of dominant mutations at the high frequency observed by Takemaru and Kamada (1972) and the peculiarity that over 72% of the mutants belong to just two phenotypes; there being 595 maturationless and 582 sporeless isolates out of the total of 1582. Takemaru and Kamada (1972) account for these frequencies with the suggestion that genes involved in development may be easy to mutate, but there are no reports of such mutations in other populations of *C. cinereus* dikaryons so this proposition is untenable. An alternative interpretation (Moore, 1981a) is that the genes which were being caused to mutate were not those involved directly in development, but rather genes which modify the dominance of pre-existing developmental variants. Dependence of dominance (or penetrance) on the modifying action of other genes is a well-established idea in genetical theory (Fisher, 1928, 1931; Sheppard, 1967; Manning, 1976, 1977), and could have considerable selective advantage in a system which imposed recessiveness on variants in genes concerned with development. This interpretation was arrived at following work showing that the penetrance of *scl* genes in heteroallelic dikaryons depended on the segregation of modifiers (Moore, 1981c). Nyunoya and Ishikawa (1979) showed that *roc* and *fis*^{C} segregated in ratios suggestive of multiple gene control and dominance modification has also been invoked to explain segregation patterns of a gene conferring resistance to *p*-fluorophenylalanine in *C. cinereus* (Senathirajah and Lewis, 1975; Lewis and Vakeria, 1977). As differentiation in basidiomycetes involves extensive protein processing (Zantinge *et al.* 1979; de Vries *et al.* 1980; Moore and Jirjis, 1981), modifiers might be involved in processing signal sequences of structural proteins. In the presence of particular modifier alleles (those which cause the change in penetrance), signal processing might lead to normal structural proteins failing to reach their correct destination, or abnormal proteins being partially corrected so that they do reach the target site, despite being defective.

Isolation of strains of *C. cinereus* which have mutations in both mating type factors (*Amut Bmut* strains) has opened up new possibilities for genetic analysis of morphogenesis in this organism. *Amut Bmut* strains are homokaryotic phenocopies of the dikaryon; i.e. they emulate the dikaryon in that their hyphae have binucleate compartments and extend by conjugate nuclear division with formation of clamp connections. Also, the cultures can produce apparently normal fruit bodies. On the other hand they are homokaryons, and are able to produce asexual spores (usually called oidia) and, most importantly, containing only one (haploid) genetic complement (Swamy *et al.*, 1984). This last feature allows expression of recessive developmental mutations and these strains have been used to study a number of developmental mutants (Kanda and Ishikawa, 1986) especially in meiosis and spore formation (Zolan *et al.*, 1988; Kanda *et al.*, 1989a, 1990; Kamada *et al.*, 1989a,b) and in the formation of fruit body primordia (Kanda *et al.*, 1989b), but no overall fruit body developmental pathway has yet emerged, nor has any information about major regulators.

5.4.4 Expression of fruiting genes

Genetic analysis of the sort discussed so far gives no guidance about the way in which genes causing developmental variants exercise their effects. Among the first enzymes identified as having an important role in morphogenesis were glucanases involved in the degradation of fungal cell walls. The concept that cell wall materials are re-utilised during morphogenesis originated with the studies made by Wessels with *Schizophyllum commune* (Wessels, 1965; Wessels and Niederpruem, 1967; Wessels and Sietsma, 1979), and has received support from work with fruit bodies of *Flammulina velutipes* (Kitamoto and Gruen, 1976) and *Coprinus congregatus* (Robert, 1977a,b) among basidiomycetes, as well as *Aspergillus nidulans* cleistothecia (Zonneveld, 1977). The latter example is important because a mutant of *A. nidulans* which lacked α-1⟶3 glucan was unable to form cleistothecia (Polacheck and Rosenberger, 1977) and mutants deficient in either cleistothecial formation or conidiation or both, confirmed the correlation (at least) between the presence of α-1⟶3 glucan, depletion of glucose, synthesis of α-1⟶3 glucanase and cleistothecial formation (Zonneveld, 1974).

Another important aspect of the sequence of studies on *A. nidulans* cleistothecium development is that it emphasises the flexibility of the developmental process by showing that if glucan reserves are low, proteins may be utilised for cleistothecium formation. The exact nature of the nutrient limitation conditions determine whether glucans or proteins are used during morphogenesis, but when circumstances demand, specific glucanase activity is replaced by specific proteinase action (Zonneveld, 1980). This sort of flexible integration of enzyme activities to suit the prevailing conditions goes some way to explaining why only a small fraction of the genome is specific to morphogenesis, and correspondingly few morphogenesis-specific polypeptides have been identified. A development-specific protein has been identified in sclerotia of *Sclerotinia sclerotiorum* (Russo *et al.*, 1982) and a polypeptide specific to fruit body (ascomatal) development has been detected in *Neurospora tetrasperma* (Nasrallah and Srb, 1973, 1977) and localised to the mucilaginous matrix surrounding the asci and paraphyses (Nasrallah and Srb, 1978). In *Sordaria brevicollis*, 17 out of over 200 polypeptides detected after pulse-labelling were found in perithecia after crossing (Broxholme *et al.*, 1991). De Vries and Wessels (1984) found only 15 polypeptides specifically expressed in fruit body primordia of *Schizophyllum commune*. Analysis of specifically-transcribed RNA also suggest that expression of only a small proportion of the genome is devoted to morphogenesis in both *S. commune* (Zantinge *et al.*, 1979; de Vries *et al.*, 1980) and *Coprinus cinereus* (Yashar and Pukkila, 1985; Pukkila and Casselton, 1991). In the latter organism, Kanda *et al.* (1986) found only four so-called 'cap proteins' which were abundant in cap cells but rare in the stem.

In situ hybridisation has been used to demonstrate reallocation of ribosomal-RNA between fruit bodies and their parental vegetative mycelium in *S. commune* (Ruiters and Wessels, 1989a) and accumulation of fruiting-specific RNAs in the fruit body has also been demonstrated (Mulder and Wessels, 1986; Ruiters and Wessels, 1989b). Dons *et al.* (1984) cloned a gene from among the fruiting-specific sequences which belongs to a family of sequences encoding hydrophobins. These are cysteine-rich polypeptides which are excreted into the culture medium but polymerise on the wall of aerial hyphae as they emerge into the air (to form fruit body initials, for example) and invest them with a hydrophobic coating (Wessels, 1992, 1996; and see Fig. 5.17). In *S. commune*, some hydrophobin genes are under control of the mating-type genes (Ásgeirsdottir *et al.*, 1995; discussed in section 5.2.2.4), and sequences coding for hydrophobins have been found in *Agaricus bisporus* (Lugones *et al.*, 1996; De Groot *et al.*, 1996), one of which specifically accumulates in the outer layers of mushroom caps (the 'peel' tissue) during fruit body development (De Groot *et al.*, 1996). However, hydrophobins have been very widely encountered in fungi (de Vries *et al.*, 1993); about 20 have been recognised by gene sequence homology. They are small secreted proteins comprised of 75–125 amino acids that have eight cysteine residues spaced in a specific pattern ($X_{2\text{-}38}$-C-X_{5-9}-C-C-$X_{11\text{-}39}$-C-$X_{8\text{-}23}$-C-X_{5-9}-C-C-$X_{6\text{-}18}$-C-$X_{2\text{-}13}$, in which C is cysteine and X is any amino acid), with a high proportion of non-polar amino acids. They are two-domain proteins (one domain hydrophilic, the other hydrophobic) capable of self-assembly at hydrophilic-hydrophobic interfaces (interfacial self-assembly) into amphipathic films which may be very insoluble (protein films formed by *S. commune* SC3 are insoluble in most aqueous and organic solvents).

The hydrophobins are a large and diverse family of proteins which contribute to the non-specific interactions which assist microorganisms to attach to surfaces (Marshall, 1991). As such, they have been suggested to have roles in spore dispersal and adhesion (particularly in pathogens) as well as during morphogenesis (Hazen, 1990; Stringer *et al.*, 1991; Wessels *et al.*, 1991; Wessels, 1992; St Leger *et al.*, 1992; Talbot *et al.*, 1993; Bidochka *et al.*, 1995a,b). In the morphogenetic context it is important to remember that there are numerous hydrophobins which may function differently and at different times. The *S. commune* SC3 hydrophobin coats aerial hyphae and hyphae at the surface of fruit bodies, the SC4 hydrophobin coats voids (air channels?) within solid fruit body tissues (Wessels, 1994b). Both confer hydrophobicity to these surfaces, but since hydrophobins form amphipathic layers, they can also make hydrophobic surfaces wettable. Teflon sheets immersed in SC3 hydrophobin become coated with a strongly adhering protein film that makes the plastic surface completely wettable (Wösten *et al.*, 1994). The hydrophobins alone suggest mechanisms which may be responsible for adherence of hyphae to each other and to other surfaces (Fig. 5.17). More generally, they indicate that the surface properties of the hypha can be controlled and

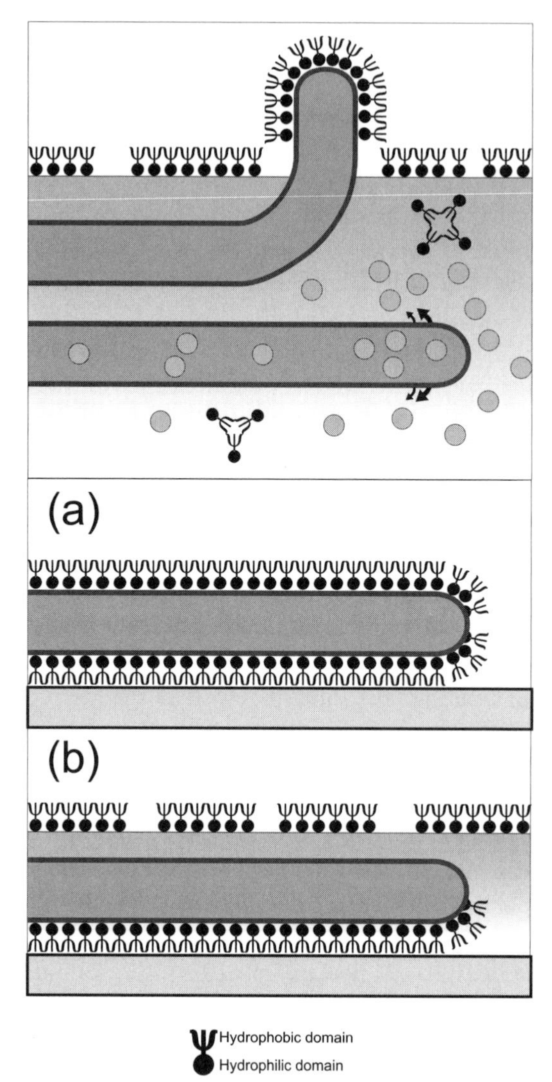

Figure 5.17 Fungal hydrophobins. Top panel depicts the excretion of hydrophobins into the culture medium by submerged hyphae. The hydrophobins assemble to form a hydrophobic coating on the surface of aerial hyphae to protect emergent structures (modified after Wessels, 1992). Bottom panel shows two models for the attachment of hyphae to hydrophobic surfaces. In (a) the hypha is growing in air over a hydrophobic substrate and secreted hydrophobin monomers assemble directly on the hyphal surface exposed to air and between the wall and the hydrophobic substratum, firmly attaching the hypha (observed with hyphae of *Schizophyllum commune* forced to grow over dry Teflon. In (b) it is assumed that free hydrophobin monomers are secreted into mucilage surrounding the hyphae; by assembling on the hydrophobic surface, the hydrophobin creates a hydrophilic surface to which the hypha can be attached (Wösten *et al.*, 1994).

manipulated to serve particular morphogenetic purposes as a result of specific gene expression.

5.5 Overview

Whatever genes are directly involved in morphogenesis, they are presumably ultimately controlled in some way by the transcriptional regulators produced by the mating type factors (section 5.2.2). Certainly, most of the recognisable developmental-specific genes seem to be transcriptionally regulated (de Vries *et al.*, 1980; Huang and Staples, 1982; Lee and Dean, 1993; Schuren *et al.*, 1993). However, the translational regulation observed in *Aspergillus* conidiation is a powerful means of relating entry into a developmental pathway to the nutritional status of the supporting mycelium. Given the prevalence of data which indicate that mycelia (i) need to develop a state of competence before they are able to undertake a developmental pathway, and (ii) can be precipitated into a developmental pathway by a variety of environmental signals, it is difficult to believe that translational triggering is not widely used as a regulator throughout the higher fungi.

Another message which comes clearly from these studies is that recessive mutations can lead both to loss and gain of the ability to form multicellular structures. As examples we can cite the *scl* mutants of *C. cinereus* which are involved in fruit body initiation and which have lost the ability to form sclerotia. Contrast these with the *fis* mutants, some of which cause monokaryotic fruiting (Uno and Ishikawa, 1971), and the *roc* gene, which causes stromatic proliferations (Nyunoya and Ishikawa, 1979) of *C. cinereus*, and the *hap*, *fi* and *fb* genes in *Schizophyllum* which confer on the monokaryon the ability to form a fruit body, a phenotype which is normally a character of the dikaryon. Attempts have been made to simplify many of these observations into a single developmental pathway (Esser and Hoffman, 1977; Esser *et al.*, 1977; Meinhardt and Esser, 1983; Fig. 5.15), yet much of the evidence points to there being a number of discrete partial pathways which can run in parallel. This appears to be reflected in the fact that variation in fruit body morphology is common in higher fungi (Watling, 1971; Chiu *et al.*, 1989) and can span generic (Bougher *et al.*, 1993) and even wider taxonomic boundaries (Watling and Moore, 1994).

Consideration of these fruit body polymorphisms has led to the suggestion that normal morphogenesis may be an assemblage of distinct developmental subroutines (Chiu *et al.*, 1989). This concept views the genetic control of overall morphogenesis as being compartmentalised into distinct segments which can be put into operation independently of one another. Thus, this model postulates subroutines for hymenophore, hymenium, stem, cap, etc., which in normal development appear to be under separate genetic control (Fig. 5.16). In any one species they are thought to be invoked in a specific sequence which generates the particular ontogeny and morphology of that species but the same subroutines may be invoked in a different sequence as an abnormality in that same species or as the norm in a morphologically different species. The model provides a unifying theme for categorising fruit body ontogeny and for clarifying phylogenetic and taxonomic relationships (Watling and Moore, 1994).

Using what is known about the few systems which are reasonably well understood (mating type factors and conidiation regulators), it is tempting to speculate on the genetic architecture which might underpin such a model. A word of warning is necessary, though, because there seem to be some major differences in gene regulation between the different groups of fungi (Jacobs and Stahl, 1995). Fortunately, there is a good catalogue of major similarities between fungi and other eukaryotes. As in other eukaryotes, fungal genes are controlled by regulatory proteins which can bind to short sequences upstream of the start of transcription. The structure of these proteins in fungi and the sequences they recognise show homologies with those in other eukaryotes. A major difference, though, is the rarity of CAAT and TATA boxes in fungi (even when present in a sequence they can be deleted usually without effect on formation of transcription complexes). Remarkably, efforts to express genes of filamentous fungi introduced into yeast have failed, and expression of ascomycetous genes in basidiomycetes has resulted, in most cases, in partial or total loss of regulation. Such observations imply that gene regulation mechanisms may be specific at a high taxonomic level; certainly beyond the family level and perhaps at phylum or subphylum level. Whilst of great interest from the point of view of evolution, where does this leave attempts to use observations made in one group for prediction and speculation in other groups?

It probably has little effect on such speculations, providing they do not attempt to explain the unknown in too much detail. The strategy of the regulation may be more similar than the tactics employed. Although the genetic structure may be different, at the level of functional expression there are many similarities. For example, the gene structures are different but mating type factors seem to serve the same function in much the same way in most fungi in which they occur. Similarly, there is an over-riding impression that the membrane and hyphal surface are crucial players in morphogenesis. Hydrophobins are now an extremely common feature throughout the fungi (Wessels, 1996) and represent the sorts of proteins which can manipulate the surface properties of hyphae. There must be many more such proteins awaiting discovery.

Development of form

In this chapter I will describe and illustrate the formation of fruit bodies and related structures from their earliest stages (fruit body 'initials') through to maturation. This will include discussion of the cell types concerned and their patterns of distribution, the development of form and the way in which tissue domains are defined, and aspects of fruit body construction that relate to, or are determined by, the mechanics and physical structure of the object. Development is a dynamic process. In a conventional book, conveying the full dimensionality of morphogenesis is extremely difficult. Illustrations are in two dimensions, but morphogenesis occurs in four – the three dimensions of space *and* the fourth dimension of time. Please remember that as you now continue.

If the activities of any organism can be described as having a purpose, then the purpose of the activities described so far in this book is to provide the fungi concerned with reproductive potential. Resources that the invasive, exploratory mycelium has won are not squandered on vegetative growth. Rather, as we have seen, from a very early stage the mycelium puts aside reserves for later use in asexual and/or sexual reproduction. The network of metabolic regulatory mechanisms provides some sort of partitioning system that diverts part of the absorbed nutrient into reserve materials. With the exception of perennating structures, like dormant spores, overwintering sclerotia, or some perennial (bracket) fruit bodies, fungi store these reserve materials for short times. Physiological and genetic data show that reproduction occurs when the mycelium reaches a competent state. Competence presumably amounts to the accumulation of sufficient reserves (strictly speaking, the internalisation of sufficient substrates; see section 4.1 above) to permit entry into the asexual reproductive pathway or to support the minimum sexual fruiting structure, depending upon other genotypic factors (the mycelium must have a particular genotype to form a fruit body), and also depending upon the environmental conditions.

In the construction of fruiting structures, the emphasis is clearly on designs which permit production of a large number of propagules, and also on dispersal of those propagules. Consequently, fruit bodies feature mechanisms and structures to expand spore production and to make dispersal more effective. However, another important emphasis is that reproduction should be guaranteed. There is no point (i.e. no selective advantage) in producing vast numbers of spores in a beautiful piece of cellular architecture which falls over and rots before the progeny are dispersed. Fungal fruiting bodies consequently manifest considerable developmental plasticity which gives them the opportunity to adapt to unusual or adverse conditions, at least partially, and produce at least some progeny spores. In the ultimate, some fruiting structures may be aborted and their substance converted and translocated to support the growth of other fruit bodies elsewhere. Regeneration of fruit body initials on the remains of old fruit body tissues (section 6.5.2), and failure of supernumerary initials in support of development of a lesser number are so commonly encountered as to be part of the normal morphogenetic strategy of most higher fungi.

In discussing the formation of fruit bodies of *Coprinus sterquilinus*, Buller (1931, pp. 127–129) described how

> Tiny rudiments of fruit-bodies ... arise simultaneously at many points everywhere over the surface of the substratum. Of the scores or hundreds of rudiments ... only one or a few can possibly survive, owing to the size of the fruit-bodies to be produced and the limitation in the quantity of nutrient substances stored up within the mycelium. ... Certain rudiments are therefore selected for further development. The selection takes place in two stages. [In] the first selection ... illuminated rudiments on the upper exposed surfaces of the substratum are prevented from growing ... by the inhibitory action of light. A second selection ... depends on nutrition and mechanical opportunity ... rudiments which happen to occupy the most favourable situations in respect to the mycelium and to free space for growth obtain the most nutriment, grow fastest, and ... survive and

attain maturity. The more vigorous development of one rudiment causes the entire inhibition of growth of all the smaller neighbouring rudiments within a certain radius.

Thus, there is a pattern in the distribution of fruit bodies which is probably determined by nutrient supply and environmental conditions (Madelin, 1956a,b; see also the introduction to Chapter 4 and section 4.3.1), but the pattern emerges from a more uniformly distributed field of fruit body initials through one or more selection processes. The same sort of selective implementation of development as a result of competition and selection is probably applied at other levels within the fruit body. It is likely that metabolic influence over a local region determines morphogenetic pattern at all dimensions from the cellular to the ecological.

6.1 Initiation of structures

Aggregation of hyphae to form tissues begins from the vegetative hyphae of a mycelium, which is itself a diverse and dynamic population comprising a variety of differentiated phenotypes (Boddy and Rayner, 1983a,b; Gregory, 1984; Rayner and Webber, 1984).

The basic ground tissue of fungal structures is described by the general term 'plectenchyma'. This word is derived from the Greek *plekein*, meaning to weave, combined with *enchyma* which means infuse or permeate. The overall meaning implies an intimately-woven tissue and adequately describes a structure formed by entwining hyphae. There are two types of plectenchyma: 'prosenchyma' (Greek *pros*, toward; i.e. approaching or almost a tissue) is a loosely organised tissue in which the components are recognisable as hyphae; and 'pseudoparenchyma' (Greek *pseudo*, false with *parenchyma*, a type of plant tissue) which is seen to be comprised of tightly packed cells resembling plant tissue in microscope sections. In pseudoparenchyma, the hyphal origin of the tissue is not obvious in individual microscope sections, though scanning electron microscopy reveals this origin well enough (Read, 1994). Read (1983, 1994) and Read and Becket (1985) have suggested that the term 'cellular element' should be used in fungi in preference to the word 'cell' because fungal cells are always hyphal compartments and consequently different from the concept of the cell applicable to plants and animals (see section 2.1). Macroscopic fungal structures formed by hyphal aggregation may be either linear organs; variously called strands, rhizomorphs and fruit body stems, but also including synnemata (erect, parallel aggregations of hyphae producing conidia); or globose masses, which would include sclerotia and the more familiar fruit bodies, as well as stromata (a mass of vegetative hyphae which often contains or bears fruiting structures).

6.1.1 *Linear organs*

Association of morphologically similar hyphae in parallel aggregates is quite common. Mycelial strands and cords develop under circumstances which require large scale movement of nutrients (and water) to and from different parts of the mycelium. They are formed in mushroom cultures, for example, and channel nourishment towards developing fruit bodies (Mathew, 1961). Some mycorrhizal fungi also form strands which radiate into the soil, greatly supplementing the host plant's root system and gathering nutrients for the host (Read *et al.*, 1989; Read, 1991). Interestingly, the pseudoparenchymatous tissue of the mycorrhizal sheath is not formed in the soil or *in vitro*. This tissue development requires a surface, oxygen and a supply of nutrients (Read and Armstrong, 1972).

Strands are also migratory organs, advancing from one food base over nutrient-poor surroundings to explore for new sources of nutrient. Strands of the dry-rot fungus, *Serpula lacrimans*, may penetrate several metres of brickwork from a food base in decaying wood (Butler, 1957, 1958; Watkinson, 1971; Jennings and Watkinson, 1982) and can grow over plastic sheets and many building materials (Jennings, 1991). Strands hasten capture of new substrate because they ensure that a large number of hyphae reach the new resource at the point of contact with it. For pathogens this is described as increasing the inoculum potential of the fungus (Garrett, 1954, 1956, 1960, 1970) but it is just as important a strategy for saprotrophs to concentrate mycelial resources on capture and consolidation of a new food base. Strands provide translocation routes in both directions (see above), and their distribution around a food base changes with time (Thompson, 1984). Resorption of hyphae of the strand, followed by redistribution of the nutrients so recovered towards active growth points enable migration of the colony from place to place (Rayner *et al.*, 1985a,b; Boddy, 1993; Rayner *et al.*, 1994).

Mycelial strands (Fig. 6.1) arise when branches of a leading hypha form at acute angles and grow parallel to the parent hypha which also tends to grow alongside other hyphae it encounters. Anastomoses between the hyphae of the strands consolidates them into a bundle. Also, narrow hyphal branches (called 'tendril' hyphae) from the older

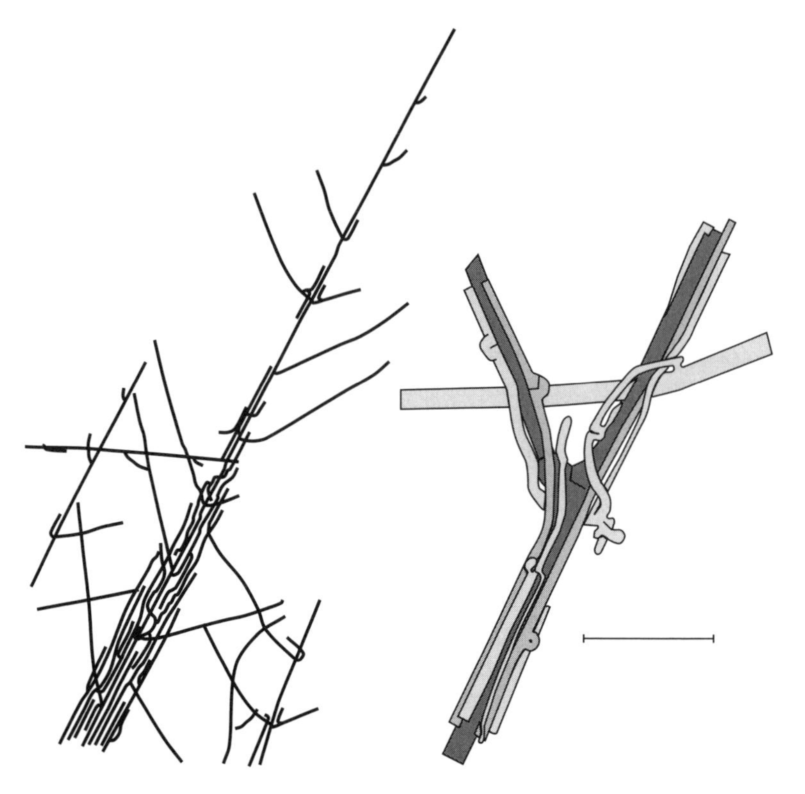

Figure 6.1 Mycelial strands of *Serpula lacrymans*. The strand is shown in a general habit sketch on the left. The drawing on the right shows tendril hyphae intertwined around main hyphae (scale bar = 20 μm). Redrawn after Moore (1995).

regions of the main hyphae intertwine around the other hyphae (Nuss *et al*., 1991). Some of the central hyphae may be wide-diameter and thin walled from the beginning (called vessel hyphae). In older strands narrow, but thick-walled, 'fibre' hyphae are formed, running longitudinally through the mature strand, and presumably strengthening it.

Strand formation tends to occur in ageing mycelium on an exhausted substrate. It has been argued that strand formation results from the limitation of new growth to the immediate vicinity of the remaining nutrient (Watkinson, 1975, 1979). When the substrate is exhausted the existing hyphae themselves are likely to be the main nutrient stores (especially of nitrogen). Consequently, new hyphae formed in those conditions are expected to grow alongside existing hyphae because the latter are the source of nutrition. For as long as it remains the case that the existing hyphae are the main supplier of nutrient the integrity of the strand will be reinforced. However, when the strand encounters a new external source of nutrients which is greater than its own endogenous supply, the stimulus to cohesive growth is lost. When that happens, the full inoculum potential of the strand is released as spreading, invasive, hyphal growth from the numerous hyphae of the strand envelops the new substrate.

Although mycelial strands contain morphologically differentiated hyphae, these are only loosely aggregated. Some fungi produce highly differentiated 'rhizomorphs' which are aggregations of hyphae with well developed tissues and a general appearance very much like plant roots (Fig. 6.2). An important example is *Armillaria mellea*, which is a pathogen of trees and shrubs and spreads from one root system to another by means of its rhizomorphs (Rishbeth, 1985). Rhizomorphs serve translocatory and migratory functions and, as with strands, translocation is bidirectional (Granlund *et al.*, 1985). The fundamental difference between strands and rhizomorphs is that the latter have a highly organised apical growing point and exhibit extreme apical dominance. The apical region of the rhizomorph contains tightly packed cells and behind the tip is a medullary zone containing swollen, vacuolated and often multinucleate cells surrounded by copious air- or mucilage-filled spaces. The medullary region forms a central channel through the rhizomorph. In mature rhizomorphs this channel is traversed by narrow fibre hyphae and wide-diameter vessel hyphae. At the periphery of the rhizomorph, the cell compartments are smaller, darker, and have thicker walls, and there is a fringing mycelium extending outwards between the outer layers of the rhizomorph, resembling the root-hair zone of plant roots (Townsend, 1954; Motta, 1969, 1971; Botton and Dexheimer, 1977; Motta and Peabody, 1982; Powell and Rayner, 1983; Cairney *et al.*, 1989; Cairney and Clipson, 1991).

Garrett (1953, 1970) and Snider (1959) described the development of rhizomorphs *in vitro*, but details of cellular interactions at the site of their inception are sparse. Rhizomorphs are usually initiated as compact masses of aggregated cells. The ultimate origin has been ascribed to locally enhanced acute-angled branching of some marginal hyphae in a mycelium; a phenomenon described as 'point-growth' (Coggins *et al.*, 1980). It is likely that these linear organs originate from originally unpolarised hyphal aggregations which somehow become apically polarised. Rayner *et al.* (1985a) suggested that all linear hyphal aggregations could be related together in a hierarchy depending on apical dominance. Mycelial strands and rhizomorphs are extremes in a range of hyphal linear aggregations. A range of other forms have been recognised and given particular names (Townsend, 1954; Butler, 1966; Garrett, 1970).

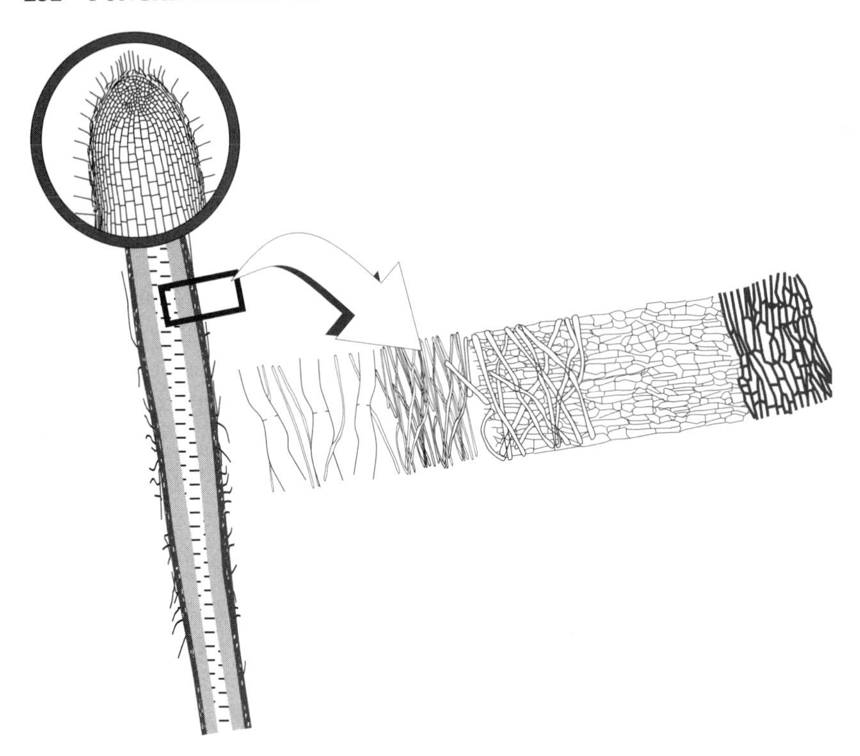

Figure 6.2 Rhizomorph structure. The diagram on the left is a sectional drawing showing general structure, with the apical region magnified to show the appearance of a growing point of tightly packed cells (redrawn after de Bary, 1887). Behind the tip is a medullary zone containing swollen, vacuolated and often multinucleate cells surrounded by copious air- or mucilage-filled spaces. The medullary region forms a central channel through the rhizomorph traversed by narrow fibre hyphae and wide-diameter vessel hyphae in mature tissues. The microscopic appearance is indicated in the drawing on the right. Redrawn after Moore (1995).

Whilst a complex terminology enables distinctions to be made between species and their lifestyle strategies, it should not be allowed to obscure the close developmental relationships which exist between the structures. Interpreting linear structures as a related group differing in apical dominance is a valuable way of drawing them together for comparison. Fruit bodies should also be included in this arrangement. In moist tropical forests aerial rhizomorphs, mainly of *Marasmius* spp., form a network which intercepts and traps freshly fallen leaves, forming a suspended litter layer (Hedger, 1985; Hedger *et al.*, 1993). Hedger *et al.* (1993) showed that the rhizomorphs have a reduced fruit body cap at their tips which may protect the apex against desiccation. Overall, these aerial linear organs are functionally analogous to soil rhizomorphs, but developmentally analogous to indefinitely-extending fruit body stems (Jacques-Félix, 1967). As mentioned above, many fruit bodies are served by radiating mycelial strands which convey nutrients towards the fruit body. In some cases these are so highly developed that the junction between strand and fruit body stem is obscure. The term 'radicating' is used to describe fruit bodies whose stems are elongated into root-like *pseudorhizas* which extend to the surface from some buried substrate. Even in species which do not normally produce pseudorhizas, they can be induced by keeping fruiting cultures in darkness (Buller, 1924) so that etiolated fruit bodies are formed (see above) in which the stem base can be extended for many centimetres so as to drive the fruit body primordium at its tip towards any source of light. Whether the linear structures are classifiable as rhizomorphs, pseudorhizas, strands or extending stems is less important than the implication that a close morphogenetic relationship underlies all fungal linear hyphal aggregations.

The similarity, at least in sections prepared for the light microscope, with the plant root has prompted the suggestion that rhizomorph extension results from meristematic activity (Snider, 1959; Motta, 1967; 1969, 1971; Motta and Peabody, 1982; Garraway *et al.*, 1991). This is such a completely wrong interpretation, unfortunately peddled in otherwise respectable books (e.g. Hawksworth *et al.*, 1995), that it is worth widening the discussion a little here to firmly dismiss the notion.

Cells of a plant apical meristem divide in various planes to give an orderly distribution of cells and allow the meristem to grow as an organised whole. Some of the cells in a true plant meristem divide by wall formation at right angles to the surface and subsequent cell enlargement consequently lengthens the organ. Longitudinal divisions are also crucial to increase the girth and divisions in other planes contribute to increase in volume. Motta (1969) claimed that cell divisions in a narrow zone about 25 μm behind the extremity of a rhizomorph of *Armillaria* were effected by walls being formed in diverse directions. His comparisons with shoot- and root-tips of Phanerogams induced him to suggest the existence of an apical meristem. However, in the fungal hypha cross-walls are formed at right angles to the long axis of the hypha (see section 2.3); true longitudinal wall formation occurs only in the very specialised basidia of Tremellales. Thus, a meristem-like structure would be totally alien to the growth strategy of the fungal hypha. The kingdom is simply not equipped with the cell biological mechanisms required by apical meristems.

The 'meristem' analogy has unfortunately also been used in two other situations. Tissue layers involved in rapid cell formation in which the hyphae run parallel to one another have been recognised in agaric fruit bodies. They frequently demarcate the major tissue layers of the fruit body and were called meristemoids by Reijnders (1977). Reijnders, though, emphasised their difference from true meristems: "These meristemoids closely resemble the meristems of the phanerogams; sometimes they are referred to as such. This is not correct because the meristemoids initiate from hyphae which together form a simple tissue; cell division, therefore, can only take place in one direction, *and the cell walls between the cells of different hyphae are double*" (Reijnders, 1977). Again, some fungi have been described as having 'meristem arthrospores' or 'meristem blastospores' (Hughes, 1953), but this is, again, an unfortunate misuse of the word.

Hughes (1971) compared cell proliferation in fungi with some lower plants and concluded that " . . . septation of fungal conidia results, not from the activity of a single [i.e. meristematic] cell, but by division of any or all of the cells. An apparent dictyoseptate condition [dictyoseptate, i.e. having both longitudinal and transverse cross-walls] of some conidia or other reproductive units in fungi may arise from the compacting of a coiled septate hypha or of hyphae which may branch repeatedly to form a more or less solid mass of cells." In the rhizomorph, ultrastructural (especially scanning electron microscopic) observations clearly demonstrate the hyphal structure of the rhizomorph tip (Botton and Dexheimer, 1977; Powell and Rayner, 1983; Rayner *et al.*, 1985a) and, as suggested by Rayner *et al.* (1985a), the impression of central apical initials giving rise to axially arranged tissues in a 'meristematic' fashion is undoubtedly *an artifact caused when compact aggregations of parallel hyphae are sectioned.* It is quite categorically the case that meristems do not occur in fungi.

6.1.2 Globose structures

As the previous section implies, there is a gradation and close relationship between different fungal structures which can sometimes make categorisation uncertain. Nevertheless it is helpful to distinguish, at least initially, between linear and globose structures. Sclerotia are representative globose structures, being hyphal aggregations in which concentric zones of tissue form an outer rind and inner medulla, sometimes with a third zone (the cortex) distinguishable between these two tissues. Sclerotia arise as hyphae ramify locally, their branches intertwining to form the multihyphal aggregate which finally may have its cells so compressed that in microscope sections the tissues look like plant parenchyma (pseudoparenchymatous fungal tissues). The term sclerotium is a functional one, describing tuber-like objects which detach from their parental mycelium at maturity. The structure can remain dormant or quiescent when the environment is adverse and then, when conditions improve, germinate to reproduce the fungus. On the basis of development, there are a number of different sclerotial types (Townsend and Willetts, 1954). Some sclerotia consist of only a few cells and are therefore of microscopic dimensions. At the other extreme, the sclerotium of *Polyporus mylittae*, found in the deserts of Australia, can reach 20–35 cm in diameter.

Sclerotia are resistant survival structures. The outer rind layer is composed of tightly-packed hyphal tips which become thick-walled and pigmented (melanised) to form an impervious and protective surface layer, so dormant sclerotia may survive for several years (Sussman, 1968; Coley-Smith and Cooke, 1971; Willetts, 1971). The bulk of the sclerotium is formed by the medulla, and its cells (and those of the cortex where present) may accumulate reserves of glycogen, polyphosphate, protein and lipid as cytoplasmic deposits or in the form of a thick secondary cell wall composed of unusually thick fibrils (Fig. 4.3; Waters *et al.*, 1972). These accumulated reserves are utilised when sclerotia 'germinate' to form mycelium, conidia or fruit bodies. The mode of germination may be a matter of size in some cases. Sclerotia of *Coprinus cinereus* which are quite small (about 250 μm diameter) germinate to produce a mycelium but many other species produce spores and fruit bodies. The giant sclerotium of *Polyporus mylittae* can form a fruit body without being supplied with water as the flesh is honeycombed with blocks of translucent tissue in which the hyphae form copious amounts of an extracellular gel which is thought to serve as both nutrient and water store (Macfarlane *et al.*, 1978).

The sclerotium developmental pathway comprises initiation, development and maturation. During initiation, mycelial hyphae begin to aggregate to form small knots of hyphae which are the sclerotium initials. These then undergo a development phase, when the initials expand and grow to full size, accumulating nutritional reserves from the parent mycelium. During maturation the tissue layers become clearly demarcated, especially the surface layer with pigmentation of its constituent cell walls, but maturation also involves conversion of the reserve nutrients to forms suitable for long-term storage (Chet and Henis, 1975). Townsend and Willetts (1954) and Willetts and Wong (1971) distinguished several kinds of development in sclerotia. In the loose type (*Rhizoctonia solani* as an example) sclerotial initials arise by branching and septation of hyphae; the cells become inflated and fill with dense contents and numerous vacuoles. The mature sclerotium is pseudoparenchymatous, but with an open structure and an obvious hyphal nature. At the periphery of this type of sclerotium the hyphae are more loosely arranged and generally lack thickened walls (Willetts, 1969). The type of development called 'terminal' is characterised by repeated dichotomous branching and cross-wall formation. It is exemplified by *Botrytis cinerea* and *B. allii*. Eventually, the hyphal branches cohere to give the appearance of a solid tissue. In *Botrytis*, a (usually flattened) mature sclerotium may be about 10 mm long, 3 to 5 mm wide and 1 to 3 mm thick. It is differentiated into a rind composed of several layers of round cells with thickened, pigmented walls, a narrow cortex of thin-walled pseudoparenchymatous cells with dense contents, and a medulla of loosely arranged filaments. *Sclerotinia gladioli* (which causes dry rot of corms of *Gladiolus*, *Crocus* and other plants) has the lateral type of sclerotium development in which sclerotial initials arise by formation of numerous side branches from one or more main hyphae, or strands of several parallel hyphae. The mature sclerotium, about 100–300 μm in diameter, is differentiated into a rind of small thick-walled cells and a medulla of large thin-walled cells.

More complex sclerotia and other types of sclerotial development have been found (Butler, 1966; Chet *et al.*, 1969) and many fungi enclose portions of the substrate and/or substratum within a layer of pigmented, thick-walled cells. The inclusions include host cells if the fungus is a pathogen, and the whole structure may be regarded as a kind of sclerotium. The enclosure results from the formation of the impervious layer of rind tissue, which is especially protective against desiccation. This may develop in other circumstances as an incomplete covering over the surface of a mycelium, forming what has been called a pseudosclerotial plate. This is quite common in Petri dish cultures, but occurs in nature, too. For example, hazel branches have been shown to be 'lashed' together by a pseudosclerotial plate being formed over the surface of hyphal aggregations of the fungus *Hymenochaete corrugata* (Ainsworth and Rayner, 1990). Similar adhesions between rhizomorphs and leaf litter in tropical forests create a litter layer suspended in the canopy some distance off the ground (Hedger *et al.*, 1993).

Pseudosclerotial plates seem to form to protect a persistent mass of mycelium, if they completely enclose the mycelium on which they are formed, the structure which results is called a sclerotium even if it also contains non-fungal material. Since a number of dissimilar structures are encompassed by the name it is thought that the different forms arose by convergent evolution, most probably evolving, in Ascomycotina, from aborted spore-forming organs like perithecia, cleistothecia and conidial masses (Willetts, 1972, 1997; Cooke, 1983; Willetts and Bullock, 1992) and in Basidiomycotina from aborted fruit body initials (Moore, 1981a). Willetts (1997) re-defined some structures formed in the Sclerotiniaceae which had previously been described as sclerotial stromata (Whetzel, 1945), making a firm distinction between sclerotia and stromata. He maintains that the stromata are 'produced within host tissue by random or haphazard aggregation of vegetative hyphae' whereas sclerotia are . . . 'produced under the control of morphogenetic factors by branching and growth of a compact tissue formed within the host or substrate.' (Willetts, 1997). The first type of resting structure to evolve is thought to have been a stroma arising by branching and interweaving of infection hyphae within host tissues. Subsequently, a variety of stromata with more compact tissues evolved from this initial loose hyphal network. Willetts and Bullock (1992) suggested that sclerotia evolved from aborted spore-forming (macroconidiogenous) tissue, but Willetts (1997) presented evidence indicating that sclerotia probably predated macroconidia. His preferred interpretation is that both evolved from a common ancestral tissue organisation; sclerotia as an adaptation to cold and wet conditions and macroconidia as an adaptation to warmer, drier conditions.

Close relationship between sclerotia and reproductive tissues is shown in many ways, including their ability to give rise to sporulating organs immediately they germinate. High humidity inhibits conidial differentiation in *Monilinia fructicola* and occasionally leads to formation of sclerotium-like stromata (Willetts, 1969; Willetts and Calonge, 1969), providing an example of the environment directly influencing the pathway of development. Another example provides evidence for the same genes being involved in both sclerotium and fruit body initiation in the basidiomycete *Coprinus cinereus* (Moore, 1981c; and see section 5.4.3 and Fig. 5.16). Sclerotia of *Coprinus cinereus* are polymorphic in their internal structure. Volz and Niederpruem (1970) reported that mature sclerotia possessed an outer unicellular rind layer, composed of cells with thickened and pigmented walls, which enclosed a medulla composed of a compact mass of thin-walled bulbous cells and accompanying hyphae. In a later study (Waters *et al.*, 1975a) the sclerotia were found to have a multilayered rind enclosing a compact medulla composed predominantly of thick-walled cells (Waters *et al.*, 1972, 1975a). Hereward and Moore (1979) showed that these different structures were expressions of a genetic polymorphism in the natural population of this fungus. The more common aerial sclerotium of the two polymorphic forms (called the Z-type) had a rind which was only one cell thick; the other form (the H-type) had a rind many cells thick which extended to at least half the diameter of the sclerotium. The Z-type proved to be the wild-type and genetically dominant form. The H-type was a naturally occurring variant caused by an allele of the gene *scl-1* which was designated *scl-1*^{H}. Though recessive to wild-type, the *scl-1*^{H} allele was dominant to strains carrying another allele of *scl-1* (designated *scl-1*^{0}) which failed to produce sclerotia. Since *scl-1*^{H} caused the formation of sclerotia having an abnormal proliferation of cells, particularly in the rind, it was suggested that the *scl-1* gene may be involved in the control of the disposition and extent of tissue layers during sclerotium development. The *scl-1*^{0} strains which were unable to make sclerotia as monokaryons, were able to form both sclerotia and fruit bodies when dikaryotised with wild type monokaryotic strains. However, dikaryons 'homozygous' for sclerotium-defective genes were unable to make fruit bodies. The conclusion was that a common initiation pathway gave rise to hyphal aggregations which, depending upon environmental conditions, either developed axial symmetry and became fruit body initials, or developed radial symmetry and became sclerotia (Moore, 1981c; Fig. 5.16).

Just as the initially formed structures may develop into sclerotia or fruit bodies in *C. cinereus*, so it is also the case that not all hyphal aggregates produced by the ascomycete *Sordaria* are protoperithecia. Some are larger, often less spherical, lacktrichogynes and are sterile (Broxholme *et al.*, 1991). At least some may be sclerotia, implying that here, too, there are two morphogenetic pathways.

6.1.3 Morphogenetic patterns

Development of any multicellular structure in fungi requires that hyphae grow towards one another to cooperate in forming the differentiating organ. Changing the fundamental growth pattern of the hyphae must involve reactions to specific chemicals but knowledge of the sex hormones of lower aquatic and terrestrial fungi and the pheromones determining mating-type specific agglutination in yeasts are about the limit of current knowledge about the control of hyphal interactions.

As pointed out by Reijnders and Moore (1985), fungi offer a unique system for study of cell-to-cell tropisms and specific cell-to-cell adhesion since the change from one state to another is part of their normal development. Once the mycelial knot or tuft (constituting the initial of the developing fruit body, for example) is established, major tissue domains are demarcated very quickly. Fruit bodies of Hymenomycetes are particularly interesting in this respect because although similar phenomena occur also in many ascomycetes, the fruit bodies formed by Hymenomycetes are in general much larger and more complex than those found in other fungi. In *Coprinus cinereus*, fruit body initials only 800 μm tall are clearly differentiated into cap and stem (Moore *et al.*, 1979) though this is only 1% of the size of a mature fruit body (Figs 6.3 and 6.4). Similarly, in *Agaricus bisporus*, 2 mm tall primordia consist of a disoriented mass of hyphae (Flegg and Wood, 1985) and yet by the time the primordium reaches 6–10 mm diameter it has differentiated into the tissue zones present in mature mushrooms. These examples emphasise that, as in plants and animals, the basic 'body plan' of the fungal fruit body is established very early in development of the 'embryonic' structure.

Development of these organised fungal structures requires that hyphae grow towards one another, cooperate and are coordinated in formation of the differentiating organ. This is the diametrically reversed character of the invasive, 'undifferentiated' mycelium. Control of this peculiar reversal in behaviour presumably depends on autotropic agents and sur-

Figure 6.3 Some basidiomycete fruit bodies. *Coprinus cinereus* (top left), *Volvariella bombycina* (top right), *Pleurotus pulmonarius* (bottom left), and a view of the underside of *Hexagonia tennis*, showing the typical polypore hymenophore. Photographs kindly provided by Prof. Siu Wai Chiu.

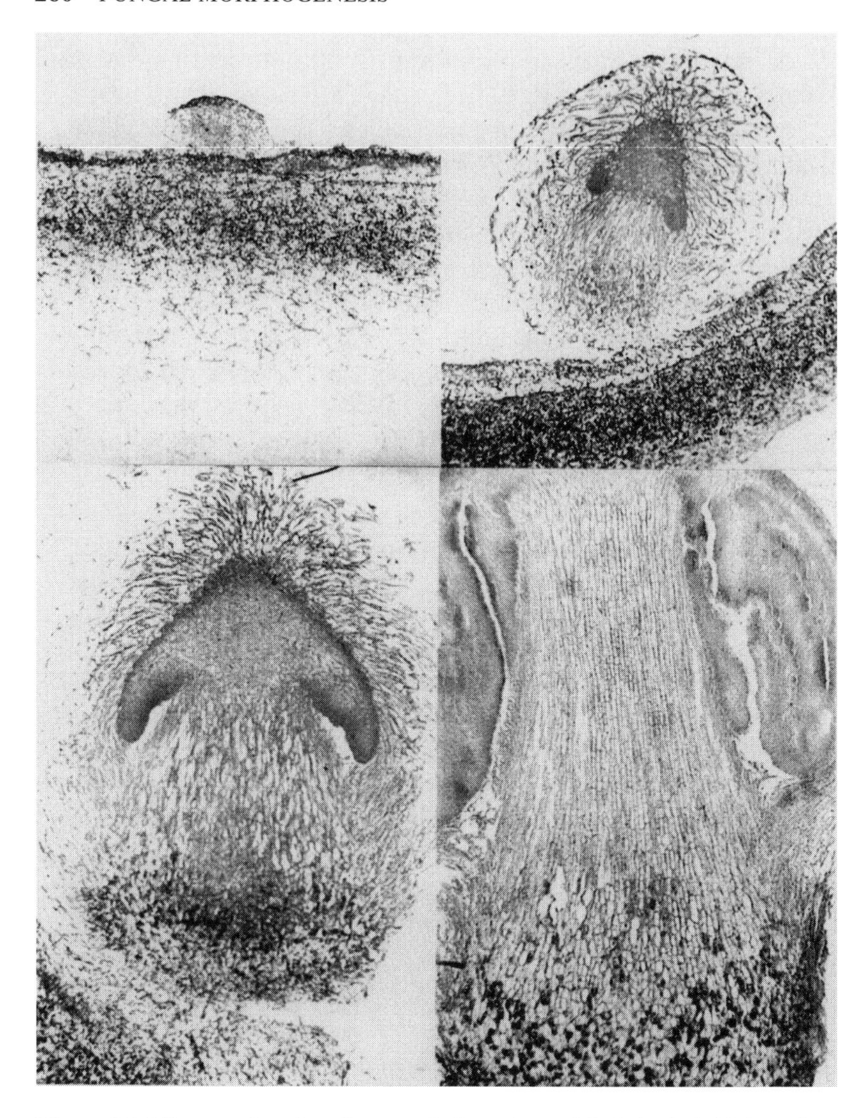

Figure 6.4 Micrographs of median vertical sections of fruit body initials and primordia of *Coprinus cinereus* to show the earliest stages in tissue demarcation. The fruit body initial at top left is approximately 200 μm tall and shows hyphal aggregation but without obvious tissue delimitation. At top right, an 800 μm initial is already clearly demarcated into veil, cap and stem tissues. Bottom left and right are 1.2 mm and 3 mm tall primordia, respectively.

face-active molecules, but nothing is known yet about their nature in higher fungi. Light microscopy does not reveal many differences between the generative hyphae (which in tissue masses have been called protenchyma) and mycelial hyphae. Generative hyphae usually have a minimum number of nuclei (i.e. increase in the number of nuclei is an aspect of cell differentiation) and may or may not bear clamp connections. Generative hyphae give rise to other (differentiated) cell types of the structure they comprise and also give rise to the hymenium (spore-bearing tissue layer) in fruit bodies. The protenchyma of the primordial fruit body displays two types of organisation: a bundle of nearly parallel hyphae, and interwoven hyphae which form a plectenchyma. Bundles of strictly parallel hyphae (called meristemoids by Reijnders (1977)) appear later in the stem and in the lateral parts of the cap.

The hyphae of these aggregates are generally embedded in a mucilaginous matrix which has been visualised in a variety of ways (Fig. 6.5; Van der Valk and Marchant, 1978; McLaughlin, 1982; Williams *et al.*, 1985; Umar and Van Griensven, 1997a). As well as in fruit body primordia, an extracellular matrix (often called a sheath) surrounding hyphal cells has been demonstrated in several fungi, including plant pathogens (Dubourdieu and Ribereau-Gayon, 1981; Benhamou and Ouellete, 1986; Chaubal *et al.*, 1991) and wood decay fungi (Daniel, 1994; Connolly *et al.*, 1995; Connolly and Jellison, 1995). Important components of these extracellular matrices are β-1$\longrightarrow$6, β-1$\longrightarrow$3-linked glucans (Dubourdieu and Ribereau-Gayon, 1981; Buchala and Leisola, 1987; Chaubal *et al.*, 1991; Nicole *et al.,* 1993, 1995) but they contain proteinaceous materials as well as polysaccharides (Palmer *et al.*, 1983). Functions proposed for the fungal extracellular matrix include: (i) recognition of the substratum, (ii) adhesion to the substratum, (iii) modification of the extracellular ionic environment, (iv) storage, concentration and retention of materials externalised by the hyphae, (v) protection against dehydration and adverse environmental changes, (vi) binding the hyphae together, and (vii) providing a medium for the transmission of enzymes or other effectors (e.g. hormones) released by the hyphae.

6.1.4 Chemoattractants in aquatic fungi

A number of water moulds included amongst the lower fungi like some of the chytridiomycetes and those distant relatives, the oomycetes, are known to produce chemoattractants involved in growth processes leading up to sexual reproduction (see reviews by Gooday and Adams, 1993;

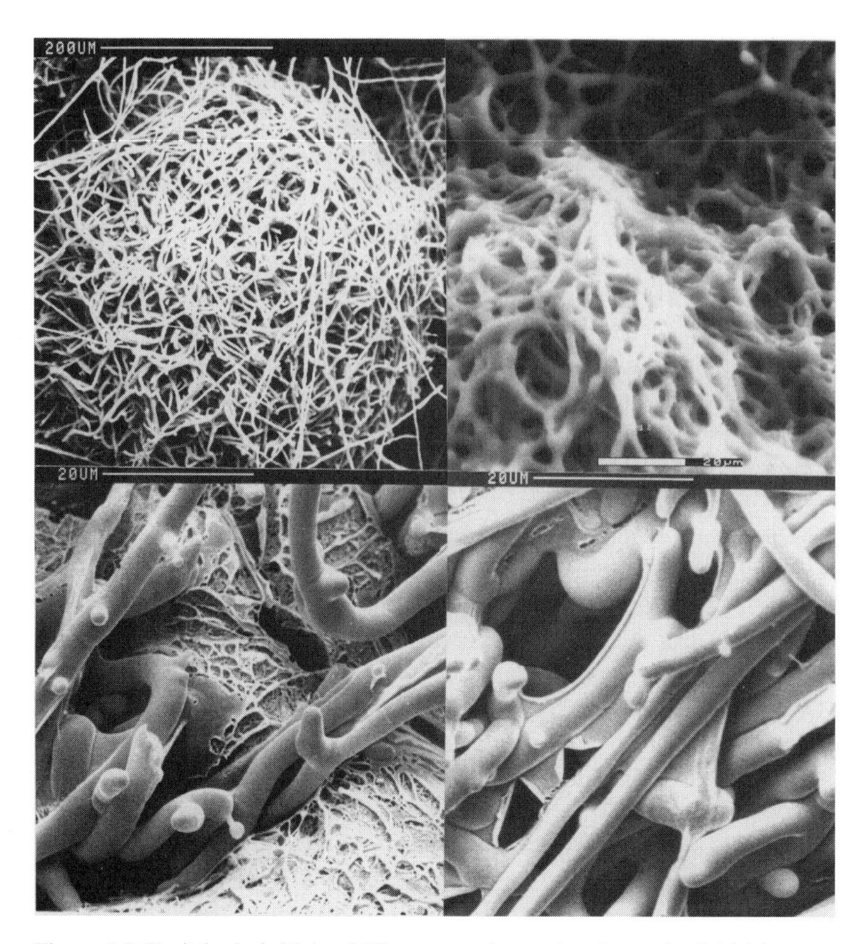

Figure 6.5 Fruit body initials of *Pleurotus pulmonarius*. An entire initial is shown at top left in a specimen prepared for cryo-scanning electron microscopy (SEM) in which surface water is sublimed away to reveal the flash-frozen hyphae beneath as a tangled mass of hyphae. Top right shows a similar specimen, but imaged using an Ectroscan environmental SEM which operates at a sufficient pressure to enable imaging of live tissue. This photograph shows that living hyphae are surrounded by a sheath of mucilage. Frozen and partially-desiccated mucilage can be imaged in conventional cryo-SEM preparations (bottom two micrographs. Photographs kindly provided by Ms Carmen Sánchez.

Gooday, 1994; Mullins, 1994). These can quite properly be described as sex hormones (Bu'lock, 1976). The aquatic chytridiomycete *Allomyces* produces uniflagellate motile gametes which are differentiated as male and female although they arise on the same haploid thallus and have the same genotype. The female gametes and gametangia produce a substance called sirenin to which the male gametes show strong chemotaxis. Sirenin is a bicyclic sesquiterpene diol (Fig. 6.6; and see section 3.15.2.1) which is probably derived from farnesyl phosphate (Nutting *et al.*, 1968; Pommerville, 1981; Pommerville *et al.*, 1988). The hormone is active at concentrations of about 10^{-10} M and it works by regulating the movement of male gametes. Female gametes are only sluggishly motile but male gametes swim in random smooth arcs interrupted by stops after which the cell swims in a different direction. Sirenin organises the direction of swimming by shortening the run between interruptions if the cell moves away from the source of hormone and diminishing the number of stops if the cell is moving towards the source. Inactivation of the hormone by the male gamete is essential to the overall activity but it is not known whether this results from enzymic breakdown or irreversible binding to some component of the cell (Machlis, 1973). This chemotaxis thus leads to cell contact which is a prelude to plasmogamy.

The female sex hormone of another water mould, *Achlya*, has also been characterised in some detail. This material, called antheridiol, is a steroid (Fig. 6.6) the activity of which can be detected by bioassay in 10^{-11} M solution (Raper, 1952; Barksdale, 1969). The mating sequence reported by Raper (1966) consisted of: development of antheridial hyphae on the male; production of oogonial initials on the female; growth of antheridial hyphae towards oogonial initials; formation of cross-walls separating oogonia and antheridia; and, after the two made contact, the antheridium grew through a lysed portion of the oogonial wall, after which its own wall was dissolved. Antheridiol is produced continuously by the female and under the influence of the hormone, branches on the putative male thallus which might otherwise grow out as vegetative branches are caused to elongate rapidly and differentiate into antheridia. The male is also induced to excrete a second hormone, oogoniol or hormone B (also a steroid), and it is in response to this that the female initiates oogonial differentiation and amplifies antheridiol levels to those which attract antheridial hyphal growth (McMorris, 1978a,b). There are thus at least two contributors to this hormonal 'conversation'; the female produces antheridiol but takes up very little itself and does not synthesise oogoniol though it does have a receptor for this hormone. On the other hand, the male makes no antheridiol but is sensitive to it, and one of the responses is to produce oogoniol to which the male is insensitive.

Antheridiol and oogoniol are thought to be alternative products of a branched biosynthetic pathway. In the male, antheridiol amplifies that branch of the pathway which leads to oogoniol synthesis and also

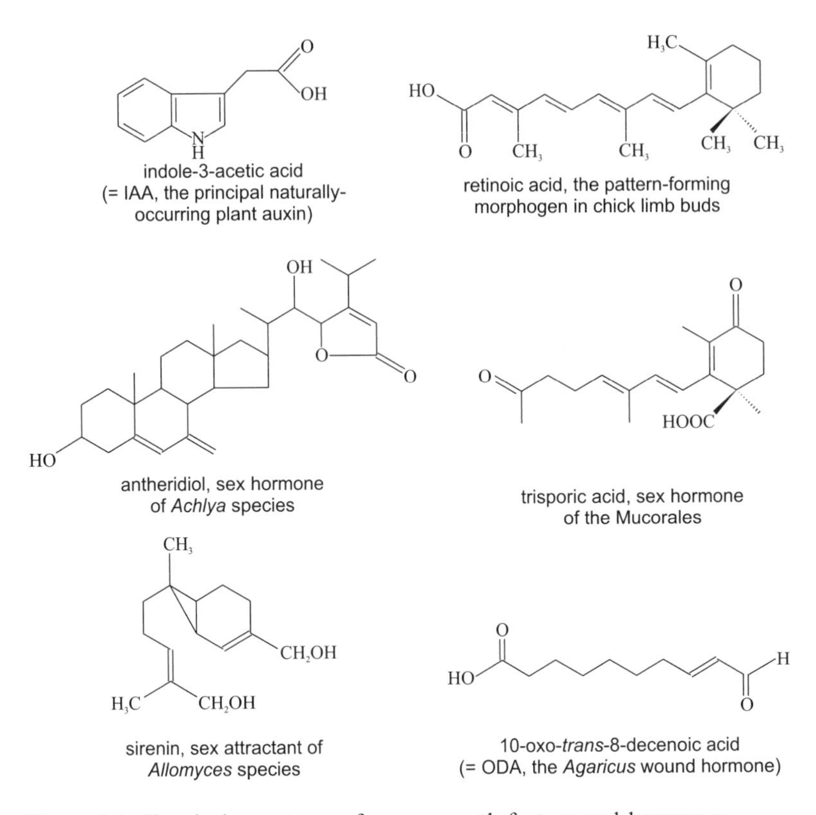

Figure 6.6 Chemical structures of some growth factors and hormones.

increases respiration, induces breakdown of glucan reserves in the cytoplasm and triggers *de novo* synthesis of cellulase. These metabolic changes contribute to processes involved in antheridium initiation, including aggregation of vesicles at the sites where initials develop (Mullins and Ellis, 1974; Horgen, 1981; McMorris, 1978a,b; Timberlake and Orr, 1984). It is likely that broadly similar responses are elicited in the female by oogoniol. It is not known how these sterols influence gene regulation but the available data seem to indicate a system very much akin to sterol regulation in animals (Horgen, 1977) where the steroid receptors are located in the nucleus and hormone-receptor complexes bind directly to DNA to regulate gene expression. However, antheridiol does not induce changes of this sort (Horton and Horgen, 1989) and the hormone binding proteins are present in the cytosol (Riehl *et al.*, 1984), though they are quite similar to other steroid hormone receptors (Riehl and Toft, 1984; Mullins, 1994).

6.1.5 Chemoattractants in the Mucorales

The only other fungi in which the activity of known hormones has been well characterised are some members of the Mucorales. These are filamentous, terrestrial, lower fungi with mycelia typically composed of unbranched coenocytic hyphae. A little way behind the advancing hyphal tips of vegetative mycelia asexual sporangiophores are produced. However, in the vicinity of a mycelium of opposite mating type, sporangiophore formation is suppressed and sexual differentiation takes place, involving formation of sexual hyphae (zygophores) which grow towards each other, fusing in pairs to form gametangia. Zygophore formation is determined by trisporic acid (Fig. 6.6); if this chemical is added to pure, unmated, cultures sporangiophore formation ceases and zygophores form instead (Gooday, 1973b, 1974b, 1994; Bu'lock, 1976). Although there are a number of trisporic acid-related compounds (and compare with retinoic acid in Fig. 6.6), some of them corresponding to intermediates in the pathway, both mating types produce and respond to the same hormone.

The trisporic acids are synthesised from β-carotene: the molecule is cleaved to retinal, a C2 fragment is lost, and then there is a series of oxidations. The complete reaction sequence occurs *in vitro* only when both plus and minus mating types are grown in mixed culture or in an experimental apparatus in which they are separated by a membrane permeable to small molecules. Both mating types have the genetic capacity to produce the enzymes of the complete pathway, but the alleles which determine the mating type repress complementary steps in the later stages of trisporic acid synthesis. Thus, in plus strains synthesis of enzymes needed to form the 4-keto group is repressed by the MT+ allele while enzymes involved in forming the 1-carboxylic acid group are repressed by MT− (Bu'lock *et al.*, 1973; Bu'lock, 1975). Each mating type thus produces a precursor which only the opposite mating type can convert to trisporic acid. The precursors diffuse between the strains and have the status of prohormones which stimulate trisporic acid synthesis. Early steps in the pathway are repressed to a rate-limiting level by a mechanism which allows activation by trisporic acid. When plus and minus strains come together, therefore, the complementary synthesis of trisporic acid consequent on the co-diffusion of the prohormone precursors leads to derepression of the early part of the pathway and amplification of overall trisporic acid synthesis (Jones *et al.*, 1981; van den Ende, 1984).

The increasing gradient of prohormone diffusing from each zygophore induces a chemotropic response. The zygophores can grow towards one another from distances of up to two millimetres (Gooday, 1975a). When the zygophores make contact they adhere firmly in a way that implies that mating type-specific and species-specific substances are formed on the zygophore surface. These features are clearly an aspect of the mating type phenotype and are necessary for completion of the mating programme, without adhesion the zygophores continue unproductive extension growth, but the nature of the substances involved is unknown.

6.1.6 Chemoattractants in yeasts

In higher fungi, mating through the activity of hormones is well documented in ascomycetous and basidiomycetous yeasts (Kelly *et al.*, 1988; Dyer *et al.*, 1992; Bölker and Kahmann, 1993; Duntze *et al.*, 1994) and there is some evidence that diffusible factors also have a role in inducing ascogonial and trichogyne formation in some filamentous ascomycetes (reviewed in Dyer *et al.*, 1992). These observations provide some idea of the sorts of molecules which might be involved in cell to cell contact. Many yeasts release diffusible sex hormones ('pheromones') as a prelude to the cell fusion that leads to conjugation (Nielsen and Davey, 1995). Mating type factors in the yeast *Saccharomyces cerevisiae* (see Chapter 5) are responsible for production of peptide hormones (pheromones called a- and α-factors; Fig. 6.7) and pheromone-specific receptors. In animals, pheromones are chemicals emitted into the environment by an organism as a specific signal to another organism, usually of the same species. Effective at minute concentrations, pheromones often have important roles in regulating social behaviour of animals, being used to attract mates, to mark territories, and promote social cohesion in communities. Yeast pheromones were named by analogy to the animal hormones because they organise the mating process. They bind to pheromone receptors on the surface of cells of opposite mating type acting through GTP binding proteins to alter metabolism and to: (i) cause recipient cells to produce an agglutinin, so that cells of opposite mating type adhere; (ii) stop growth in the G1 stage of the cell cycle; (iii) change wall structure and consequently cell shape. Both pheromones cause their target cells to elongate into projections but have no effect on cells of the same mating type or on diploids. In *Ustilago maydis*, conjugation tube formation in the yeast-like sporidia is induced by mating-type-specific pheromones released by haploid cells. These pheromones are short lipopeptides: 11–15 amino acids with a C-terminal cysteine residue to which a farnesyl

α-factor

H_2N-Trp-His-Trp-Leu-Gln-Leu-Lys-Pro-Gly-Gln-Pro-Met-Tyr-COOH

a-factor

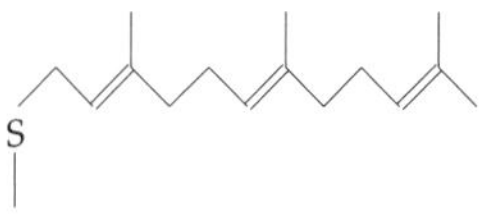

S

H_2N-Tyr-Ile-Ile-Lys-Gly-Val-Phe-Trp-Asp-Pro-Ala-Cys-$COOCH_3$

Figure 6.7 Simplified chemical structures of yeast pheromones. From Chiu and Moore (1998).

group (a 15-C isoprenyl moiety, see section 3.13.2.1) is attached. The latter makes the pheromone extremely hydrophobic.

These pheromones prepare the cells for conjugation and contribute to the recognition of different mating types. However, the major step in the recognition of compatible cell types involves macromolecules on the cell surfaces which cause cells to agglutinate. Some of these are constitutive (i.e. cells agglutinate immediately the different clones are mixed) while others are inducible, the cells only acquiring the ability to agglutinate after growth in mixed culture. In both *Hansenula wingei* and *Saccharomyces cerevisiae* there is evidence that the molecules directly involved in agglutination, the agglutinins, are probably glycoproteins. In *H. wingei* one of the agglutinin components consists of 28 amino acids and about 60 mannose residues. The agglutinins seem to be located on surface filaments external to the cell wall. The function of the agglutinins is to bring cells of opposite mating type together. They do this by virtue of their ability to bind in a complementary manner, the agglutinin of one mating type binding specifically to that produced by the compatible mating type. Following this adhesion of yeast cells by complementary binding, protuberances grow out from the cell walls and when these meet cytoplasmic communication is established by dissolution of the walls. These phenomena (reviewed by Duntze *et al.*, 1994) are obviously specifically part of the mating process; yet they clearly demonstrate that fungal cells are capable of producing surface glycoproteins which, by a sort of antigen – antibody reaction, can achieve a very specific adhesion. It is exactly this sort of specific cell binding which one might expect to be part of the cell-to-cell communication which contributes to the construction of differentiated multicellular structures. The activity of pheromones is also a model for the way extracellular signals can affect proliferation and differentiation of eukaryotic cells. The signals are detected at the plasma membrane in *S. cerevisiae* by a family of protein kinases known as the mitogen- or messenger-activated (MAP), or extracellular signal-regulated (ER) protein kinases (Bardwell *et al.*, 1994). This class of enzyme is highly conserved in eukaryotes and as there are at least five physiologically distinct MAP kinase signalling pathways in yeast (Levin and Errede, 1995), it is likely to be involved in transmitting a range of signals through the cytoplasm and into the nucleus to generate changes in metabolism and gene transcription. The pheromone-responsive MAP kinase cascade interacts with a protein required for establishing cell polarity. So spatial information can also be conveyed through regulation of the rearrangement of the actin cytoskeleton (Leeuw *et al.*, 1995).

6.1.7 Signalling in filamentous fungi and their fruit bodies

In higher fungi anastomosis between homokaryotic hyphae (including incompatible ones) occurs freely and cell fusion between compatible homokaryons is thought to occur without the need for specific mating hormones (Bölker and Kahmann, 1993). However, recent analysis of mating type factor B in *Schizophyllum commune* has shown that its sequence shares homology with the pheromone and pheromone receptor sequences of *Saccharomyces cerevisiae* (Kües and Casselton, 1992). Since mating type locus *B* controls nuclear migration and clamp cell fusion it seems likely that targeting of the clamp cell apex on the parent hypha depends on a pheromone-regulated signalling mechanism.

Apart from this example, how hyphal tips find each other and then have their branching pattern changed in a coordinated way to create multihyphal structures is unknown. Homing reactions have been described in which hyphal tips tend to grow towards germinating spores of the same species (e.g. Kemp, 1977; and see discussion in Moore, 1984b). The mechanism of this reaction between different individuals of the same species is unknown, as is its relevance to cooperation between hyphae of the same individual to produce a fruiting structure assembled from contributions of a number of cooperating hyphal systems. Hyphal cooperation is so fundamental that it can lead to the formation of chimeric fruit bodies. Kemp (1977) described fruit bodies of *Coprinus* consisting of two different species, *C. miser* and *C. pellucidus*. The hymenium comprised a mixed population of basidia bearing the distinctive spores of the two species but the chimera extended throughout the fruit body as both species could be recovered by outgrowth from stem segments incubated on nutrient medium. Bringing hyphal tips together and coordination of branching patterns are obvious possibilities for control by hormones or growth factors but very few chemicals which may function in these ways have been identified (reviewed by Uno and Ishikawa, 1982; Wessels, 1993; Novak Frazer, 1996).

The search for fruit-inducing substances (often called 'FIS') and for growth hormones in higher fungi is considerably complicated by the extensive exchange of metabolites between the mycelium and its substratum and between the mycelium and its fruit body and between the different tissues of the fruit body itself. Interplay between cap and stem in mushrooms, for example, must involve such extensive exchange of 'general purpose' metabolites that some are bound to have quite fortuitous effects on growth patterns if extracted. Distinguishing these from true growth hormones is a task which has not yet been accomplished.

The experimental approaches which have been used usually rely on 'capturing' the active chemical in some way and then demonstrating directed growth in a bioassay. Examples are the absorption of a trichogyne attractant of *Neurospora crassa* into agar blocks containing activated carbon (Bistis, 1983). The blocks were simply placed onto mycelium of one mating type for 18 h and then transferred to cultures of opposite mating type. In 24 out of 27 transfers from mating type *A* culture to a culture of mating type *a*, trichogynes grew towards the block of agar, apparently attracted by some material previously adsorbed from the A culture. The reciprocal test, a transfer of blocks from *a* to *A*, was less effective (2/27 attracted trichogynes). No attraction occurred in transfers to cultures of the same mating type. Similar work has suggested hormones are involved in induction of ascogonial development in *Ascobolus stercorarius* (Bistis, 1956, 1957; Bistis and Raper, 1963), attraction of trichogynes to conidia in *Nectria haematococca* (Bistis and Georgopoulos, 1979) and *Bombardia lunata* (Zickler, 1952). In none of these cases has the active principle been identified.

A variety of chemicals and extracts have the ability to induce or enhance fruiting in higher fungi. These include zearalenone (Nelson, 1971; Wolf and Mirocha, 1973), extracts from *Agaricus* (Rusmin and Leonard, 1978) and cerebrosides (Kawai and Ikeda, 1982) which induce fruiting in *Schizophyllum commune*, phenolic lactones extracted from sexually deficient strains of *Aspergillus nidulans* (Champe and El-Zyat, 1989), and extracts of *Pyrenopeziza brassicae* which influence ascocarp development in that organism (Ilott *et al.*, 1986; Siddiq *et al.*, 1989). The mode of action of these compounds is unknown and, indeed, their true significance must remain in doubt until their chemistry is properly established. Application of even simple nutrients like ammonium salts can induce fruiting in some species (Morimoto *et al.*, 1981) so it is important to know the effect exerted by fruit-inducing extracts on the medium as well as their direct effect on the organism. Another example of a morphogenetic effect which might be chemically mediated is the interaction between *Armillaria mellea* and *Entoloma abortivum* in which the former disturbs the developmental pattern of the latter, resulting in development of the agaric fruit bodies of *E. abortivum* being stopped at various stages, producing carpophoroids, which are aborted, hypertrophied or otherwise abnormal fruit bodies (Watling, 1974). Presumably *A. mellea* either produces or destroys some extracellular chemical signal which is required for normal development of *E. abortivum* to proceed.

There is some evidence for the involvement of specific chemicals in sexual differentiation of some filamentous ascomycetes (e.g. production of linoleic acid in *Ceratocystis* spp., *Neurospora crassa*, *Nectria haematococca* and mycosporines in *Pyronema omphaloides*, *Morchella esculenta*, *Nectria galligena* (reviewed by Dyer *et al.*, 1992)) but whether these compounds act as morphogens or metabolites is not established. Serious doubts about the relevance of all of these compounds to hormonal/growth factor control of growth and development arise from three features of the published researches. First, for the most part the compounds involved have proved to be basic metabolites which seem to lack the specificity expected of hormones. Second, the concentrations used have greatly exceeded (often by many orders of magnitude) those expected of signalling molecules. Third, bioassays usually depend on all or nothing response to the chemical when it is added to the culture in some way; there is no evidence for reaction to a gradient in a way which might support the idea that the chemical in question is a morphogen which determines differentiation patterns within a morphogenetic field.

Chemicals which have been claimed to be directly involved in fruit body induction in basidiomycetes include the following, although in no case is there any evidence about what mechanism(s) might be involved in their claimed activity:

- cAMP and AMP in *C. cinereus* (Uno and Ishikawa, 1971, 1973a,b, 1982);
- an unidentified fruit inducing substance (FIS) in *S. commune* (Leonard and Dick, 1968);
- an unidentified low molecular weight compound from *Agaricus bisporus* causing fruiting induction in *S. commune* (Rusmin and Leonard, 1978);

- an unidentified, diffusible factor (or factors) in *Phellinus contiguus* (Butler, 1995);
- cerebrosides in *Schizophyllum commune* (Kawai and Ikeda, 1982).

A fruit body inducing substance extracted from *A. bisporus* is interesting because of its cross-reactivity to the taxonomically unrelated *Schizophyllum commune*. In addition, its general chemical similarity to substances with growth factor-like properties extracted from *Flammulina velutipes* (Table 6.1) and *C. cinereus* (see below) suggests that there may be a family of chemically related factors or hormones responsible for mediating primordium initiation and fruit body elongation.

At intervals over many years there have been claims of experimental evidence for growth hormones in mushroom fungi. Most of these reports lack numerical data and adequate controls. Failure to establish any consistent mechanistic model from even the best documented of these accounts is very disappointing, especially in comparison with the success achieved in the parallel (and contemporary) search for, isolation, characterisation and commercial exploitation of plant growth hormones, particularly the auxins. Many of the experimental approaches used with mushrooms echo those used in the plant work and have relied on the observation that partial removal of the cap often results in curvature of the stem with the greatest amount of stem growth occurring under the remaining cap. Consequently, the one consistent conclusion which has been drawn from all of this work is that growth of the mushroom stem depends upon the continued presence of the cap. This has always been taken to imply that cap tissues produce growth hormone(s). Much of the literature has been reviewed by Gruen (1982) and Novak Frazer (1996), and the latter author summarised potential sites for hormone action as shown in Fig. 6.8.

Hagimoto and Konishi (1959) found an association between gill segments remaining on a surgically modified cap and the bends of the stem in *Agaricus bisporus*. Gruen (1963) removed halves of caps of young fruit bodies of *A. bisporus* and found the stems bent through 135° in 8 days. The bend was always directed towards the removed part of the cap. Gruen concluded that gills are the centre of regulation of the growth of the stem, possibly producing a 'growth factor' which is not formed in other parts of fruit bodies. In more extensive experiments of the same type with the same species, Hagimoto (1963) showed that there is a reliable correlation between the position of the gills and bending of the stem and claimed that there are two independent centres of regulation of stem bending: one in the gills, the other in the stem.

Table 6.1. *Growth factor-type substances, extracted from basidiomycetous fruit bodies, with fruiting-inducing, stem elongation and/or cap expansion promoting properties. From Novak Frazer (1996), with permission.*

Source organism	Activity	Active substance	Stability	Solubility	Authors
Agaricus bisporus *Flammulina velutipes* *Lentinus edodes* *Pleurotus ostreatus*	Mycelial growth; fruiting induction	Whole fruitbody extract, active principle unknown, but ninhydrin +ve, reducing sugar +ve, organic acid −ve, phosphate −ve, fatty acid −ve	Heat, acid, base stable	Water, aqueous methanol (insoluble in absolute methanol, chloroform, petroleum, benzene)	Urayama (1969)
Coprinus cinereus	Fruiting induction	cAMP, 3′-AMP, theophylline			Uno & Ishikawa (1971, 1973a,b)
Coprinus cinereus	Fruiting induction	Whole culture extract, active principle unknown		Water	Uno & Ishikawa (1971, 1973a,b)
Coprinus cinereus	Hyphal aggregates (fruiting induction)	cAMP			Matthews & Niederpruem (1972)
Agaricus bisporus	Fruiting induction (in *Schizophyllum commune*)	Fruitbody tissue extract, active principle unknown but < 12,000 molecular weight	Heat, acid, base stable	Water, 50% ethanol, 50% acetone	Rusmin & Leonard (1978)
Schizophyllum commune	Fruiting induction	Cerebrosides (glycosphingolipid)	Unknown	Acetone	Kawai & Ikeda (1982)

Phellinus contiguus	Fruiting induction	Mycelial extract, active principle unknown	Heat stable	Water	Butler (1995)
Agaricus bisporus	Stem elongation, cap expansion	Gills, active principle unknown	Unknown	None	Hagimoto & Konishi (1959)
Agaricus bisporus *Coprinus macrorhizus* (= *cinereus?*) *Armillaria matsutake* *Hypholoma fasciculare*	Stem elongation, cap expansion	Gills, active principle unknown. *(contains IAA, but inactive)	Heat, acid, base stable	Ether, acetone, ethanol, water (insoluble in petroleum ether, benzene)	Hagimoto & Konishi (1960)
Agaricus bisporus	Stem elongation; converts tryptophan to IAA	Gills, active principle unknown but < 12,000 molecular weight. *(contains IAA, but inactive)	Unknown		Konishi & Hagimoto (1961)
Agaricus bisporus	Stem elongation	Gills, active principle unknown. Glutamic acid, leucine, cysteine, glycine, serine, asparagine, glutamine, threonine, tyrosine, valine, proline, arginine, $(NH_4)_2SO_4$ & NH_4Cl all tested +ve in the bioassay	Unknown	Ether, acetone, ethanol (petroluem ether, benzene insoluble)	Konishi (1967)

(cont.)

Table 6.1. (*cont.*)

Source organism	Activity	Active substance	Stability	Solubility	Authors
Agaricus bisporus	Stem elongation	Gills, active principle unknown	Unknown	None	Gruen (1963)
Flammulina velutipes	Stem elongation	Gills, active principle unknown	Unknown	None	Gruen (1969)
Coprinus radiatus	Stem elongation	Gills, active principle unknown	Unknown	None	Eilers (1974)
Flammulina velutipes	Stem elongation	Gills, active principle unknown	Unknown	None	Gruen (1976)
Flammulina velutipes	Stem elongation	Gills, active principle unknown, but $<12{,}000$ molecular weight	Heat stable	None (diffused into agar blocks)	Gruen (1982)
Coprinus congregatus	Stem elongation	Cap inhibitor, active principle unknown, but $<12{,}000$ molecular weight	Unknown	None	Robert & Bret (1987)
Coprinus congregatus	Stem elongtion	Cap inhibitor, and stimulator, active principle unknown	Unknown	Unknown	Robert (1990)
Agaricus bisporus	Mycelial growth, stem elongation	Wound hormone, ODA (10-oxo-*trans*-8-decenoic acid), enzymatic degradation product of linoleic acid	Unknown	Extraction patented	Mau *et al.* (1992)

Coprinus micaceous	Unknown	*Cytokinin-like activity in fruit bodies	Unknown	Unknown	Szabo *et al.* (1970) (cited by Gruen, 1982)
Lentinus tigrinus	Unknown	*Auxin-like, gibberellin-like, cytokinin-like activities in stems and caps	Unknown	Unknown	Rypacek & Sladky (1972, 1973; cited by Gruen, 1982)
Agaricus bisporus *Boletus elegans* *Grifola frondosa* *Phallus impudicus* *Phellinus pomaceous*	Unknown	*Gibberellin-like activity in caps	Unknown	Unknown	Pegg (1973)

*Indicates the presence of plant growth regulators which have no stem elongation or cap expansion properties on the fungal fruit bodies tested (see Gruen, 1982).

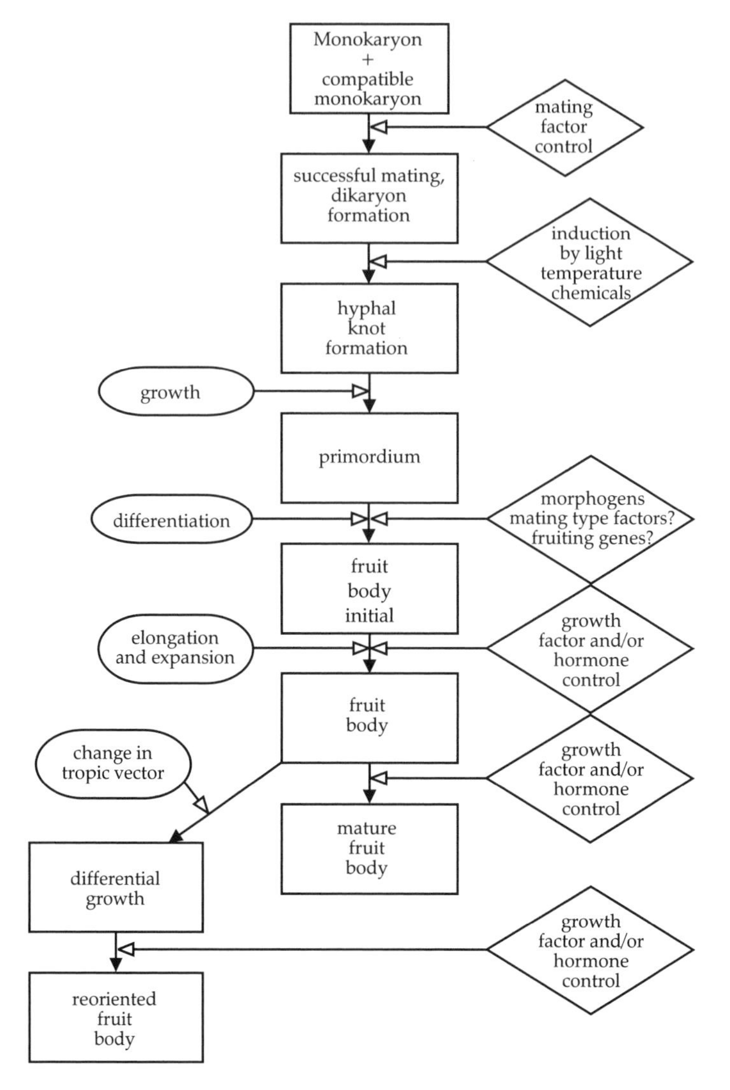

Figure 6.8 Potential target sites for growth factors and/or hormones during mating, dikaryon formation, primordium induction, fruit body differentiation and maturation of mushroom fruit bodies. From Novak Frazer (1996).

The hymenomycetes can be divided into two groups according to whether growth of the stem depends on the cap. In *A. bisporus* and *Flammulina velutipes* the influence of the cap extends through the whole period of development. Removal of the cap or gills stops growth of the stem, and its gravitropic reactions are weakly expressed (Gruen, 1963, 1969, 1982; Hagimoto 1963). Experiments with *Coprinus* spp.

(Borriss, 1934; Eilers, 1974; Gooday, 1974a; Cox and Niederpruem, 1975) suggest that the cap is necessary only at the initial stages of fruit body development. After the stem reaches one eighth to one quarter of its mature length, it can continue growth and show gravitropic responses without the cap. Gooday (1974a) claimed that stem elongation in *C. cinereus* did not require connection with the cap suggesting that elongation is autonomous and endotrophic. However, Hammad *et al.* (1993a) demonstrated that intact fruit bodies elongated about 25% more than decapitated ones (amounting to 2–3 cm greater length), so although the stem may not be *dependent* on the cap, its elongation certainly benefits considerably from the presence of the cap.

Numerous studies seem to imply that extracts or diffusates of the cap can stimulate growth of the stem. Disappointingly, the only chemical candidates which have been extracted are various amino acids and ammonium salts (Konishi and Hagimoto, 1962; Konishi, 1967). Compounds of this sort have been shown to have specific inhibitory effects on sporulation (Chiu and Moore, 1988a,b), but by diverting growth between different routes of differentiation rather than enhancing or inhibiting growth. It is difficult to believe that such simple components of primary metabolism would be employed as *specific* growth hormones by a group of organisms which exhibit enormously versatile and varied ability to synthesise secondary metabolites (Claydon, 1985; Moss, 1996; and see Chapter 3). Remember that some plant pathogenic fungi synthesise plant growth hormones.

Yet, the only mushroom hormone which has been characterised is 10-oxo-*trans*-8-decenoic acid (called ODA; Fig. 6.6) which is produced as a result of wounding in *A. bisporus* (Tressl *et al.*, 1982) as an enzymatic breakdown product of linoleic acid (Mau *et al.*, 1992). ODA doubled the percentage elongation of excised stems at all concentrations tested down to 10^{-8} M. Mau *et al.* (1992) described the compound as a hormone because "ODA stimulated mushroom mycelial growth and stipe elongation." These are not sufficient reasons and ODA must still be counted as an uncertain candidate as a true hormone. Very little attention has been given to the compounds with hormone or growth-factor-like activity described in earlier studies of extracts from *A. bisporus*, *F. velutipes* and other basidiomycetes mentioned above (Hagimoto and Konishi, 1959, 1960; Konishi and Hagimoto, 1961; Gruen, 1963, 1969, 1976; Konishi, 1967; Eilers, 1974; Robert and Bret, 1987; Robert, 1990).

Novak Frazer (1996) catalogued the attempts that have been made to purify and identify fungal growth substances which might regulate fruit body development as shown in Table 6.1. In most of these studies the emphasis was on some demonstration of presence or absence of hormone-like or growth-factor-like activity. Active ingredients found in fruit body extracts were not chemically identified; only their capacity to promote stem elongation and/or cap expansion was demonstrated. Konishi (1967) partially purified a substance from *A. bisporus* caps which appeared to enhance stem elongation in *Agaricus*. The 'growth factor' was claimed to be comprised of various amino acids. Whether these were functioning as separate growth factors or as nutrients is not known. The concentration used was 10^{-4} M, which is a respectable concentration for a nutrient but is probably 4–6 orders of magnitude greater than would be expected of a hormone/growth factor. No efforts have been made to purify the active ingredient further.

Direct evidence for the existence of fungal growth factors is very fragmentary and derives from experiments involving very different species and extraction methods, so it is very surprising that the various extracts (including those analysed for their ability to promote fruit body induction/formation) have exhibited similar activities and chemical properties (Novak Frazer, 1996). The active principles in extracts from *A. bisporus*, *Coprinus macrorhizus*, *Hypholoma fasciculare*, *Armillaria matsutake* (Hagimoto and Konishi, 1960; Konishi, 1967; Urayama, 1969), *Lentinus edodes*, *F. velutipes*, and *Pleurotus ostreatus* (Urayama, 1969; Gruen, 1982), which all cause stem elongation and cap expansion, are $< 12{,}000$ molecular weight, heat stable, acid/base stable and soluble mainly in polar solvents including water. The uniformity in characteristics and cross-reactivity (e.g. *A. bisporus* extracts which induce fruiting in *S. commune* (Rusmin and Leonard, 1978)) suggest the active compounds may be members of a family of hormones or growth factors which have similar but slightly different chemical structure in each species. This would be similar to the auxins, a family of compounds based on indole-3-acetic acid which are active on a very wide variety of plants (Jacobs, 1979; Salisbury and Ross, 1985).

Though the majority of the substances listed in Table 6.1 stimulate extension, those isolated from early primordial stages of *C. congregatus* by Robert and Bret (1987) and Robert (1990) have inhibitory activities. Thus, both inhibitory and stimulatory substances are produced in fruit bodies.

Novak Frazer (1996) points out that, " . . . by definition, growth factors (especially those controlling small morphogenetic fields) must be unstable, either intrinsically or through active destruction, as one of the ways to establish the concentration gradient", which is needed to regulate the morphogenetic field. This consideration raises a serious doubt about stability of the putative growth factor(s) during the lengthy bioassays (24–72 h) reported in the literature. Consequently, the evidence concerning hormones in higher basidiomycetes remains confused, conflicting and fragmentary. Lilian Hawker wrote in 1950: "It is desirable that research should be directed towards an interpretation of tropisms in fungi based on the study of growth-regulators. At present nothing is known of any mechanism in fungi comparable to the redistribution of auxins in the higher plants". Sadly, this is still true more than 40 years later.

6.1.8 Tropisms as morphogenetic changes

Fungi exhibit a variety of tropic reactions in response to gravity, light and some other external stimuli (Moore, 1991). Tropisms result from differential growth and consequently represent a true morphogenetic process. However, since exposure to the tropic stimulus is under the experimenter's control, tropisms provide an excellent model system for study of the control of differential growth and morphogenetic changes.

Fungi exhibit a variety of tropisms. The very smallest fruit body initials grow perpendicularly away from the surface on which they arise independently of the direction of light or gravitational signals (Buller, 1905, 1909; Plunkett, 1961; Schwantes and Barsuhn, 1971). Similar behaviour has been claimed for the gills of *Lentinus lepideus* (Buller 1905) and pores of *Polyporus squamosus* (Buller, 1909) which first grow perpendicularly to the hymenophore surface but in the course of further development show a pronounced gravitropism. These reactions are probably analogous to the avoidance reactions of *Phycomyces* sporangiophores (Johnson and Gamow, 1971; Lafay *et al.*, 1975); that is, the very young structures are simply growing away from their support. They may be based, therefore, on negative chemotropisms to products of mycelial metabolism or to water activity (Gamow and Böttger, 1982). As they develop further, the stems of ground agarics (gilled mushrooms) seem generally to be non-phototropic but show a marked negative gravitropism whereas lignicolous and coprophilous hymenomycetes are often both phototropic and gravitropic (Plunkett, 1961). Anemotropism can also be demonstrated (Badham, 1982). In normal morphogenesis in nature, an initial period of light-seeking growth in the earliest stages of development is followed by negative gravitropism. Buller (1909, p.48) ascribed this to "a remarkable change in the physiological properties of the stem"; that is, from being primarily positively phototropic, the growth control apparatus is rendered almost insensitive to light but negatively gravitropic.

Plunkett (1961) concluded that "geotropism is only dominant under conditions of low light intensity when it may be presumed that the phototropic mechanism is understimulated." This seems to imply an expected relationship between photo- and gravitropism. Certainly, it is clear that in sporangiophores of *Phycomyces* phototropic and gravitropic responses interact so that the direction of growth at equilibrium is at "a compromise angle between gravi- and phototropic tendencies" (Varjú *et al.*, 1961).

The natural and experimental switches between dominating tropisms shows that they are different and interactive. In some instances (as in Plunkett, 1961) they can be shown to interact in ways dependent on simple physical factors like being shaded from incident illumination. In others (as in Badham, 1982) a morphogenetic change in one tissue seems to be at least coordinated with a change of tropism in another. The vegetative hyphae of ascomycete and basidiomycete cultures which give rise to gravitropic, phototropic and anemotropic fruit bodies do not themselves show such tropisms though their genetic descendants within those structures do. Thus, the tropic response is an attribute of hyphae in multi-hyphal structures which can be regulated. It is significant that phototropically abnormal mutants of *Phycomyces* fall into seven complementation groups, two of which are also affected in gravitropic and avoidance (chemotropic) responses (Bergman *et al.*, 1973). If *Phycomyces* can be realistically used as a comparison for higher fungi, then this adds weight to the belief that one tropic *response* can be regulated by sensitivity to different tropic effectors.

It seems to be generally assumed that tropic bending of such structures as the mushroom stem are conditioned by uneven distribution of growth regulators, an idea which was probably introduced by Buller (1934) and Borriss (1934). When a mushroom stem which has been laid horizontally bends upwards the coordination of growth rates on the upper and lower sides of the stem is most readily understood in terms of an asymmetrically distributed growth hormone. This is not a unanimous interpretation, though. Jeffreys and Greulach (1956) reached the conclusion, from their experiments in which they tested the effect of auxin, that "The results suggest, not only that auxin is not involved, *but also that no other hormone is involved.* [my emphasis] It seems likely that each hyphal strand responds individually to environmental factors. Because the strands are aggregated, this results in a unit action by the stem." There are two other papers (Banbury, 1962; Gorovoj *et al.*, 1989) which contain statements which indicate that their authors were convinced that the apparently coordinated expression of gravitropic response is in truth a common but independent response by the individual component hyphae of the structure concerned. Moore (1991) discussed the possibility that linear hyphal systems may conduct mechanical stresses particularly well, and differential mechanical loading might orchestrate differential growth. Gravitropism in *Coprinus cinereus* has been used to test this (Greening *et al.*, 1993) but application of even large lateral loads failed to affect either the direction or the magnitude of gravitropic bending.

The kinetics of the gravitropic response in the two most studied basidiomycetes, *C. cinereus* and *Flammulina velutipes* (reviewed in Moore *et al.*, 1996), reveal evidence for the presence of growth factors controlling differential growth. In *C. cinereus*, gravitropic curvature begins at the apex of the stem and proceeds in a basipetal direction (Kher *et al.*, 1992); this implies that some sort of bending signal (growth factor?) migrates towards the base. Again in *C. cinereus*, the initial gravitropic response is due to the lower hyphae (in the lower half of a horizontal stem) elongating faster than the upper hyphae (the upper half of the same horizontal stem). Light microscopic studies reveal that lower hyphae increase in length by 4–5-fold without increase in girth, and growth studies indicate that the lower surface of the stem elongates at a faster rate than the upper surface to generate the gravitropic bend (Greening and Moore, 1996, Greening *et al.*, 1997; see section 6.2.2.5). Clearly, there must be a mechanism by which differential growth is coordinated in a way which ensures that cell length, but not diameter, is increased.

Potential growth factors have been extracted from pre- and post-meiotic fruit bodies of *Coprinus cinereus* and two different activities found, possibly corresponding to two different substances (Moore and Novak Frazer, 1993, 1994; Novak Frazer, 1996). One substance (Fungiflex 1) was produced by both immature and mature fruit bodies of *Coprinus cinereus* but the other (Fungiflex 2) was only produced by mature fruit bodies. Fungiflex 2 was found to be between 10–100 times more abundant in the cap than in the stem. Both substances were < 12,000 molecular weight and heat stable. Fungiflex 1 has a faster action (results seen after 1 h exposure) and has inhibitory properties, similar to substances extracted from *C. congregatus* (Robert and Bret, 1987). Fungiflex 2 has a slower action (results seen after 6 h exposure) and has stimulatory properties, similar to substances described in earlier studies (Table 6.1).

6.1.9 Pattern formation

Since there is no direct experimental evidence for the existence of morphogens in the differentiating primordium it is useful to (i) make comparisons with other organisms in which more information exists about the coordination of developmental processes; (ii) review the *a priori* 'evidence' for chemical control of pattern formation; and (iii) consider the theoretical aspects of pattern formation in biological systems.

Prime examples of pattern forming growth factors are retinoic acid and plant auxins (Fig. 6.6). Retinoic acid, a derivative of vitamin A, is an endogenous signalling substance which specifies the position of mesenchymal cells within the embryonic chick limb bud. Limb development depends on the gradient of retinoic acid in the interstitial space around cells (Brickell and Tickle, 1989; Tickle, 1991); the concentration gradient establishes the cell-to-cell communication required for the formation of digits in the correct orientation. Digits in the wing are formed in reaction to the position-dependent dose of the morphogen. Retinoic acid binds to nuclear retinoic acid receptors which regulate the expression of several homeobox genes (Tickle, 1991). The regulated genes specify the position of digits in the limb bud. Thus, retinoic acid acts as a morphogen (through dose-dependent specification of position of cells) and regulates gene expression (through homeobox genes). Auxin is instrumental in establishing polarised differentiation of vessels and cell shape during vascular differentiation in plant embryos via a concentration gradient (Sachs, 1991). Exogenously added auxin activates transcription of specific mRNAs in plant tissues undergoing either cell elongation or cell division (Key, 1989; Guilfoyle *et al.*, 1993) and auxin binds to nuclear receptors (Löbler and Klämbt, 1985; Guilfoyle *et al.*, 1993; Ulmasov *et al.*, 1995 and references therein). Auxin-regulated mRNAs show distinct patterns of organ-specific, tissue-specific and development-specific expression (Guilfoyle *et al.*, 1993).

The important features of morphogen activity thus seem to be control of specific gene expression via binding proteins (receptors). The result is that morphogenesis becomes polarised. Polarisation involves control of gene expression in a gradient but as transcriptional control tends to be on/off, polarised development must be expressed as a coordination of events of many cells (Sachs, 1991). Another feature of the retinoic acid and auxin examples is that the response to the morphogen is dependent on the reacting tissue; i.e. one morphogen molecule can elicit different responses from different cells within its morphogenetic field. No fungal parallels of these examples are known. Yet it is abundantly clear that cell and tissue patterns are established during development of fungal multicellular structures. Many of these patterns are akin to others found in animals and plants and, like them, tissue patterns in fungal fruit bodies are probably best interpreted as arising through the activity of 'morphogens' acting at a distance (Moore, 1984b; Reijnders and Moore, 1985), which may therefore be viewed as short-distance growth hormones. The origin of organising centres for the gill plates in *Coprinus cinereus* has been shown to depend on their spatial arrangement. Two neighbouring centres are able to inhibit the formation of a third between them until their radial growth into the expanding fruit body separates them by a distance which releases undifferentiated tissue from the inhibitory influence emanating from the existing organising centres. There is, in other words, a morphogenetic field around the gill organising centre within which new centres cannot form (Rosin and Moore, 1985a; Moore, 1987; discussed in detail below).

Another example arose from analysis of the distribution of the specialised cystidial cells in the spore-bearing hymenial tissues of *C. cinereus* which was shown to be non-random (Horner and Moore, 1987). Cystidia are large, inflated cells (see Figs 6.28 and 6.30) which are readily seen in microscope sections so their relationships are open to numerical analysis. Horner and Moore (1987) examined sections of *C. cinereus* gills prepared for the light microscope and categorised cystidia spanning the gill cavity as 'distant', having other cells separating them, or 'adjacent', with no intervening cells. In either case, neighbouring pairs of cystidia which emerged from the same hymenium were described as *cis*, or as *trans* when they emerged from opposite hymenia (Fig. 6.9). They argued that if the distribution of cystidia was entirely randomised there should be an equal number of *cis* and *trans* in both the distant and adjacent categories. However, quantitative data showed a distinct shortage of adjacent *cis* cystidia (Fig. 6.9), suggesting that formation of a cystidium lowers the probability of another being formed in the immediate vicinity (Horner and Moore, 1987). The plot shown in Fig. 6.9 suggests that inhibition extends over a radius of about 30 μm and is limited to the hymenium of origin. On the basis of these observations, the region around a cystidium was identified as a morphogenetic field controlled by the cystidium at its centre.

Implicit in this interpretation is that the distribution pattern of cystidia might be dependent on interplay between activating and inhibiting factors. In this instance (as with the determination of gill development mentioned immediately above) the patterning process is open to interpretation using the activator–inhibitor model developed by Meinhardt and Gierer (1974; Meinhardt, 1984) which supposes that morphogenetic pattern results from interaction with an activator which autocatalyses its own synthesis, and an inhibitor which inhibits synthesis of the activator. Both diffuse from the region where they are synthesised, the inhibitor

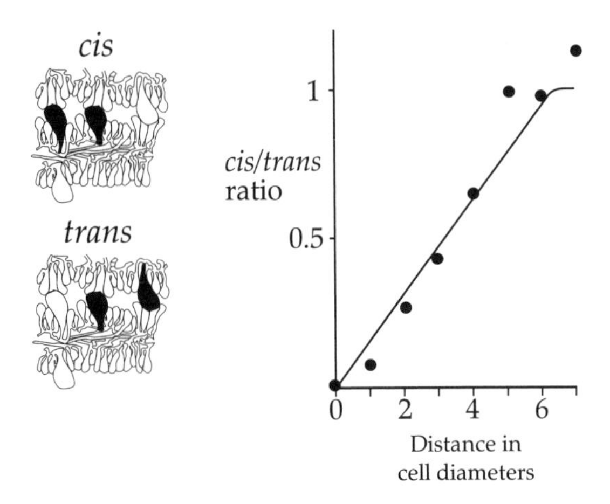

Figure 6.9 Cystidium distribution in the *Coprinus cinereus* hymenium. The drawings on the left show the categorisation of neighbouring pairs of cystidia in micrographs as either *cis* (both emerge from the same hymenium) or *trans* (emerging from opposite hymenia). The accompanying graph compares the frequencies of these two types over various distances of separation and shows that closely-spaced *cis* neighbours are less frequent than closely-spaced *trans* neighbours, implying some inhibitory influence over the patterning of cystidia emerging from the same hymenium.

diffusing more rapidly and consequently preventing activator production in the surrounding cells. This results in long-range inhibition and short-range activation. Increase in the activator is eventually arrested (limited, for example, by diffusion) and stable activator and inhibitor concentration profiles are established. A wide variety of patterns can be generated in computer simulations by varying diffusion coefficients, decay rates and other parameters (Meinhardt, 1984). The model readily accounts for stomatal, cilial, hair and bristle distributions, and has been applied successfully to simulate leaf venation and phylotaxy (Meinhardt, 1984).

As patterning depends on two compounds, the basic model consists of two equations describing changes in concentration of activator (a) and inhibitor (h), each varying with both position (x) and time (t):

$$\frac{\partial a}{\partial t} = \frac{ca^2}{h} - \mu a + D_a \frac{\partial^2 a}{\partial x^2} + \rho_0$$

In this equation the first term represents production of activator (a), which is proportional to a^2 (representing autocatalysis) and inversely proportional to inhibitor concentration (rate constant c). The second term represents decay, with first order rate constant μ. The third term describes changes due to diffusion, with diffusion coefficient D_a.

Production of inhibitor is described by the second equation and is also dependent on activator concentration, but is not inhibited, it also decays, with rate constant v, and diffuses with a diffusion coefficient D_h.

$$\frac{\partial h}{\partial t} = ca^2 - vh + D_h \frac{\partial^2 h}{\partial x^2}$$

The mechanism in these equations leads to a polar concentration profile, with a peak of activator concentration at one end and no activator at the other end.

Expansion of the model to include substrate concentration (s) and a substance whose concentration, y, is high in the differentiated state produces four equations which describe the formation of net-like structures from a field of undifferentiated cells within which a branching filament of differentiated cells develops (Meinhardt, 1976). Two of the equations are similar to the above, but modified to make activator and inhibitor production dependent on substrate concentration, with a constant basal rate of production of both. The other two equations describe changes in s and y. Substrate (s) is assumed to be produced at constant rate in each cell, to decay with first order rate kinetics, to be removed by differentiated cells and to diffuse. Concentration of the 'differentiation substance' (y) increases in proportion to activator concentration, decays and also undergoes positive feedback (after production of a particular concentration of y by a, further production of y is independent of a). Reiteration through these equations over a series of time intervals leads to formation of activator maxima (and, therefore, 'differentiation') in a file of neighbouring cells so that a filament of differentiated cells arises from the uniform field which is the starting point for the model.

Since the model makes inhibitor production dependent on activator concentration and the highest activator concentration occurs at the tip of the filament, this provides a mechanism for apical dominance within the filament. Inhibitor eventually diffuses away from the tip. A lateral branch can then arise as new activator maxima form, first within the filament and then laterally where substrate concentration is still high. These equations make useful predictions about mycelial growth, but they need to be recast because as originally constructed the model deals with differentiation of existing undifferentiated cells. That is, the branching filament which is the analogue of a fungal mycelium is modelled to emerge within a pre-existing field of cells. This is clearly not applicable to hyphal growth and branching. Consequently, these equations need to be combined with the descriptions of hyphal growth kinetics outlined in Chapter 2. The effort should be worthwhile because the Meinhardt models, based on

Turing's (1952) original analysis, are very effective in predicting the formation of periodic structures like stomata on leaves, cilia on embryos and distribution of hairs and bristles on insects. It will be obvious that such a model would have potential in describing the distribution of cystidia, gill and pore spacing and other tissue patterns in fruit bodies as well as distribution of structures in a developing mycelial colony. Green and Smith (1991) analysed application of these models to animal systems and concluded that the answer is 'probably yes' to the question 'do gradients and thresholds of growth factors acting as morphogens establish body plan?'

A *belief* that the same may be true for analogous patterning phenomena in fungi outlined in this book is implicit in the way in which they are described here. Unfortunately, this is no more than a belief. Differentiation of a cystidium being influenced by pre-existing cells of the same type implies that a morphogenetic field exists around cystidia, extending to a radius of about 5 cell diameters. This is not a unique example, similar considerations probably also apply to the non-random distribution of the hyphae which give rise to sphaerocysts in the stem of *Lactarius rufus* (Watling and Nicoll, 1980). These observations imply that localised patterning of tissues might be *explained* in terms of the asymmetric distribution of some chemical (the morphogen or inducer) which thereby extends the influence of the morphogenetic centre over the surrounding hyphae. In the specific cases referred to above the morphogenetic fields extend only over short distances in the 20–100 μm range. It is important to appreciate that this is only a plausible explanation. In no case has a morphogen been extracted and characterised. So although it is abundantly clear that coordination of developmental processes is successfully achieved in fungal multicellular structures, the evidence for chemicals able to perform the signal communication involved is sparse and disappointingly unconvincing. As indicated above, there are very few clues even to the nature of the morphogens which might be the activating and/or inhibiting growth factors in these phenomena. Until we have more detailed knowledge of potential morphogens it is impossible to judge whether the characteristics assumed for inclusion in the modelling process (diffusion constants, decay rates, etc.) are realistic. It is also necessary to take account of two structural features which are particular to fungi. One is that *lateral* contacts between fungal hyphae are extremely rare, being represented only by lateral hyphal fusions. The constituent cells of plant and animal tissues are interconnected laterally by frequent plasmodesmata, gap junctions and cell processes. The absence of similar structures connecting adjacent hyphae suggests that any morphogens which do exist are likely to be communicated exclusively through the extracellular environment (Reijnders and Moore, 1985). Transmission exclusively outside the cell is likely to influence the chemical nature and physical characteristics of any morphogen chemicals which might be produced.

The other structural feature which might influence the progress, and even the nature, of signalling systems in fungal tissues is that in most tissues there is a repetitive substructure comprised of a central hypha (which remains hyphal) and an immediately-surrounding family of hyphae which differentiate in concert. These hyphal aggregations were identified and termed hyphal knots by Reijnders (1977, 1993). The indications are that the central hypha induces the differentiation of its surrounding family. If this is the case, then any control exerted by morphogens must be imposed on the central induction hypha which may not differentiate itself, but simply relay the message to its dependent family. This two-stage process may influence the physical characteristics of the morphogen(s), but it might also influence their number. If the induction hyphae determine the terminal differentiation of their surrounding family, the morphogen(s) may simply need to instruct a differentiation process to occur, without defining the nature of that differentiation. The latter might be determined by local environmental, physical or nutritional factors.

Computational models which have proved successful in describing and simulating the morphology, growth and development of animals and plants need to be adapted to take account of specifically fungal characteristics. A start is being made on this. Regalado *et al.* (1996) describe a mathematical model which generates fractal structures that can be effectively applied to assessing the role of internal and external factors in formation of heterogeneities in mycelial growth. However, this model is a long way from what has been achieved in simulating plant growth and development (Prusinkiewicz and Lindenmayer, 1996). Using an entirely different approach, Stočkus and Moore (1996) succeeded in simulating the bending kinetics of mushroom stems in response to gravitropic stimuli. However, the simulations were not perfect. The mathematical expressions used were derived from work with plants and evidently require additional terms to take account of some details of the specifically fungal response. Another limitation is that the fungal modelling is so far exclusively two-dimensional. Much further work remains to be done before three-dimensional modelling, of the sort accomplished with plants (Korn, 1993), can be undertaken.

6.2 Cell differentiation

The observations, referred to immediately above, which enabled Reijnders (1977, 1993) to recognise the hyphal aggregations he called hyphal knots were an extension of an approach to the description of tissue construction in mushrooms and toadstools which is called hyphal analysis. This is a procedure whereby the range and type of differentiated cell types are catalogued and used as taxonomic criteria. Hyphal analysis was introduced by Corner (1932a,b), who coined the terms monomitic, dimitic and trimitic to describe tissues consisting of one, two or three kinds of hyphae. Later, the words sarcodimitic and sarcotrimitic were used to describe fruit bodies having two or three types of hyphae of which one is inflated and has thickened walls (Corner, 1966; Redhead, 1987). Hyphal analysis in this sense is an almost entirely descriptive study, aimed at establishing structural features as taxonomic criteria. The approach has spawned a range of technical terms which, except for the specialist, tend to obscure rather than enlighten. Many of the cell types which hyphal analysis usefully identifies are named for their morphological features alone (the essence of taxonomic description), yet the names carry functional overtones (words like 'generative' and 'skeletal' are used) though this is entirely a matter of presuming a function without proof or even evidence beyond the morphology. Walls which are seen to be unusually thick by light microscope observation are almost always assumed to be mechanically strengthening (skeletal or ligative hyphal characters). Yet, as I have discussed above (sections 4.2.1 and 6.1.2), fungal wall structure is modified both chemically and physically to serve as a transient nutrient store. Thus, wall thickness is not a reliable guide to wall function.

The functional and morphogenetic purposes of the hyphal differentiation which hyphal analysis describes have only rarely been considered. Listing the presence or absence of a cell type is considered adequate. Consideration of its function, how it arises, how its position may be regulated, and its prevalence do not seem to be of great concern to most practitioners. The first *quantitative* hyphal analysis was done by Hammad *et al.* (1993a,b) who showed that enumerating cell types at different stages of development (in the fruit bodies of *Coprinus cinereus*) is a powerful way of revealing how the macroscopic aspects of fruit body structure emerge during morphogenesis as a result of changes in hyphal type and distribution. This work will be described in detail below. Initially, I will describe the 'classic' sort of hyphal analysis, relying heavily on the review by Pegler (1996), but attempting to translate (and then avoid!) most of the arcane technical terms. The taxonomic importance of hyphal analysis is immense (Pegler, 1996). It can even be applied *in vitro* to identification of fungi (especially wood rotting species) in culture (Nobles, 1958, 1965, 1971; van der Westhuizen, 1958, 1963, 1971; Stalpers, 1978; Rajchenberg, 1983; Lombard, 1990; Nakasone, 1990). My interest, however, is in the contribution it makes to our appreciation of two features. First, the analysis demonstrates the range of functional differentiation of which the hypha is capable. Second, the varied appearance of the hyphal types and the intergradations between them illustrates the adaptability, and versatility, the fungi express in constructing their fruit bodies.

6.2.1 Hyphal analysis

Corner (1932a,b) pioneered study of the hyphal components of the fruit body, developing the micromorphology of their hyphal components as a new taxonomic approach. Current understanding is that gross morphology depends on the hyphal system present and as the system is constant between fruit bodies it can be used to delimit genera. Initially, work on hyphal types was restricted to the 'polypores' (Corner, 1953; Cunningham, 1954), fruit bodies in which the spore-bearing hymenophore is located in pores rather than on gills (Fig. 6.10). However, hyphal analysis has been a major contributor to the realisation that some of the fungi which produce mushrooms with gills (so-called 'agarics') have close relatives among the polypores. The Friesian tradition which sees toadstools which have pores as being sharply separated from mushrooms with gills is not a natural classification (see below).

Pegler (1996) traced the origins of hyphal analysis to Patouillard (1887, 1900) and Fayod (1889). The former proposed new genera based on the presence or absence of clamp connections, the presence of thick-walled hyphae, lactiferous hyphae (hyphae containing, and perhaps transporting, fungal 'latex', the modified fatty acid, lactarinic acid, described in Chapter 3) and gelatinisation of the 'context', the main flesh of the fruit body cap. Fayod (1889) examined the structure of agaric fruit bodies and identified two types of tissue. The first tissue he described was '*tissu fondemental*' which formed the basic structure of the fruit body, being composed of regularly arranged hyphae lacking anastomoses, mainly parallel in the stem. The other tissue type, called the '*tissu connectif*', was composed of thin, irregular hyphae with clamp connections. Ames (1913) catalogued differences in the nature of hyphae in different

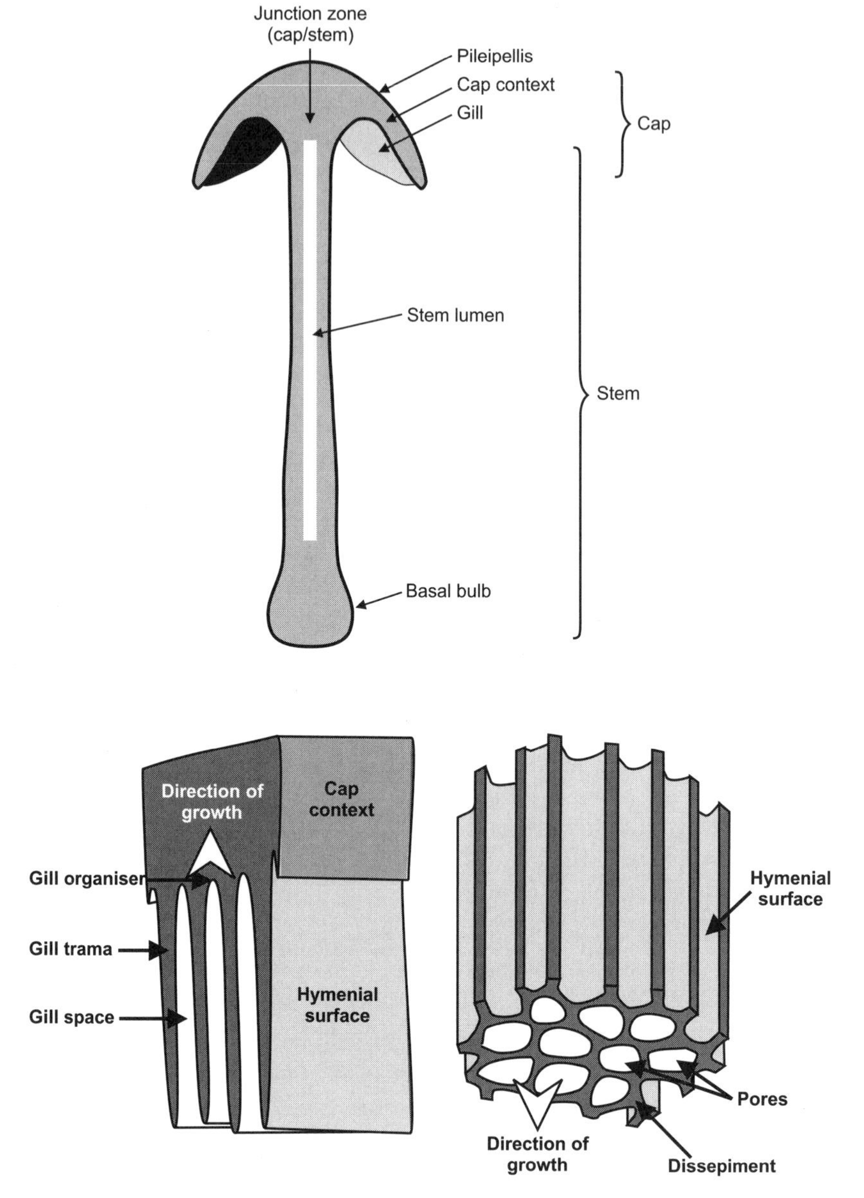

Figure 6.10 Diagrams of the main structural features of mushrooms and toadstools.

parts of polypore fruit bodies, stating that differences in wall thickness and fluid content of the hyphae conferred special qualities on the tissues. Interestingly, although Ames (1913) found that the hyphal patterns were fairly constant between fruit bodies, she did not recognise hyphal systems.

Corner's analysis (1932a,b) used manual dissection of fruit body tissues as a complement to examination of (comparatively thick) microscope sections. In his first study (of the stemmed polypore *Microporus xanthopus*), he found four types of hypha contributing to fruit body construction (Fig. 6.11):

1. unbranched, thick-walled, aseptate hyphae which he called *skeletal hyphae*;
2. branched, thin-walled, septate hyphae with clamp connections were called *generative hyphae*;
3. branched, thick-walled, aseptate hyphae which interwove the other hyphae were called *binding hyphae* (Pouzar (1966) suggested the term 'ligative' for these hyphae, better to indicate their connective function);
4. longitudinally arranged, branched, thick-walled, aseptate hyphae were called *mediate hyphae*.

Corner (1932a) argued that skeletal hyphae formed the basic framework structure, connected together by binding hyphae to establish a cylinder in the stem or a layer over the hymenophore. The different types of hyphae were thought to differentiate from a basic arrangement of hyphae with uniform structure, the generative hyphae described by Corner (1932b); skeletal hyphae arising in the growing region and binding hyphae (with limited and intricate growth) developing behind the growth zone (Fig. 6.12). Mediate hyphae were thought to be a transition stage in formation of the skeletal type from the generative.

The three basic hyphal systems originally proposed were:

1. *monomitic*, in which there is only one type of hypha (generative);
2. *dimitic*, which has both generative and skeletal hyphae, but lacking binding hyphae;
3. *trimitic*, which has all three hyphal types, generative, skeletal and binding.

Corner (1950) monographed *Clavaria* and related genera, finding that while most genera were monomitic (generative hyphae only) there were

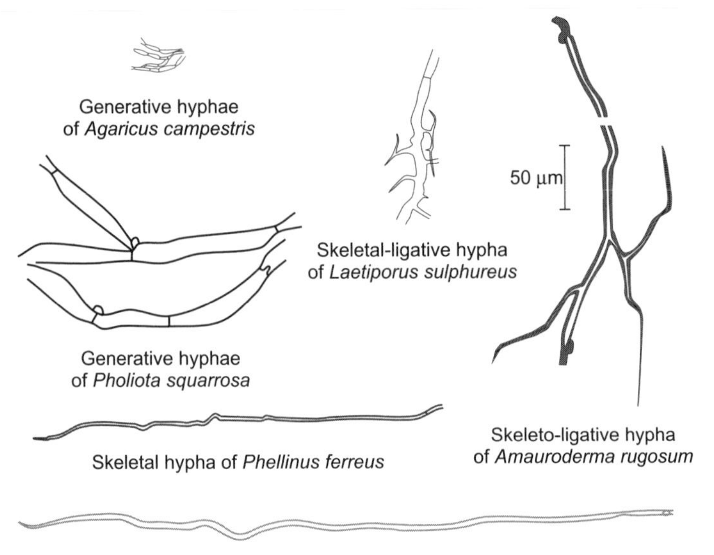

Figure 6.11 Types of hyphae recognised in hyphal analysis. Note the range of hyphal differentiation provided by variation in the length of the hyphal compartment, degree of inflation, amount and nature of branching and wall characteristics. Revised and redrawn to the same scale after Pegler (1996).

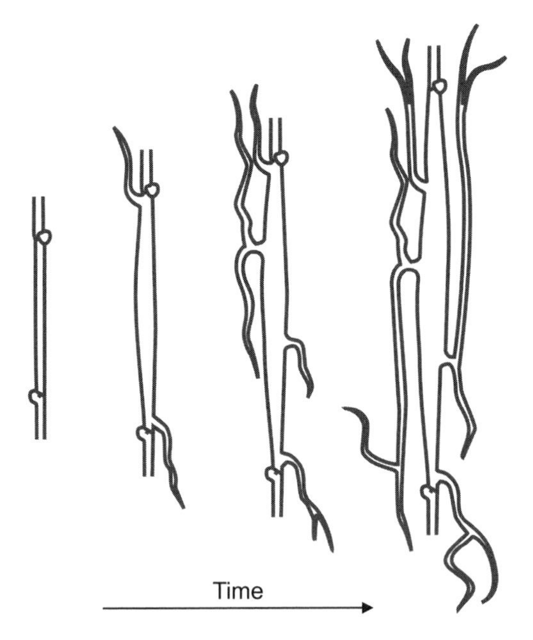

Figure 6.12 Diagrams illustrating the formation of a skeleto-ligative hypha. Note that this is a complex differentiation process involving elongation and inflation of an intercalary hyphal compartment coupled with intercalary branching to produce branches of a specific (ligative) morphology. Redrawn after Pegler (1996).

many variations. Some generative hyphae were inflated with clamp connections, some inflated but without clamp connections; some not inflated; some contained repeatedly branched sterile cells (dichophyses) in the hymenium; others had gelatinised cells. Corner's (1953) summary of the hyphal construction of polypores made it clear that the dimitic classification included tissues which possessed either skeletal hyphae or binding hyphae. Cunningham (1947) had shown that skeletal hyphae may branch freely, with the branches arising from a relatively short main stem (termed 'bovista-type hyphae'; Fig. 6.13). Teston (1953a,b) found a continuous series of intermediates between skeletal and binding hyphae, indicating that hyphal types were not as clear cut as many had believed, and suggesting that hyphal systems are no more natural a classification system than any other aspect of morphology.

Pinto-Lopes (1952) criticised Corner's mitic terminology and proposed a different way of describing hyphal differentiation, separating them first in terms of the presence or absence of clamp connections. As described by Pegler (1996), the Pinto-Lopes (1952) categorisations are:

1. Secondary hyphae with clamp connections:
   (a) tertiary hyphae colourless, with clamp connections and walls unthickened or only slightly thickened;
   (b) tertiary hyphae colourless, with clamp connections and thickened walls;
   (c) tertiary hyphae colourless without clamp connections, walls more or less thickened;
   (d) tertiary hyphae yellow or brown, without clamp connections and walls more or less thickened.
2. Secondary hyphae without clamp connections:
   (a) tertiary hyphae colourless, septate and walls slightly thickened;
   (b) tertiary hyphae yellow, septate, and walls slightly thickened;
   (c) tertiary hyphae colourless, aseptate, walls much thickened;
   (d) tertiary hyphae yellow or brown, aseptate and walls much thickened.

In turn, Corner (1954) criticised the Pinto-Lopes categorisation for the rigidity of its definitions, non-use of nuclear states of the hyphae and consequent implications for the mycelial nature of homothallic species.

In a long series of papers Cunningham (summarised in Cunningham, 1954, 1956, 1958) applied Corner's concept of mitic systems to taxonomy of basidiomycetes of Australia and New Zealand which have poly-

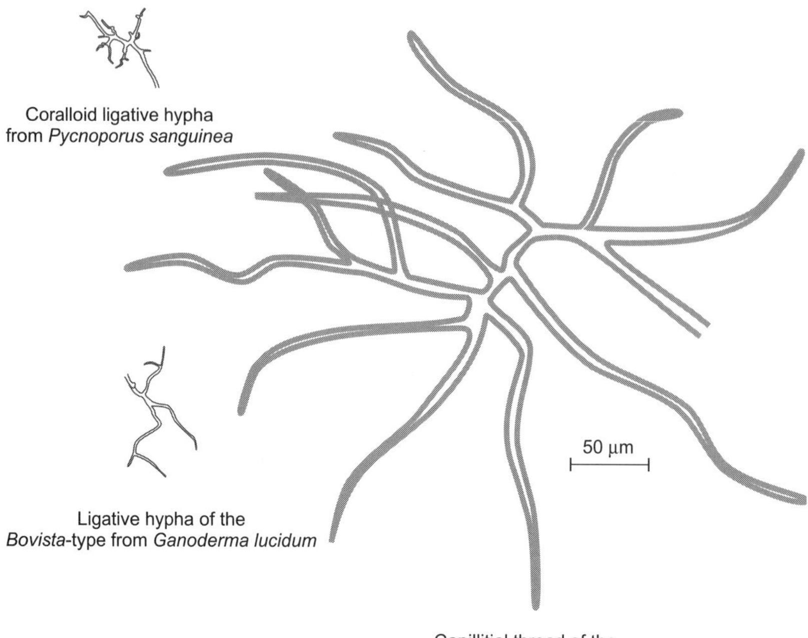

Figure 6.13 Ligative hyphae and a capillitial thread. The capillitium is the name given to the sterile hyphal elements which occur amongst the spores of gasteromycetes. Note how similar morphology extends across an enormous size range. Revised and redrawn to the same scale after Pegler (1996).

poroid, toothed (hydnoid) or resupinate (flattened to the substratum) fruit bodies. Many other studies applied Corner's approach (references in Pegler, 1996) and by 1964 the value of hyphal systems was widely accepted so that Donk (1964, 1971) was able to use the hyphal characteristics of fruit bodies in defining aphyllophoroid families. Aphyllophoroid is a term encompassing basidiomycetes in which the hymenophore (spore-bearing tissue) covers a flattened (resupinate), club-like (clavarioid), toothed (hydnoid), tubular (polyporoid), and occasionally lamellate (gilled, but tough and not fleshy as in agarics) fruit body. This analysis indicated that skeletal hyphae probably evolved independently in several of the families in this Order. The approach was further expanded to other orders and families of the Basidiomycotina by Jülich (1981).

In his analysis of the genus *Trogia*, Corner (1966) recognised the *sarcomitic construction*, which features long and inflated cells (500–3000 × 10–30 μm), mostly unbranched and with relatively narrow septa (Fig. 6.14). He argued that these took the place of skeletal hyphae, and enabled the fruit body to enlarge by cell inflation, like an agaric, their thickened walls stiffening the tissue. In company with generative hyphae these form a dimitic hyphal system (sarcodimitic). A sarcotrimitic construction in the context of *Trogia stereoides* arose as generative hyphae became thick-walled, interweaving with inflated skeletal cells, rather like binding hyphae but being septate. Polypore taxonomy progressed in a range of studies using hyphal analysis as a primary criterion for defining genera (discussion and references in Pegler, 1996). In studies of *Lentinus,* a widespread genus closely related to *Polyporus* (Corner, 1981; Pegler, 1983), skeleto-ligative hyphae (Fig. 6.11) were recognised. These developed from intercalary or terminal compartments of generative hyphae, which lengthened, became thick-walled, and formed branches of determinate growth as tapering 'ligative' processes. Many specialised hyphal systems have been identified and some have been associated with particular features of the fruit body. For example the asterodimitic context (*Asterostroma* and *Asterodon*; Corner, 1948; Hallenberg, 1985) is composed mostly of asterosetae (thick-walled, bristle-like hyphae) which give it a spongy texture composed of minute air-pockets, avoiding water absorption from the fruit body surface.

The range of hyphal modifications encountered in gasteroid Basidiomycotina (puffballs, earthstars and stinkhorns) has been summarised by Pegler *et al.* (1995). Specialised cell types occur, including the capillitial threads (Fig. 6.13), which form a tangled mass of hyphae (the capillitium) among the spores. These are specialised skeletal hyphae. They are often stellate, with many short, tapering branches from a central stem. They resemble skeleto-ligative hyphae and presumably contribute to forming or maintaining the powdery spore mass (*gleba*).

Hyphal analysis has also been applied to agarics (Kühner, 1980; Singer, 1986). Singer stressed the range of conducting systems evident in otherwise monomitic hyphal systems in the Agaricales. These are presumed to be involved in transportation and include: (i) lactifers, which contain fungal 'latex'; (ii) oleiferous hyphae, containing a resinous substance; (iii) gloeo-vessels (Singer, 1945), or gleoeoplerous hyphae, attached to gloeocystidia, all having colourless or yellowish but highly refractile contents; (iv) chryso-vessels, which are similar to the gloeo-vessels but contain an amorphous, golden yellow substance; and (v) coscinoids (Singer, 1947), which have a pitted, sieve-like surface and link to coscino-cystidia. Singer used context structure to separate Agaricales from the Russulales. The terms which feature here (introduced by Beck-Mannagetta, 1923) are homoiomerous (composed only of unre-

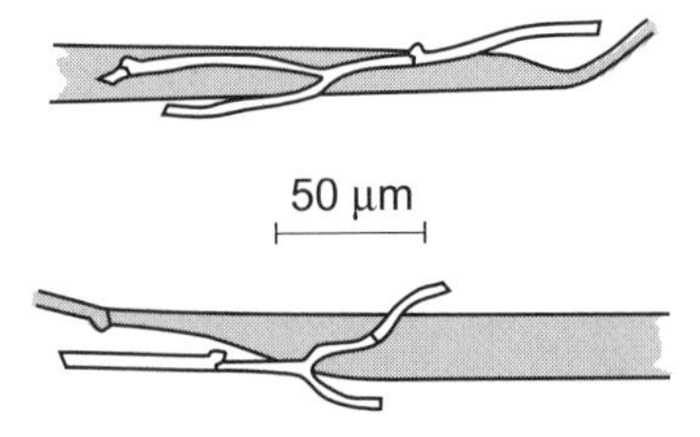

Figure 6.14 Sarcomitic hyphal system of *Trogia anthidepas*, showing the apex and base of an inflated generative hypha with narrow generative hyphae for comparison. Compare Fig. 6.16. Redrawn after Pegler (1996).

markable hyphae) and heteromerous. Heteromerous context or trama (the hyphal layer in the central part of the gill, see Fig. 6.10) is characterised by sphaerocysts, first depicted by Corda (1839), which are inflated cells situated in a ring surrounding a central ('induction') hypha (Reijnders, 1976; Watling and Nicoll, 1980). In *Lactarius,* the rosettes are simple but in *Russula*, several simple rosettes can unite to form large complexes. Recently, these have all been examined in detail by Reijnders (1993) who has come to the general conclusion that cell structures which are considered peculiar to each of the specialised trama types can be found in some form (either less well developed or restricted to a particular developmental stage) in many other, unrelated taxa.

Very similar aggregations of hyphae, termed hyphal knots (Reijnders, 1977) have been observed in a wide range of species (Reijnders, 1993). The common features of Reijnders' hyphal knots seem to be a central hypha (which remains hyphal) and an immediately-surrounding family of hyphae which differentiate in concert. Some hyphal knots do not show conspicuous differentiation, remaining as systems of tightly interwoven hyphae of uniform structure, but at the extreme, swollen cells in a ring or cylinder around a central hypha may be formed in species taxonomically far removed from the Russulaceae. Hyphal knots are found particularly frequently in plectenchymatous tissue and occur in bulb, stem, veil and cap as well as tramal tissues. Reijnders (1993) discusses the impact of their widespread occurrence on taxonomic and phylogenetic arguments about the Agaricales. I wish to point out that their analogues (perhaps, even, homologues) might occur much more widely than that because elements of Reijnders' description of hyphal knots are detectable in descriptions of many fungal multihyphal structures, from strand formation (Butler, 1958) through development of sclerotia (Townsend and Willetts, 1954), and on to the descriptions given above of paraphysis distributions around basidia as well as the influence exerted by cystidia on their surroundings in the *Coprinus* hymenium (Fig. 6.9, and the possible relationships between narrow and inflated hyphae in the *Coprinus* stem (see below).

Perhaps, in all multihyphal fungal structures, the ultimate morphogenetic regulatory structure is the Reijnders hyphal knot: a little community comprising an induction hypha (or hyphal tip, or hyphal compartment) and the immediately surrounding hyphae (or tips, or compartments) which can be brought under its influence. Larger scale morphogenesis could be coordinated by 'knot-to-knot' interactions.

6.2.2 Quantitative hyphal analysis

None of the hyphal analyses described above attempted any objective numerical analysis of the cell types they described. It seems that the best one can hope for from the types of study these represent is some subjective comment like 'infrequent' or 'numerous'. However the technology required for detailed and extensive studies of cell size and cell distribution is readily available and there is no excuse for continued reliance on the sort of subjective interpretations of over-thick microscope sections which result in mycological journals still being peppered with line drawings which look like freely-interpreted piles of spaghetti. The essential steps of the preferred procedure start with the use of thin sections cut from specimens embedded in resin (e.g. glycolmethacrylate). Resin blocks readily yield 5 μm thick sections when cut with glass knives on a suitable microtome. Thinner sections can be obtained, but 5 μm thick sections provide good contrast when stained with conventional histological staining procedures. The second step is to observe the light microscope preparations using a video camera attached to the microscope and to pass the video images to a personal computer. The third step then involves digitising the video frames with a suitable digitiser card and software. For numerical analyses, image analysis software can be applied to the captured video images to provide automatic or semi-automatic sizing and enumeration of components of the microscope image.

This approach was pioneered by Hammad *et al.* (1993a,b) who studied development of the fruit body in *Coprinus cinereus*. They discovered that the stem comprises two cell populations and that cell inflation is accentuated in cells occupying a specific zone of the stem. Differential expansion of cells in this zone readily explains how the stem changes from a solid cylinder to a hollow tube during its development.

6.2.2.1 Narrow and inflated hyphae in the fruit body stem of Coprinus cinereus

Although *Coprinus* is one of the most frequently studied fungal genera there is surprisingly little information concerning the structure of the stem of its fruit body other than that the stem is composed of greatly inflated and elongated cells. Lu (1974) seems to have given the most comprehensive description of stem structure of the time as: "The stipe includes a central column of dikaryotic hyphae and a cortex of giant multinucleate cells." Further, Gooday (1975b) noted the presence of narrow hyphae showing apical labelling with *N*-acetylglucosamine (a precursor of chitin, see Fig. 2.1, section 2.2.1) rather than the uniformly distributed labelling shown by the inflated hyphae. Most observers, therefore, considered that the main body of the stem was basically uninteresting, being made up exclusively of inflated cells. Where narrow hyphae have been observed they have been dismissed as fragments of the population of generative hyphae from the young primordium (A. F. M. Reijnders, personal communication). Hammad *et al.* (1993b) showed that this is not the case.

Low magnification images of transverse sections of stems of any fruit body primordium more than a few mm tall were dominated by highly inflated cells (Fig. 6.15) but also had a scattering of very much narrower hyphal profiles, more evident at higher magnification (Fig. 6.15). The narrow hyphae were also clearly visible in longitudinal sections and SEM images (Fig. 6.16). Narrow hyphae stained selectively with numerous standard histological stains, including Mayer's haemalum, the periodic acid-Schiff reagent (PAS), toluidine blue, Delafield's haematoxylin, 0.5% (w/v) acid fuchsin in 0.5% (v/v) acetic acid, alcian blue/safranine, Luxol fast blue, and 1% aqueous acid fuchsin. Most staining reactions were differential in that only some narrow hyphae were stained in any one transverse section and adjacent compartments in longitudinal sections could stain differently.

Hammad *et al.* (1993b) measured cross-sectional areas of hyphal profiles in 5 μm thick sections with an image analysis program in transverse sections cut from stems of fruit bodies of a range of developmental ages. For each section of each piece of stem, the area of every cell within two randomly chosen radial transects 12 μm wide was measured. Individual cells in a transect were measured in strict order, starting from the exterior of the stem and ending at the lumen. Fig. 6.17 shows an example transect and Figs 6.18 and 6.19 the graphical plots derived from it. Analysis of the size spectrum of the cells was done by grouping their cross-sectional areas

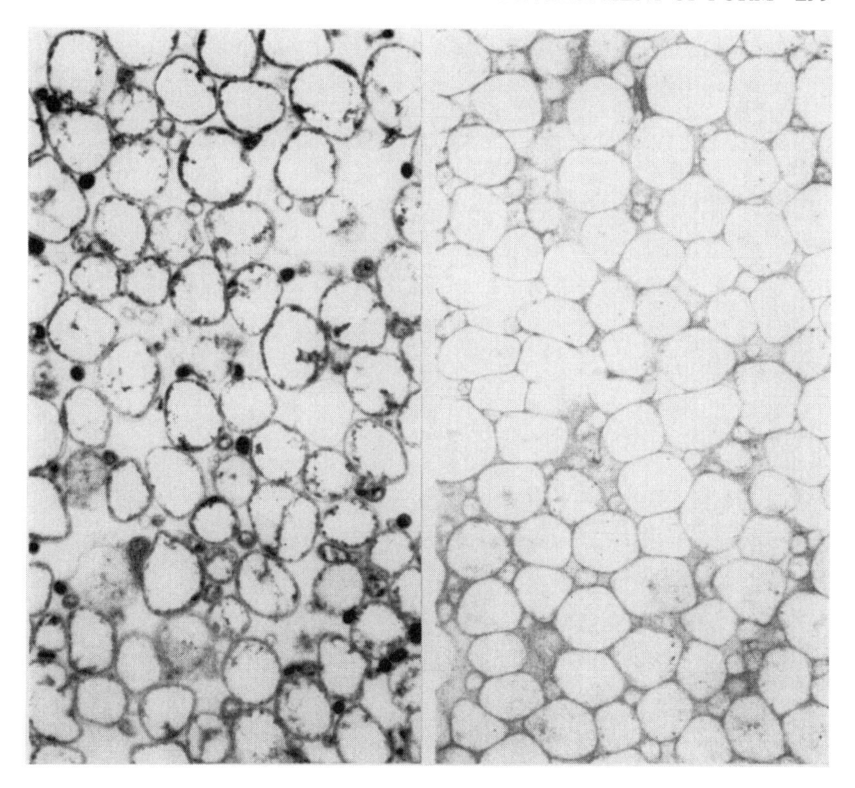

Figure 6.15 Transverse sections of stems of a fruit body of *C. cinereus*, showing highly inflated cells and a scattering of very much narrower hyphal profiles. Note that in the left hand image, in particular, some of the narrow hyphae are unstained, suggesting that they may have different cytoplasmic contents (×500).

in frequency classes of 10 μm^2. In all transects the 0–10 and 10–20 μm^2 classes were the most numerous and the cell area of larger hyphae was widely dispersed (Figs 6.20 and 6.21). The frequency distribution, in 5 μm^2 classes, of the combined data for 3794 cells is shown in Fig. 6.21. On the basis of these frequency distributions two distinct populations of hyphae were identified and categorised as narrow hyphae, with cross-sectional area $< 20\ \mu m^2$, and inflated hyphae, with cross-sectional area $= 20\ \mu m^2$. Narrow hyphae constituted 23% to 54% of the cells in transverse sections of the stem but only contributed 1% to 4% of the overall cross-sectional area (Table 6.2).

All the available evidence indicates that narrow hyphae in stems of *C. cinereus* have diverse functions. They tend to be particularly concentrated at the exterior of the stem where they may serve as an insulating layer, and as a lining to the lumen where they may excrete material into the

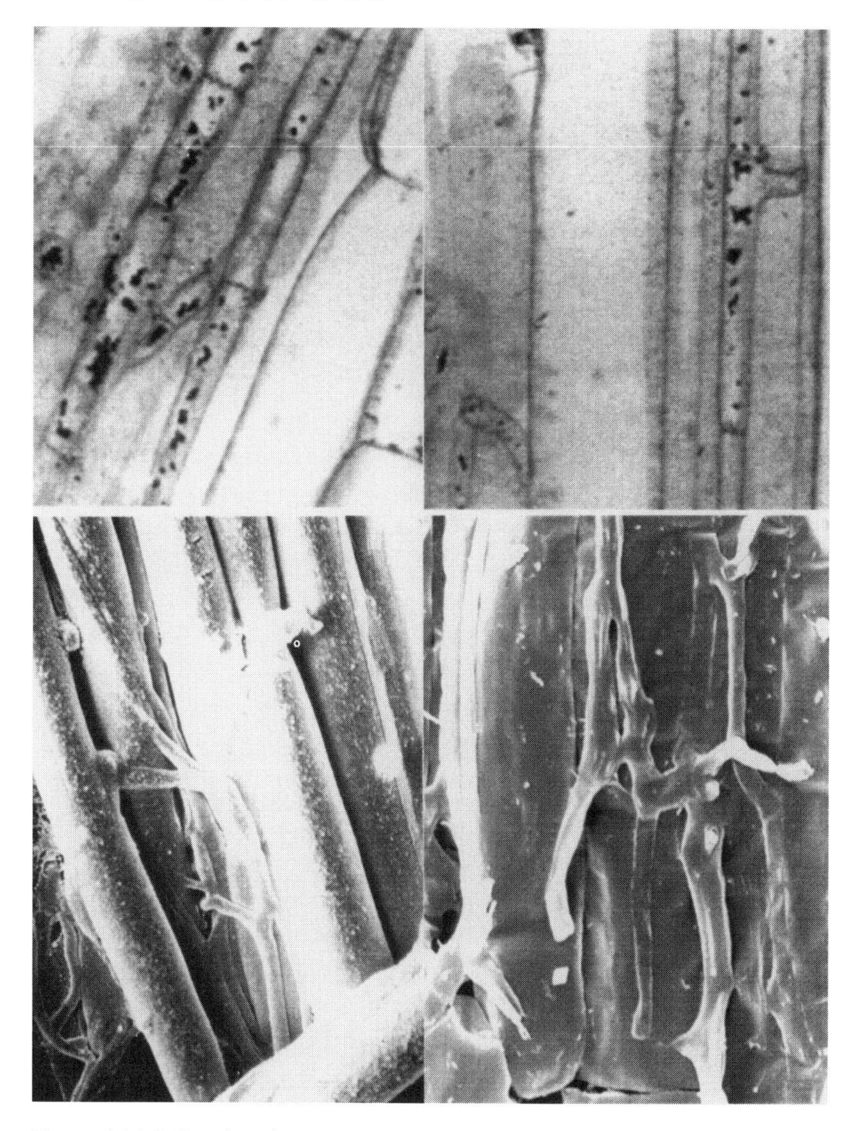

Figure 6.16 Inflated and narrow hyphae in longitudinal sections (top, ×3000) and SEM images (left, ×500; right, ×1000) of stems of a fruit body of *C. cinereus*.

cavity or represent the remnants of the initially central core of dikaryotic hyphae. Cox and Niederpruem (1975) referred to a brown gel in the lumen which disappeared as the stem extended. This gel might be produced by the narrow hyphae or consist of their degradation products if the lumen is produced by degradation of the initial central core.

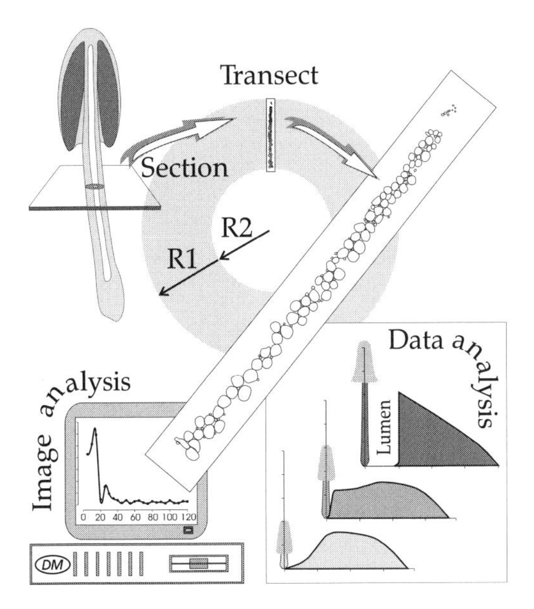

Figure 6.17 Determination of cell population distributions. The diagram shows how glycolmethacrylate sections of the stem were used for image analysis of cell cross-sectional areas. R1 and R2 refer to the radial extent of stem tissue and stem lumen, respectively, which are shown in Table 6.2. Transects were routinely 12 μm wide; a wider transect is shown here for illustrative convenience. Redrawn after Moore (1995).

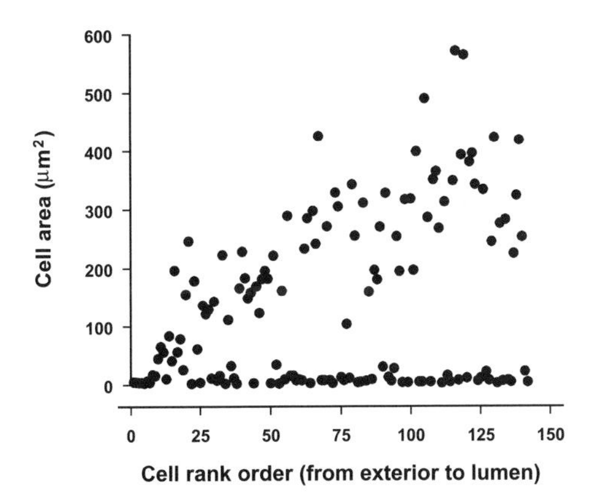

Figure 6.18 Plot comparing cell cross-sectional area with position of the cell in the transect (rank order) for the transect illustrated in Fig. 6.17. Data from Hammad *et al.* (1993b). In this and other plots showing rank order, remember that the *x*-axis shows a stem radius and that the outermost cell is arbitrarily put as position zero, subsequent cells being numbered in strict sequence towards the lumen (or towards the centre of solid stems).

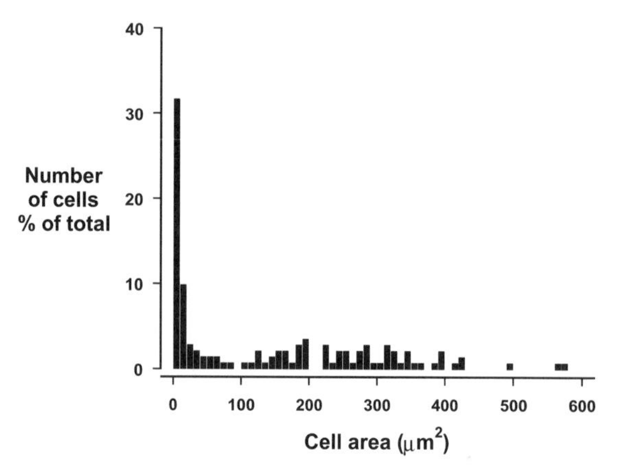

Figure 6.19 Plot showing the cell size distribution of the transect illustrated in Fig. 6.17. Data from Hammad *et al.* (1993b).

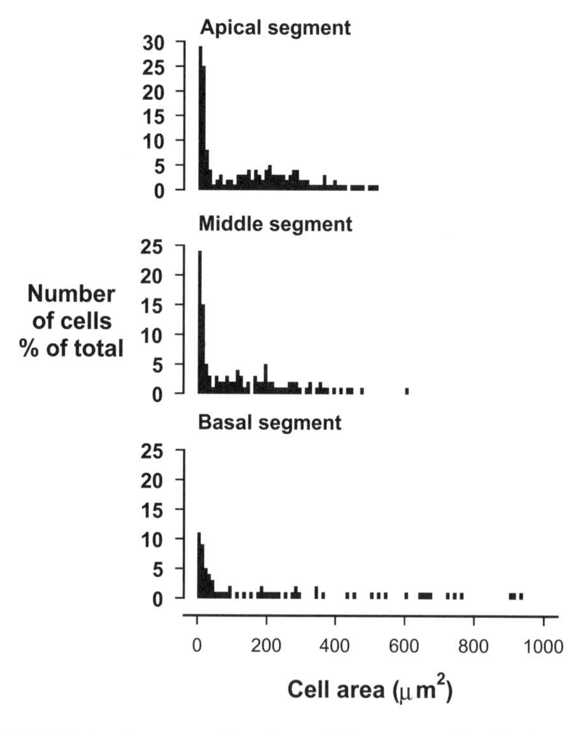

Figure 6.20 Cell size (cross-sectional area) frequency distributions of cells in transects of sections cut from the apical, middle and basal zones of a 45 mm long stem. Note that throughout the fruit body narrow hyphae (up to 20 μm^2 in cross-sectional area, represented by the first two categories in the frequency histograms) were a major component of the hyphal population. Data from Hammad *et al.* (1993b).

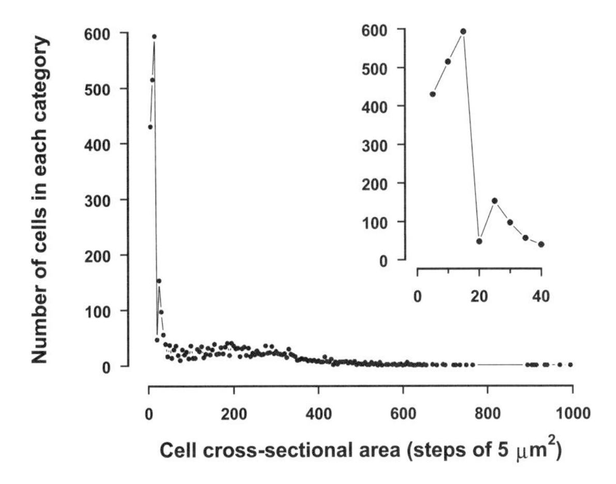

Figure 6.21 Frequency distribution of cell cross-sectional areas in the cumulated data from all the transects analysed. A total of 3794 cells are represented here. Inset shows the clear demarcation between cell populations at the 20 μm^2 category. Data from Hammad *et al.* (1993b).

Narrow hyphae stained densely with Mayer's haemalum, toluidine blue and aniline blue/safranine and revealed especially strong, particulate, staining with the periodic acid-Schiff reagent for polysaccharide. However, not all narrow hyphal profiles in a transverse section and not all hyphal compartments belonging to any one narrow hypha in longitudinal sections stained equally. The reason for this differential staining is not known, but it might reflect differential function among the narrow hyphal population or, since narrow hyphae may be important in translocation of nutrients through the stem, it may simply reflect variation in the vertical distribution of cytoplasmic materials in course of translocation. The narrow hyphae seem to form networks independent of the inflated hyphae; they were seen to be branched and to be fused laterally with other narrow hyphae but inflated hyphae were neither branched nor associated in networks.

What makes some hyphae become inflated and multinucleate while others remain morphologically similar to the vegetative mycelial hyphae is not known although the even (i.e. non-random) distribution of the former (see section 6.2.2.2) implies some form of organisational control. This differentiation occurs at an extremely early stage as both narrow and inflated hyphae can be seen in primordia 3 mm tall. However, during stem elongation, the numerical proportion of narrow hyphae decreased (Table 6.2) implying that approx. 25% of them are recruited to the inflated category as the fruit body develops and this seems to be a fourth way

Table 6.2. *Comparison of the numbers of narrow hyphae and the area they contributed to the total area of cells in the transect for fruit bodies of* Coprinus cinereus *at different stages of development and at different positions within each fruit body.*

Stem length (mm)	Zone	Narrow hyphae (% of total)	% Area contributed by narrow hyphae	Number of cells in the transect	R1* (µm)	R2* (µm)
6	Middle (full diameter)	47.2	4.2	496	1664	546
27	Apical	48.2	3.4	164		
	Apical	50.3	3.2	145	832	416
	Upper mid-region	45.1	3.4	173		
	Upper mid-region	46.6	3.0	189	966	312
	Lower mid-region	41.3	3.3	104		
	Lower mid-region	31.3	2.3	80	989	884
	Basal	45.2	3.9	146		
	Basal	40.4	4.4	146	839	780
27	*Mean values*	*43.6*	*3.4*			
45	Apical	38.8	2.7	165		
	Apical	32.6	2.2	132	1173	832
	Middle	34.5	1.7	87		
	Middle	40.2	3.9	117	973	1768
	Basal	27.8	1.4	54		
	Basal	37.1	1.5	62	875	ND
45	*Mean values*	*35.2*	*2.2*			
70	Apex (full diameter)	53.8	4.0	318	691	748
	Apical	23.8	1.1	105		
	Apical	28.7	1.6	108	856	1300
	Upper mid-region	25.0	1.4	104		
	Upper mid-region	23.3	1.4	103	821	1716
	Lower mid-region	30.8	1.7	117		
	Lower mid-region	29.9	1.6	107	834	1872

Table 6.2 (*cont.*)

Stem length (mm)	Zone	Narrow hyphae (% of total)	% Area contributed by narrow hyphae	Number of cells in the transect	R1* (µm)	R2* (µm)
	Upper basal	38.7	2.5	119		
	Upper basal	37.8	2.0	111	948	1664
	Lower basal	43.6	2.6	117		
	Lower basal	50.0	2.8	152	1153	1092
70	*Mean values*	*33.2*	*1.9*			

The 6 mm fruit body was at a pre-karyogamy stage. Sporulation was occurring in the 27 mm long fruit body and had been completed in the 45 and 70 mm long fruit bodies. Entries in the table show total cell numbers and narrow hypha proportions for radial transects except for the 6 mm and 70 mm apex transects which were full diameters.
*R1 and R2 refer to the actual length of the radii spanning stem tissue and lumen respectively. Data from Hammad *et al.* (1993b).

of expanding the fruit body. Already well documented are the heteromerous trama of Russulales, with inflating rosettes of sphaerocysts, sarcodimitism in some of the tougher pleurotoid/mycenoid agarics, and the dimitism of the aphyllophoralean type seen in such genera as *Pleurotus* and *Lentinus*. The Coprinoid type of fruit body expansion described by Hammad *et al.* (1993b) may be an adaptation to rapid development on ephemeral substrates.

Normal extension in vertical stems of *C. cinereus* occurred in two phases. The first phase was a period of relatively slow extension rate, in the region of 10 µm min^{-1}, which was followed by a rapid period of extension with a rate up to 110 µm min^{-1}. During the most rapid phase an average stem may elongate by 80 mm in less than 12 h (Moore and Ewaze, 1976). The kinetics of this elongation have attracted a great deal of attention since Buller (1924) examined *C. sterquilinus*. Another *Coprinus* species (probably *C. radiatus*) was studied by Borriss (1934, using the name *C. lagopus*), Hafner and Thielke (1970), and Eilers (1974); *C. congregatus* by Bret (1977); and *C. cinereus* by Gooday (1974a), Cox and Niederpruem (1975, using the name *C. lagopus*) and Kamada and Takemaru (1977a,b, 1983, using the name *C. macrorhizus*).

In all these species the upper half, and generally the upper mid-region, was the most active zone of elongation. Eilers (1974) found that stem elongation in *C. radiatus* was accompanied by a 68-fold increase in cell length and a doubling of the cell number. Although the DNA content of the stem of *C. cinereus* has been found to increase abruptly just before the most rapid phase of elongation (Kamada *et al.*, 1976) and stem cells become multinucleate (Lu, 1974; Moore *et al.*, 1979; Stephenson and Gooday, 1984), stem elongation has been attributed solely to cell elongation (Gooday, 1975b; Kamada and Takemaru 1977a).

This view was supported by the cell measurements of Hammad *et al.* (1993a) in *C. cinereus*. These showed that there was little increase in cell (longitudinal) sectional area between a 3 mm fruit body and an 8 mm tall fruit body (Table 6.3) both of which were at pre-meiotic developmental stages. Presumably any stem elongation occurring at these stages is due primarily to cell proliferation rather than cell elongation. By contrast, there was a large increase in cell sectional area between the stems of the 8 mm fruit body (pre-meiotic) and that of a 25 mm fruit body undergoing meiosis. Initially the cells in the basal and middle regions of the stem inflated. The apical cells did not expand to the same extent, and even in a fully elongated fruit body apical cells were considerably shorter than other cells. Cells in the extreme basal and apical regions were always shorter than those in other regions of elongated stems. For example, cells near the cap/stem junction at the extreme apex of an 83 mm fruit body (fully elongated) had a typical longitudinal sectional area of 3000 μm^2 compared with an average for the whole of the apical section examined (about 10 mm long) of 6258 μm^2. The most elongated cells were found in the upper mid-region of the stem. The length/width ratios of about 2 in pre-meiotic stems (3 mm and 8 mm fruit bodies) increased greatly after meiosis, particularly in the upper middle regions, to 10, 20 and approximately 35 in 48 mm, 55 mm and 83 mm tall fruit bodies.

Overall, therefore, stem extension of *C. cinereus* involves increase in length and cross-sectional area of inflated hyphae and recruitment of narrow hyphae into the inflated population.

6.2.2.2 Spatial distribution of inflated and narrow hyphae

Spatial distributions of cells in transverse sections of the stems is illustrated by plotting cell area against cell rank order (where cell number one is deemed to be at the exterior of the stem and the last cell is at the lumen end of the radius (see Figs 6.17 and 6.18). The data set in Fig. 6.18 was representative of other transects and other fruit bodies (Fig. 6.22). Narrow hyphae coated the outside of the stem and lined the lumen but

Table 6.3. *Sectional area and length/width ratios of cells in longitudinal sections of stems of* Coprinus cinereus.

Fruit body height (mm)	Section	Mean stem cell area (μm^2)	Length/width ratio
3	Middle	148	1.9
8	Middle	211	2.0
25	Apex	292	1.8
	Middle	3857	11.5
	Base	2705	6.2
48	Apex	3184	9.0
	Upper middle	6813	12.7
	Lower middle	5735	10.3
	Base	3449	10.6
55	Apex	9243	11.5
	Upper middle	9496	18.1
	Lower middle	10522	19.9
	Base	6533	13.8
83	Upper apex	6258	12.6
	Lower apex	11960	22.0
	Middle region 4	12894	26.0
	Middle region 3	11672	30.2
	Middle region 2	10448	35.1
	Middle region 1	5538	20.4
	Upper base	4785	14.1
	Lower base	2681	8.8

Each entry represents the mean of 50 measurements.

were also dispersed throughout the remaining tissue. Where no lumen was present, the central region was occupied exclusively by narrow hyphae. This was the case in fruit bodies <6 mm tall in which the lumen had not yet developed (Fig. 6.23), and at the extreme apex of mature fruit bodies above a well developed lumen (Fig. 6.24).

Distributions of these populations were judged by the nearest neighbour method. Inflated hyphae, regardless of the age of the fruit body (from 27–70 mm tall) and regardless of position within the stem, were very strongly evenly distributed (in an even distribution the presence of one individual lowers the probability of another of the same sort occurring near by). On the other hand, the spatial distribution of narrow hyphae was random in almost all cases (in a random distribution the presence of one individual does not affect the probability of another of the same sort occurring near by), differing significantly from randomness

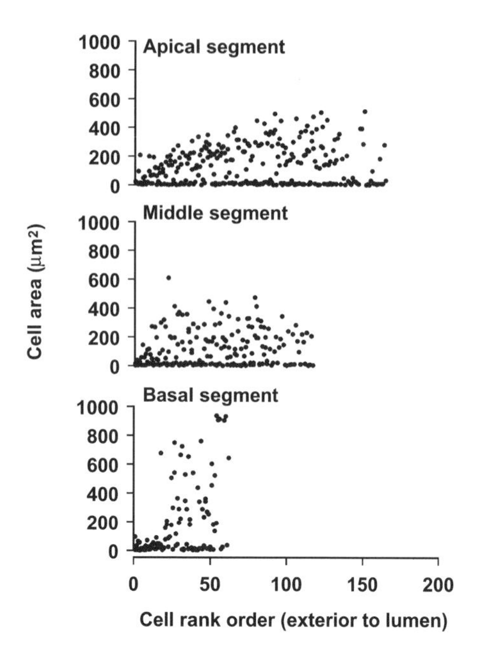

Figure 6.22 Rank order plots for transects of a 45 mm fruit body (same data as Fig. 6.20). Note that narrow hyphae were present throughout the fruit body. Data from Hammad *et al.* (1993b).

only in the upper middle region of the 27 mm tall fruit body and the upper apical region of the 70 mm tall fruit body where there were slight tendencies towards even distributions. Generally, narrow hyphae were interspersed with inflated hyphae across the full radius of all stems irrespective of position along the length of the stem and irrespective of the developmental age of the stem (compare Fig. 6.23, from a 6 mm tall fruit body, Fig. 6.22, from a 45 mm fruit body, and Figs 6.18 and 6.24 which come from a 70 mm tall fruit body). Significantly, narrow hyphae are present at all positions in the fruit body (from base to apex) and at all stages of development. The population of narrow hyphae was reduced by about 25% as size increased from 27 to 70 mm, presumably due to approx. 25% of narrow hyphae becoming inflated. This fraction might therefore be considered to be that fragment of the primordial generative hyphae preserved as a reserve of hyphal inflation capacity to support final maturation growth; but it is only a minority fraction of the narrow hyphal population. Other members of that population have other functions. If inflated hyphae do arise by expansion of the randomly distributed narrow hyphae, the even (i.e. non-random) distribution of inflated

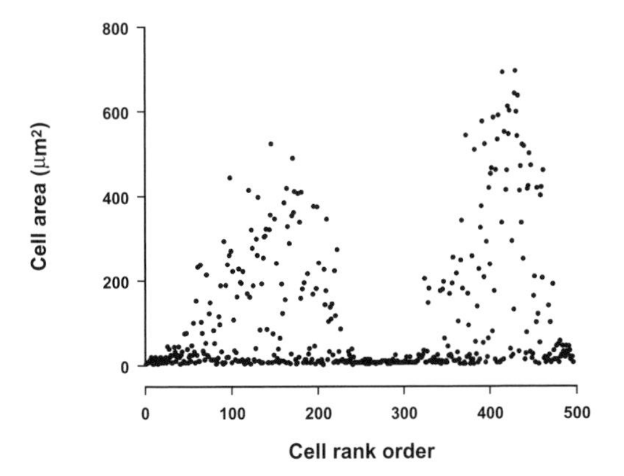

Figure 6.23 Rank order plot of a transect across the full diameter of the middle region of a 6 mm tall stem. Narrow hyphae occupy the central column of tissue in the stem. Data from Hammad *et al.* (1993b).

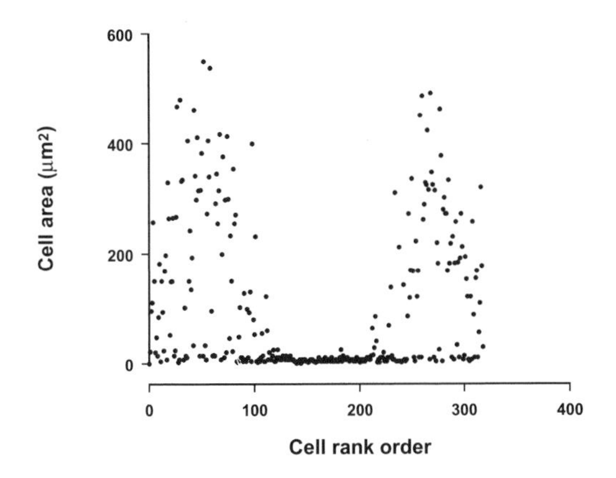

Figure 6.24 Rank order plot of a transect across the full diameter of the extreme apex of a 70 mm tall stem. The section passed above the lumen and narrow hyphae occupy the central region of the stem. Data from Hammad *et al.* (1993b).

hyphae implies that a pattern forming process determines their differentiation.

6.2.2.3 Developmental changes in the distribution of hyphal sizes within the stem

Comparing data from all transects between fruit bodies of different size revealed a progressive change in the distribution of inflated hyphae (Fig. 6.25). In 6 mm and 27 mm tall fruit bodies the inflated hyphae increased

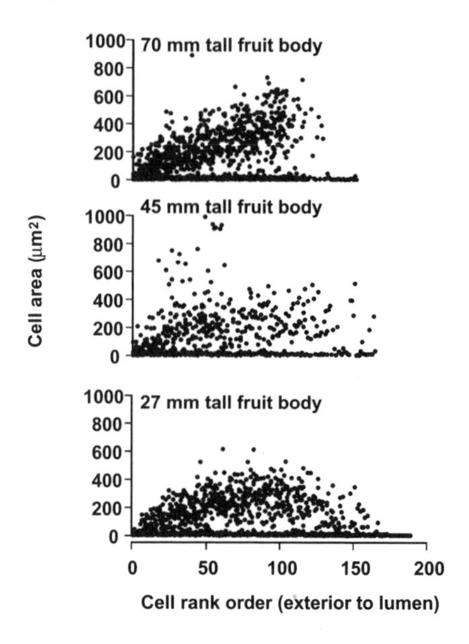

Figure 6.25 Cumulated rank order plots of 27, 45 and 70 mm stems. Note how the cell size distribution across the radius of the stem changes with increasing stem length. Data from Hammad *et al.* (1993b).

in cross-sectional area up to halfway across the cortex but then their size declined towards the lumen (Figs 6.23 and 6.25). In the 45 mm tall fruit body the cross-sectional area of inflated hyphae increased gradually from the exterior to the lumen and this pattern was even more pronounced in the 70 mm tall fruit body, the peak cell area being adjacent to the lumen rather than in the mid-cortex (Fig. 6.25).

The data obtained by Hammad *et al.* (1993b) show that expansion of the stem is mainly due to increase in cross-sectional area of inflated hyphae in the region between the mid-cortex and the lumen. Inflated hyphae around the periphery of the stem do not enlarge much. The mechanical consequences of this pattern of cell inflation are obvious. Increase in cross-sectional area of inflated hyphae in the middle of the cortex will (i) result in the central core being torn apart, leaving its constituent cells as a remnant around the inner wall of the lumen so created; and (ii) stretch, reorganise and compress the tissues in the outer zones of the stem (Fig. 6.26). Consequently, formation of the mature stem as a cylinder with outer tissues under tension and inner tissues in compression (the optimum mechanical structure for a vertical cylindrical support) is entirely a result of the pattern of cell inflation within the stem as the stem develops. This specific pattern of inflation must be

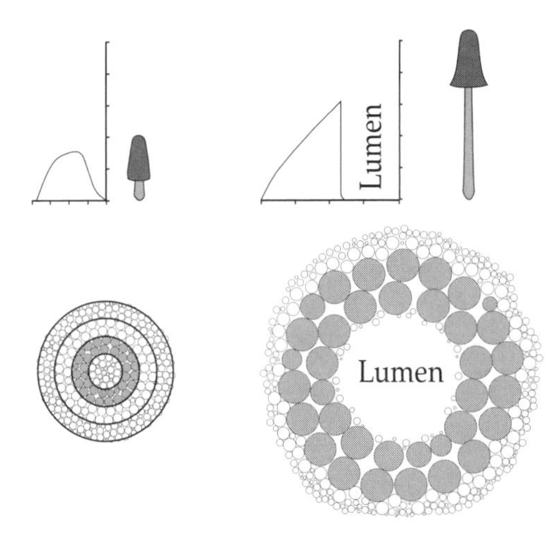

Figure 6.26 Interpretation of the geometrical consequences of the cell size changes during development revealed by the data of Fig. 6.25. The graphs in the upper part of the figure show the lines of best fit for the scatter plots (Fig. 6.25) of 27 and 70 mm stems, together with scale drawings of the fruit bodies. The diagrams in the lower part of the figure are transverse sections of the stems, drawn to scale. The diagrammatic transverse section of the stem of a 27 mm tall primordium (on left) is composed of solid tissue which is divided into four zones corresponding to the zones in the radius in the graph above. The central (zone 4) and outermost (zone 1) zones are comprised of rather smaller cells than the two cortical zones. During further growth the most dramatic cell inflation occurs in the cells of zone 3 which are here shown shaded. Growth from 27 mm to 70 mm in height is accompanied by a 3.6-fold increase in cell area in zone 3 and a 1.6-fold increase in zone 2. The cells in the other zones have to be rearranged to accommodate the inflation of zone 3 and a major consequence is that a lumen appears in the centre of the stem. Changes in area are shown to scale in these diagrams, though only a total of 327 cell profiles are illustrated. As this is only a tiny proportion of the cells involved *in vivo*, the diagrams inevitably distort the apparent relationship between cell size and stem size. Narrow hyphae have been ignored in these diagrams, though they are distributed throughout the tissue and their conversion to inflated hyphae contributes to stem expansion. From Moore (1996).

organised by signalling molecules which determine differential cell inflation across the stem radius.

These observations concentrate attention on the integration of control of cell expansion throughout the fruit body, and this is dealt with in the next section. However, they also demonstrate that not all narrow hyphae are identical; some stain where others do not, some are recruited to become inflated, others are not. Similarly, inflated hyphae differ; some inflate much more than others, but the distinction is based on position, not cell morphology. So, returning to the topic discussed in section 6.2.1, one might well ask whether the *C. cinereus* stem should be described as monomitic (inflated hyphae being the most obvious component even though in a numerical minority in some stages), or as dimitic (narrow + inflated), or perhaps trimitic (narrow + inflated + very inflated, or generative narrow + specialised narrow + inflated) or even some higher mitic multiple (generative narrow + specialised narrow + inflated + very inflated). Perhaps the safest conclusion is that categorisation of cell types on morphological grounds is bound to reveal a disappointingly small part of the story unless it is complemented by detailed analysis of function, distribution and how these change during development.

6.2.2.4 Coordination of cell inflation throughout the maturing fruit body

Cell inflation in a specific pattern like this raises the question of whether, and if so how, inflation of different parts of the fruit body is integrated during development. Most of the changes in shape during fruit body development in basidiomycetes depend on two types of cell inflation: a slow process typical of young primordial stages and a more rapid one characteristic of maturation. Both, but particularly strong inflation, are aspects of differentiation (Reijnders and Moore, 1985). Reijnders (1963) showed that the different zones of fruit bodies of a great many species enlarge proportionally, so that different tissues mature without being impeded, compressed or distorted by the growth of other parts. Such coordination of cellular differentiation in fruit body primordia would have to operate over distances of many millimetres at least. Local cell inflation during fruit body development has often featured in studies of, for example, stem growth (especially) or cap expansion (e.g. Gooday, 1974a, 1982, 1985; Moore *et al.*, 1979; Rosin *et al.*, 1985 with *Coprinus cinereus*; Bret, 1977 on *C. congregatus*; Wong and Gruen, 1977; Gruen , 1982; Williams *et al.*, 1985; Gruen, 1991 on *Flammulina velutipes*; Bonner *et al.*, 1956; Craig *et al.*, 1977 on *Agaricus bisporus)*. However, the only holistic account of inflation over the whole fruit body has been done by Hammad *et al.* (1993a).

Fundamental to this study was an objective, physiologically-relevant timescale. Morphological events have usually been used to define the temporal stages in development of whole fruit bodies (Madelin, 1956a; Takemaru and Kamada, 1972; Matthews and Niederpruem, 1973; Morimoto and Oda, 1973; Moore *et al.,* 1979). The smaller species of *Coprinus* are particularly useful because meiosis is well synchronised in all of the basidia. Pukkila *et al.* (1984) defined these stages in terms of

Table 6.4. *Developmental stages of basidiome development in* Coprinus cinereus *designated by Pukkila* et al. *(1984).*

			Other designations in the literature*			
Stage	Time (h)	Predominant feature	a	b	c	d
1	0–12	Basidial differentiation	Stages 1–2	Stages 1–3	Day 8	
2	13–24	Karyogamy and meiotic prophase	Stage 2	Stage 2		−2 Days
3	25–36	Meiotic divisions and spore formation	Stages 3–4	Stage 4	Day 9	−1 Day
4	37–48	Stem elongation and spore release	Stage 5			Day 0

*References: (a) Moore *et al.*, 1979; (b) Morimoto & Oda, 1973, 1974; (c) Kamada *et al.* 1978; (d) McLaughlin, 1982.

progress through meiosis and spore formation in the hymenium of *C. cinereus*. The 48 h which encompass meiosis and sporulation were divided into four 12 h periods (Table 6.4).

Hammad *et al.* (1993a) expanded on this approach by examining a large sample of fruit bodies to determine the exact timing of major meiotic and sporulation stages. This established a timeline to which other processes can be referenced simply by microscopic examination of a sliver of cap tissue (Fig. 4.5). Using standardised culture conditions, Hammad *et al.* (1993a) found that fruit body primordia were formed 5 days after inoculation and developed into mature fruit bodies within 2–3 days, during which the basidia underwent a sequence of morphologically and physiologically distinctive stages. The dikaryotic basidioles underwent nuclear fusion (karyogamy) and then entered meiosis I, followed immediately by meiosis II. After completion of meiosis, four sterigmata emerged followed by formation of four basidiospore initials, nucleus migration into the young spores, maturation and discharge of the spores. All these events were completed within 18 h. However, basidial development was not totally synchronous. For example, nuclear fusion, meiosis I and meiosis II, or the stages meiosis II (basidia containing four nuclei), sterigmata formation and expansion of spores could be seen in a single (*ca.* 1 mm^2) fragment of gill tissue. The overlap between different stages varied through development. In a 'meiosis I specimen' about 60% of the basidia had two nuclei; at meiosis II about 70% had four nuclei; and when sterigmata first appeared about 90% of the basidia were at the same stage. Defining karyogamy as time zero, basidia took 5 h to reach meiosis

I (Fig. 4.5) and meiosis II was completed after a further hour. From meiosis II to the emergence of sterigmata required 1.5 h, spores emerging 1.5 h after that. Spore formation continued for 1 h and then nuclear migration started. Spore pigmentation commenced 1 h after spore formation and spores matured and were discharged 7 h later. This timeline is objective in the sense that it depends upon processes which are endogenously controlled. It is reliable because these processes are central to fruit body function and it is versatile since by examining slivers (*ca.* 1 mm^2) at known time intervals the effects of any change in cultivation conditions or culture genotype become apparent. The most rapid phase of stem elongation occupied the 5 h prior to spore discharge and cap autolysis and started 8 h after karyogamy. Cap expansion started as spores matured, about 14 h after karyogamy. Expansion of the different cell types in the cap as well as inflation of cells of the stem began immediately post-meiotically (Table 6.5). Such coordination may be achieved by some sort of signalling system that 'reports' the end of meiosis to spatially distant parts of the fruit body. The route such a signal might take is not clear, but primary gills are attached to the stem, with their tramal regions in full hyphal contact with stem tissues, so the connection between tissues undergoing meiosis and the upper (most reactive) regions of the stem may be fairly direct.

6.2.2.5 Specific wall synthesis drives tropic bending

Another example of the tight control of cell inflation in *Coprinus cinereus* is the regulation of cell extension during the tropic bending. The use of tropic responses of mushroom fruit bodies, especially gravitropism, as non-invasive means of investigating morphogenesis has been outlined in (section 6.1.8). Greening *et al.* (1997) carried out cytological studies of gravitropically bending stems of *C. cinereus* to determine the morphometric patterning which achieved the gravitropic curvature. Stem material was taken from gravistimulated stems at the region which showed the greatest degree of gravitropic bending (invariably in the upper 20 to 30% of the stem). The outer and inner flanks of the bent stem were sectioned separately). Cell measurements were made with image analysis software in the same way as described by Hammad *et al.* (1993a). Cell cross-sectional areas were not significantly different, nor were there any differences in the number of narrow hyphae or in packing density of hyphae between the inner and outer flanks of the bend. In fact, the only difference found was in cell length (Table 6.6). Hyphae of the inner flank were short with cytoplasm localised around the septa. In contrast, hyphae of the outer flank were very long, containing large vacuoles which main-

Table 6.5. *Sectional areas (μm^2) of hymenial cells measured in longitudinal sections together with pooled data for cells of the stems of the same specimens.*

	Fruit body height (mm)							
Cell type	0.3	0.6	0.8	2.5	2.7	4.8	5.5	8.3
Apex				292 ± 25		3184 ± 174	9243 ± 548	9109 ± 390
Middle	148 ± 7		211 ± 13	3857 ± 194		6274 ± 214	10009 ± 434	10138 ± 308
Base				2705 ± 181		3449 ± 200	6533 ± 474	3733 ± 182
Veil cells	238 ± 14	242 ± 11	276 ± 16					
Basidia					151 + 3	181 + 3	177 + 3	138 + 3
Paraphyses					193 + 7	244 + 9	253 + 7	215 + 7
Spores					39 + 1	48 + 1	43 + 1	45 + 1
Cystidia					1194 ± 28	1423 ± 44	2495 ± 93	1391 ± 42
Cystesia					305 + 7	303 + 11	387 + 17	260 + 8

Each entry for the hymenial cell types is a mean of 50 measurements. The data for the cells of the stems are means of 50 to 400 measurements.

Table 6.6. *Cell morphometric analysis of sections of gravitropically-responding stems of* Coprinus cinereus *at the point of maximum curvature.*

	Lower region of the bend (inner flank)		Upper region of the bend (outer flank)	
	n	mean ± SEM	n	mean ± SEM
Mean cross-sectional	626	9.82 ± 0.60	881	9.43 ± 0.49
area of narrow	803	8.93 ± 0.79	625	8.95 ± 0.43
hyphae (μm^2)	775	9.15 ± 0.78	538	9.58 ± 0.88
Mean cross-sectional	1422	176.08 ± 6.51	1442	183.00 ± 7.62
area of inflated	1152	185.96 ± 9.24	1202	178.55 ± 9.39
hyphae (μm^2)	1195	184.29 ± 11.74	1203	189.71 ± 15.38
Mean width of	20	29.90 ± 3.64	20	19.85 ± 3.01
hyphae (μm)*	20	18.74 ± 4.67	20	16.25 ± 3.06
	20	18.73 ± 3.47	20	18.08 ± 4.42
	20	19.25 ± 3.85	20	18.86 ± 5.71
	20	17.13 ± 2.97	29	18.74 ± 4.91
		(range)		(range)
% Narrow Hyphae	3 stems	30.5–39.1	3 stems	28.8–41.5
Packing density	3 stems	0.44 ± 0.02	3 stems	0.47 ± 0.28
Cell length of inflated	34	542 ± 35.0	34	116 ± 7.4
hyphae in three	34	534 ± 37.9	34	170 ± 12.4
different stems (μm)	34	698 ± 50.6	34	107 ± 5.7
Cell length of inflated	20	263.12 ± 50.5	20	141.0 ± 24.0
hyphae in five	20	439.74 ± 153.2	20	133.16 ± 22.1
different stems* (μm)	20	299.35 ± 114.9	20	122.70 ± 18.5
	20	385.30 ± 107.8	20	137.36 ± 27.5
	20	271.45 ± 77.2	20	127.57 ± 24.3

*Measurements made 1 year later than the others, using diffferent analysis methods. Stems were fixed 3 h (*4 h) after reorientation to the horizontal and embedded in glycol methacrylate (*Spurr's resin). Sections were cut 2–4 μm thick with glass knives on an ultramicrotome and observed by light microscopy, video images being processed by

tained the cytoplasm at the periphery of the cells. Hyphae of the outer flank were significantly longer than those of the inner flank of the bend in a ratio of between 4 and 5:1.

In *Coprinus*, production of the tropic bend is a result of the outer flank extending more than the inner. Overall, the outer flank extends more than the inner in a ratio of 3:2. Production of the tropic bend in *C. cinereus* is generated as a direct and unique result of an increase in hyphal length only in the outer flank. Evidently, the intercalary wall growth which drives the 4–5-fold increase in length that is solely responsible for bending, is somehow regulated to lengthen the cells without increasing their girth. In this respect it may be relevant that in stem cells of *C. cinereus* the presence of left handed chitin helices in the wall is unique (Kamada *et al.*, 1991), suggesting that helicity of the wall structure is of considerable importance. It may be that the growth pattern which characterises tropic bending is to extend the helices without increasing their diameter.

Whether this specific speculation is borne out or not will depend on further research into the cellular nature of growth during tropic bending. However, there is clear evidence in these analyses that growth in length and growth in girth of inflated cells in the stem are separate events which are under separate control. Further, those controls are affected differentially by the response pathway to a tropic perception. Overall, the morphometric analyses of Hammad *et al.* (1993a,b) and Greening *et al.* (1997) show that growth of the *C. cinereus* stem depends upon:

- differentiation of narrow hyphae into inflated hyphae;
- increase in compartment length of inflated hyphae;
- increase in diameter of inflated hyphae.

Morphogenesis of the stem depends upon:

- differential control of hyphal inflation on the basis of position of the hyphae within the whole structure;
- differential control of hyphal wall synthesis to distinguish between increase in length and increase in girth on the basis of position in relation to an externally applied signal.

6.3 Tissue domains

Section 6.2.2.5 describes how inflation of particular cells in a specific region of the stem is able to create a change in morphology. The new morphology develops as a consequence of the mechanical impact of cell inflation being controlled and emphasised in a specific group of hyphae in a particular region. This example brings out two important generalisations. First, tissues with distinctly different functions may not be very different morphologically. Second, tissue inflation is a major morphogenetic factor in fungal multicellular structures.

6.3.1 Defining tissues

Tissues must be defined in terms of function but they may need to be recognised at stages before their function becomes apparent. Developmental studies are therefore essential in establishing the cell lineage which leads to a particular tissue. In the majority of cases in the literature, tissues are identified on the basis only of observation of the mature structure. This is an adequate definition of function, but without the developmental timescale, it is an incomplete description of morphogenesis. Two features emerge from the developmental studies which have been made. Different tissues may be defined initially on the basis of position, rather than morphology, and tissues which have similar function in the mature structure may have very different origins. Regionalised cell expansion in the *Coprinus* stem is an example of the first of these features. The whole structure of the stem is changed by inflation of a particular group of cells. Yet, before the inflation occurs those cells are not differentiated from their fellows in any obvious way. Rather it is their position, in an annulus located at the central half of the radius, which defines them as a particular population of cells. Even after maturation, they are different from their fellows only in the extent of their inflation and no other differentiated feature marks them out. Nevertheless, this population of cells serves a specific function which contributes to development of a crucial feature of the mature fruit body. They can rightly be termed a tissue different from other tissues in the stem on these criteria, even though cell morphologies are uniform. The different origins of tissues of similar function will be discussed on a tissue level in section 6.4. However, it is also the case that similar functions are served by cells of different origins. This is a relatively new observation which may not be obvious so it is worth discussing here.

Possibly the most unexpected examples come from analyses of cell lineages in mushroom hymenia. These tissues may be thought of as being of simple function (spore-bearing), yet they also serve mechanical and structural purpose. A hymenium is essentially an epidermal layer of closely packed hyphal tips. Ultimately the most important hymenial cell is the ‘meiocyte’, which is the cell in which meiosis occurs and in basidiomycetes this cell also extrudes the four basidiospores and is called a basidium. A cell described as a basidium is a hyphal tip that is quite clearly characterised by karyogamy, meiosis and the formation of basidiospores. In other words, application of the nomenclature involves consideration of the past and future behaviour of the cell, i.e. its development over time. Usually, other descriptive terms, like basidiole, paraphysis, sterile element or cystidium, are applied on the basis of the immediate morphology and/or position of the cell without reference to its ontogeny or fate, yet these are important considerations. Understanding the mechanisms of differentiation and morphogenesis requires that the descriptions of developmental pathways must be precise. A hyphal branch from the central tissue (the trama) of the gill which becomes a hymenial cell may be pluripotent initially, but it must be channelled to follow one developmental pathway in particular. The varied pathways of differentiation open to a tramal branch involve commitment to expression of particular (probably different) sets of characters at different stages in morphogenesis so that the hymenium is provided with cells that have appropriate functional characteristics.

Some cell lineages are unable to express morphologies developed by others. As in most other agarics, the *A. bisporus* hymenium lacks differentiated structural cells. Instead, the hymenium is made up of young basidia (basidioles) which are continually produced to replace those which mature and sporulate, thereby enabling the fruit body to produce spores gradually over the 57 days that it is viable (Buller, 1922; pp. 405–435). In such a 'classic' agaric hymenium, successive generations of basidia are seen as serving a structural function during their young stages. Allen *et al.* (1992) showed that this is achieved through meiotic arrest causing a prolonged meiotic prophase. These observations also showed that far more basidioles enter meiosis than can ever complete it, so that, for a very large number of basidioles, the state of arrest in meiotic prophase represents the terminal stage of differentiation. Basidioles with a single fusion nucleus rapidly became the majority class and remained so right through to senescence of the fruit body. Clearly, release from meiotic arrest occurs only rarely. This is not wastage of reproductive potential but use of one differentiation pathway to serve two distinct but essential functions: reproduction on the one hand and structural support on the other.

Many agarics are like *Agaricus* in having hymenia in which morphological differentiation is lacking and the less obvious functions have to be deduced from other features (Manocha, 1965; Wood *et al.*, 1985). The gill of *Coprinus* illustrates the other extreme, being comprised of highly differentiated cells with what appear to be distinct and recognisable functions, the morphological differentiation signalling the functional differentiation. In particular, *Coprinus* has a highly differentiated cell type, the paraphysis, which is used to construct the epidermal pavement. However, the mature structure of the hymenium is arrived at through a defined developmental sequence. In young (primordium) fruit body caps the hymenium consists of a loosely organised layer of hyphal tips of broadly similar size, though occasional unusually large cells may be identified as presumptive cystidia (Fig. 6.27). About 8% of these tramal hyphal branches become cystidia; the rest become basidia (Horner and Moore, 1987) which proceed to karyogamy and initiate the meiotic cycle ending with sporulation. During the early stages of hymenium development it is noticeable that adjacent basidia arose at the apex of sister branches of parental tramal hyphae. This is the normal mode of basidium proliferation in agarics (Oberwinkler, 1982) and contrasts with proliferation at sub-basidial clamp connections described in *Schizophyllum commune* (Niederpruem *et al.*, 1971; Niederpruem and Jersild, 1972). Thus, all cells of the hymenium layer proper in young primodial specimens, originate as branches from the hyphal tissue of the subhymenium and remain in contact with their subhymenial neighbours through dolipore septa. The presence of clamp connections and paired nuclear profiles shows that dikaryon growth is maintained during formation of the hymenium. The subhymenium is a much more open tissue and the cells of which it is composed contain large accumulations of glycogen granules and are usually multinucleate. Generally, throughout the primordial gill only a small proportion of the cell volume is occupied by vacuoles (Rosin and Moore, 1985b).

The more mature specimen contrasts starkly with this description. At maturity the hymenium is differentiated into three very distinct cell types: the basidium, the cystidium and the paraphysis. The basidium is only slightly enlarged in comparison with primordial hymenium cells, and contains fairly dense cytoplasm with few, generally small, vacuoles. The paraphyses are enormously inflated in comparison with the primordial hymenium cells, and the bulk of the cell volume is occupied by a single vacuole; the cytoplasm being restricted to the periphery. Inflation of the paraphyses produces a hymenium in the form of a closely adpressed and apparently interlocked paraphyseal pavement containing interspersed basidia (Moore *et al.*, 1979). This is quite distinct from the uncrowded structure seen in the primordium and also differs greatly from the structure of the mature hymenium of *Agaricus bisporus* (Craig *et al.*, 1977; Wood *et al.*, 1985). A particular feature of the cell inflation which generates this pavement formation is that it is entirely restricted to the hymenium. The subhymenium remains essentially hyphal, and despite their considerable enlargement, continuity with the hyphae of the subhymenium has been demonstrated for cystidia, paraphyses and basidia (Moore *et al.*, 1979). The paraphyses actually arise after the young basidium (basidiole) population commits to meiosis, as branches which emerge from hyphal compartments beneath the basidia (Rosin and

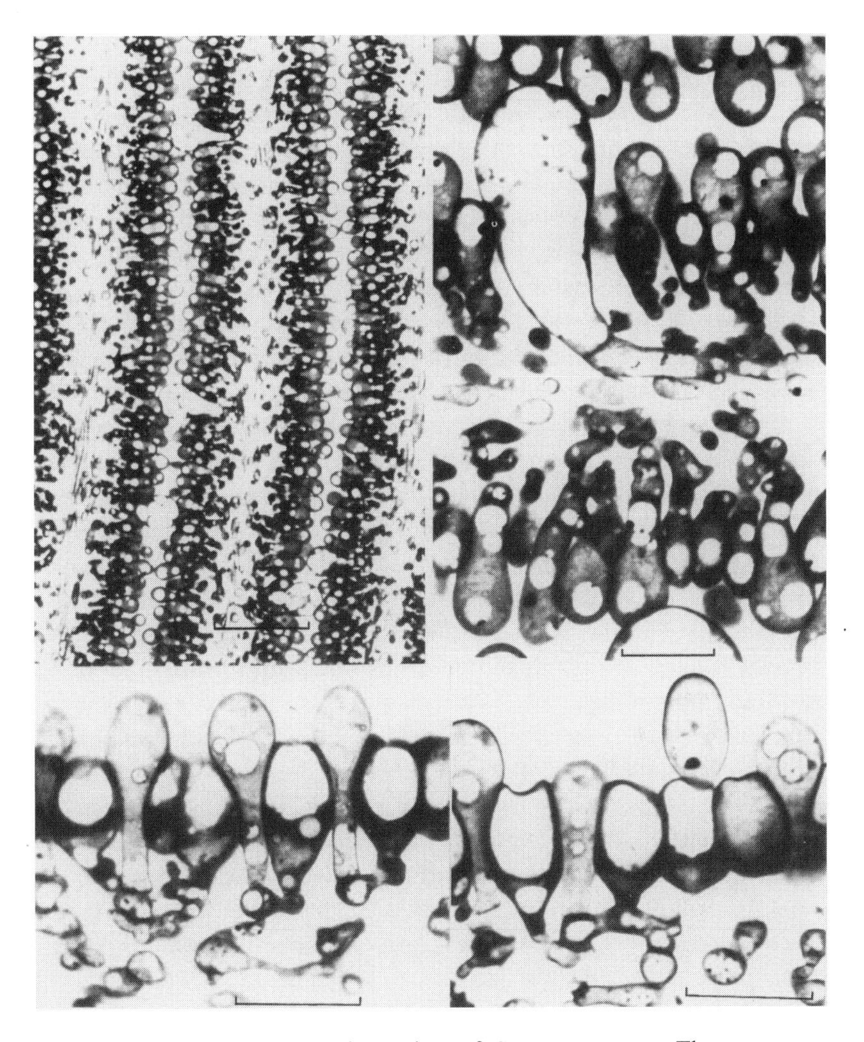

Figure 6.27 The differentiating hymenium of *Coprinus cinereus*. The top two images show the hymenium of a primordium 1 to 2 mm tall, similar to that shown in the bottom left photograph of Fig. 6.4. They reveal that at this stage the hymenium consists of a closely packed layer of young basidia, with occasional large, but still young, cystidia. Paraphyseal branches from sub-basidial cells are evident as more densely-stained hyphal tips, beginning to force their way into the basidial layer. The two lower photographs are sections of a later stage in which paraphyseal insertion has been accomplished and paraphyses have begun to inflate. Their connection to sub-basidial cells is still evident. Scale bars: top left = 20μm, top right = 5μm, bottom = 10 μm.

Moore, 1985b). The tips of these branches force their way into the hymenium. They grow to about two-thirds the height of the basidia but then cease apical extension growth and initiate spherical growth (uniform, non-polarised wall growth) which causes them to expand. This inflation causes them to take up all remaining intercellular space and form a mechanically-integrated pavement (Fig. 6.27). About 75% of the paraphysis population is inserted before the end of meiosis, the rest insert at later stages of development (Rosin and Moore, 1985b).

At maturity, when the *Coprinus* hymenium is composed of basidia embedded in a pavement of paraphyses, individual basidia are surrounded by about five paraphyses (Rosin and Moore, 1985b); thus, more than 80% of the hymenial cells in *Coprinus* serve a structural function. *Agaricus* and *Coprinus* hymenophore tissues reach essentially the same structural composition by radically different routes, using very different cell types. This is an example of different cells serving the same function. In another example the same final morphology serves different functions. Both *Coprinus cinereus* and *Volvariella bombycina* have inflated cells (called cystidia) on both the main surface of the gill (facial or pleurocystidia) and over the margin of the gill (marginal or cheilocystidia; Figs 6.27 and 6.28). Cystidium morphology and distribution are widely used in taxonomy. Unfortunately, much of the literature dealing with cystidia considers them only from the taxonomic or purely morphological viewpoints (e.g. Lentz, 1954; Price, 1973; Erikssen *et al.*, 1978), reflecting assiduous attention to fine distinctions of nomenclature at the expense of appreciation of the remarkable developmental plasticity revealed by their occurrence and form. According to Smith (1966), cystidia "occur haphazardly in the hymenium, depending on the species, and vary from abundant to absent"; a description of a cell distribution pattern which, in terms of developmental biology is totally useless. Brefeld (1877, cited in Buller, 1909) concluded that cystidia are metamorphosed basidia, a view summarised by Corner (1947) in the phrase "cystidia represent sterile basidia which become overgrown." Certainly, young basidia and young cystidia both originate as the terminal compartments of branches from the hyphae of the subhymenium, but cystidia are *not* overgrown basidia and such dismissive statements ignore the fact that the mature cystidium is a cell that is highly differentiated for its particular function.

Both facial and marginal cystidia in *V. bombycina* are established when the hymenium is first laid down on the folded gills and, apart from location, their differentiation states and ontogeny appear to be identical. Facial cystidia in *C. cinereus* are also established as components of the very first population of dikaryotic hyphal tips which form

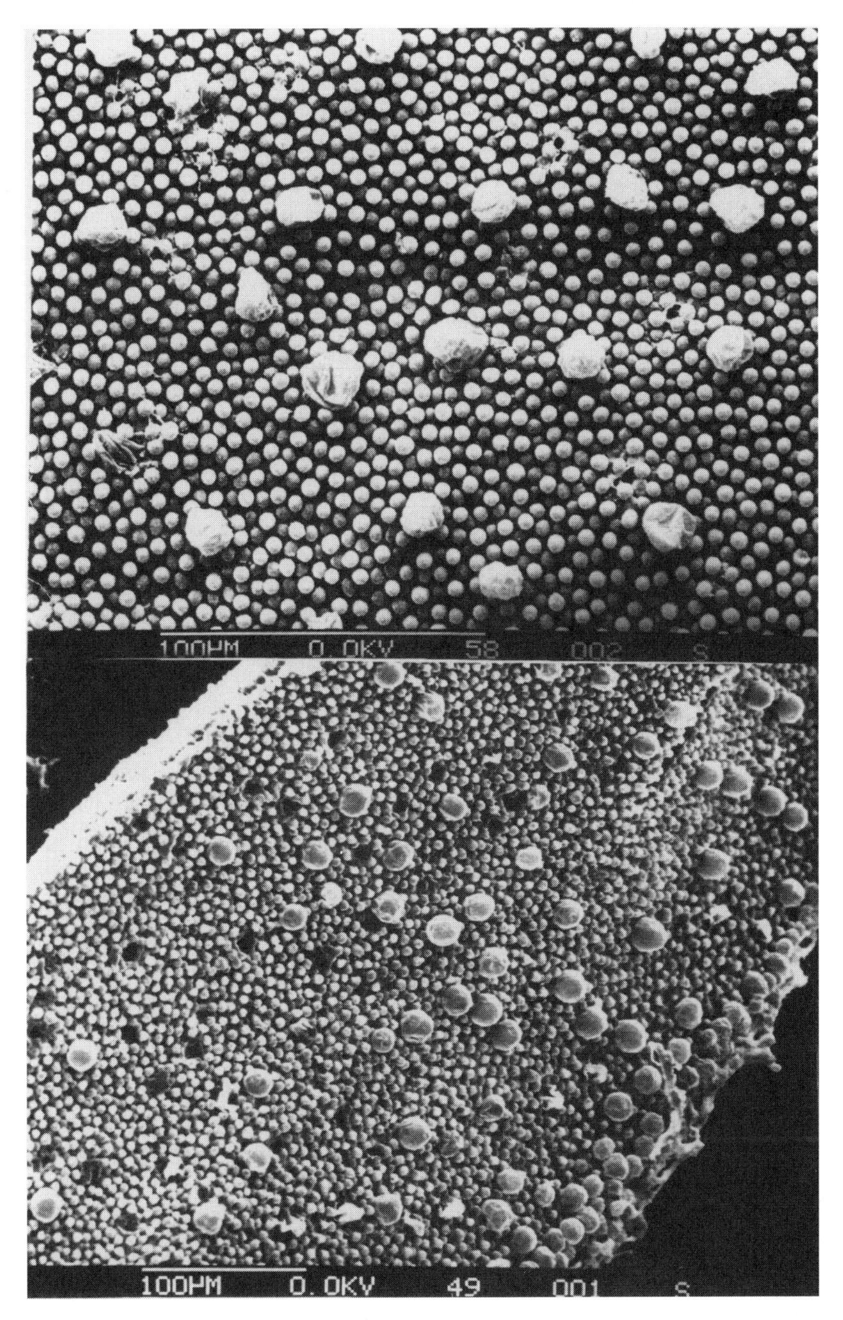

Figure 6.28 SEM views of the surface of the gill of *Coprinus cinereus*, showing the distribution pattern of facial and marginal cystidia (the largest cells visible here).

hymenial tissue (Rosin and Moore, 1985b; Horner and Moore, 1987) and are mostly binucleate as a result. Marginal cystidia in *C. cinereus* are the apical cells of branches from the multinucleate gill trama, which become swollen to repair the injury caused when primary gills pull away from the stem; marginal cystidia retain the multinucleate character of their parental hyphae but have the same morphology as facial cystidia (Chiu and Moore, 1993). Facial cystidia in *V. bombycina* have been observed with aqueous droplets adhering to them (Fig. 6.29). They are, therefore, secretory cells, releasing water vapour and possibly other volatiles into the atmosphere of the gill space immediately around the basidia (Chiu and Moore, 1990b). I believe that facial cystidia of *Coprinus* serve a mechanical, almost engineering, function which depends on their growing right across the gill cavity and firmly attaching to the opposing hymenial layer.

At early stages in growth of the cystidium across the gill cavity the cell(s) with which it will collide in the opposing hymenium are indistinguishable from their fellow basidioles. However, when the cystidium contacts the opposing hymenium, the cells with which it collides (called cystesia) develop a granular, vacuolated cytoplasm, more like that of the cystidium than that of their neighbouring basidioles. The contact triggers the differentiation, and results in the apex of the cystidium being firmly joined to the apex of the cystesium, probably by formation of a joint wall layer. The outcome is that the cystidium bridges the gill cavity and is firmly attached to both hymenia (Fig. 6.30). Why? Early mycologists had no doubts. The following quote comes from Buller (1909), but Chow (1934) expressed similar opinions:

> According to Cooke [M. C. Cooke, *Introduction to the Study of Fungi*, published in London in 1895] "The usual interpretation of the function of cystidia is, that they are simply mechanical contrivances projecting from the hymenium and thus keeping the gills or lamellae apart." Possibly this view is correct for certain species of *Coprinus*, for example, *C. micaceus*, where large cystidia are found on the gill surfaces; but where cystidia coat the swollen gill edges, as in *C. comatus*, we may regard them as packing cells. They form cushions where the gills are in contact with each other and the stem, and they probably facilitate the separation of these structures on the expansion of the cap. (Buller, 1909)

I would not take issue with these views with regard to the packing cell function of marginal cystidia, but for facial cystidia my view is that these cells mechanically tie adjacent gills together, rather than keeping them apart as cushions or buttresses facilitating their separation. One argument for this interpretation is that if cystidia serve as buttresses to keep the gills apart, then the buttress must have a firm foundation to carry the

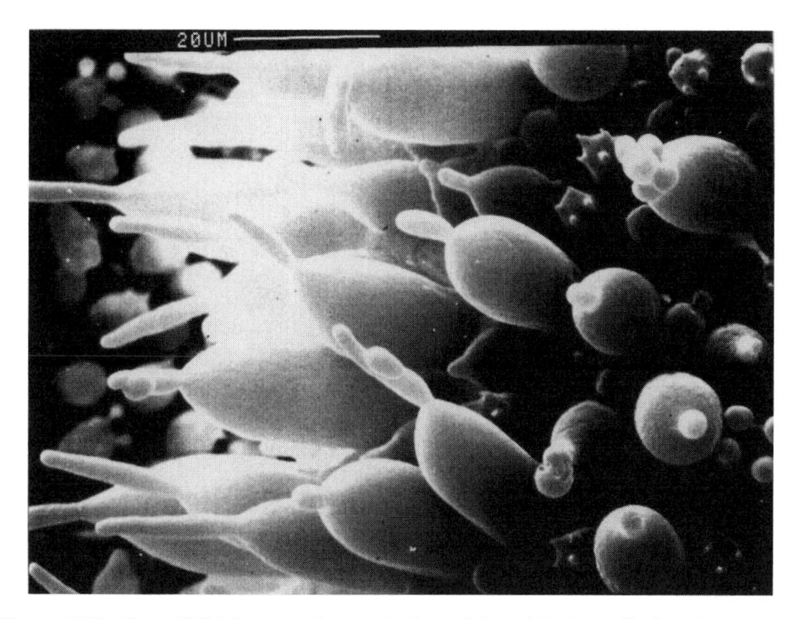

Figure 6.29 Cryo-SEM image of marginal cystidia of *Volvariella bombycina*. The flash-freezing has preserved droplets of mucilage on and around the extended necks of some of the cystidia.

mechanical load. Yet the trama of the gill is a very open structure, with many intercellular spaces (Fig. 6.27). There is no evidence in this structure of firm support for the base of the cystidium. Another argument is that a key feature of the *Coprinus* facial cystidium is that it is very firmly attached to the *opposite* hymenium (Fig. 6.30), but a buttress does not need firm attachment. A lintel can simply rest upon a column for support. Firm adhesion to cystesia in the opposing hymenium suggest that cystidia function as tension ties between neighbouring gills, rather than as buttresses. Indeed, cellular reinforcement of cystidial function, in the form of basal clasping cells (Fig. 6.30), also seems to emphasise the mechanical ability of the whole arrangement (cystesia, cystidium and clasping cells) to resist *tension* along its long axis rather than compression. Consequently, facial cystidia and the cell structures associated with them in *Coprinus* are specialised as tension ties connecting the gills together to resist forces which would otherwise pull the gills apart. The origin of those forces and the ways in which they are put to use for morphogenetic purposes will be discussed in the next section.

The examples considered in this section make it clear that definition of tissues is often far from easy in fungi. Despite their alleged lowly position in the scheme of things, fungi exhibit even long-range coordination of

Figure 6.30 Cystidia of *Coprinus cinereus*. The panel of three images on the left show a cystidium spanning the gill space and fused with cystesia in the opposite hymenium. In the top micrograph the fluorescence of calcofluor white reveals newly-synthesised chitin in the walls by which the cells adhere. The middle image is a bright field micrograph of the same cell; and the bottom of the three is a critical-point-dried SEM specimen showing the junction line between cystidium and cystesium. The image on the right is a young cystidium with a basal collar of clasping cells. Thus, a cystidium spanning between two hymenia has its connections to both hymenia reinforced. Scale bar = 20 μm.

cellular activities (section 6.2.2.4). This should leave no doubt that true tissues, as communities of cells of concerted function, do exist in fungal multicellular structures. However, fungi have such a high degree of developmental plasticity that the cell types which serve specific functions and/or the functions of specific cell types may well differ from case to case. One result of these features is that a community of cells which quite realistically makes up a distinctive tissue may not be easily recognisable even after it has performed its function because differences in cell morphology are too subtle to demarcate the tissue without detailed analysis. The tissue which determines stem structure in *Coprinus* would be an example of this (section 6.2.2.3). Another result is that errors may arise when cell function is assigned on the basis of morphology. Basidium function in *Agaricus* (the majority serving as structural rather than reproductive elements) and different functions of facial cystidia between *Volvariella* and *Coprinus* (this section) serve as examples here. The key features which appear to determine cell patterning in the *Coprinus* hymenium are the establishment of a cell layer comprised of basidia with a scattering of cystidia, and cessation of basidial growth which seems to remove a constraint on branching and lateral proliferation of the sub-basidial cells so that paraphyseal branch formation can take place. This latter is accompanied by an accumulation of carbohydrate in the sub-basidial cells as these become the active growth zone of the hymenium. These processes seem to require a sequence of major controls: specification of the distribution of cystidia; determination of the apical differentiation of tramal hyphae into basidia and, as a consequence of basidial differentiation (including arrestment of extension growth of the basidial hyphal apex), emergence of paraphyses as sub-basidial branches which insert between the basidia before arresting apical wall growth and initiating spherical wall growth. With regard to later stages, the structure and function of the gill tissue in relation to basidiospore development have been studied in detail by McLaughlin (1974, 1977). The results suggest that carbohydrates accumulate in Golgi vesicles in basidia, cystidia and basidiospores and that the accumulations are related to wall synthesis. It was concluded that Golgi vesicles migrated from the Golgi cisternae to the cell wall and that during migration the contents of the vesicles underwent progressive change in preparation for their contribution to wall growth. At all times, however, the contents of the vesicles were demonstrably different from glycogen; and whereas glycogen and Golgi vesicles co-existed in cystidia, the vesicles were absent from paraphyses and subhymenial cells although these cells contained large deposits of glycogen (McLaughlin, 1974). A subsequent study of the early ontogeny of basidiospores concentrated on the control of spore shape and the parts played by the spore wall and cytoplasmic components in basidiospore development (McLaughlin, 1977).

6.3.2 Inherent properties of cell mosaics

An inherent property of a system of linked polygons, is that the average number of cell sides is exactly six, and that the area of a cell is proportional to the number of sides of the cell and the nature of this proportionality can identify different cell populations (Lewis, 1943). These considerations apply just as much to a mosaic of cells which are compressed together as to non-biological systems. In cell sheets, cell faceting is a mechanical consequence of their compression; the phenomenon is discussed in detail by Dormer (1980). Rosin and Moore (1985b) analysed the distribution of paraphyses and basidia geometrically using a method based on that of Lewis (1931, 1943). The principles for geometrical analysis of tissues were derived for layers of dividing animal and plant cells. Application of them to fungal hymenia is dependent on the assumption that the insertion of hyphal tips (in this case specifically to form paraphyses) into a hymenium layer is formally equivalent to cell division within the layer. Geometrically this seems to be valid providing the argument does not depend on constraints on the plane of division. Measurements were made of maximum cell diameter (replacing area) and the number of sides for randomly chosen basidia and paraphyses in photographs of the hymenium surface of fruit bodies in different stages of development. The analysis showed that basidia and paraphyses fell into two distinct populations. Basidial diameter remained approximately constant irrespective of the number of sides, while for paraphyses the number of sides increased with size.

The failure of many-sided cells to attain the size required to comply with the expected progression has been explained in terms of the division of cells around a non-dividing cell (Lewis, 1943). In the context of our present knowledge of the ontogeny of the hymenium in *Coprinus cinereus*, the analogous statement for this tissue would be that additional paraphyses are inserted around the basidia. Thus the geometrical analysis of basidia and paraphyses reinforces the view that the basidial population is numerically static. Only paraphyses insert into the hymenial layer after its initial differentiation from the protenchyma (Rosin and Moore, 1985b).

6.3.3 Tissue expansion as a morphogenetic factor

It is obvious that inflation plays a central role in differentiation of the cells and tissue which make up the hymenium, but it seems likely that the significance extends beyond this level of structure and that cell inflation has a profound bearing on the morphogenesis of the fruit body as a whole (Reijnders and Moore, 1985). Indeed, most of the changes in shape which characterise the later stages of fruit body maturation depend on cell expansion. It follows from this that the distinction between cell division and cell expansion (in terms of their contribution to development) is an important one, though we have very little information about it.

Another important point is that of the differentiation of cells (i.e. hyphal compartments) corresponding with their location. It is often observed that the cells of one recognisable hypha become abruptly different because they are influenced by another morphogenetic factor (the nature of which is obscure). Thus in *Leucocoprinus cepaestipes* the lower cells of hyphae at the surface of the cap form an 'epidermal' layer called the pileodermium, and the narrower upper continuations of the same hyphae merge into the universal veil (Reijnders, 1948). Similarly, in *Coprinus poliomallus* the narrower hyphae of the gill trama suddenly widen and form isodiametric cells when they pass into the lipsanenchyma (primordial tissue, other than the veil which covers the hymenium; Reijnders, 1979). Such differences in adjacent cells of continuous hyphae, leading to completely different patterns of differentiation either side of a single dolipore septum, can be detected everywhere in primordial fruit bodies. The differentiation of hymenial elements from the tramal hyphae is particularly notable (Fig. 6.31), but the same feature can be observed in vegetative structures, such as the distinction between thick-walled and thin-walled cells in sclerotia of *Coprinus cinereus* (Fig. 6.31). The nature of the signal(s) involved in directing such differentiation and how they are localised in a hyphal structure are unknown. Clearly, though, the septum, which is capable of extremely rapid response to experimental stress (Todd and Aylmore, 1985), must be involved in partitioning these sorts of regulatory signals between cell compartments in the same hyphal strand. The structure of the dolipore/parenthesome septum is sufficiently complex for us to anticipate quite sophisticated involvement in localising differentiation signals relative to the long axis of the parent hypha. But even though the hypha is the very basis of the sorts of structures considered here, control of the longitudinal communication of organisational signals is probably insufficient to account for the regulation of the levels of differentiation which can be observed.

The scope of the differentiation seen in fruiting and vegetative fungal structures alike is every bit as complex as that commonly studied in plant and animal systems. Yet a consequence of the fungal dependence on hyphal organisation is that lateral communication between cells contributing to the same tissue must involve export and import of control signals. Longitudinal secondary walls are not found in fungi; so lateral communication between adjacent cells must take place across two hyphal membranes and two mature hyphal walls. Lateral communication must

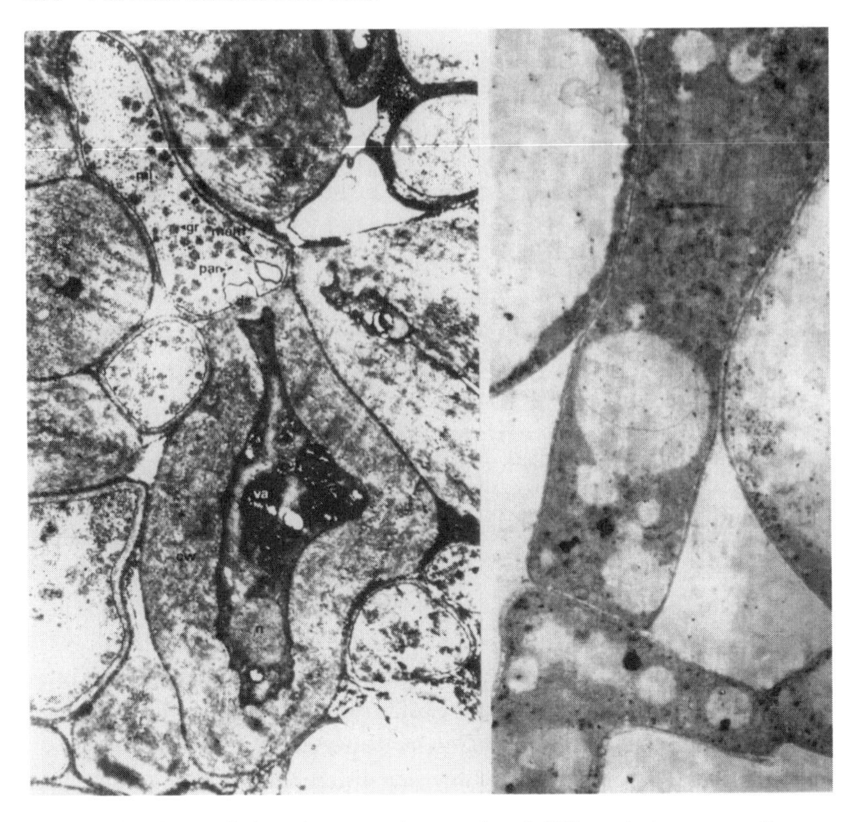

Figure 6.31 Transmission electron micrographs of differentiation across the dolipore septum in *Coprinus cinereus*. At left is a section of the medulla of a sclerotium in which the thick-walled and thin-walled phenotypes occur in hyphal compartments either side of a septum (×9000). At right is a section of an immature hymenium in which the enormously inflated paraphysis shares a dolipore septum with a cell below which is of more normal hyphal appearance (×4200).

play an important role in defining the positional information on which tissue differentiation in fruiting and vegetative structures depends. Yet there appears to be no general evidence for anything akin to plasmodesmata or gap junctions in the lateral walls of fungi. A few reports exist of channels through, or pits within, thickened cell walls (in peridioles of *Nidularia* (Reijnders, 1977), and sclerotia of *Penicillium* (Lohwag, 1941). Hawker and Beckett (1971) describe plasmodesmata which maintain cytoplasmic continuity between gametangia and suspensors during zygospore formation in *Rhizopus sexualis*, even after considerable secondary thickening of the septa. Though they demonstrate that cytoplasmic continuity can be provided for, these examples are in newly-formed cross-walls. In general, therefore, a cytoplasmic route for signals capable of conveying positional information between physically different hyphae would seem to be excluded (Rosin *et al.*, 1985).

Cell inflation is an aspect of differentiation of the cell and is a mark of specialisation. The generative hyphae which make up the basic, undifferentiated, tissue of the primordium never have inflated cells. Large cells occur principally in fleshy fungi and are characteristic of Agaricales, Clavariaceae and some Gasteromycetes. Primordia of the most specialised genera of the Agaricales, such as *Mycena*, *Coprinus*, *Conocybe*, *Bolbitius*, and *Pluteus* have remarkably inflated cells even at very young stages. Cell elongation in primordia of Agaricales generally starts immediately after their formation, and this is also true for vegetative structures like sclerotia and rhizomorphs. Reijnders (1963) related this early inflation to the general shape of the primordium; showing that the different zones of the primordium enlarge proportionally; they do not impede the growth of other parts and compressed tissues are seldom observed. Following this continuous and slow inflation, there is usually a period of rapid expansion, which constitutes the final maturation phase of the structure during which most of its characteristic shape is established.

In early primordia of *Coprinus*, when the gill hymenium consists essentially of a cell layer of basidia, the entire gill is about 100 μm wide at the cap margin. About 100 paraphyses had been inserted into a transect across the hymenium of gills at about the middle of development, prior to the onset of meiosis, when gill width was 300–400 μm. In the mature gill after meiosis, approx. 140 paraphyses were counted along such transects and they had an average width of 8 μm in gills about 1500 μm wide at the cap margin (Rosin and Moore, 1985b). Paraphyseal enlargement proceeded continuously during maturation and Rosin and Moore (1985b) claimed that insertion and expansion of paraphyses was sufficient to account for the entire increase in gill width. Increase in gill length will, of course, depend on the same factors but is further dependent on continued differentiation of gill tissues from the protenchyma at the apex of the cap (Rosin and Moore, 1985a). Nevertheless, increase in gill area is clearly dependent on an enormous increase in the volumes of the cells which make up the hymenium and the combined effect of these increases in cell volumes is to drive the change in shape of the fruit body as it matures.

During maturation, the cap of a *Coprinus* fruit body passes through a change in shape of such magnitude that the initially vertical orientation of the gills is transformed to a horizontal orientation (Buller, 1924, 1931). In *C. cinereus* this process is completed in about 6 h and is achieved by a gradual revolution of the edge of the cap in concert with radial splitting of cap and gill tissues so as to allow the cap to open like an umbrella. In describing and accounting for cap expansion in different species of *Coprinus,* Buller (1931) stressed the part played by the cap tissue. He distinguished clearly the smaller species (*C. curtus*, pp. 27–28; *C. plicatilis*, pp. 40–42) in which the accent is on differential growth in the central disc of cap flesh which ensheathes the stem from the larger species (*C. comatus*, p. 73) in which differential growth of cap flesh bounding the gills is supposed to account for the changing shape of the gills. Any role played by the paraphyses is minimised in Buller's accounts; they are described as "the elastic elements of the hymenium" which "perform an important mechanical function as the cap opens", though it is clear that this author considers their role in cap structure to be a secondary one (Buller, 1931, p.28), their major function being to separate adjacent basidia.

My view is quite different. Based on observations on *C. cinereus*, I believe that the paraphyses are the major driving force of cap expansion (Moore *et al.*, 1979). Measurement of the profile contour length of the cap at successive stages during expansion shows that it remains constant in the final autodigestive phase (Fig. 6.32); and even though spent gill tissue is removed by autodigestion a considerable portion of the gill remains intact until very late in the expansion process. Yet as the paraphyses differentiate their inflation must increase the area of the gill. In the early stages this appears to be accompanied by extension of the cap flesh (as judged from the measured length of the profile; Fig. 6.33). However, in the later stages the expanding gill is bounded by a flexible but evidently inextensible band of cap tissue, a combination which will inevitably lead to an outward curvature even without the active participation of the cap flesh (Fig. 6.34). The cap flesh will become an active (supporting) participant after the autodigestive process has removed the bulk of the gill, but initially paraphyseal inflation provides the major expansionary force.

In reference to *Coprinus comatus*, Buller (1931) remarked on the small amount of cap flesh in relation to the amount of gill tissue in coprinoid caps and commented on the mechanical problem of supporting the gills as they expand (his assumption being that the gills were passively hanging from the cap tissue). This is not a problem if inflation of the paraphyses enables them to contribute both to the support and the load. If the cap flesh is assumed to act as an inextensible but flexible outer border to a gill plate which is attempting to increase in area then the mechanics of the combined structure become integrated. By generating the packing force within a stressed skin (the stressed skin = the cap flesh), the paraphyses make each gill a self-supporting plate which is hanging from its connec-

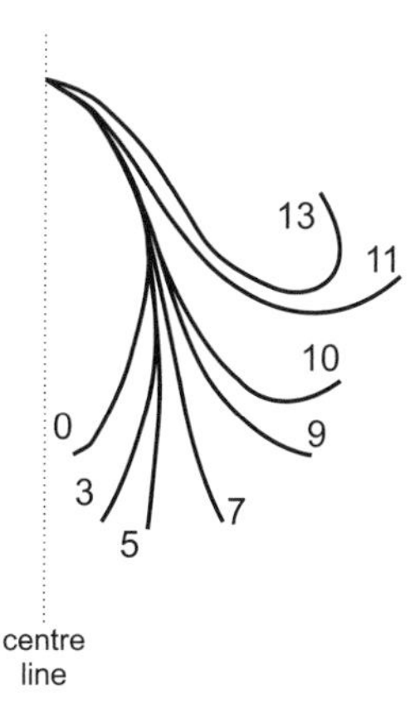

Figure 6.32 Development of the fruit body cap in *Coprinus cinereus*. Diagrams recording the way in which the profile of the cap changes during its final phase of expansion. Consecutive photographs were made of the same fruit body as it matured and tracings were then made of the upper edge of the cap profile. Numbers show hours elapsed since observation started at time zero. During the course of the sequence shown the stem extended from 25 to 93 mm, spore discharge commenced about 5 h from the start of observation. Redrawn after Moore *et al.* (1979).

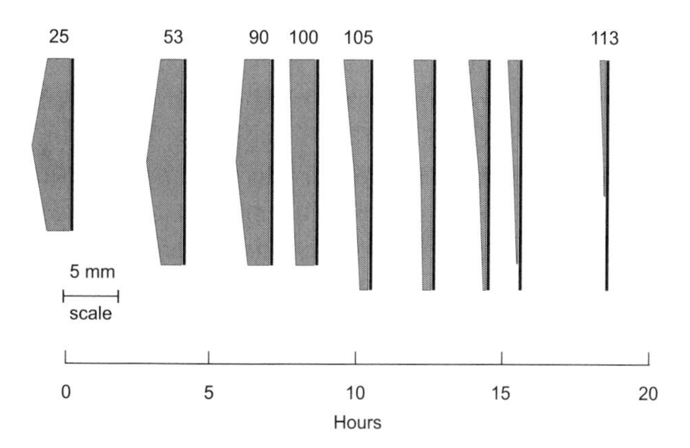

Figure 6.33 Development of the fruit body cap in *Coprinus cinereus*. Scale drawings of gill growth and autolysis. At regular intervals in the final stages of fruit body development a small segment was surgically removed from an otherwise undisturbed fruit body and preserved in formalin. Subsequently, the outer contour length (length of cap flesh or pileipellis) was measured and the depth of gill tissue measured at the top, centre and bottom of a suitable primary gill in each segment. In these diagrams the shape of the cap is ignored (for which see Fig. 6.40); the vertical bars represent the contour length and the shaded area represents the gill lamella. The numbers above the diagrams record the length (in mm) of the stem at the time the sample was taken. In this specimen, spore discharge began at about the 7th hour. These observations show that the cap flesh remains intact and extended throughout the entire period of gill autolysis, and that for the final period of cap development the length of the cap flesh is unchanged. Redrawn after Moore *et al.* (1979).

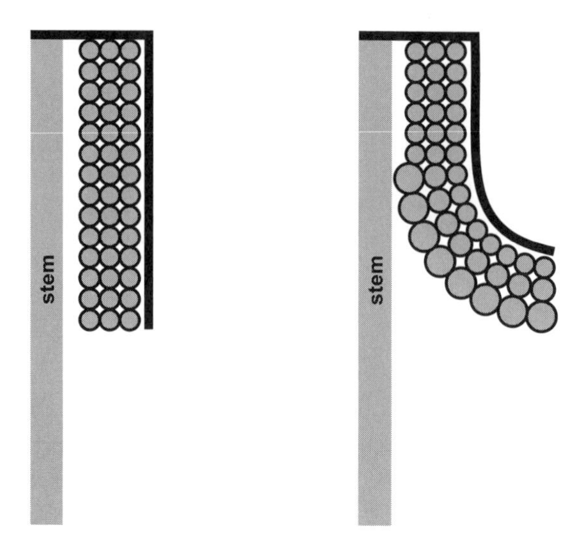

Figure 6.34 Development of the fruit body cap in *Coprinus cinereus*. Diagram to show how cell expansion in the hymenium can account for the outward bending of the gills (as indicated in the profiles shown in Fig. 6.40) if it is assumed that the cap flesh (pileipellis) acts as a flexible but inextensible outer boundary (shown as a thick black line). Redrawn after Moore *et al.* (1979).

tion to the stem at the cap apex. Since all the gills are laterally connected by facial cystidia (see section 6.3.1) the whole of the cap forms an integrated mechanical structure which is stressed into shape by inflation of its constituent cells. Since the majority of those cells are paraphyses, it follows that paraphyseal enlargement is the main driving force for reflection and erection of the gills.

The inflated cells do not have strengthened walls but they do have much enlarged vacuoles, and it is worth emphasising that the cell enlargement observed as the cap expands is produced by the osmotic influx of water (Table 6.7). In the cap of *Coprinus cinereus*, this is a response to urea accumulation, and to provide for this urea accumulation glutamate dehydrogenases, glutamine synthetase, ornithine acetyltransferase and ornithine carbamolytransferase (and presumably other enzymes) are coordinately derepressed during fruit body development (see section 4.2.3). Enhanced operation of the urea cycle, usually associated with disposal of excess nitrogen, would correlate well with the upsurge in activity of enzymes involved in autodigestion (see section 6.5). In *C. lagopus* and *C. comatus* increases in the activities of glucosidase, protease and chitinase have been noted during cap expansion (Iten, 1970; Iten and Matile, 1970; Bush, 1974) and glucanases have been identified in auto-

Table 6.7. *Comparison of fresh weight and dry weight of individual primordia and mature fruit bodies of* Coprinus cinereus.

		Stage 2 primordium (n = 13)	Stage 5 fruit body* (n = 19)
Cap	Fresh weight (mg)	136 ± 44	485 ± 236
	Dry weight (mg)	16 ± 6	34 ± 16
Stem	Fresh weight (mg)	174 ± 82	354 ± 179
	Dry weight (mg)	16 ± 7	17 ± 9
	Length	15 ± 2	75 ± 19

Fruit body stages are described in Table 6.4. These mature (Stage 5) fruit bodies were harvested just prior to spore discharge to ensure that they had an undisturbed crop of basidiospores. Data from Moore *et al.* (1979).

lysates of *C. cinereus* (Schaeffer *et al.*, 1977); so the degradation of the substance of the spent gill tissue may well provide the substrates which intermediary metabolism could utilise to support the expansion of the still intact regions. It seems quite clear that these aspects of cap metabolism are not a novelty in terms of the reaction sequences involved; much of what occurs in the cap can be shown to occur at other times, and in other tissues. However, the cap does present a very special regulatory picture. What is specific to the cap is that enzymes involved in these aspects of metabolism are endogenously up-regulated so that the cap cells become metabolically quite distinct from those of either the stem or the parent mycelium.

Urea is used as an osmoticum in this way in several fungi (see section 4.2.3), but even in *Coprinus cinereus* the stem of the fruit body uses a different metabolic scheme to take up water osmotically. Rather than nitrogenous metabolites, the emphasis in the stem is on a mixture of sugars (see discussion in section 4.2.1). *Agaricus bisporus* and *Lentinula edodes* fruit bodies have accumulations of mannitol close to 50% on a dry weight basis (see Table 4.10). One of the roles ascribed to mannitol is that of osmoregulation or turgor regulation, that is the maintenance of a high osmotic or suction pressure inside the cell to establish turgor and create an inflow of water, especially when the substrate is of high osmolality. Thus, mannitol is thought to provide for support and expansion of the fruit body in both these organisms. If mannitol does serve an osmoregulatory function in both *A. bisporus* and *L. edodes*, it is particularly noteworthy that the two species differ in the way in which the fruit body expands: in *A. bisporus* this occurs by hyphal inflation, whilst in *L. edodes* hyphal proliferation is responsible. Thus, the two very different basidiomycete strategies for tissue expansion, cell inflation in Agaricales and cell proliferation in Aphyllophorales, can be facilitated by the same metabolite.

Fruit body expansion in the Russulales is achieved by columns and rosettes of hyphae expanding in an orchestrated way (Reijnders, 1976) whereas other agarics have simple, gradually elongating and inflating hyphae, accompanied in *Coprinus* by narrow hyphae (Hammad *et al.*, 1993a,b) which resemble the inducer hyphae in *Russula* and *Lactarius* (Reijnders, 1976; Watling and Nicoll, 1980). In Amanitaceae the formation of the gills is depends on what is called an acrophysalidic tissue (Bas, 1969) by which flesh hyphae through massive inflation of individual cells allow the fruit body to expand. In some members of the Tricholomataceae a further modification is found leading to a two-component flesh termed sarcodimitic (see section 6.2.1; Corner, 1966, 1991) and, in the belief that this is a fundamental method of fruit body expansion, Redhead (1987) has adopted the family Xerulaceae for agarics exhibiting the character.

Bracket fungi achieve massive spore production by increasing the longevity of the fruit body. They do this by producing spores over several months (e.g. *Polyporus*), or by producing a perennial fruit body in which growth is renewed at regular intervals (e.g. *Fomes*). Examination of the flesh of these fruit bodies reveals an intricate pattern of hyphae which may branch profusely to bind adjacent hyphae together (section 6.2.1) or elongate, thicken, lose their living contents to form strong tubes which act as structural members like cylindrical girders in architectural constructions. These hyphae increase in number as the fruit body grows and ages. The major group with this type of hyphal construction have correspondingly woody fruit bodies and a poroid hymenium (polypores) but, as indicated in the previous paragraph, it can be found in a rudimentary way in some fungi with gills, for example, *Lentinulla* and *Lentinus*. It is now considered that *Lentinus* is more closely related to the polypores than to the other agarics. The presence or absence of mixtures of hyphal types (dimitic or trimitic forms, see section 6.2.1) correlates with other characters used in classification of polypores (Ryvarden, 1992).

The pioneering microscopical analyses of Fayod (1889) showed that in the homoiomerous agarics (gill trama composed only of hyphal tissue) the tissue between the hymenial surfaces might be one of four kinds. Although modified as more observations have been made in the past 100 years, the basic idea still holds true: in mature specimens, bilateral (divergent), convergent, regular and irregular patterns of hyphal arrangement can be found. In each case they correlate with other characters which together define taxonomic families.

6.3.4 Embryonic gills are convoluted

A sinuous, labyrinthiform hymenophore appears to be a normal 'embryonic' stage in fruit body development in agarics, yet a regular radial arrangement of the gills is characteristic of the mature fruit body. How this is achieved is a function of the expansion of the maturing primordium generating stresses between tissue layers which stretch or inflate the convoluted gills into strict radii. Stretching is the effective force in *C. cinereus*; the cystidia being critical elements in the communication of the formative stresses around the fruit body. Inflation occurs in *V. bombycina*: the tightly appressed cells of the hymenium forming a tensile layer containing the compression generated by cell inflation in the gill trama.

Chiu and Moore (1990a,b) were the first to show that in both *V. bombycina* and *C. cinereus,* gills are formed as convoluted plates. In both species the first-formed gills were radially arranged, but as the cap expanded more gills were formed. In *V. bombycina*, new gills were formed in two ways (Chiu and Moore, 1990a). First, by bifurcation of an existing gill near its free edge. Initiation of the folding which produced bifurcations on existing gills was localised and irregular, resulting in sinuous, contorted gills (Fig. 6.35). The formation of two daughter gills depended on completion of the bifurcation along the entire edge of the parental gill.

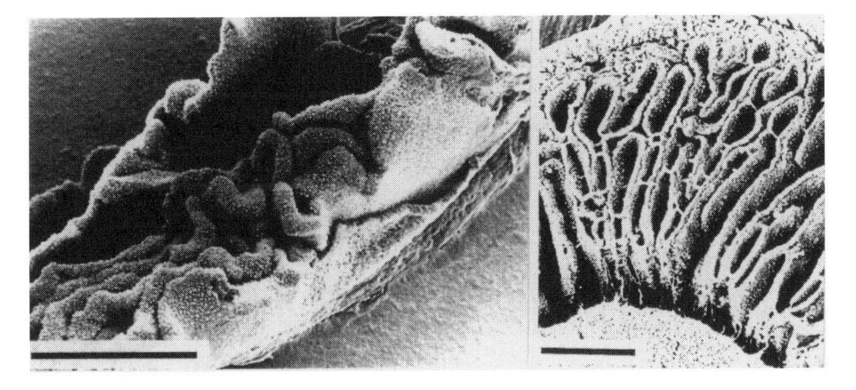

Figure 6.35 'Embryonic gills' in both *Volvariella bombycina* (left) and *Coprinus cinereus* (right) are clearly seen to be convoluted in these SEM preparations (scale bars = 0.5 mm).

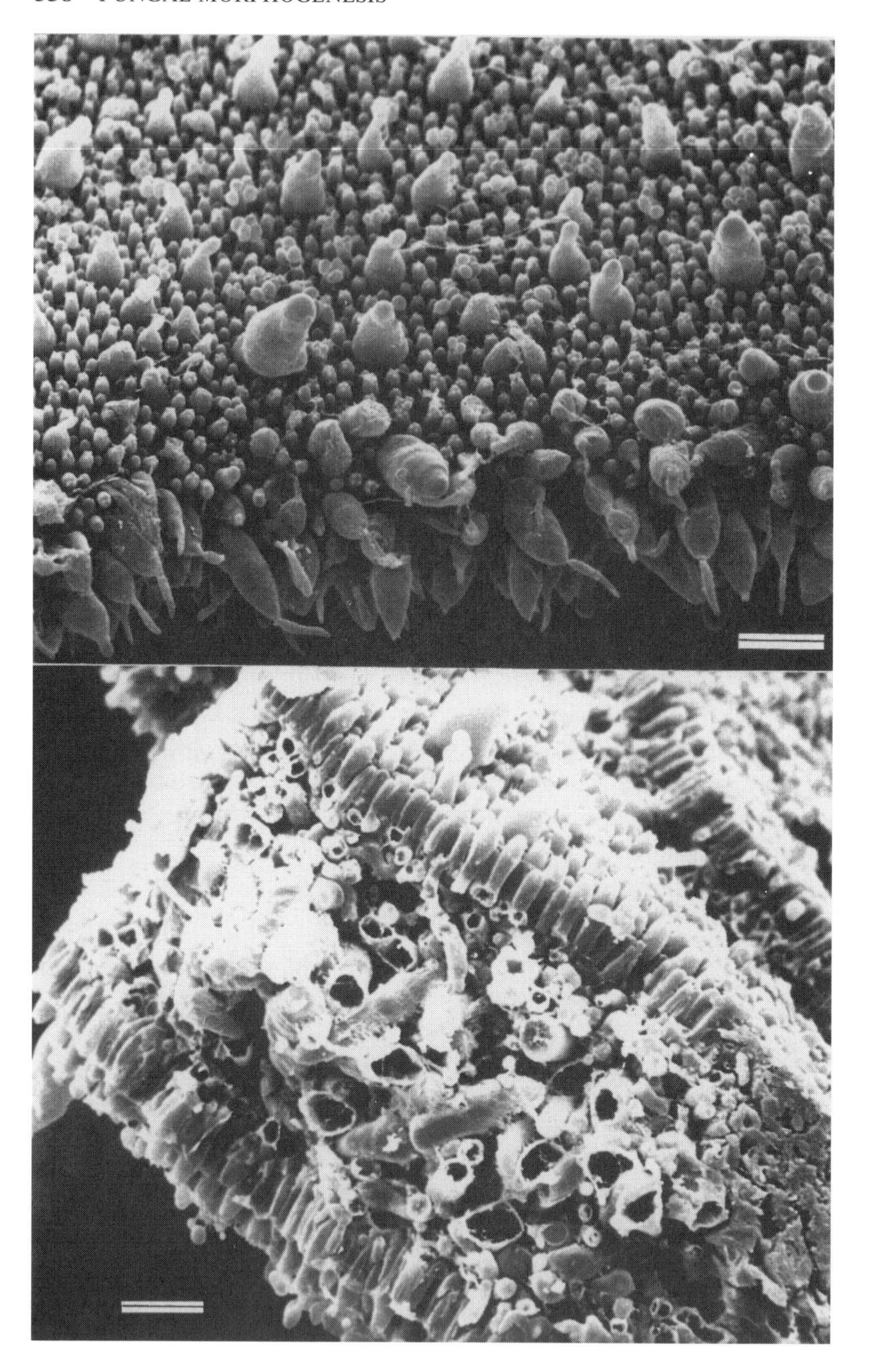

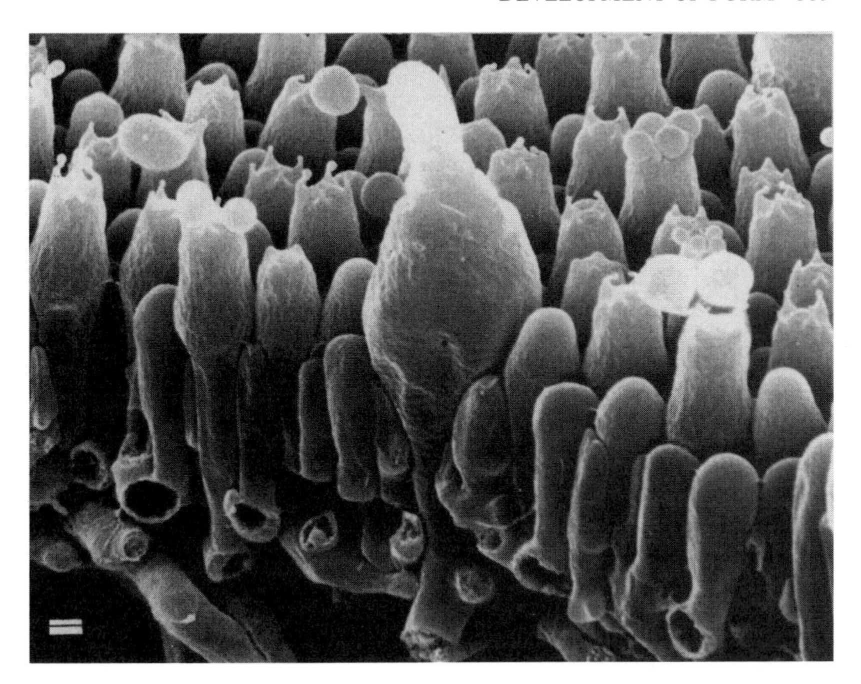

Figure 6.36 The mature gill of *Volvariella bombycina* in SEM images. Top, on the facing page, is a surface view (scale bar = 20 μm), and beneath it is a view of the exposed trama of a gill broken whilst frozen (scale bar = 20 μm). The image immediately above is a magnified view of the hymenium (scale bar = 5 μm). Note that the hymenial cells are closely packed and interlocked to form a coherent layer stretched over a trama which contains a mass of greatly inflated cells. The gill is inflated into shape and is a stretched-skin construction which owes its structural strength to the combination of compression in the trama and tension in the hymenium.

New generations of gills appeared as ridges in the region between existing gill roots, creating new folds on the cap context representing the free edges of new secondary or tertiary gills, the gill spaces on either side extending into the cap context as the gill grew by its root differentiating from the context.

The sinuous 'embryonic' gills of a *Volvariella bombycina* fruit body are inflated into their final shape by the inflation of the cells of the trama. The hymenium of *V. bombycina* is a skin-like layer of tightly appressed cells (Fig. 6.36), and the trama of the gill becomes filled with greatly inflated cells as maturation proceeds. These features suggest that expansion of tramal cells in gills enclosed by the hymenial 'epidermis' generates compression forces which effectively inflate, and so stretch, the embryonic gills to form the regularly radial pattern of the mature cap. When complete, this stressed-skin construction exhibits great similarity to the stretched skin construction which is used for light-weight, high-strength engineering structures. The comparison is not so fanciful. Requirements for efficient operation combined with minimum expenditure on unnecessary weight are common both to critical engineering structures and to successful organisms.

In *C. cinereus*, more gills are added as the fruit body enlarges by bifurcation of existing gills either on one side or at the stem-gill junction (Fig. 6.35), and by division of gill organisers at the roots of existing gills. Consequently, *Coprinus* gills are also formed as convoluted plates (Fig. 6.35). Subsequently, tensions generated by growth of other parts of the

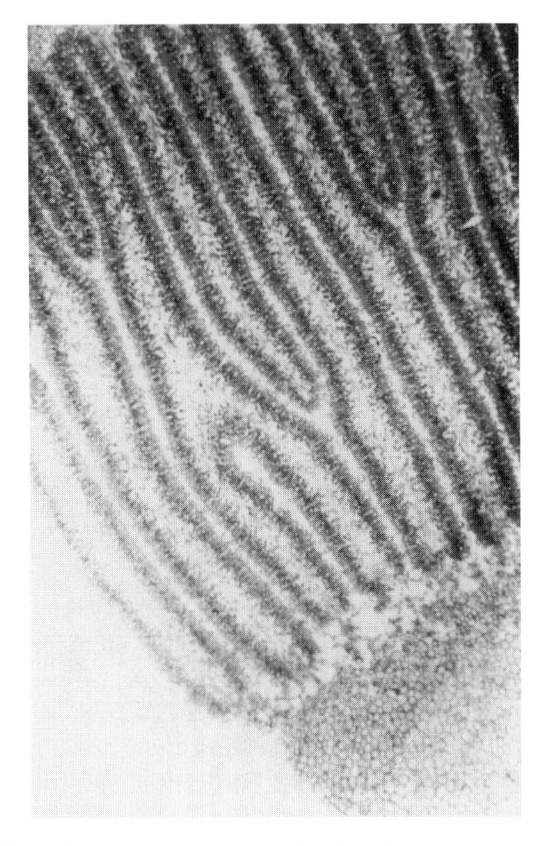

Figure 6.37 Part of a transverse section of a fruit body cap of *Coprinus cinereus*. Stem tissue is in the bottom right corner. All of the gills which have their tramal tissue continuous with the stem are primary gills. Note that one of these shows a Y-shaped profile, but both arms of the Y are still connected to the stem. This indicates that primary gills proliferate by formation of a new gill space within the trama of a pre-existing primary gill (×120).

fruit body place geometrical stress on the 'embryonic' gills, like a folded cloth being straightened by stretching. Such a mechanism requires that the folded elements (in this case the gills) are anchored. The connection of primary gills to the stem provides the initial anchorage; subsequently cystidium–cystesium pairs interconnect gill plates around the stem. Tensions generated by expansion of the cap will then be communicated and balanced throughout the structure. Cystidium–cystesium pairs act as tension elements whose function is to hold adjacent hymenia together as cap expansion pulls the gills into shape. The strength of the adhesion between cystidia and the opposing hymenium (see discussion on p. 486 in Horner and Moore, 1987) is essential to this mechanical function; when they do not exist, as in the *revoluta* mutant (Chiu and Moore, 1990b) the gills remain convoluted. This is why cystidia must be seen to be tension elements and not buttresses. An engineering comparison here, would be the bracing (tensioning) wires which are strung between the wings of a biplane and which serve to keep the wings together in flight.

As in *Volvariella*, the stress which drives the straightening of the convoluted gills is a function of the expansion of the maturing primordium. A crucial aspect of understanding how the final structure of the fruit body is attained is appreciation of the geometrical consequences of the differential growth of the primordium. For example, as a typical fruit body of *C. cinereus* grows from 1 to 34 mm in height, the circumference of the stem increases 9-fold and the circumference of the outer surface of the cap increases 15-fold; this latter corresponds to more than a 3000-fold increase in volume (Fig. 6.38) The differential growth which generates primordium enlargement exerts enormous mechanical effects on relationships between tissue layers which are often concentrically arranged. Mechanical forces themselves generate many of the patterns which characterise the form and structure of the mature mushroom fruit body.

6.3.5 Mushrooms make gills

The spore-bearing gills of 'mushroom' fungi (whether agaric or aphyllophoralean) are plates suspended from the fruit body cap tissue. Intuitively one might expect such plates to develop and extend by 'downward' growth of the distal edge of the gill (that is, the edge which is eventually exposed) but this is not the case. The direction of gill development has been a matter of controversy for many years. Schmitz (1842) was one of the first to observe the presence of a general annular gill cavity around the stem, the roof of which was lined with a continuous palisade

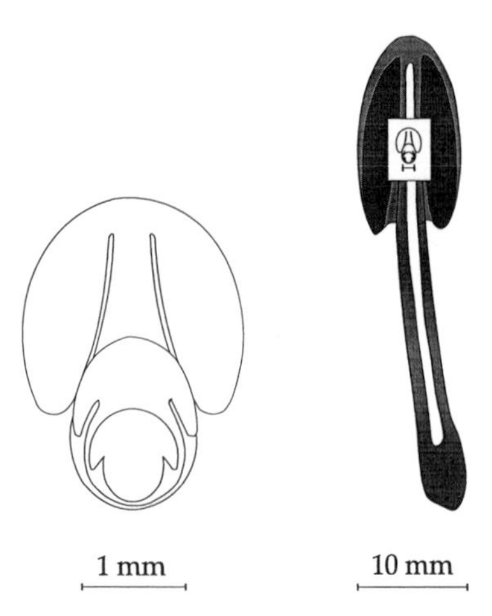

Figure 6.38 Scale diagrams showing the size relationships of primordia and fruit bodies of *Coprinus cinereus*. The outline diagrams on the left were traced from micrographs of sections shown in Fig. 6.4 and nested together to illustrate the steady outward expansion of the tissue layers. These diagrams are superimposed (to scale) on a median diagrammatic section of a mature fruit body in the panel on the right to demonstrate the full extent of the outward movement of tissue boundaries.

layer of hyphal tips which formed the young hymenophore. These observations were confirmed in Brefeld's study on *Coprinus lagopus* (1877), Hoffman's work on *C. fimetarius* (1860), and Atkinson's descriptions of *Agaricus* spp. (1906, 1914). A second mode of development was originally thought to be of limited applicability (Atkinson, 1914) until Levine (1914) reported it in *C. micaceus* and then in other species and concluded that this course of development prevailed in most agarics. According to Levine no annular prelamellar cavity is found. Instead, the palisade layer develops a series of ridges which then elongate, split, and halves of adjacent ridges unite to form the lamellae. Levine maintained that the protenchyme tissue between the ridged groups of palisade cells is continuous with the underlying stem tissue from the earliest stage. Atkinson (1916) refuted Levine's work in his treatise on *C. comatus*, *C. atramentarius* and *C. micaceus*. He maintained that, in agarics, there is first a general annular cavity with a continuous palisade layer which grows 'outward in a centrifugal direction over the under-surface of the cap, following the centrifugal growth of the latter' (p. 122). According to Atkinson the unequal growth of areas of the palisade layer gives rise to folds which are the fundaments of the lamellae: as these widen in the annular cavity, they reach the underlying stem or the fundamental plectenchyma surrounding the stem. The stem tissues and the gill trama therefore come to appear continuous. Atkinson argued that what he saw as a secondary attachment of the gill trama to the stem was what led Levine to his 'erroneous' conclusions. Similarly, Chow (1934) reported palisade pockets with the interlying tissues continuous with the stem, but, again, attributed this feature to the growth of the gills into the stem. Reijnders' observations (Reijnders, 1963, 1979) indicate that certain species do not have a continuous palisade layer or a general annular cavity and that cap and stem hyphae in those species are intimately intermingled in the regions between the palisade ridges. However, he still discusses gill development in terms implying that gills grow at their margins, and grow towards the stem.

Establishing the direction of development is crucial because the active growing point is the location of the developmental controls which establish the cell differentiation and tissue demarcation which create the gills. We need to know the direction of development in order to experiment sensibly with gill morphogenesis. Fortunately, we now have some compelling experimental evidence from work with both *Volvariella bombycina* and *Coprinus cinereus* which establishes the direction of gill development. It is diametrically opposite to that claimed by the early observers referred to above.

6.3.5.1 Volvariella gills

In fruit body development of *Volvariella bombycina*, primary gills arise as ridges on the lower surface of the cap, projecting into a preformed annular cavity, described as a levhymenial mode of development. The gills clearly project into the annular cavity, but the question is, do they *grow* into it? I believe that the answer to this question is no, and see gill development as dependent upon outward progression of the organised hyphal branching which creates the hymenium on either side of the foot (or base) of the gill where it is joined to the cap tissue (Fig. 6.10). To understand this it is essential to appreciate the scale of the changes which occur during development. Fig. 6.38 illustrates the relationships between primordial fruit bodies (in which tissue formation is initiated) and their mature counterpart. Clearly the depth of the gill even in a small species like *Coprinus cinereus* increases by several millimetres, in some larger species the increase in depth may be many millimetres or even centimetres. These gills differentiate from the undifferentiated tissue of the cap and as gill differentiation progressively 'uses up' undifferentiated tissue in the inner regions of the existing cap, so further undifferentiated tissue is formed in the outer regions of the cap. Consequently, the cap is continually expanding outwards and the roots of the gills are continually differentiating from the cap tissue (cap context) along outwardly directed radii. Effectively, therefore, the gill margins (the edges which arise as projections into the cavity when the under-surface of the cap first becomes folded) remain positionally fixed in space while the gill cavities enfold and extend around them.

The primordium of *Volvariella bombycina* is usually enclosed within a universal veil which protects the gills but obscures them from view. Occasionally, primordia with exposed gills arise. Chiu and Moore (1990a) used young fruit bodies of this sort to trace the relative growth rates of the different parts of the gill by painting black ink marks on the tissues. During further fruit body development, ink marks placed on the cap margin and those placed on the edges of the gills *remained at the margin or the gill edges, respectively* (Fig. 6.39). The growth increments in this experiment were very large. The radius of the cap increased from 0.5 to 2.5 cm and the depth of the gills from 1.5 to 5 mm. If growth of the cap and gill margins resulted from apical growth of the hyphal tips which occupied the margin when the ink was applied, then ink particles placed on those hyphal tips would be left behind as the hyphal apices extended (Fig. 6.40). Indeed, such an approach has been used to study growth of sporangiophores of *Phycomyces* (Castle, 1942) and *Aspergillus giganteus* (Trinci and Banbury, 1967). Both of these papers present time-lapse photographic sequences showing the sporangiophore tip growing beyond externally applied markers (*Lycopodium* spores in Castle (1942), starch grains in Trinci and Banbury (1967)). In the *Volvariella* experiment, extension at the margin would consequently have resulted in the ink marks being left at their original absolute positions, being buried beneath 4–20 mm of newly formed tissue by the end of the experiment (Fig. 6.40). It follows, therefore, that gills of *V. bombycina* extend in depth by growth of their roots into the cap context and by insertion of hymenial elements into their central and root regions. Similarly, the hyphal tips which form the cap margin when it is established at the very earliest stage of development remain *at* the margin. They do not continue to grow apically to extend the margin radially, nor are they overtaken by other hyphae; instead they are ‘pushed’ radially outwards by the press of fresh growth behind, and they are joined by fresh branches appearing alongside as the circumference of the margin is increased.

Because so much stress is placed on apical wall growth, there is a popular misconception that growth occurs only at the hyphal tip. On this basis the above interpretations may be judged as radical by some people. Yet the hyphal tips at cap and gill margins are growth-limited,

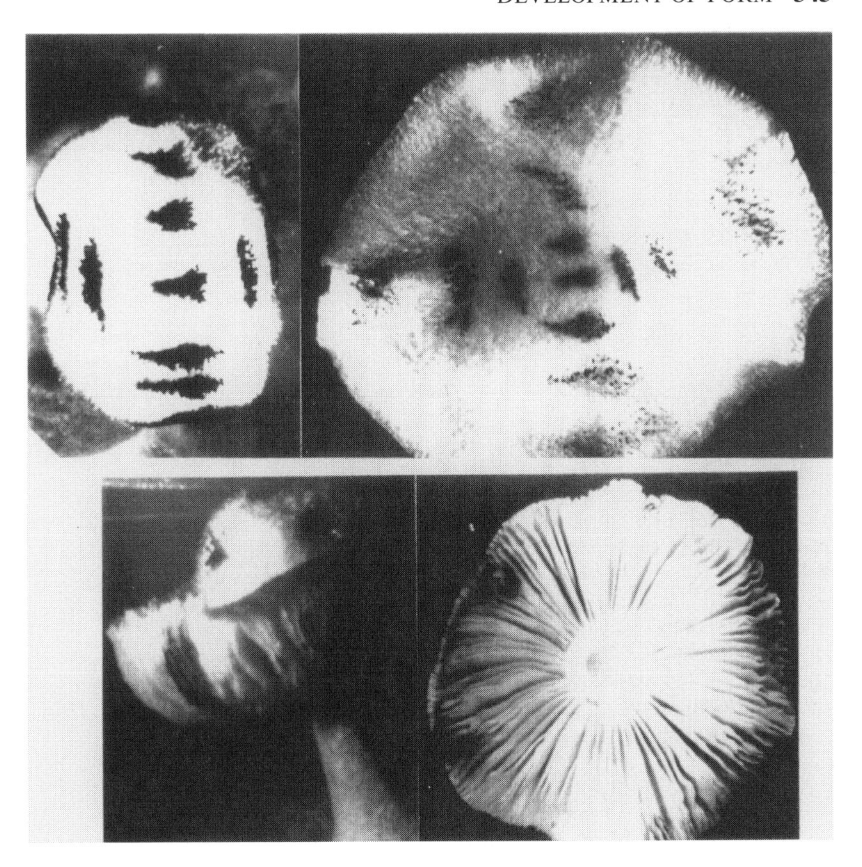

Figure 6.39 Marking experiment on a fruit body of *Volvariella bombycina*. Ink marks were placed on the top surface (upper photographs) and gill margins (lower photographs) of a 10 mm diameter cap. The photographs on the right show the disposition of the marks at maturity, when the cap diameter had increased to about 50 mm. Central marks on the top surface were unchanged, but peripheral marks became further apart circumferentially and more diffuse, though they remained close to the cap periphery. Marks on gill margins remained visible at the gill margins and cap periphery. The important observations here were that the ink marks were not submerged by hyphal growth, nor overtaken by tissue growth. Thus over a 5-fold increase in size, hyphae at the gill margin remain at the gill margin, and hyphae at the cap periphery remain at the cap periphery.

differentiated cells, so it should come as no surprise that growth of the structure they comprise is concentrated *behind* them. This fact has been appreciated for over 30 years, as even an early edition of the *Dictionary of the Fungi* defines 'inflated hypha' as "one in which cells behind the growing apex enlarge and cause the apparent rapid rate of growth characteristic of most agaric and gasteromycete fruit bodies." (Ainsworth, 1961).

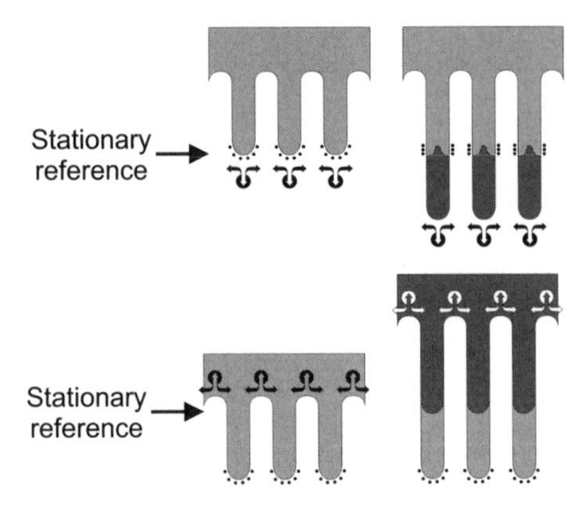

Figure 6.40 Gill formation in *Volvariella bombycina*. Diagrammatic interpretation of the experiment shown in Fig. 6.39. Top and bottom images show effect of different strategies for gill growth on the disposition of ink marks. If the gills grow at their margin (upper drawings), hyphal tips will grow beyond the ink marks and the ink will become buried (and hidden from view) within the gill spaces. If the gills extend by growth of their roots into the cap context (lower drawings), the original margins remain intact as the gill organisers 'excavate' the gill spaces. Since the ink marks remained *in situ* through a 5-fold increase in size (Fig. 6.39), it is concluded that *Volvariella* gills extend at their roots.

6.3.5.2 Coprinus gills

In *Coprinus*, the cap of the fruit body primordium encloses the top of the stem and gills are formed as essentially vertical plates arranged radially around the stem. Transverse sections show the cap as an annulus concentric with the stem, the inner circumference of the annulus being the surface of the stem and the outer the surface of the cap tissue (Fig. 6.41). There are two types of gill. Primary gills from formation have their inner, tramal tissue in continuity with the outer layers of the stem (Fig. 6.41). In secondary (and lesser ranked) gills, the hymenium is continuous over the gill edge (Fig. 6.41; and see Reijnders, 1979; Rosin and Moore, 1985a; Rosin *et al.*, 1985; Moore, 1987). Tramal tissues of primary gills remain connected to the stem until the final expansion of the cap pulls them away from the stem (see section 6.3.3; Rosin and Moore, 1985a; Moore, 1987). This frees the margins of primary gills, the injury to their margins being repaired with marginal cystidia (section 6.3.1).

Whilst intact, primary gills are connected with cap tissue at their outer edge and with the stem at their inner edge; since the circumference of the stem increases so much during maturation why doesn't the gill thickness increase by the same extent? The answer is that the tendency to widen as

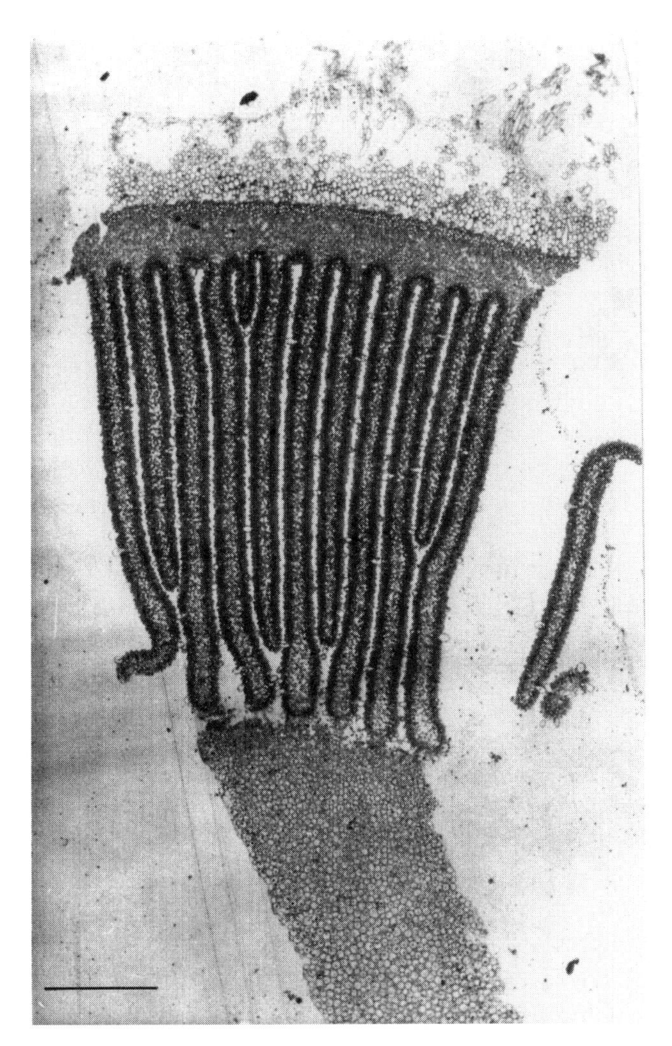

Figure 6.41 Part of a transverse section near the periphery of a fruit body cap of *Coprinus cinereus*. This shows the formation of hymenium over the margins of primary gills which are thereby becoming freed from the stem (at the bottom of this section). Note secondary and tertiary gills, and the regular radial array of gills. Scale bar = 100 μm.

the stem circumference increases is compensated by gill replication, and specifically by formation of a new gill cavity and its bounding pair of hymenia within the trama of a pre-existing gill. This forms a Y-shaped structure (Fig. 6.37). Observation of fruit body sections shows that these Y-shaped gill structures are oriented exclusively as though the new gill space originates at the stem circumference (Moore, 1987) so that the crotch of the Y-shape moves outwards towards the cap. This clearly sets the direction of development as outwards from the stem.

Consideration of the distribution of cystidia over the hymenial surface also leads to the same conclusion. There is an increased number of cystidia at the margin of secondary gills and this higher density of cystidia at the margin shows categorically that the margin remains stationary in space at its point of formation, that is, it does not grow towards the stem. The reason for this conclusion is that if the margin were responsible for growth of the gill and grew towards the stem, then the zone behind the margin (and the zone behind that, etc) would once have been the location of the margin and would therefore exhibit at least the cystidial density characteristic of margins. There is no evidence for cystidial death or removal at this formative stage and the cystidia are contributed by the two hymenia either side of the secondary gill, so if the margin extends then the body of the gill, being made up of every previous margin, can have more cystidia than the margin but cannot have fewer. Since observation shows that the general body of the hymenial surface does have a much lower density of cystidia than that which characterises the margin, it follows that the margin does not extend and the secondary gills (and gills of lesser rank) grow at their roots.

Gill development in most *Coprinus* species is highly synchronised and many developmental processes can be seen to proceed along axes which are polarised such that the 'wave' of differentiation migrates from the stem towards the pileipellis across the width of the gill, and from the cap margin towards the cap apex along the length of the gill (Rosin and Moore, 1985a). This developmental polarity is the same as that seen during the later stages of maturation. In *Coprinus cinereus* it is the tissue at the cap margin which first reaches the stage of development at which meiosis is initiated (Raju and Lu, 1970). Chow (1934) notes that the maturation of the basidia follows the same general order in *Coprinus* species and begins at the interior-inferior margin of each gill, and it is a matter of simple observation that the production and pigmentation of spores, as well as autolysis of the gill, are all initiated at the edge of the gill closest to the stem and that pigmentation and autolysis proceed from that edge towards the outer edge of the cap and from the cap margin towards the apex. These observations indicate that all events associated with maturation progress in the direction from the cap margin to the apex, and across the gill, from the inner edge (adjacent to the stem) to the outer. Thus the developmental polarity established when the gill tissues are first delimited is maintained throughout the fruit body maturation process.

These observations indicate that gills in the *C. cinereus* fruit body develop radially outwards, their roots extending into the undifferentiated tissue of the cap context. The developmental vector is directed *away from the stem*, as it is in *Volvariella*, and in both the schizohymenial and levhymenial modes of agaric gill development the gill grows *at its root and not at its margin* (Moore, 1987).

Surprisingly, the same also seems to be true for 'polypores with gills' like *Pleurotus*. When gill edges of *P. pulmonarius* and *Lentinula edodes* were marked with the vital stain Janus Green, the edge remained intact even when further growth and development was quite considerable (Fig. 6.42). Evidently, these gills also grow at their root. However, outward growth of the cap margin was not the same as in agarics. In agarics the cap margin advances outwards as a result of the press of growth behind and marks made on the margin remain at the margin as the cap develops. In *P. pulmonarius* and *L. edodes* the margin proved to be a region of active growth, reference marks made on the margin were overtaken by growth of the tissue and the new cap margin was free of the staining added at the primordial stage (S. W. Chiu, personal communication). Presumably these different growth patterns are consequences of the suite of developmental strategies which, like cell inflation versus cell proliferation in particular, create the morphogenetic distinction between agarics and polypores (discussed in sections 6.3.3 and 6.3.6). Certainly, the radially outward growth of cap tissue evident in *Pleurotus pulmonarius* (Fig. 6.42) is functionally similar to the lateral extension of the fruit body in the resupinate polypore *Phellinus contiguus* (Butler and Wood, 1988; and see section 6.3.6). It seems, therefore, that 'gilled polypores' maintain the ancestral direction of development for the structural tissue of the fruit body, but abandon the ancestral direction for development of the hymenophore. Pores extend at their free edge, but gills extend at their base.

6.3.5.3 Making gills

A fundamental 'rule' during the very earliest stages of agaric gill formation seems to be: if there is room, make a gill; without reference to the exact spatial orientation of the gill tissue so formed. This results in the convolutions in embryonic gills. The mechanics of fruit body expansion compensating for any meandering in the direction taken during gill formation (section 6.3.4). The formative element which directs the development of undifferentiated tissue of the cap into gill tissue is an organiser in the tissue at the extreme end of the gill cavity. This gill organiser is responsible for the progression of the gill cavity radially outwards,

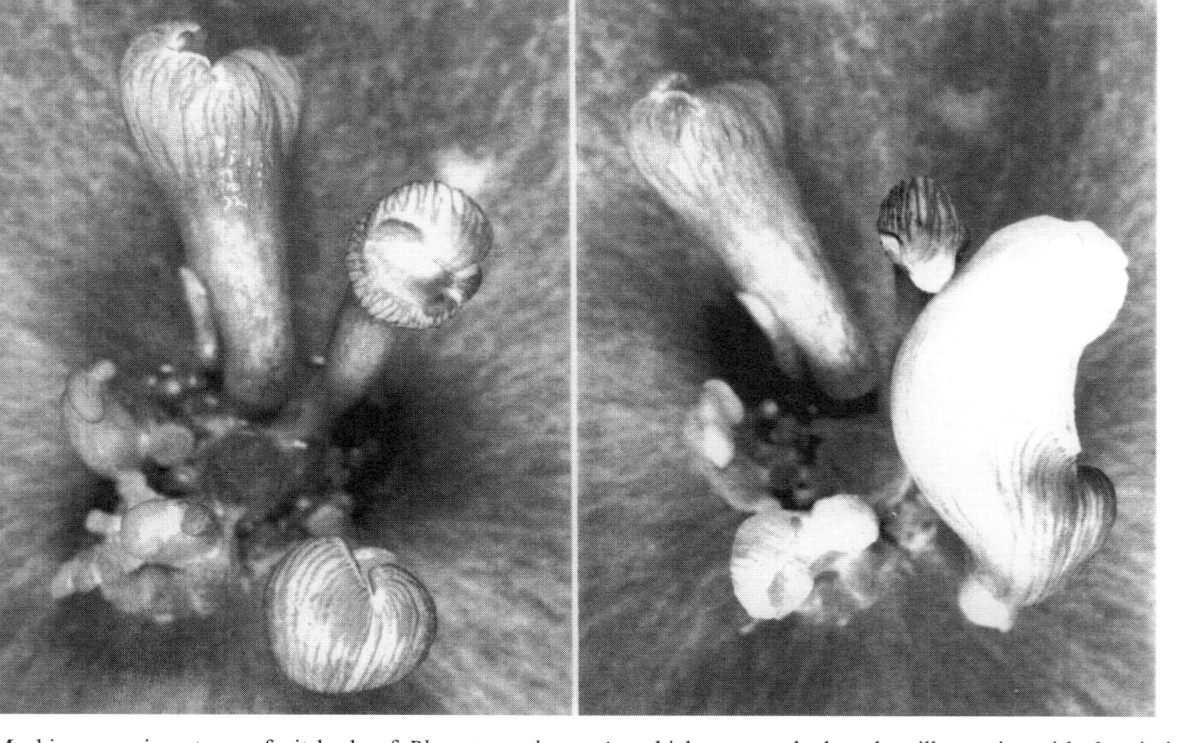

Figure 6.42 Marking experiment on a fruit body of *Pleurotus pulmonarius* which was marked at the gill margins with the vital stain Janus Green (left hand image). After further incubation the originally-marked gill margins remained intact although further growth of the cap periphery involved production of entirely new (unstained) tissue which extended the cap periphery radially outwards. Evidently, these gills also grow at their root. However, in *P. pulmonarius* the cap margin proved to be a region of active growth, reference marks made on the margin were overtaken by growth of the tissue and the new cap margin was free of the staining added at the primordial stage. Photographs kindly provided by Ms Carmen Sánchez.

away from the stem. It directs the 'undifferentiated-to-differentiated' transition. Presumably this is largely an increase in branch frequency to produce branches of determinate growth which are mutually 'attracted' so that they form the opposing young hymenia on either side of what will become the gill cavity. Cap expansion separates the two protohymenia, thus extending the gill cavity deeper into the tissue of the cap, a process which is called cavitation (Fig. 6.43; Moore, 1987, 1995, 1996). The organiser responsible for this is moving radially outwards penetrating, as the primordium grows, successive generations of undifferentiated prosenchymatous tissue in the cap context which lies between the gills and the 'epidermis' (pileipellis) of the cap (Rosin and Moore, 1985b). Since it is a radial progression (as well as progression into the depth of the gill), neighbouring organisers become further and further separated from one another as development proceeds (Fig. 6.44). As the distance between neighbouring organisers increases a new one can arise between them (Rosin and Moore, 1985a). When a new gill organiser emerges, the margin of a new (but 'secondary') gill is formed. It is extended not by growth of its margin, but by continued radial outward progression of the two gill organisers which bracket it into the undifferentiated prosenchyma of the cap context.

Moore (1995, 1996) suggested that diffusion of some sort of activating signal along the fruit body radius assures progression of the gill organiser along its radial path. He also suggested that each organiser may produce an inhibitor which prevents formation of a new organiser within its diffusion range (i.e. the gill organiser uses this inhibitor to control its morphogenetic field). As radial progression into the extending cap context causes neighbouring organisers to diverge, a region appears between them which is beyond the range of their inhibitors. At this point a new organiser can arise in response to the radial activating signal. Interaction between the diffusion characteristics of the activator and the inhibitor is all that is necessary to control gill spacing, gill number, gill thickness, and the radial orientation of the gill field.

Evidently, gill organisers can also come into existence within the gill trama (Figs 6.37 and 6.57). In conventional agarics this will initially produce gill folds (Fig. 6.57). If the organiser continues to penetrate towards the cap tissue and through the entire length of the gill, then two gills will result from division of the first. However, the process may not be completed. When it is not completed, folded gills and/or bifurcated gills will persist into maturity and be a part of the species description for that subject. Indeed, even pore-like structures might result if penetration of gill organisers into the gill trama or cap context is localised to punctate

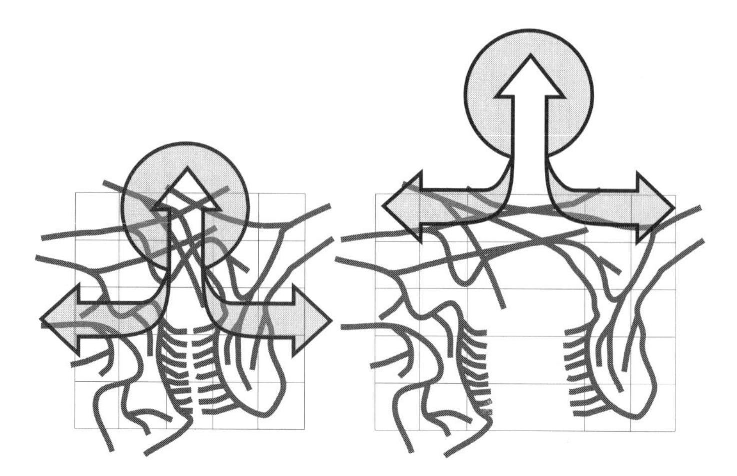

Figure 6.43 Gill organisers are thought to promote the branching pattern which enables the hymenium to be formed as an 'epidermal plate' of laterally-arrayed hyphal tips. As the organiser migrates radially outwards into newly formed (undifferentiated) cap context the internal expansion of the primordium can separate the hymenia to extend the gill space.

Figure 6.44 Diagram showing how the outward migration of gill organisers could be driven by a signal radiating from the stem and their integrity determined by a morphogenetic field governed by a growth factor they themselves produce.

regions rather than spreading radially. A range of hymenophore types has been recorded in *Lentinus* spp. (Hibbett *et al.*, 1993a) which might well reflect this sort of ontogeny. Thus, the generation of gill organisers on the basis of the rule 'where there is space, make gill' is able to account for most gill morphologies. Perhaps it is this essentially simple rule which has enabled similar gill morphologies to arise repeatedly by convergent evolution (Hibbett and Vilgalys, 1991, 1993).

6.3.6 *Toadstools make pores*

Few developmental studies of poroid fruit bodies have been published. Corner (1932b) described how pores and dissepiments arose in an area on the underside of the young part of the cap of *Polystictus xathopus*, a region he called the 'pore field'. Pores, of course, are the tubular holes which are lined with the hymenium. The tissue which separates the pores is called the dissepiment (Fig. 6.10). This word literally means a partition and is specifically assigned to the tissue between the pores of a polypore. Described in this way, it is implied that the pore is an entity in its own right. However, like the space between the gills of gilled-fungi, the pore is defined by the pattern-forming growth and branching processes which take place in the hyphae of the dissepiments. The pore is the place where the dissepiment does not grow. Corner (1932b) defined the 'pore field' as an annulus near the margin on the underside of the cap, in which localised development of differentiating hyphal growth resulted in initiation of dissepiments as protruding ridges, which branched and united around non-extending areas. The latter became the pore bases. Continued growth of the dissepiment away from the supporting cap tissue delimited the pore. In *Fomes levigatus* the pore and dissepiment areas could be distinguished before the first basidia appeared (Corner, 1932a) and similar early demarcation of hymenium in pore areas has also been described in *Ganoderma lucidum* (Mims and Seabury, 1989). Corner (1948) described a spine field corresponding to the pore field of polypores in the resupinate hydnoid fruit body of *Asterodon ferruginosus* fruit body, in which spines arose as localised areas of downgrowth 200–300 μm in diameter in which the hyphal tips were positively gravitropic. Little attention has since been paid to hymenophore differentiation in polypores. Most subsequent studies have been directed towards the contribution of hyphal analysis to taxonomy (discussed in section 6.2) and to tropisms of polypore fruit bodies (Plunkett, 1956; Kitamoto *et al.*, 1974b). An exception is a study of development of *Phellinus contiguus*, which produces a poroid fruit body which is flat on the substratum (that is, a resupinate fruit body). Butler and Wood (1988) described an isolate obtained from decayed timber which regularly produced resupinate pores over extensive parts of colonies cultured in Petri dishes and in a series of papers the morphogenesis of those pores has been described.

In this strain of *Phellinus contiguus,* pore initials can be recognised with the naked eye and were formed 34 mm behind the colony margin. Fruit bodies formed in agar cultures of *Phellinus contiguus* were generally similar to naturally-occurring specimens; even the pore size and density were within the range found in nature. Butler (1988) distinguished two developmental processes in pore morphogenesis which she called island initiation and aerial fascicle growth (a fascicle is a little group or bundle of hyphae). Fruit body tissue arose as islands in the narrow initiation zone immediately behind the mycelial margin. The island initials were formed in scattered regions of mainly submerged mycelium, implying that the pattern formation which defines the pattern of initials is related more to position than to hyphal lineage (Reijnders and Moore, 1985) and is mediated through the agar medium. New pores were formed by lateral extension from the island initials. The later developmental process, involving aerial fascicle growth, was characterised by a distinction between localised regions of loose, quite rapidly extending hyphae and lateral regions which either did not extend or extended more slowly. Although dissepiment hyphae originated beneath the agar surface of the culture (see below), this localised extension of the established dissepiment was primarily an aerial process. Thus, coordination of hyphal behaviour to form pores presumably involves communication using some volatile metabolite. The final pore density was not defined in the original pore field. Secondary pore formation occurred both by differential growth within thick dissepiment areas and by subdivision of primary pores. However, cooperation between adjacent dissepiment initials resulted in pores of uniform height.

Pore development continued on isolated disks of mycelium and after other surgical operations required for observations of living material. Butler (1988) stresses the importance of such observations to the recognition of aerial spatial relationships between adjacent structures, which would be lost easily during fixation and embedding. On the basis of these observations, Butler (1988) interpreted all features of later development, that is delimitation of primary and secondary pores, vertical dissepiment elongation and lateral extension of the fruit body area, in terms of localised aerial fascicle growth. Such an interpretation places the nature of the fascicle, its mode of growth and coordination between fascicles in a central position in determining pore morphogenesis. This notion is amplified in the discussion of Reijnders' hyphal knots in section 6.1.9, and it is also worth comparing the description of island initials above with the progressive appearance of NADP-GDH activity in hymenia of *Coprinus cinereus* illustrated in Fig. 4.9 and discussed in section 4.2.3.

Another point worth stressing is that there was no clear separation between a period of lateral diagravitropic growth, resulting in pore delimitation, and a period of vertical positively gravitropic growth, resulting in dissepiment elongation. The two processes were coincident. Butler and

Wood (1988) showed that fruit body formation was positively gravitropic; pores were formed only in inverted cultures (agar upwards). Disorganised growth of the hymenophore occurred in dishes with the agar lowermost, showing that some hyphae that contribute to hymenophore construction are not positively gravitropic, though extension of the dissepiments clearly is. The effect of inversion in *Phellinus contiguus* contrasted with that described for *Polyporus brumalis* (Plunkett, 1956, 1961) in which dissepiment growth was not gravitropic and inversion resulted in continued coordinated growth of the dissepiments in an upward direction. Sensitivity to gravity varied between strains of *P. weirii*; two out of eight strains formed fruit bodies when inverted (Li, 1979).

Other environmental factors which affected initiation and final form of fruit bodies in *P. contiguus* were aeration and illumination. Reduced aeration inhibited fruiting, but as this was counteracted by KOH it seems likely that CO_2 was the effective agent (Raudaskoski and Salonen, 1984). A single short light exposure prompted initiation of pores over a large area of the colony and pore formation continued on parts of the mycelium formed in the dark. Light and aeration also promote cap development from the stem of *Polyporus brumalis* (Plunkett, 1956) and *P. ciliatus* (Akimova, 1982). Light also induces hymenophore development in *Favolus arcularius* (Kitamoto *et al.*, 1974b).

Butler (1992a) identified convergent growth and differentiation of aerial skeletal hyphae as indicators of the first differentiation of dissepiment-forming hyphae. The first skeletal hyphae occurred more than one day later than the first hymenial cells. Hymenial elements were the first to differentiate, basidia and setae (a seta is a stiff hair-like sterile hyphal end with thick walls, projecting from the hymenium) arose directly from 3-day-old divergently-growing mycelium, setae differentiating first. In the pore bases this sparse discontinuous hymenium became more continuous as more basidia, and also setae, differentiated in parallel with dissepiment growth. Consequently, three kinds of hyphae, namely setae, basidia and dissepiment hyphae arise in the pore field from divergently growing mycelial hyphae.

Butler (1992b) showed that setae, basidia and dissepiment building hyphae all differentiated on explants. Setae and basidia developed directly from hyphal compartments at or close beneath the agar surface though dissepiments were aerial structures (Butler, 1992a). In explants, sites of differentiation of hyphal types of the fruit body occurred on the outer surface and on cut surfaces exposed to air. The first basidia and dissepiment building hyphae occurred close to the sites of setae that differentiated earlier. Older regions of the outer surface were not competent to differentiate into these cell types. Instead, the hyphae in older regions often developed a golden colour, suggesting that they had differentiated into a different hyphal form. In contrast with the outer surface, internal hyphae retained competence to differentiate into fruit body hyphae but this was only expressed at exposed surfaces. Retention of competence for fruit body differentiation by internal hyphae but not by older surface hyphae could explain fruit body initiation in response to injury or surgery (see section 4.3.2). What it means is that the internal hyphae have a capacity to regenerate which the surface hyphae do not. Differences in this degree of commitment may account for the differences in tropic responses which have been categorised by Gorovoj *et al.* (1987). Using gravitropism as the most obvious and readily observed tropic response, Gorovoj *et al.* (1987) identified three types of mechanisms occurring in different groups of fungi, as follows.

The first pattern occurs in most of the gill and polypore fungi with soft quick-growing fruit bodies the gills or the walls of the tubular hymenophores have a longitudinal section in the form of a wedge-shape narrowing downward. Apart from a few exceptions like *Panaeolus*, spore formation and release in such fungi occur simultaneously over the whole surface of the hymenophore for a relatively long period. The wedge-shaped section and strictly vertical orientation of gills and tubes permit the spores to fall out freely. In these fungi the hymenophore may be the main centre regulating gravitropic growth, as on the one hand the hymenophore has its own orientation mechanism assuring positively gravitropic growth, while on the other hand there is evidence that the hymenophore may influence growth of the stem of the fruit body.

The second pattern is exhibited by species of *Coprinus* and their relatives in which the gills are parallel-sided and spore formation and release are synchronised over a relatively short period of time. Spore formation is accompanied by autolysis of those parts of the gill which have released spores, and by the advancement of the spore maturation zone from bottom to top (i.e. margin to apex) of the cap (Rosin *et al.*, 1985). Buller (1922, p. 240) says that the gills of *Coprinus* species are not gravitropic and if this is truly the case then it is the stem alone which is responsible for spatial orientation of these fungi.

Finally, the third pattern occurs in bracket fungi which have a very tough, solid consistency. After their position has been changed, for example by a fall of the tree on which they grow, the direction of growth changes in the growing zones and only the newly formed parts of the fruit body assume a vertical position. The fruit body as a whole, though, is not gravitropic. The fruit body primordium is an undifferentiated spherical mass. If this emerges on a vertical tree trunk it grows out horizontally to form the 'bracket'. If it forms on the lower side of a branch the fruit body develops a circular form with a central attachment to the tree; rather like an inverted mushroom. This gravimorphogenetic effect is evident when the branch is disturbed sufficiently to reorient the fruit body drastically. When this happens the existing fruit body tissues cannot adjust; instead, new tissues are formed so that a fresh crop of hymenial pores can be directed downwards (illustrated in Buller, 1922, p. 110). Thus, in this instance the 'tropic' signal (gravity in the specific example cited) exerts a morphogenetic effect, determining the overall shape and form of the fruit body as it reacts (Ingold, 1953). These different reaction mechanisms must reflect differences in cell biology of the organisms concerned and the observed differences in level of commitment between surface and internal hyphae in *Phellinus contiguus* are likely to be a part of this.

Butler (1995) found that fruiting could be promoted by one or more metabolic products found in extracts of *P. contiguus* cultures. The fruiting-inducing factor was produced by colonies of fruiting size which had been grown in either light or dark. Production of the factor did not require differentiated fruiting tissue and its concentration was maintained during formation of fruit body tissue. The factor did not replace the requirement for light. The concentration of fruiting-inducing factor below the margins of small, 6-day-old, colonies was significantly lower than from an equivalent position below larger, fruiting-size colonies. This suggests that production of the fruiting-inducing factor is responsible for the minimum size of colony capable of fruiting in the light. Thus, accumulation of fruiting-inducing factor may be a necessary prerequisite for the light response. In developmental terms, in the presence of sufficient fruiting-inducing factor, colonies are competent to fruit in response to light. The requirement for a period of vegetative growth before the colony is competent to initiate fruit body differentiation in response to light is of general occurrence in the basidiomycetes (Ross, 1985). The extract from *P. contiguus* resembles the fruiting inducing substance (FIS) of *Schizophyllum commune* described by Leonard and Dick (1968) in its diffusibility in water and heat stability. It has not been characterised any further than this (see section 6.1.7).

Gillian Butler's work on *Phellinus contiguus* provides an ideal focus for comparing developmental concepts between agaricoid and poroid forms. We have already seen that the two differ fundamentally in the way in which they increase the bulk of the fruit body: agarics employ cell inflation, polypores use cell proliferation (section 6.3.3). Also, as discussed immediately above, mushrooms and toadstools react to environmental signals by tropic readjustment of existing tissues, whereas bracket (and presumably resupinate) polypores react by a morphogenetic regrowth process which produces new tissue on top of the old. Yet, it is evident that the major cell biology controls are much the same between the two groups. In both there is evidence for cell differentiation into a (small) range of cell types, in a specific sequence (temporal control of development) and in specific places (spatial control of development). The ability of internal hyphae of *P. contiguus* (but not surface hyphae) to form fruit body hyphal types is an expression of differential commitment which is mirrored in the different reactions of hymenial and other cells of *Coprinus cinereus* to explantation. Obviously, in both cases some cells become committed, others do not. The need to reach a state of competence, which is then unleashed into a morphogenetic change by a specific signal, is also a common theme (in *P. contiguus* accumulation of a specific effector molecule may be involved). Similarly, what might be called 'island patterning' is common to the two groups. Scattered cells react, communicate and/or amplify the message to their neighbours to build an island of common reaction around the initiator and then the islands interact and unite into a broader ('continental') reaction. The overall conclusion which seems appropriate is that the different groups of fungi share a common cell biology but this is combined into different strategies appropriate to the lifestyle and behaviour pattern of the organisms concerned. Different edifices are constructed from the same bricks.

6.4 Strategies of basidiomycete fruiting

6.4.1 Shape as a taxonomic criterion

The shape and form of the final structure is clearly of crucial importance to the developmental biologist, but in higher fungi, especially mushrooms and toadstools and their relatives, shape became such a dominant factor in classification that it strangled systematics through most of the 20th century (Watling and Moore, 1994). Fungal classification was developed by Fries (1821) on the basis of the earlier work of Persoon (1801) and was an arrangement founded primarily on the shape and form of the spore-producing tissue, the hymenium, and on the hymenophore, the structure on which the hymenium was borne. In this approach, fruit bodies with hymenophores in the shape of plates (gills) hanging beneath an umbrella-shaped cap were assigned to the 'agarics' irrespective of any other aspect. This group was contrasted with those fungal fruit bodies which had pores (tubes) in a layer beneath the cap (boletes and bracket fungi) and those with teeth or spines hanging down from the cap (hydnoids). The latter groups included species in which the 'fruit body' is formed as a sheet of fertile tissue on some supporting substratum (resupinate forms). Other major groups included those bearing spores externally on a club-shaped (clavarioid) or coral-like (coralloid) structure and those in which spores developed inside the fruit body (gasteroid species).

The earliest studies of development (Fayod, 1889) indicated that the Friesian groupings were artificial. Unfortunately, these observations had to wait forty years for support and recognition (Rea, 1922; Heim, 1931; Singer, 1936, 1951). Indeed, it was not until the turn of the century that the microscope became a regular tool in identification and classification (Patouillard, 1900). As a result, there was more than 100 years of stagnation, during which superficial characters were the basis of fungal classification, before anatomical and developmental information was correlated with chemical and morphological data to produce a rational scheme of classification. Consequently, notions of homology (features that have the same evolutionary origin but have developed different functions), analogy (features that are superficially similar but have evolved in different ways), neoteny (retention of the juvenile body form, or particular features of it, at maturity) and the like have come late to mycology.

The function of the fruit body is to produce as many spores as the structure permits and there are many ways in which spore production is optimised. Although the structure of the fruit body tissue may be fundamentally similar in different species, it differs in detail from one group of larger fungi to another. Fayod (1889) showed this for different agarics and it has been demonstrated for bracket fungi (Corner, 1932a, 1932b; Cunningham, 1965), clavarioid fungi (Corner, 1950) and hydnaceous fungi (Maas Geesteranus, 1971, 1975). The hyphal analysis which emerged from some of these studies and its further developments are discussed in section 6.2. Likewise, some of the different ways in which fruit body expansion is achieved are discussed in section 6.3.3.

Reijnders (1948, 1963, 1979) has been instrumental in formalising the descriptive terminology of fruit body development on the basis of extremely extensive observation. He stressed the importance of three sets of features:

1. development and nature of the veil and pileipellis (the 'epidermis' of the cap) in relation to covering the developing hymeno-

phore (the hymenophore carries the hymenium, a cell layer responsible for eventually producing the spores);
2. the sequence of development of the stem, cap and hymenophore, which are the major functional zones of the fruit body;
3. the mode of development of the hymenophore.

The terminology is discussed by Watling (1978, 1985, 1996) and illustrated in Figs 6.44 to 6.51. These observations show that there are at least ten different ways by which the familiar mushroom shape can be formed, a shape which is excellently designed to give protection to the developing hymenia in exposed environments. Naturally, they have all been given different names! These range from those with naked development which is called gymnocarpic (Fig. 6.45), to several where hyphae emerge from one or more parts of the developing primordium to envelop the developing hymenium in a veil (Figs 6.46–6.50). Veils are considered to be protective, allowing the hymenia to develop in an environment which can be controlled by the organism itself; for example by regulating exudations by cystidia (section 6.3.1 and Fig. 6.29). The shape of the gills is strongly tied to the constraints of this environment within the developing fruit body. Reijnders described as rupthymenial those cases where gill differentiation appears to proceed away from the stem; as levhymenial where gills appear to differentiate towards the stem; and as schizohymenial instances where gills differentiate from the undifferentiated tissue by a splitting process which forms the gill space. The manner in which the gill is attached to the stem apex, of which there are at least eight different types, is a valuable feature in identification.

Quantitative aspects of gill development and distribution have been analysed by Pöder (1992) and Pöder and Kirchmair (1995). These authors used mathematical models to make simulations of hymenophores as well as measurements of natural fruit bodies to compare the effectiveness of gill and poroid systems. Because the widest part of the basidium must project in towards the centre of the pore, tubular hymenophores impose different adaptive pressures on size and form of basidia than do gills. Generally, the physical constraints of a poroid structure restricted maximisation of the hymenophore surface, resulting in their producing fewer basidia per unit area than gills. Furthermore, more sterile tissue is needed to construct pores (simply on geometrical criteria) than is needed to construct gills. In poroid systems, optimisation of the relationship between thickness of pore walls (dissepiments) and width of the pore is crucial to surface maximisation. For a given wall thickness even small deviation from the optimal pore diameter causes significant loss of inter-

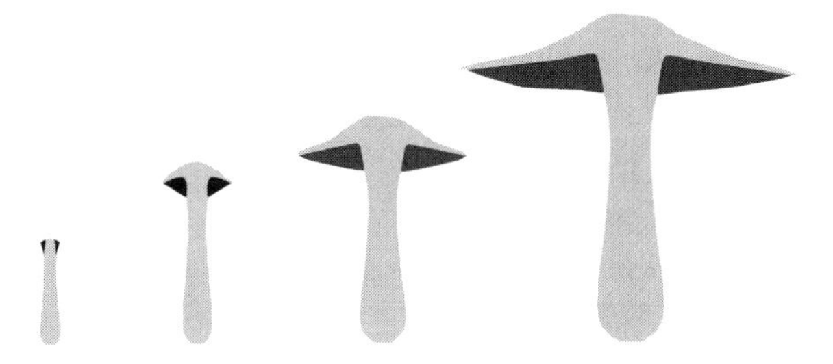

Figure 6.45 Schematic diagrams showing gymnocarpic development, where the hymenium (dark shading) is naked at first appearance and develops to maturity on the fruit body surface.

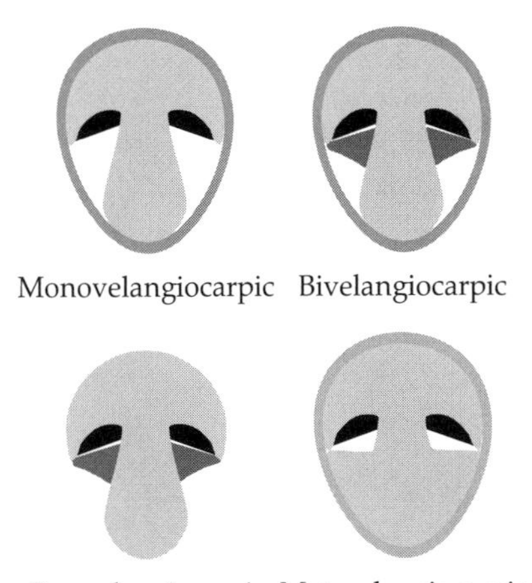

Figure 6.46 Four types of velangiocarpic development shown as diagrammatic vertical sections through immature fruit bodies. In velangiocarpic fruit bodies the hymenium (shown black in the diagrams) is protected by veil tissue (shaded grey) until the spores mature and it then becomes exposed. Monovelangiocarpic types have a single (universal) veil enveloping the primordium. In the bivelangiocarpic type there is an inner (partial) veil (dark grey) in addition to the enveloping universal veil. In paravelangiocarpic types the veil (dark grey) is reduced and often lost at maturity. In the metavelangiocarpic type a union of secondary tissues emerging from the cap and/or stem forms an analogue of the universal veil.

nal surface area. Maximum surface area is provided by thin-walled narrow tubes but a consequence of this strategy is that the long tubes must be kept exactly vertical for efficient spore release. In nature, relatively long tubes seem to occur preferentially in the perennial fruit bodies of bracket fungi. Annual fruit bodies tend to have shorter (1–5 mm long) and wider pores and compensate for their suboptimal geometry by con-

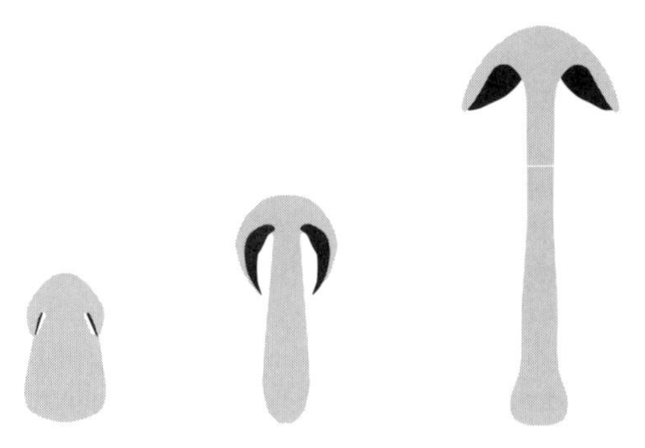

Figure 6.47 During gymnovelangiocarpic development the hymenium is protected by a very reduced veil, seen only during adolescence, which is formed between the stem and the closely applied cap.

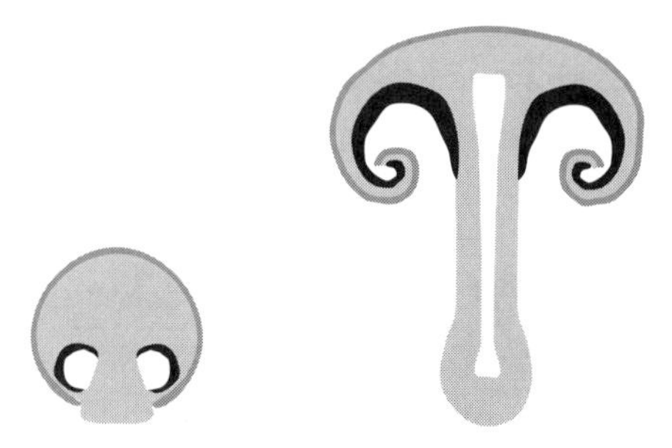

Figure 6.48 In pilangiocarpic development, the hymenium (black in the diagram) is protected by tissue extending downwards from the margin of the cap.

tinuous growth of the dissepiments with only the lower part of the continuously growing tubes being used for sporulation.

In gill systems, the possible number of gills is limited by the minimum thickness required to construct gills and the necessary distance between them for successful spore release. Nevertheless, maximising hymenophore surface depended on optimising the ratio between cap and stem diameters. A relatively slender stem (high cap to stem ratio) permits longer primary gills and allows space at the cap margin for secondary and lesser ranked gills. However, stem size is the crucial factor. Insertion of a series of tertiary gills gives a relative gain in surface area of about 2%, but the reduction of stem size to change the geometry to allow tertiary gills to be inserted itself increases surface area available for hymenium by about 40%. Obviously, there are limits to such adaptations which are set by the mechanical problems encountered by slender stems. Pöder and

Figure 6.49 Stipitoangiocarpic development describes an arrangement in which the hymenium is protected by tissue extending upwards from the stem base, but this does not enclose the primordium.

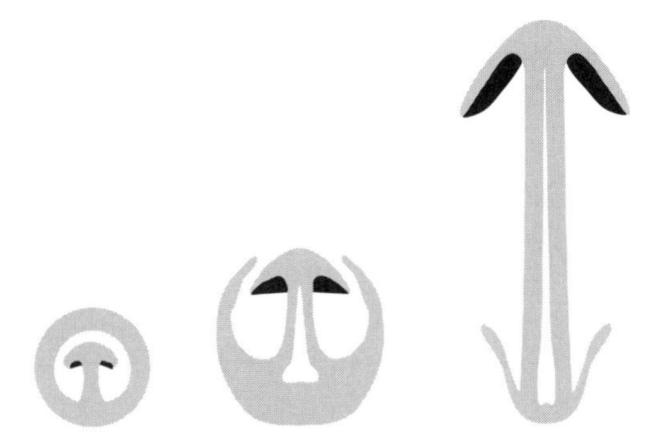

Figure 6.50 In bulbangiocarpic development the tissue protecting the hymenium is largely derived from the basal bulb of the stem and initially completely encloses the primordium.

Kirchmair (1995) stress that simple geometry is only part of the story. Their analysis goes far beyond earlier studies (Ingold, 1946; Bond, 1952) which were limited to measurements obtained from published illustrations of agarics and, of course, were limited to two dimensions. Pöder and Kirchmair (1995) point out that in nature the relation between form and function is the result of evolutionary adjustment of the whole fruit body in relation to its ecological status. Other aspects of fruit body biology may exert higher selective pressure, influence its architecture and affect the optimisation of hymenophore area.

There is a large group of fungi which have a completely enclosed fruit body (Fig. 6.51), their basidiospores being confined within the fruit body. When this continues to maturity they are called the gasteromycetes. They

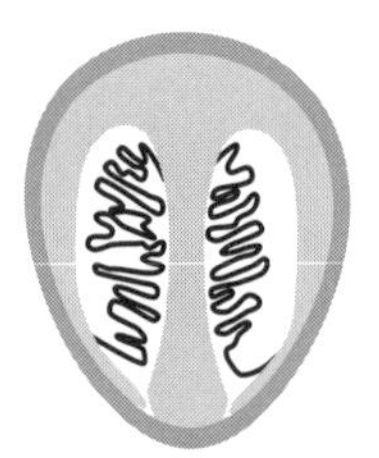
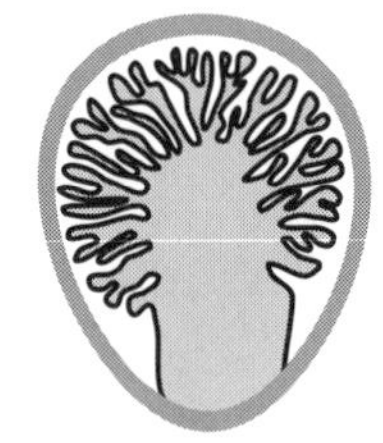
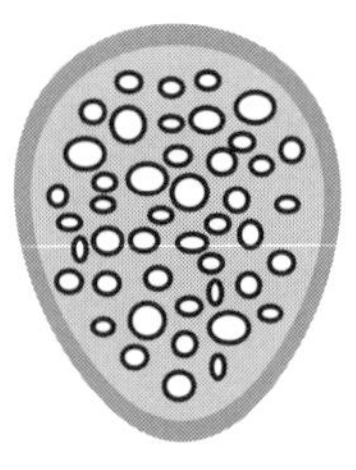

Figure 6.51 Endocarpic development produces a fruit body in which the mature hymenium is enclosed or covered over. Three patterns of this gasteromycetous form of fruit body are shown here as diagrammatic vertical sections through mature fruit bodies.

include puffballs, earth-stars, earth-balls, in all of which the fruit body has become an epigeous or hypogeous sack of spores. In addition, the stinkhorns are included in this group. These parallel the agarics in gross morphology, but the hymenium is adapted for insect dispersal of the spores (as opposed to wind dispersal in agarics). There are corresponding adaptations in shape, colours, and penetrating smells to attract insects. Similarly, the basidiomycetous hypogeous fungi, as with their ascomycete cousins the truffles, are attractive to animals, often having a distinctive odour. In some taxa this resembles the male pheromone of pigs. Many quite unrelated groups have developed hypogeous taxa, which retain the internal form of the ancestor although looking alike from outside, something that early mycologists failed to appreciate.

Current taxonomy relies on suites of characters being the key to natural classification (Bondartsev, 1963), makes full use of fruit body anatomy, and is making increasing use of developmental studies, bringing together fungi which were formerly thought dissimilar (Reijnders, 1963; Reijnders and Stalpers, 1992). With a clearer view of the range in fruit body form which can be represented in fungal taxa, we can begin to appreciate, and assess the significance of, the plasticity in fruit body form which is often seen within a single species.

6.4.2 *Variation in shape and form*

Variation occurs at different levels and for different reasons. There is variation (more properly called plasticity) in the shape and form of fruit bodies produced by a particular strain which can be demonstrated by *in vitro* culture; there is variation between collections of what might, on other grounds, be judged the same species; and there is variation at the supraspecific level. The first of these is an epigenetic plasticity: instances where, for some reason, the development of a normal genotype is disturbed, but without change to that genotype. Such plasticity in fruiting morphogenesis may be a strategy for adaptation to environmental stress. The 'rose-comb' disease of the cultivated mushroom, *Agaricus bisporus,* in which convoluted growths of hymenium develop over the outer surface of the cap, is caused by mineral oil fumes in mushroom farms (Lambert, 1930; Flegg, 1983; Flegg and Wood, 1985). Viral infections have been causal in some instances, for example, in *Laccaria*, *Armillaria* and *Inocybe* (Blattn *et al.*, 1971, 1973), and fungal attack in others. For example, Buller (1922) showed that gill-less fruit bodies of *Lactarius piperatus* were caused by parasitism by *Hypomyces lactifluorum* and Watling (1974) showed that primordia of *Entoloma abortivum* can be converted to a puffball structure by interaction with *Armillaria mellea.*

This sort of fruit body polymorphism, or developmental plasticity, has been reported in various fungal species (Buller, 1922, 1924; Keyworth, 1942; Singer, 1975), but thorough studies have only been made on *Psilocybe merdaria* (Watling, 1971; Reijnders, 1977), *Agaricus bisporus* (Worsdell, 1915; Atkins, 1950; Reijnders, 1977; Flegg and Wood, 1985), *Coprinus cinereus* (Chiu and Moore, 1990b) and *Volvariella bombycina* (Chiu *et al.*, 1989). In *Agaricus bisporus* the developmental variants reported include: carpophoroids (sterile fruit bodies; Singer, 1975), forking (where a single stem bears two or more caps; Atkins, 1950; also called bichotomy by Worsdell, 1915), proliferation (additional secondary caps arise from cap tissues; Worsdell, 1915), fasciation (a bundle of conjoined fruit bodies; Worsdell, 1915), and supernumerary hymenia (the fruit body has additional hymenia on the upper surface of the cap; Langeron and Vanbreuseghem 1965). All of these forms have also been observed in *Volvariella bombycina* (Chiu *et al.*, 1989) and in *Psilocybe merdaria* (Watling, 1971). In addition, in both *V. bombycina* and *P. merdaria* other fruit body forms, including morchelloid (also observed in *Coprinus cinereus* by Chiu and Moore, 1990b) and gasteroid fruit bodies, arose spontaneously alongside the normal agaric fruit bodies. Early developmental stages of agaricoid and gasteroid forms of *Lentinus tigrinus* were indistinguishable (Hibbett *et al.*, 1994b), the enclosed fruit body resulted from proliferation from the margins of the developing gills.

The fact that an agaric fungus might be able to produce both morchelloid (remember that *Morchella* is an ascomycete) and gasteromycete-like fruit bodies alongside its normal mushroom fruit bodies has significance for the use of morphology in systematics. To put this into the context of an approximate animal counterpart (and assuming the same weight for the systematic divisions), the parallel would be for cats to be able, quite normally, to give birth to litters containing the odd kitten looking like an aardvark, dolphin or even iguana.

Transition of the agaric hymenial pattern to the morchelloid one, particularly the position of the hymenium on the upper surface of the cap, has been suggested to be due to reversion or atavism to a fruit body organisation seen in ascomycetes (Worsdell, 1915). Similarly, the gasteroid forms of *Volvariella bombycina* might, by the same logic, be taken to reflect some phylogenetic relationship with the so-called gasteromycete genus *Brauniella* which Singer (1955, 1963, 1975) has, on other grounds, suggested to be ancestral to some species of *Volvariella*. However, making phylogenetic points on the basis of morphological variants is inherently dangerous, and carries the particular danger of raising the old notion of ontogeny recapitulating phylogeny. Animal biologists have battled through this stage in the development of evolutionary ideas. In the first quarter of this century animal evolution was thought of as resulting mainly from modification of adult form and development was seen as a recapitulation of previous mature stages. This was encapsulated in "the individual in its development recapitulates the development of the race" in MacBride's *Textbook of Embryology* (1914). Walter Garstang's views were diametrically opposed. He coined the term paedomorphosis which he summed up as "Ontogeny does not recapitulate Phylogeny: it creates it" (Garstang, 1922).

In *V. bombycina* these teratological forms arose spontaneously. In two different strains in various media tested, in addition to the normal hemiangiocarpous mode of development, fruit bodies developed either angiocarpously, with enclosed hymenium, or gymnocarpously, with exposed hymenium. In gymnocarpous fruit bodies the genus-specific basal volva might be absent or abruptly bulbous in contrast to the normal membranous form. Fruit bodies, either solitary or in aggregates, could be sterile carpophoroids, gasteromycetoid forms, normal agaricoid forms, agaricoid forms with upturned/inverted caps or supernumerary hymenia, or morchelloid forms. Heterogeneity in the hymenophore ranged from the sinuous, labyrinthiform hymenium resembling that of a *Morchella* or *Tremella* fruit body to the normal radial gills of a typical agaric. Some fruit body variants showed irregularities in the hymenium: spore tetrads enveloped in mucilage, occasional asynchrony of spore maturation, variation in number of spores borne by a basidium from 2 to 4, and variation in spore shape from oval to heart-shaped. This spontaneous fruit body polymorphism observed in two independent strains implies that an ontogenetic programme is a sequence of subroutines which can be modulated independently. Importantly, all hymenia in these forms were functional in the sense that they produced at least some apparently normal basidiospores. The function of the plasticity in fruiting morphogenesis seems to be to maximise spore production and favour dispersal of spores even under environmental stress caused by a range of extrinsic factors.

Fruiting is a complex polygenic process (see section 5.4.4) further modulated by environmental factors (Manachère *et al.*, 1983; Raudaskoski and Salonen, 1984; Manachère, 1985, 1988; Leatham and Stahmann, 1987 and see Chapter 4), and it is against this sort of background that fruit body variants like those of *V. bombycina* must be interpreted. When parasitism, disease and toxic reactions can be ruled out, it seems that the variation in forms of fruit bodies reflects a natural developmental plasticity and the key feature seems to be that even the most extreme morphological abnormalities were accompanied by normal sporulation. Thus, these developmental variants are actually or potentially functional as meiospore production/dispersal structures. The manner in which they arise implies that normal fruit body development is comprised of a sequence of independent but coordinated morphogenetic subprogrammes, each of which can be activated or repressed as a complete subroutine (Chiu *et al.*, 1989). For example, this interpretation would claim there may be a 'hymenium subroutine' which, in an agaric, is normally invoked to form the 'epidermal' layer of the hymenophore (gill lamella); but if it is invoked aberrantly and additionally to form the upper epidermis of the cap, it forms, not a chaotic travesty of a hymenium, but a functional supernumerary hymenium. Similarly, the 'hymenophore subroutine' produces the classic agaric form when invoked on the lower surface of the cap, but if wrongly invoked on the upper surface, it produces, not a tumorous growth, but a recognisable inverted cap.

Chiu *et al.* (1989) hypothesised that the development of fungal structures in general depends upon organised execution of such subroutines; the sequence and location in which they are invoked determining the ontogeny and form of the fruiting structure. Invocation of these developmental subroutines may be logically equivalent to the 'mode switches' between different mycelial states discussed by Gregory (1984) and Rayner and Coates (1987). Some of the subroutines can be identified with specific structures, such as basal bulb, stem, cap, hymenophore and hymenium, but others are rather subtle, affecting positional or mechanical morphogenetic features such as the subroutine (seemingly inactive in morchelloid forms) which ensures that gill lamellae are strictly radially arranged or a 'grow to enclose' capability, possibly associated primarily with the veil subroutine but perhaps expressed in the stem base to generate pilangiocarpic fruit bodies. Essentially the same subroutines could give rise to morphologically very different forms depending on other circumstances. For example, the agaric gill hymenophore subroutine seems to be expressed with the rule "where there is space, make gill" (Chiu and Moore, 1990a,b). When this is combined with mechanical anchorages the contortions initially produced by this rule are removed as the gills are stretched along the lines of mechanical stress (see section 6.3.4). Where such anchorages are absent the expansion forces are not communicated through the gills and the labyrinthine structure remains, as in morchelloid forms.

Very few of the subroutines have so far been identified. Indeed, very few studies have been made in sufficient detail to identify their possible genetic basis and how they might be assembled into the different developmental programmes which give rise to the characteristic structures of the major fungal groups. Hibbett *et al.*, (1993a,b) provide some relevant observations about how change in ontogeny can be related to phylogeny, but much more information is required.

6.5 Commitment, regeneration and senescence

It is a common expectation among mycologists that vegetative mycelial cultures should be recoverable readily from the tissues of fruiting (and other multicellular) structures collected in the field. The expectation is more often fulfilled than not; usually with quite simple media and frequently with the ability to reform the fruiting body given appropriate environmental and nutritional conditions. Neither plant scientists nor animal scientists can contemplate such routine preparation of cell cultures from excised slivers of fully differentiated tissues, still less the regeneration of the whole organism. So the question arises whether fungal multicellular structures consist of cells as fully committed to a differentiated state as are their plant or animal counterparts. Animal embryologists distinguish a number of different types of commitment (see the Introduction to Chapter 1 and refer to Slack, 1991, for detailed discussion) but the two which are applicable to fungi are the successive steps, specification and determination. A tissue explant is said to be specified to become a particular structure if it will develop autonomously into that structure after isolation from the embryo, subject to its being provided with appropriate conditions. An explant is determined to become a particular structure if it develops autonomously into that structure irrespective of the conditions into which it is explanted.

6.5.1 Commitment

Commitment in the *Coprinus* hymenium has been demonstrated quite effectively in *C. cinereus* by McLaughlin (1982), and in *C. congregatus* by Bastouill-Descollonges and Manachère (1984). However, these authors did not discuss their experiments from this viewpoint, placing more stress in the former case on sterigma formation, and in the latter on the potential for renewed fruiting from excised lamellae. This is a regeneration phenomenon related to, but distinct from, developmental commitment. Chiu and Moore (1988a) repeated and extended these observations using *C. cinereus*. *C. congregatus* and *C. cinereus* are similar in certain respects: both have synchronised meiotic divisions in their basidia, so that progress through meiosis can be employed as an objective marker for the physiological age of a fruit body. Also, both have similar response to light induction of primordium formation (Morimoto and Oda, 1973; Ross, 1985), have dark-inhibitory and dark recovery processes in the maturation of caps (Durand, 1983; Kamada *et al.*, 1978), and show tissue-specific accumulation of glycogen (Moore *et al.*, 1979; Ross, 1985). The species differ in at least one aspect: stem elongation in *C. congregatus* is regulated by the cap (Manachère *et al.*, 1983) while excised and decapitated stems of *C. cinereus* continue to elongate (Gooday, 1974) though not to the same extent as intact fruit bodies (see section 4.2.1; Hammad *et al.*, 1993a).

The approach used by Chiu and Moore (1988a) was to remove gill lamellae from caps at various stages of development and place them onto the surface of nutrient agar explantation medium or 1% water agar. Their further development was then observed during incubation at 27°C. In the same series of experiments the potential for renewed fruiting was studied using gills and 10 mm lengths of dark-grown 'stems' (pseudorhizal stem bases; Buller, 1931, pp. 112–117) and of normally-grown stems as inocula for nutrient agar medium. The ability of these fruit body segments to form new fruit body primordia during further incubation at 27° C was then scored. In scoring the fruiting pattern, only the first-flush primordia were taken into consideration and could be of two sorts: (i) direct fruiting when primordia formed directly on the inoculum, and (ii) indirect fruiting when primordia were formed on the outgrowing mycelium away from the inoculum. Mixed fruiting is a combination of both of these patterns (terminology of Bastouill-Descollonges and Manachère, 1984). The physiological age of all fruit body tissues was related to the stage attained in meiotic division (Fig. 4.5).

Cytological examination of 16 specimens of explanted lamellae taken at the dikaryotic stage (prior to meiosis) showed that only very few probasidia in some samples proceeded to prophase I, even after 2 days incubation on explantation medium. The majority of probasidia in such samples remained at the stage which they had reached at the time of explantation despite the fact that surrounding cells (at this stage of development, largely tramal hyphae) formed hyphal outgrowths (Fig. 6.52). Probasidia of samples taken at or after prophase I all completed meiosis and sporulation after explantation (25 specimens were tested). In contrast, paraphyses, cystidia and tramal hyphae in the same samples reverted to hyphal growth by formation of one to many hyphal apices (Fig. 6.52).

Lu (1972) suggested that the 10-h period before prophase I was the stage programmed for initiation of karyogamy and chromosome pairing, and Raudaskoski and Lu (1980), using hydroxyurea to arrest DNA synthesis in the dikaryotic stage, showed that this treatment stopped further development of fruit bodies. Lu (1982) suggested that duplication of DNA at the dikaryotic stage immediately prior to meiosis commits the basidia to genetic recombination. These explantation experiments of Chiu and Moore (1988a), however, demonstrate that such tissues are specified but not determined for sporulation. Determination to sporulation (i.e. full commitment) was evident only in material explanted at prophase I or later. This is similar to the situation in *Saccharomyces cerevisiae*, where commitment to recombination does not inevitably lead to commitment to meiotic division, the latter requiring duplication of the spindle pole body which occurs early in the first meiotic division (Berry, 1983; Dawes, 1983). Raju and Lu (1973) found that the spindle pole body duplicated at diplotene in *C. cinereus*. Thus, *S. cerevisiae* and *C. cinereus* share similar requirements for the attainment of competence and commitment to recombination and meiotic division. There is a difference, in that *S. cerevisiae* cells removed from sporulation medium after commitment to recombination but before commitment to sporulation can return to mitotic vegetative growth (Berry, 1983). In *C. cinereus*, though, all the isolated lamellae explanted at the dikaryotic stage maintained their hymenial structure even after two days incubation; the majority of hyphal outgrowths penetrating through the hymenium from below (Fig. 6.52). Thus, although such young probasidia are unable to continue development on explantation, they are somehow inhibited

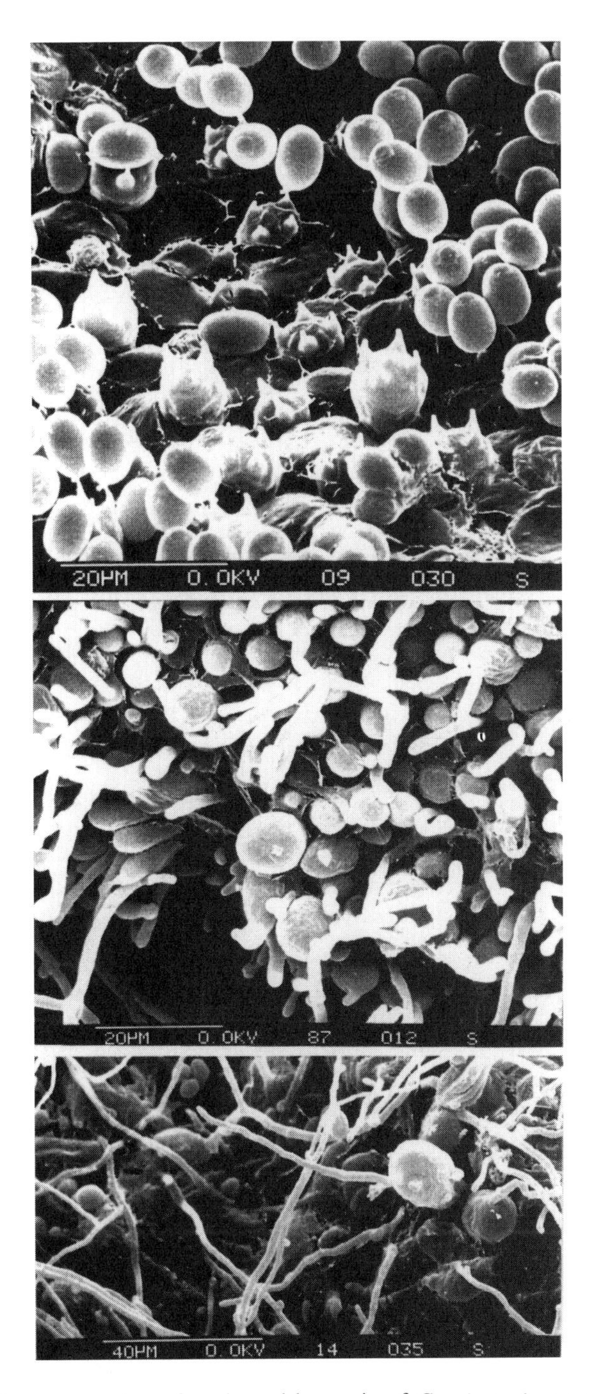

Figure 6.52 Regeneration of explanted hymenia of *Coprinus cinereus*. Top image shows SEM surface view of a normal mature hymenium. Middle and bottom image show hymenia of gills explanted to agar medium regenerating to form hyphal outgrowths. Note the cystidium forming numerous outgrowths in the bottom image.

from reversion to the vegetative state; they are specified as meiocytes but not yet determined for sporulation. Paraphyses, although highly differentiated by being much swollen (Fig. 6.31), retain the ability to revert immediately to (dikaryotic) vegetative growth on explantation.

Although basidia appeared to be specified for meiosis and sporulation, these processes were slowed in the explants. Some gills isolated at prophase I had formed only sterigmata after one day, producing spores after two days. Raju and Lu (1970) claimed that basidia required 11.5 h to complete meiosis and 8 – 10 h for sporulation; in the Meathop dikaryon, karyogamy through to spore maturation occupies 10 h (Moore *et al.*, 1987; Fig. 4.5). Thus, explantation and isolation from the environment of the fruit body cap slows the rate of maturation quite considerably. A similar effect was noted by McLaughlin (1982) in experiments on the effects of applied electrical fields on sterigma formation by isolated gills of *C. cinereus* floating on a liquid nutrient solution. In *C. congregatus*, Bastouill-Descollonges and Manachère (1984) also demonstrated that isolated lamellae carried out meiosis and sporulation at a retarded rate. All the evidence suggests that prophase I is the critical stage at which basidia become determined for the division programme. In the experiments of Chiu and Moore (1988a), similar results were obtained whether water-agar or nutrient-agar was used as the explantation medium. Thus, meiosis in basidia, once initiated, is endogenously regulated and proceeds autonomously. Only gross interference with events such as DNA or protein synthesis arrests nuclear division and sporulation (Lu, 1982). This autonomous, endotrophic phenomenon and the synchrony of nuclear division in *C. cinereus* make isolated lamellae an ideal material to study cell differentiation.

Apart from 'scattered' cystidia and cystesia (Horner and Moore, 1987), the mature hymenium of *Coprinus* comprises the two highly differentiated cell types, basidia and paraphyses. The basidia produce the spores, but the more numerous paraphyses provide the structural foundation for the gill, and their inflation produces the characteristic expansion of the fruit body. Yet, it seems likely that only the basidial (meiocyte) morphogenetic pathway is one to which a true developmental commitment is made. An important implication of the ready ability of paraphyses (and tramal hyphae) to revert to vigorous hyphal growth on explantation is that this growth mode must be actively and continually inhibited *in vivo* to ensure the orderly formation and development of the fruit body. Thus, it is extremely important that a differentiating 'environment' is maintained within the intact tissue, possibly via morphogens, to ensure fulfilment of development. Primordia are often enveloped in a mucilage which could serve as the medium through which morphogens could maintain the differentiated state and avoid the dedifferentiation which so readily occurs upon transplantation.

6.5.2 *Renewed fruiting as a regeneration phenomenon*

All basal stem portions of normally-grown fruit bodies (16 samples) and all parts of the pseudorhizas of dark-grown fruit bodies (10 samples) made direct fruiting within 4 days. This compares with cultures inoculated with vegetative dikaryon which, under the same conditions, required 10–14 days to form fruit bodies (Chiu and Moore, 1988a). The other (middle and apical) portions of stems of normally-grown fruit bodies (16 samples) produced (in 9–12 days) all types of fruiting pattern, unpredictable from their physiological age or physical size at the time of inoculation. The types of fruiting pattern observed on cultures inoculated with isolated lamellae are summarised in Table 6.8. This shows a clear dependence of the type of fruiting pattern on physiological age. Tissues explanted prior to karyogamy showing a preponderance of direct fruiting, those explanted during or after meiosis showing a minimum of direct fruiting. Exactly similar results were obtained for *C. congregatus* by Bastouill-Descollonges and Manachère (1984). Unusually rapid formation of fruit body primordia upon segments of fruit bodies used as inocula for cultures (i.e. direct fruiting) is not uncommon, but the high degree of developmental synchrony characteristic of the smaller *Coprinus* species permits assessment of its dependence on the physiological developmental state of the tissue. Such tissues are clearly rather more competent to initiate fruiting than the average vegetative dikaryon inoculum, but it is unlikely that this indicates a morphogenetic phenomenon akin to commitment to a developmental pathway.

The only physiological aspect of normal fruit body development with which a correlation is evident is the disposition of accumulated glycogen. The developing fruit body of *C. cinereus* accumulates large quantities of glycogen, which appear first in the stem base and later in the subhymenial regions of the gills (Moore *et al.*, 1979; Gooday, 1985). Further, the greatly reduced frequency of direct fruiting in cultures initiated with lamellae explanted during or after meiosis correlates with the rapid, immediately post-meiotic, utilisation of glycogen (Moore *et al.*, 1987). The implication, therefore, is that direct fruiting occurs on a tissue with a high glycogen content. Brunt and Moore (1989) have, indeed, claimed that intracellular glycogen stimulates

Table 6.8. *Fruiting pattern of cultures inoculated with gills of* Coprinus cinereus.

Physiological age at explanation	Number	Fruiting pattern observed* (% of cultures)		
		Direct	Mixed	Indirect
Dikaryotic (prekaryogamy)	19	63	5	32
Prophase I	38	24	39	37
Meiotic division	25	8	60	32
Sporulation	39	18	56	26

*Direct fruiting means that the new primordia were formed only on the original inoculum; indirect, that primordia form only on the outgrowing mycelium; mixed, combines the two. Data from Chiu & Moore, 1988a.

renewed fruiting in *Coprinus cinereus*. They used fragments of fruit body cap, stem base and stem top as inocula for a standard culture medium (containing malt extract, yeast extract and glucose). The yield of indirect fruit bodies was not correlated with the glycogen content of the culture inoculum, but yield of direct fruits showed a positive correlation coefficient of 0.88. This was despite the fact that the amount of glycogen contained in the fruit body inoculum fragments represented only a small nutritional supplementation in these experiments (estimated as 0.024% (w/w) of total carbohydrate). Supplementing the medium with this proportion of purified rabbit liver glycogen did not enhance direct fruiting. These observations suggest that glycogen within the inoculum improves fruiting in *Coprinus cinereus*, and apparently out of proportion to its value as a nutrient. On the other hand, in *Pleurotus sajor-caju* (*P. pulmonarius*), no correlation could be found between glycogen content and fruiting capacity in experiments featuring *in vivo* fruiting on media with different carbon sources and *in vitro* renewed fruiting of excised stems (Chiu and To, 1993). Subsequent detailed analysis of the relation between glycogen concentration and fruit body development in *C. cinereus* concluded that the carbohydrate cannot be linked exclusively, or even predominantly, with any one of the several processes during fruit body maturation pathway (Ji and Moore, 1993). Therefore, the endogenous glycogen level does not represent a fruiting signal. Bastouill-Descollonges and Manachère (1984) used the phrases " this potential for direct regeneration remains 'memorized' in the inocula" and "the competence of hymenial lamellae to sporulate in an autonomous way", clearly implying that the fruit body tissues used as inocula in such experiments are developmentally committed in some way to fruit body construction. Whether this represents a developmental commitment is doubtful, but the reason why fruit body tissues are predisposed to form a new generation of fruit body primordia remains unknown.

6.5.3 Breaking commitments

As indicated above (section 6.5.1), most fungal tissues produce vegetative hyphae very rapidly when disturbed and 'transplanted' to a new 'environment' or medium. Bastouill-Descollonges and Manachère (1984) and Chiu and Moore (1988a) demonstrated that basidia of isolated gills of *Coprinus congregatus* and *C. cinereus*, respectively, continued development to spore production if removed to agar medium at early meiotic stages. Evidently, basidia are specified irreversibly as meiocytes and they become determined to complete the sporulation programme during meiotic prophase I. Clearly, then, even if only to a limited extent, commitment to a pathway of differentiation some time before realisation of the differentiated phenotype can occur in these fungi.

The explantation technique of Chiu and Moore (1988a) has been developed into a rapid small-scale bioassay which can be used to study the effects of exogenous compounds on the progress of differentiation of basidia after removal from their parent fruit body. Chiu and Moore (1988b) showed that ammonium ions and glutamine halt meiocyte differentiation. Sporulation is terminated and vegetative hyphae emerge from those parts of the basidium which are in active growth at the time of exposure to the inhibitor.

In these experiments, gill lamellae were excised from fruit bodies and explanted to media containing a wide range of compounds at different concentrations. Many compounds had no effect on differentiation *in vitro* at concentrations less than 150 mM. This category included potassium chloride, potassium sulphate, D-glucose, D-fructose, sucrose, mannitol, urea, L-arginine, the glutamine analogue albizziine (L-2-amino-3-ureidopropionic acid), L-citrulline, L-ornithine, L-glutamate, L-asparagine, D-glutamine and 2-oxoglutarate. Some compounds were partially effective as they were not inhibitory to differentiation at 50 mM but inhibited at concentrations of 100 mM and above. This category included potassium nitrate, L-proline, methylamine, and the glutamine analogues L-ethionine, L-glutamic acid monohydroxamate, L-methionine sulphoxi-

Table 6.9. *Minimum effective concentrations (mM) of some inorganic salts for inhibition of sporulation of* Coprinus cinereus *in an* in vitro *assay.*

	Cation						
Anion	Ammonium	K	Na	Rb	Cs	Ca	Mg
Chloride	50	> 150	100	75	> 150	100	100
Nitrate	25	100	25	50	75	75	50
Sulphate	25	100	75	75	75	ND	100
Acetate	25	100	100	ND	ND	ND	ND
Tartrate	25	100	100	ND	ND	ND	ND

ND = not determined. Data from Chiu and Moore (1990c).

mine, azaserine, and 6-diazo-5-oxo-L-norleucine. Tests at various pH values and ammonium (chloride and sulfate) concentrations showed that highly alkaline pH values inhibit gill development, but at permissive pH values (6–8) ammonium concentrations of 50 mM were inhibitory (Table 6.9). Neither potassium chloride nor potassium sulphate had any effect. Ammonium salts injected into the caps of young fruit bodies with a microsyringe also terminated further development. Very young primordia (prekaryogamy) were not able to withstand the damage caused by injection and in most cases aborted. From the meiotic division stage onwards, very small volumes could be injected without causing non-specific damage. Injections of 2.5 μl of 1 M ammonium salt solutions (buffered to pH 7) were effective in locally suppressing sporulation if injected in post-meiotic and early sporulation stages, causing the occurrence of white zones around the point of injection as the rest of the cap matured and produced its crop of blackened spores. Similar injections of water or buffer had no visible effect on fruit body maturation. Ammonium inhibited the meiocyte development pathway *in vitro* when applied at any time during meiosis (stages prophase I through to the second meiotic division were tested). When applied at similar stages *in vivo*, ammonium retarded the rate of progress through meiosis but did not suppress sporulation. When applied at later sporulation stages (sterigma formation, spore formation, spore pigmentation), ammonium arrested sporulation completely both *in vivo* and *in vitro*.

Cytological examination of gills excised at prophase I and explanted to ammonium medium for 24 h showed a range of responses. Some were arrested at prophase I, others continued to metaphase I and some even completed the meiotic division, but no sporulation was observed. Although meiosis is well synchronised in *Coprinus*, synchrony is not perfect and the different stages at which development was arrested presumably reflect a combination of variation in exact time of exposure to ammonium and variation in stage reached by the time of exposure. Samples which were explanted at later stages suffered ammonium-arrest at correspondingly later stages. Tissue taken during meiosis (prophase I, meiotic divisions I and II) showed basidia arrested in later meiotic stages and in early sporulation stages. However, tissue explanted during those early sporulation stages seemed to become arrested immediately. Thus, exposure to ammonium causes termination of the normal developmental sequence of the basidium; the meiotic process shows some sensitivity to ammonium arrest but by far the most obvious ammonium-sensitive stages are the post-meiotic sporulation processes of sterigma and spore formation.

Some ammonium-treated basidia were apparently merely arrested, but some reverted to hyphal growth. It should be noted that this behaviour is quite unusual for basidia which characteristically continue sporulation if explanted to buffer agar after karyogamy, and remain arrested but without reverting if explanted before karyogamy (Chiu and Moore, 1988a). Thus a further direct effect of ammonium treatment is the rapid and regular promotion of reversion to hyphal tip growth among the basidial cells. This constitutes a breakdown of the commitment normally shown by these cells to their developmental pathway.

The pattern of reversion was also interesting as the new hyphal apices were not distributed randomly; rather they formed at sites expected to be involved in active wall synthesis during the normal progress of development. When the tissue exposed to ammonium treatment was in post-meiotic and early sporulation stages the reversion hyphae grew out at the sites of sterigma formation; if the basidia had formed sterigmata, hyphae, instead of basidiospores, grew from their apices; if spores were in process of formation, exposure to ammonium caused termination of spore formation and outgrowth of hyphal tips (Fig. 6.53). Hyphae also emerged from basal regions of the basidium.

Since ammonium ions or L-glutamine cause basidia, the only committed cells of the hymenium, to abort sporulation and revert to hyphal growth, normal sporulation may require some form of protection from the inhibitory effects of metabolic sources of these metabolites. Significantly, the ammonium assimilating enzyme NADP-GDH is derepressed specifically in basidia, being localised in microvesicles associated with the cell periphery (Elhiti *et al.*, 1987) where it could serve as a detoxifying ammonium scavenger. Such a function might also be ascribed to the glutamine synthetase which is derepressed coordinately with NADP-

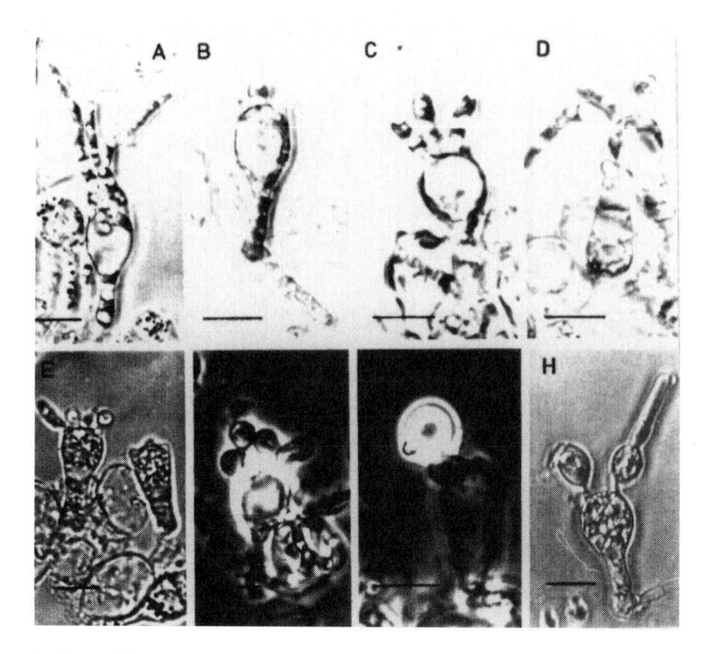

Figure 6.53 Basidium development disrupted by application of ammonium salts. Instead of forming sterigmata and spores, ammonium-treated basidia immediately dedifferentiate to produce vegetative hyphal tips. Scale bars = 10μ*m*.

GDH (Moore, 1984a; Moore *et al.*, 1987), though enhanced synthesis of glutamine in the basidium seems to be inconsistent with the inhibition of sporulation caused by L-glutamine when applied *in vitro*. However, since the NADP-GDH is localised to a particular microvesicle, glutamine synthetase and/or its product may also be vesicular, so protecting the basidium from the sort of inhibitory effect caused by exogenous glutamine. It is not known how ammonium or glutamine inhibit sporulation in *Coprinus*. The fact that the ammonium analogues hydroxylamine and methylamine and some glutamine analogues are also effective may indicate that it is the molecular structures themselves which are active, rather than some metabolic product. However, inspection of the structures of effective and ineffective structural analogues of L-glutamine listed above does not help interpretation of the sporulation inhibition process. Presumably, detection of the stereospecificity of differentiation inhibition is obscured by other events such as differential uptake and/or compartmentalisation. It is important to emphasise that though the impact of the inhibitors on sporulation is dramatic and rapid, there is no inhibition of growth. Rather, ammonium and glutamine cause diversion of effort from the highly regulated assembly of the sporulation architecture (basidium, sterigmata and spores) which must involve close control of wall growth, towards the more basic organisation of the vegetative hyphal apex.

In a later paper Chiu and Moore (1990) used the *in vitro* differentiation assay to examine the effects of metal ions, membrane-depolarising agents, ionophores, exogenous cAMP, and inhibitors which can alter fungal cell wall composition. Ammonium, glutamine and alkali metal salts were effective in tissues exposed at any time after meiotic division I; whereas ionophores, cAMP and wall synthesis inhibitors were effective only if applied during meiosis. This implies that the cell is prepared for sporulation during meiosis so that by the end of the nuclear division sporulation can proceed despite the presence of some metabolic inhibitors. On the other hand, the sporulation process must involve essential components that are sensitive to excess metal ions, ammonium and glutamine, but which do not contribute to the nuclear division. Since both types of inhibitor cause hyphae to grow from positions corresponding to sterigmata, these sites must be specified early in differentiation, and in a way which survives treatment with many different inhibitors. In view of the wide range of compounds which have been used, from substrate-analogue enzyme inhibitors like nikkomycin to inorganic salts like $NaNO_3$, it is remarkable that where inhibition of differentiation does occur the response is uniformly one of outgrowth of hyphal tips at the basidial apex and often from positions corresponding to those expected to produce sterigmata. Thus the pattern of sterigma-sites must be specified in a way which survives many different catastrophic metabolic interventions. Of the *in vitro* treatments reported, only the chilling and electrical field experiments of McLaughlin (1982) consistently altered the number and location of sterigma-sites. The sites at which sterigmata will form must be specified from within the plasmalemma and yet they must be firmly located on the basidial wall. A structure something like the focal contacts by which animal cells adhere to their substratum would satisfy these requirements (see Fig. 2.2).

Although the diverse activities of the agents tested preclude identification of their targets at the moment, it is clear that they fall into two groups. The first group comprised ammonium and glutamine and their analogues and those other inorganic anions and cations which were effective. These inhibited further differentiation but promoted hyphal outgrowths in tissues exposed at any time after meiotic division I. In contrast, the second group (ionophores, cAMP and wall synthesis inhibitors) were effective only if applied during meiosis. The differential sensitivity of basidia between meiotic and sporulation stages implies that during the nuclear division the cell is prepared in advance for sporulation, rather like the egg cell of an animal or plant, so that by the end of the cytologically recognisable nuclear division sporulation can proceed despite treatment with ionophores and wall synthesis inhibitors.

Similarly, the effects of ammonium (etc.) are presumably exerted against essential components of the sporulation process which either do not exist during meiosis or do not contribute to the nuclear division, since in tissue exposed to such compounds after meiotic division I nuclear division progressed although reversion to hyphal growth was later observed at the sterigma-sites. Basidial differentiation in *Coprinus cinereus*, therefore, can be seen as a sequence of integrated steps of commitment (Fig. 6.54) consisting of the following landmarks:

- commitment to recombination (requires completion of DNA synthesis; Lu, 1982);
- commitment to meiosis (at prophase I; Lu and Chiu, 1978; Chiu and Moore, 1988a);
- commitment to sporulation (at or after meiotic II division; Raudaskoski and Lu, 1980; Chiu and Moore, 1988a,b);
- commitment to maturation, which is assumed to be a specific step since ammonium treatment causes hyphal outgrowth from partly formed spores (Chiu and Moore, 1988b).

6.5.4 *Fuzzy logic*

Discussion of differentiation in fungi often involves use of words like 'switch' in phrases which imply wholesale diversion at some stage between alternative developmental pathways (e.g. Fig. 5.16). There are now many examples, some of which have been described above, which suggest that fungal cells behave as though they assume a differentiation state even when all conditions for that state have not been met. Rather than rigidly following a prescribed sequence of steps, these fungal differentiation pathways appear to be based on application of rules which allow considerable latitude in expression. What might be thought of as 'decisions' between developmental pathways are made with a degree of uncertainty, as though they are based on probabilities rather than absolutes. For example, facial cystidia of *C. cinereus* are generally binucleate, reflecting their origin and the fact that they are sterile cells, yet occasional

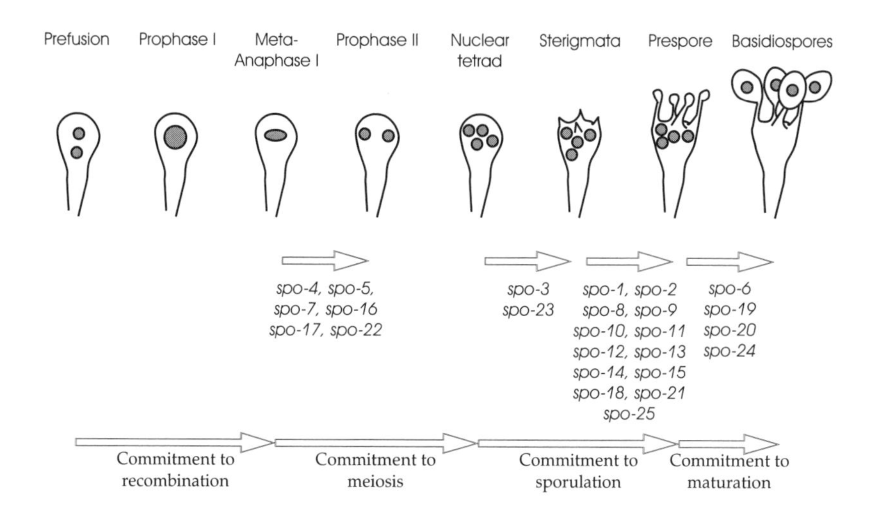

Figure 6.54 Genetical and physiological milestones in the basidial differentiation pathway of *Coprinus cinereus*. Adapted from Bourne *et al.* (1996).

examples can be found of cystidia in which karyogamy has occurred (Chiu and Moore, 1992) or of cystidia bearing sterigmata.

Development of a cystidium represents expression of a perfectly respectable pathway of differentiation and commitment of a hyphal tip to the cystidial as opposed to the basidial pathway of differentiation. The commitment must occur very early in development of the hymenium because young cystidia are recognisable in the very earliest stages (Rosin and Moore, 1985a,b). The controls which determine formation of a cystidium, instead of a basidium, by a particular hyphal apex need to be established. It is certainly the case that the basidial developmental pathway (in *Agaricus bisporus*) can be interrupted to allow this cell type to serve a structural rather than spore-producing function (Allen *et al.*, 1992), though this is clearly arrested meiosis, not sterility. Conceptually, this is similar to the cystidia of *Coprinus* which show evidence of entry into meiosis (Chiu and Moore, 1993) and which suggest that entry to the cystidial pathway of differentiation does not totally preclude expression of at least part of the meiocyte differentiation pathway. Similarly, the fact that a large fraction of the *in situ* basidial population of *A. bisporus* remains in arrested meiosis indicates that entry to the meiotic division pathway does not guarantee sporulation; a fact also demonstrated with excised gills of *C. cinereus in vitro* (section 6.5.1; Chiu and Moore, 1990a).

Further examples can be found in the literature. Watling (1971) observed some cystidia bearing hyphal outgrowths looking like sterigmata in a spontaneous fruit body variant of *Psilocybe merdaria*, while Schwalb (1978) reported that basidia of a temperature-sensitive mutant of *Schizophyllum commune* not only aborted meiosis but also produced elongated sterigmata at the restrictive temperature. A spore-deficient mutant of *Lentinula edodes* (*Lentinus edodes*) produced some abnormal basidia bearing both a hyphal outgrowth and basidiospores (Hasebe *et al.*, 1991). Similar abnormalities in basidia have been induced in *Coprinus cinereus* by transplanting gills to agar medium containing some metabolic inhibitors (Chiu and Moore, 1988a,b; 1990a; and see section 6.5.3). These explantation experiments have been discussed mainly for their value in understanding commitment to the basidium differentiation pathway, but it is equally important that all other cells of the hymenium and hymenophore showed no commitment. When explanted they reverted immediately to hyphal growth. This implies that all differentiated cells except the meiocyte have an extremely tenuous grasp on their state of differentiation. So tenuous that when removed from their normal tissue environment they revert immediately to the vegetative hyphal growth mode. That these cells do not default to hyphal growth *in situ* implies that their state of differentiation is somehow continually reinforced by some aspect of the environment of the tissue which they comprise. Interestingly, although cystidia of *Coprinus* reverted to hyphal growth when excised, cystidia of excised gills of *Volvariella bombycina* were arrested and did not show reversion to vegetative growth suggesting they are another differentiated hymenial cell type (Chiu and Moore, unpublished).

Chiu and Moore (1993) discuss the possibility that fungal differentiation pathways exhibit what would be described as 'fuzzy logic' in cybernetic programming terms. The phrase has a particular meaning, which has been equated with 'computing with words' by Zadeh (1996). It is a methodology which is useful for dealing with situations which must tolerate imprecision; effectively, where the programming term 'either/or' must become 'maybe'. It is probably easiest to illustrate by contrasting a graph based on conventional real numbers, which describes what is called a crisp function, with a fuzzy graph (Fig. 6.55). Clearly, the imprecision of the input to the fuzzy graph greatly expands the fuzzy constraints beyond those defined by the crisp function. It is this notion of tolerance of imprecision which can be applied to fungal morphogenesis.

Instead of viewing fungal cell differentiation as involving individual major 'decisions' which switch progress between alternative developmen-

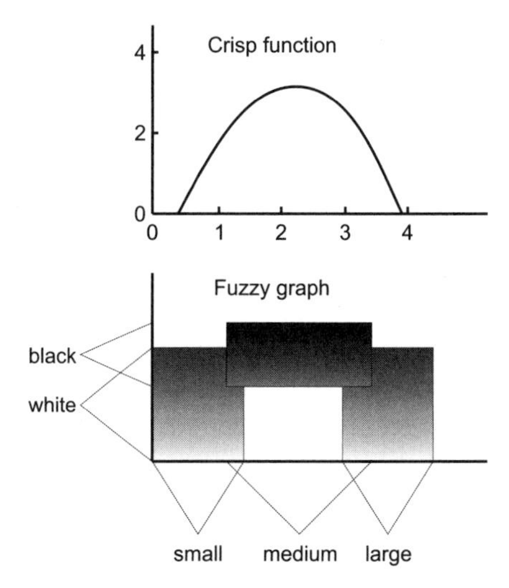

Figure 6.55 Comparison of a crisp function (top) with a fuzzy graph (bottom). Adapted from Zadeh (1996).

tal pathways which lead inevitably to specific combinations of features, this idea suggests that the end point in fungal differentiation depends on the balance of a number of minor 'decisions'. Observation shows that developmental decisions between pathways of differentiation are able to cope with a degree of uncertainty, allowing fungal cells to assume a differentiation state even when all conditions of that state have not been met. So, rather than rigidly following a prescribed sequence of steps, fungal differentiation pathways must be based on application of rules which allow considerable latitude in expression of fuzzy constraints.

6.6 Degeneration, senescence and death

Fungi are modular organisms with repetitive growth (Harper *et al.*, 1986; Andrews, 1995; Carlile, 1995) so birth and death have different meanings to those with which we, as animals, are most familiar. Death comes to all of us, though it has different manners of arrival. Esser (1990) pointed out that the various phases in the lifespan of a discrete organism (growth, reproductive phase and ageing) follows much the same pattern as the growth cycle of a batch culture of a microorganism (balanced phase, storage phase and maintenance phase). If a continued supply of substrates can be found, the mycelium is capable of unlimited growth.

However, the clone which makes up a modular organism will have regions of different age which are at different stages in the colony's developmental cycle. So even in these circumstances, in most mycelia there will be particular structures which have outlived their usefulness and must be removed. For example, a fruit body that has released its spores has served its purpose and is of no further value. It suffers a degeneration process, and its constituent cells die. The mycelium, of course, survives, to produce another crop of fruit bodies either immediately or in the next annual season. Death of the mycelium does not seem to be part of the reproductive life strategy of fungi (Griffiths, 1992) but death of cultures is regularly encountered in the laboratory. In general the processes leading up to death are referred to as senescence and in the context of the mycelium and mycelial structures, senescence is the progressive loss of growth potential which culminates in death. Some instances of this sort have been studied systematically and have provided insight into hyphal senescence. Organismal senescence has been studied in relation to the lifespan of mushrooms and the shelf life of harvested mushrooms. Another aspect of cell death is its constructive use during development. Morphogenesis can require the removal of tissue as well as tissue growth and the cell death responsible for this must be controlled in time and position. This is programmed cell death.

6.6.1 Organismal death: hyphal and fruit body senescence

Fungal senescence was first intensively studied in *Podospora anserina*, in which all isolates from nature die (Esser, 1990). The onset of senescence occurs after a short period of vegetative growth so the senescence phenotype of a strain can be measured by quantifying the growth of cultures or through serial transfers. Different isolates have different lifespans, the onset of senescence depending on nuclear genes. However, the senescence phenotype is the expression of an extrachromosomal genetic element. This is a plasmid which is integrated into the mitochondrial genome in juvenile cells where it exists as an intron in a gene which codes for subunit 1 of cytochrome oxidase. Deletion of the intron produces long-lived strains. Since the senescence DNA corresponds to an intron, the RNA splicing mechanism excises it from the primary transcript and it seems that, in some unknown way, this releases the sequence as a self-replicating DNA plasmid. When liberated from the mitochondrium the plasmid propagates autonomously and spreads through the mycelium. The plasmid transforms juvenile cells it infects immediately to the senescence phenotype (Esser, 1990; Sainsard-Chanet *et al.*, 1994). Thus, senescence in *Podospora* results from instability of the mitochondrial DNA. Growth studies suggest that onset of senescence is a randomised event, yet senescence is under nuclear genetic control.

In *Aspergillus amstelodami*, vegetative death mutants have been described which exhibit a senescent phenotype. In one, less severe, version of these mutants a specific region of the mitochondrial DNA was amplified and although the nature and size of the amplified segment differed between mutants, all shared a common core sequence. Laboratory strains of *Neurospora* are capable of indefinite growth but senescent isolates have been recovered from natural populations (Griffiths, 1992). Senescent strains of *Neurospora intermedia* were found at a frequency of about 30%, among isolates from burnt substrates on the Hawaiian island of Kauai. This senescent phenotype was called Kalilo, which is an Hawaiian word meaning 'dying, or hovering between life and death'. Kalilo strains have deficiencies in cytochromes and insertions of a linear DNA plasmid (called *kalDNA*) into the mitochondrial DNA. As senescence progresses more and more plasmid insertions occur. When death occurs intact mitochondrial DNA is barely detectable. The *kalDNA* shows no homology with either nuclear or mitochondrial DNA so is a true extragenomic plasmid. The plasmid replicates autonomously during juvenile phases, inserting into mitochondrial DNA and initiating senescence when its copy number equals or exceeds that of the mitochondrial DNA. A worldwide survey for senescence in isolates of *Neurospora crassa* discovered some from India with similarities to Kalilo strains (Griffiths, 1992). Senescent *N. crassa* strains are called Maranhar, this being a Sanskrit word meaning roughly the same as the Hawaiian Kalilo. Maranhar strains also contain a linear plasmid, *marDNA*, which can exist free or inserted into mitochondrial DNA and the insertion parallels senescence and loss of mitochondrial function.

It is not known whether these senescence processes in *Podospora* and *Neurospora* have biological significance other than as pathologies; that is, plasmid infections. It is not even clear that the amount of growth of a mycelium of these genera in nature would be sufficient to permit expression of even these senescence phenotypes. Neverthess, these situations provide interesting models for senescence and ageing in higher eukaryotes. They may also indicate how senescence might be controlled by the organism through regulated release of plasmid-like molecules targetted upon particular cell compartments or organelles. The *Podospora* example, whereby an intron is released into autonomous replication under some sort of nuclear control, is particularly interesting as it could be a model for 'self-destruct' mechanisms throughout the cell (see section 6.6.2).

There is only one instance in the literature in which any attempt has been made to assess senescence (in the strict sense) in a multihyphal structure and this is the analysis of the lifespan and senescence of *Agaricus bisporus* done by Umar and Van Griensven (1997b). In this investigation, the authors found that the lifespan of fruit bodies of *A. bisporus* was 36 days when grown in a cultivation environment which protected the culture from pests and diseases. Senescence first became evident around day 18, with cytological indications of localised nuclear and cytoplasmic lysis. These changes were followed by increased permeability of the cytoplasmic membranes and by structural changes to the cell wall. Remains of the lysed cells aggregated around and between the remaining hyphal cells. Most of the stem hyphae became empty cylinders. Other cells within the fruit body collapsed irregularly. Electron microscopy showed that most of the cells throughout the fruit body were severely degenerated and malformed after 36 days, yet a number of basidia and subhymenial cells cytologically remained alive even on day 36. When mushrooms were cultivated using conventional mushroom farming procedures, about 50% of the fruit bodies were found to have been infected by *Trichoderma harzianum* and/or *Pseudomonas tolaasii* by day 18. All such fruit bodies died on day 24 due to generalised severe bacterial and fungal infections leading to tissue necrosis and decay of the caps and stems.

In harvested *A. bisporus* fruit bodies (stored under various conditions) diffuse cell wall damage was observed first, this only later being accompanied by cytoplasmic degeneration. Consequently, Umar and Van Griensven (1997b) emphasise that the morphological changes which occur in naturally-senescent and post-harvest fruit bodies of *A. bisporus* are different. Post-harvest physiology and morphology of mushrooms is of paramount importance for mushroom marketing and has been extensively studied (e.g. Nichols, 1985; Braaksma *et al.* 1994, 1996; McGarry and Burton, 1994; Burton *et al.* 1997), but post-harvest behaviour is usually described as senescence or as an ageing process. It is quite clear from the detailed analysis of Umar and Van Griensven (1997b) that this is not the case. The harvested mushroom has suffered a traumatic injury and its post-harvest behaviour stems from that. A major factor must be that it has no way of replacing water lost by evaporation. Consequently, exposed surfaces become desiccated and are damaged first. Thus, in what might be called 'post-harvest stress disorder', further damage is inflicted on the cell inwards, from the outside. In complete contrast, during the senescence which accompanies normal ageing, the damage is inflicted first on the genetic architecture (this to include nuclear and organelle genomes) and subsequently on cytoplasmic integrity, so that cell wall damage occurs as an aspect of the eventual necrosis suffered by the lysing cell. That is, in senescence the damage starts inside the cell and proceeds outwards, from the inside.

6.6.2 Autolysis and programmed cell death

If the death of cells can be regulated and, more importantly in a hyphal system, if the death of hyphal compartments can be strictly localised, then cell death could be used as a morphogenetic process. Such a mechanism occurs in animal development and there is probably a need for a similar mechanism in fungal morphogenesis. In animals there are two patterns among dying cells (Sen, 1992). Traumatic or necrotic death occurs when the cell is suddenly confronted with extreme non-physiological conditions and loses control of its ionic balance. As a result calcium enters the mitochondria (causing swelling and dilation) and the cytoplasm becomes hypertonic. Uncontrolled water influx causes the cell and its organelles to swell and lyse. On the other hand, apoptotic or programmed cell death occurs in physiological conditions, often in response to effectors which are not lethal to other cells in the vicinity. Indeed, some mammalian cells seem to be programmed for apoptotic suicide unless suppressed by signals from other cells. Apoptotic death is relatively slower than necrosis and involves a programme of well regulated processes (including synthetic ones) which lead to internal cell degeneration and eventual removal of the dying cell by phagocytosis. Apoptosis is of enormous importance in organ development during embryogenesis where cell elimination is a key feature in morphogenesis (though only a very small minority of cells, less than 1%, may be undergoing the process at any one time).

Apoptotic cell death is also important in higher animals as a mechanism whereby autoimmunity can be avoided or minimised. Necrotic cell death releases all of the cell machinery to become potential antigens. In apoptotic death the components of the dying cell are digested within its membrane prior to phagocytosis so in most cases no antigens escape. Obviously, this last point is not a consideration in fungi. But it is particularly interesting and significant that Umar and Van Griensven (1997b) identified two modes of cell death in *A. bisporus* fruit bodies; one impacting as a trauma from the outside which can be likened to necrotic death, and the other which is more like a true, internally driven, senescence. Can this latter be considered a fungal type of programmed cell death?

Lu (1974, 1991) claimed that the gill cavities in *Coprinus* arise as a result of a cell disintegration process which he called a programmed cell death. However, Moore (1995) has pointed out that the evidence presented for this conclusion was weak and largely depended on observation of cytoplasmic inclusions which other studies had dismissed as fixation artifacts. Because of the large scale of the inflation which takes place as fruit body primordia develop, it is possible to account for the origin of spaces of the kind which become gill cavities by a process of organised branching to produce 'fracture planes' rather than cell disintegration (a mechanism called cavitation, see section 6.3.5.3). However, there is other evidence for programmed cell death.

The most dramatic of such evidence is the autolysis which occurs in the later stages of development of fruit bodies of many species of *Coprinus*. Buller (1924, 1931) described these in detail and interpreted the autolysis to be part of the developmental programme. The well-known grosser aspects of gill autolysis serve to remove spent gill tissue from the bottom of the cap upwards to avoid interference with spore discharge from regions above. That is, when the hymenium has discharged its basidiospores, autolysis is initiated to destroy that part of the gill so that it will not be a physical barrier to discharge of spores by the remaining part of the gill. The enzymology of this gross autolysis has been examined in detail in *Coprinus cinereus* and found to be due to the release of chitinase, acid and alkaline proteinase, RNA-ase, phosphatase and β-glucosidase enzymes which had previously been localised in intracellular vacuoles (Iten, 1970; Iten and Matile, 1970). Vacuoles containing acid and alkaline proteinase, RNA-ase, phosphatase and β-glucosidase were found in vegetative mycelium as well as fruit body gill tissues, so they appear to be part of the normal turnover-metabolism of the cells. However, vacuoles with chitinolytic activity were newly formed shortly before spore release was initiated. Iten and Matile (1970) suggested the chitinolytic enzymes were passively released by cells whose metabolic activity had ceased. They noted that this final stage in maturation of the fruit body was accompanied by a rapid decrease in respiratory enzyme activity an observation which might be significant in view of the role of mitochondrial degeneration in senescence in *Podospora* and *Neurospora* discussed in section 6.6.1.

Taken together these observations show that autolytic modification of the fruit body in later stages of its development in *Coprinus* involves specific production of new enzymes in particular cells at a particular time. The autolytic destruction of those cells is clearly part of the morphogenesis of the fruit body; it *is* a programmed cell death. The process cannot be dismissed as simply the final step of some other developmental programme. Buller (1924) also described a much more localised autolysis of cap flesh immediately above the gill trama in small species of *Coprinus* which enabled the gills to split and their two hymenia to be stretched apart as the cap opened like an umbrella. In this case the same autolytic programme is being exercised at an earlier stage in development and in cells which are specifically placed to achieve a particular fruit body morphology.

Umar and Van Griensven (1997c) have indicated that there may be a more general involvement of a fungal type of programmed cell death in fruit bodies of higher fungi. It is shown that in very early primordia of *Agaricus bisporus* the first gill spaces are formed as a result of cell death. The authors point out that the exact timing (prior to basidial differentiation) and exact positioning (in an annulus close to the junction of cap and stem) implies that cell death is genetically programmed as part of the morphogenetic process. It would seem that fungal programmed cell death plays a role at many stages in development of many species (Umar, personal communication). A need for such a mechanism is evident from the morphology of fruit body initials (Fig. 6.5). These appear to be tangled masses of long hyphae, yet very early in fruit body development compact structures emerge with clearly demarcated surfaces. In the initials, hyphae extend in every direction so demarcated surfaces can only arise if pre-existing hyphae which cross the boundary before it is established can be severed. The cytological evidence seems to indicate that individual hyphal compartments can be sacrificed (Fig. 6.56) in order to trim hyphae to create a particular shape. Programmed cell death is used, therefore, to sculpture the shape of the fruit body from the raw medium provided by the hyphal mass of the fruit body initial and primordium (Fig. 6.57).

Significant here also, is that this is not simply a matter of killing cells at particular times and in particular places. The cell contents which are released when the cells die seem to be specialised to particular functions too. In the autolysing *Coprinus* gills the cell contents released on death contain heightened activities of lytic enzymes. The cell deaths which carve out the shapes of fruit body primordia release mucilaginous materials. It seems, therefore, that this cell death programme includes a subroutine which causes the cells to synthesise large quantities of extracellular matrix for release when the cell lyses (Umar, personal communication).

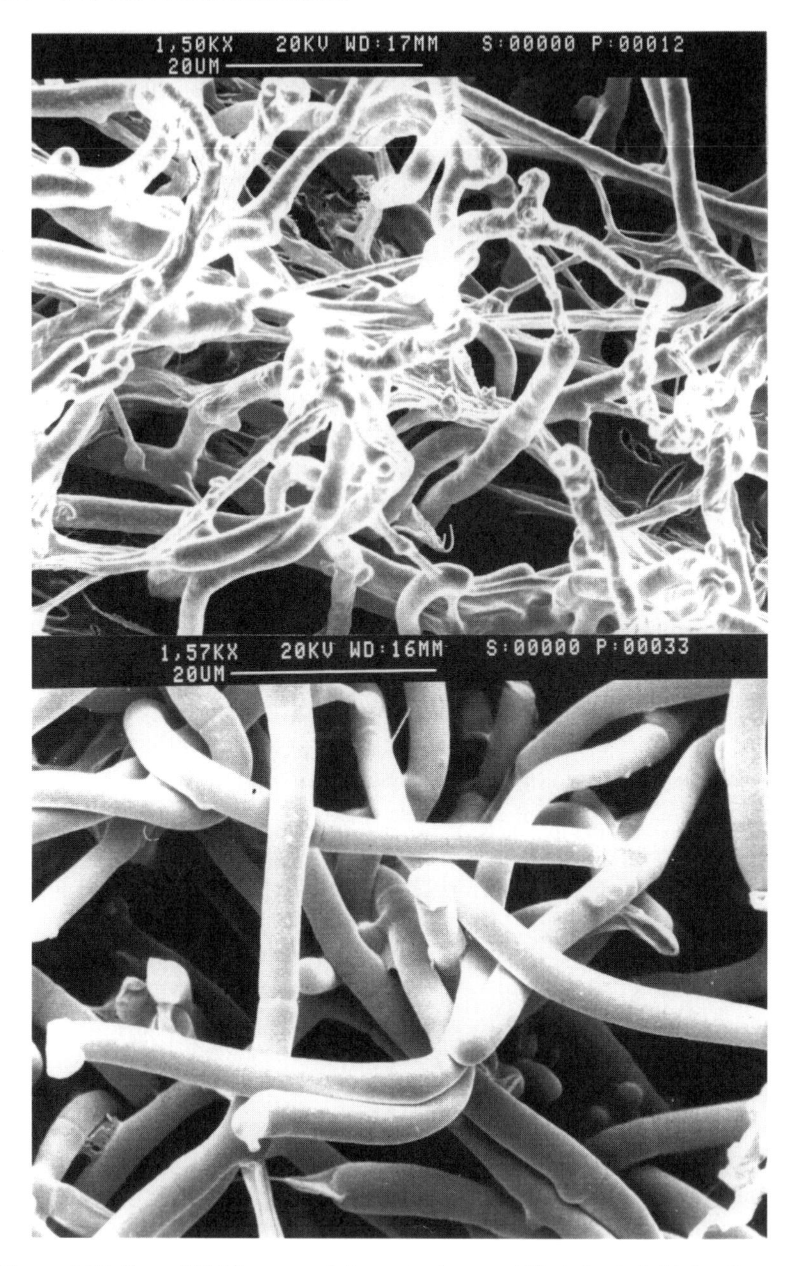

Figure 6.56 Cryo-SEM images of the outer layers of fruit body initials of *Pleurotus pulmonarius*. Many of the hyphae were alive when flash-frozen and remained fully inflated. However, cells undergoing programmed cell death tend to collapse. In the lower image a number of the hyphae show abrupt terminations where the hyphal filament has been severed by programmed death of one or more hyphal compartments. Photographs kindly provided by Ms Carmen Sánchez.

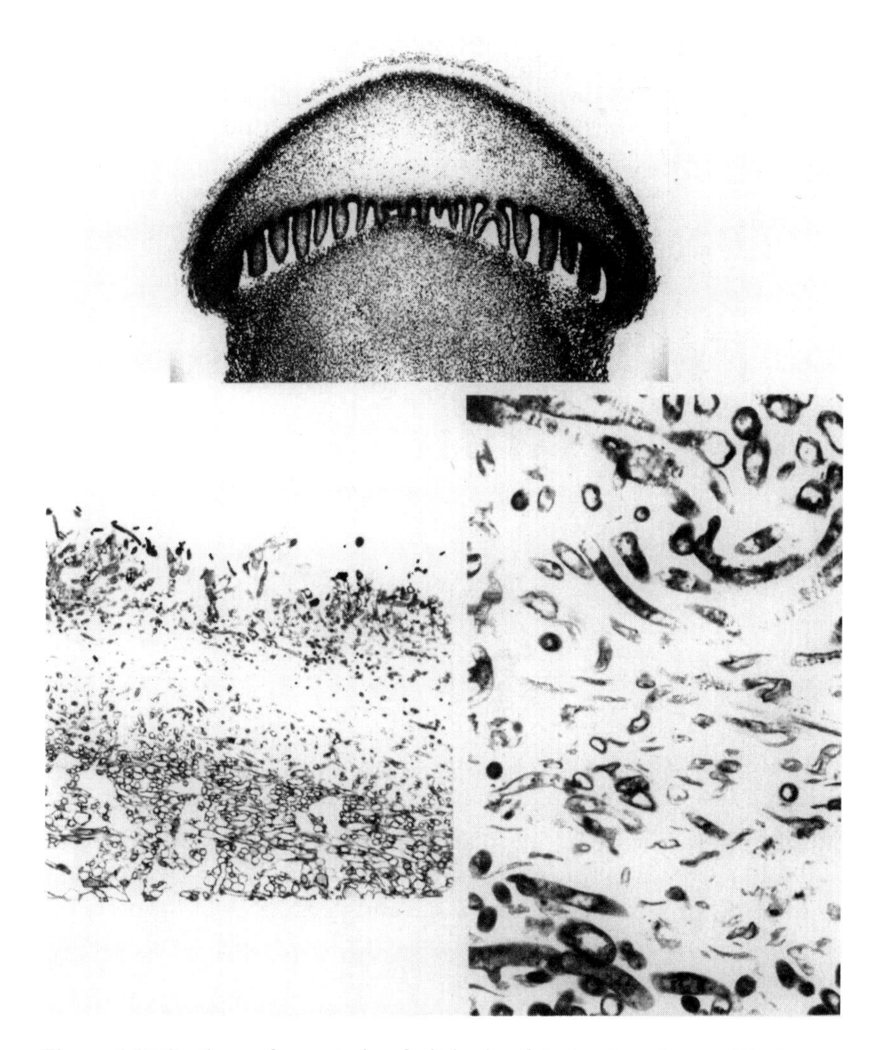

Figure 6.57 Sections of a maturing fruit body of *Psilocybe cubensis*. The top image shows a tangential longitudinal section in which there is a clearing zone demarcating the upper surface of the cap from the surrounding veil. This is a region in which hyphal cells are undergoing programmed cell death to shape the surface of the fruit body. The two lower images show progressive magnifications of this zone in which fungal programmed cell death of the 'mucoid' sort results in release of large quantities of extracellular matrix as the dying cells burst. Note the Y-shaped gill profile in the top micrograph. Photographs kindly provided by Dr M. Halit Umar.

7

The keys to form and structure

I will now attempt to summarise the discussion and present an interpretation of plausible mechanisms for control of fungal morphogenesis. Because the data are so fragmentary, this will involve extensive generalisations and potentially outrageous extrapolations. I am aware of the dangers of over-interpretation. Generalisations from a few specific cases which combine incomplete information from organisms of arguable comparability does not make the best science. At the moment there is not much choice. My motive is to produce a developmental model which can be tested by further research. So this chapter comes with the challenge: if you don't like the ideas presented here, then fill the gaps in the knowledge. Do the research to prove me wrong.

7.1 The nature of morphogenetic control

The first lesson to learn from other organisms bears on the question of the ultimate level of morphogenetic control. It is clear that in animals and plants morphogenesis is not simply a matter of playing out a predefined genetic program and it is likely that this will be true for fungi also. Goodwin (1984) states that the "developmental process involves principles of organisation which are not caused by genes in the way that computer output is caused by a program." Instead, he argued that macromolecular and cellular structures, such as the cytoskeleton and endomembrane system, provide a more promising foundation for describing the forces which define morphogenetic fields during animal embryogenesis. Those fields of force, thought to include electrical, chemical and mechanical effects, are modified by, and themselves modify, the expression of specific gene products.

Trewavas (1986) developed a similar argument about the involvement of growth substances in plant development. Rather than the abstraction 'fields of force', Trewavas (1986) likened plant development to a network of metabolic processes. Interactions between metabolic pathways create the network and the form of the tissue or organ changes in response to introduction of new components and/or new links within the network. Trewavas (1986) was keen to argue against the common notions that growth substances in plants serve as limiting factors which act in direct 'cause and effect' pattern. He argued that these ideas are best replaced by the concept of sensitivity to control. This sees the network as reaching a state which is able to react to the growth substance (or the inverse interpretation: the growth substance is such that it alone can disturb an otherwise stable network). Edelman (1992) emphasised three observations which are key to his own morphoregulator hypothesis:

1. expression of developmentally important genes is epigenetic and place-dependent, relying on previously-formed tissue structures;
2. the driving forces of morphogenesis are cellular in origin, namely cell division, movement (Edelman is concerned with animal embryogenesis), adhesion and death;
3. the links between gene expression and the mechanical driving forces are molecular, proteins, growth factors and adhesion molecules.

Edelman's (1992) morphoregulator hypothesis is essentially a network in which he places particular importance on adhesion molecules (cell-to-substratum and cell-to-cell adhesion) and cell junctional molecules. These molecules are the products of differentially-expressed (i.e. developmentally regulated) genes, but the gene products interact epigenetically to modulate the cell surface. The cell surface phenotype which results may then itself induce further differentiation and/or development-specific changes. In other discussions the part played by mechanical tensions in animal morphogenesis have also been stressed. Bard (1990) provides a general discussion, but Van Essen (1997) concludes his review of morphogenesis of the mammalian central nervous system with this paragraph:

> Morphogenesis entails an intricate choreographing of physical forces that cause differential tissue growth and displacement. Does this require an elaborate set of developmental instructions, transcending those needed to regulate the processes of neural proliferation, migration, axonal pathfinding, and synapse formation? If morphogenesis is driven largely by tension, the answer is no. Instead, the specificity of shape changes would largely be a by-product of factors that dictate the connectivity and topology of the underlying neural circuitry. This constitutes an efficient strategy for sharing the instructions that guide neural development.

Even though he places his emphasis on mechanical tension, Van Essen's argument is clearly fundamentally epigenetic and of exactly the same sort as those featured earlier in this section.

It is interesting that the embryologist, Edelman (1992), emphasises externalised molecules involved in adhesion, whilst the botanist, Trewavas (1986), emphasises interactions between internal metabolic pathways (albeit influenced by growth substances from outside the cell). Goodwin (1984) is more inclusive, mentioning mechanical and electrical as well as chemical fields as components of the 'organised context' within which developmental genes operate. Despite these cultural differences, all of these interpretations stress the role of complex interactions in specifying morphogenetic stages. In such contexts, properties are shared throughout the network because of the interconnections. Thus, change in one part results in a response by the whole. The more complex the interactions, the more stable the state. However, the concept also envisages that introduction (or removal) of a component will have sufficient influence to alter the state of the network. This helps to explain how changes in morphogenesis can depend on discrete events such as differential expression of just a few genes, or one specific environmental variable. Genes which can be characterised as developmental genes are those whose products have sufficient epigenetic impact to shift the network to a new state. Similarly, a network sensitised by its intrinsic structure to a particular environmental variable will shift in response to that variable but will be insensitive to other variables.

Again, the common observation of numerous (sometimes small) changes characterising an altered state of differentiation can also be accommodated. The assumption in this case is that these phenotypic features comprise a collection of characters which result from the interactions within the network. When the network shifts in response to a signal to which it is sensitive, all those interactions change. As a consequence, all of the characters in the collection which defines the present state of differentiation are likely to alter as the network shifts to a new state of differentiation. The differentiation signal may be correlated with many of the morphogenetic changes it induces, but may not cause them directly. Their cause lies within the network of interactions. If the signal and the morphogenetic change are studied in isolation, away from the network, it may not be possible to demonstrate any connection between them at all.

Fungi are more often excluded from than included in these sorts of generalised discussion. For example, in their review of complexity in multicellular organisms Bell and Mooers (1997) include 'fungi', but are content to rely on information which is 60 years old and covers five species, only one being a basidiomycete! Nevertheless, if these field or network interpretations are applicable to animals and plants, then they must also apply to fungi. Amongst mycologists, however, the cultural viewpoint is less advanced and more usually categorises fungal morphogenesis as a series of changes from one state to another with the implication that sequential gene activity controls the entire process. Harold (1990, 1991, 1995) has made some inroads into countering this simplistic approach by emphasising that the spatial arrangement of biochemical processes must form a foundation for understanding morphogenesis. Ion flows across membranes are the starting points for these arguments as they have the potential to provide directional references. Harold's own research has resulted in an appreciation of the potential importance of electrical processes as well as chemical processes (Harold, 1994).

Even where epigenetics is discussed in relation to fungal development, the more restricted interpretation of epigenetic modification of the genome applies, again with the implication that gene expression takes pride of place in morphogenesis. This comment is not intended to underestimate any role played by such mechanisms. There are a number of processes that silence genes in filamentous fungi (Irelan and Selker, 1996). One of these is transcriptional silencing of genes by methylation of cytosines in the DNA which seems to be common in fungi and plants (Meyer, 1996). Pre-meiotic methylation occurs during the sexual cycle in *Ascobolus immersus* and *Coprinus cinereus* to reversibly silence repeated sequences. The mechanism involves homologous pairing to sense the presence of duplications. Differences in methylation patterns have been detected between yeast-phase and filamentous growth in some dimorphic fungi (Reyna Lopez *et al.*, 1997). It is this reversible differential DNA methylation which is thought to be related to fungal morphogenesis. This is an attractive proposition in general theoretical terms (Bestor *et al.*, 1994). However, there is also evidence that methylation plays a role in the control of chromosome behaviour during meiosis in *Neurospora crassa* (Foss *et al.*, 1993). This specific possibility argues against DNA methylation being involved generally in morphogenesis. A similar conclusion is indicated by the facts that methylation occurs to a lesser extent in the mushroom *Coprinus cinereus* than it does in *Ascobolus* or *Neurospora* (Freedman and Pukkila, 1993), and that there was no evidence for development-related differences in methylation in *Agaricus bisporus* (Wilke and Wach, 1993).

Neither methylation nor any other single mechanism provides an adequate foundation for a developmental network. My preference is for a more holistic approach to identifying factors with the potential to influence morphogenesis. Some of these can be identified from the discussions in the earlier chapters of this book.

7.2 Fungal morphogenesis

Key words at each stage of development in fungi seem to be competence, induction and change. Competence is repeatedly encountered. Hyphae must be able to initiate the next step, but the next step is not inevitable. Competence may be genetic (e.g. mating types) but is primarily a physiological state. Induction is the process by which the competent tissue is exposed to conditions which overcome some block to progress and allow the next stage to proceed. Change occurs when the competent tissue is induced. The next stage always involves change in hyphal behaviour and physiology. Usually quite drastic and representing an additional property to those already expressed. That is, each developmental step takes the tissue to a higher order of differentiation. In relation to earlier discussion in this chapter, the state of competence represents the establishment of a network in the sense of Trewavas (1986), but expanded to include the full range of physical and physiological components.

Differentiated hyphal cells require reinforcement of their differentiation 'instructions'. This reinforcement is part of the context within which they normally develop (that is, it is part of their network), but when removed from their normal environment most differentiated hyphae revert to the mode of differentiation which characterises vegetative hyphae. Hyphal differentiation is consequently an unbalanced process in comparison with vegetative hyphal growth. In most hyphal differentiation pathways the balance must be tipped in the direction of 'differentiation' by the *local* microenvironment which is, presumably, mainly defined by the local population of hyphae.

Another common feature is that morphogenesis is compartmentalised into a collection of distinct developmental processes (called 'subroutines'). These separate (or parallel) subroutines can be recognised at the levels of organs (e.g. cap, stem, veil), tissues (e.g. hymenophore, context, pileipellis), cells (e.g. basidium, paraphysis, cystidium) and cellular components (e.g. uniform wall growth, growth in girth, growth in length, growth in wall thickness). They are distinct genetically and physiologically and may run in parallel or in sequence. When played out in their correct arrangement the morphology which is normal to the organism under consideration results. If some of the subroutines are disabled (genetically or through physiological stress), the rest may still proceed. This partial execution of developmental subroutines produces an abnormal morphology. Homologous subroutines can be recognised in different fungi, and gross differences in morphology can then be related to the different ways in which homologous subroutines are executed. The flow chart in Fig. 7.1 summarises these notions.

The flexibility in expression of developmental subroutines allows the fruit body to react to adverse conditions and still produce a crop of spores. It also illustrates that tolerance of imprecision is an important attribute of fungal morphogenesis. The ultimate flexibility, of course, is that the differentiation process can be abandoned in favour of vegetative hyphal growth and reversion to the invasive mycelium ensues. A lesser level of flexibility may be that a particular function is carried out by an incompletely adapted cell type.

When it comes to searching for mechanisms which might control fungal morphogenesis there is no shortage of candidates. Homologues and analogues of all of the mechanisms known in animals and plants can be found in fungi. For control at the genetic level the mating type factors (section 5.2.2) provide prime examples of transcriptional control elements able to regulate specific morphogenetic subroutines. The regulation involves transcriptional activation and repression and further 'complication' can be introduced, if necessary, by using intrachromosomal recombination to interchange regulatory cassettes.

Given the prevalence of data which indicate that hyphal systems (i) need to develop a state of competence before they are able to undertake a developmental pathway, and (ii) can be precipitated into embarking upon a particular morphogenesis by a variety of environmental signals, it is difficult to believe that translational triggering and feedback fixation

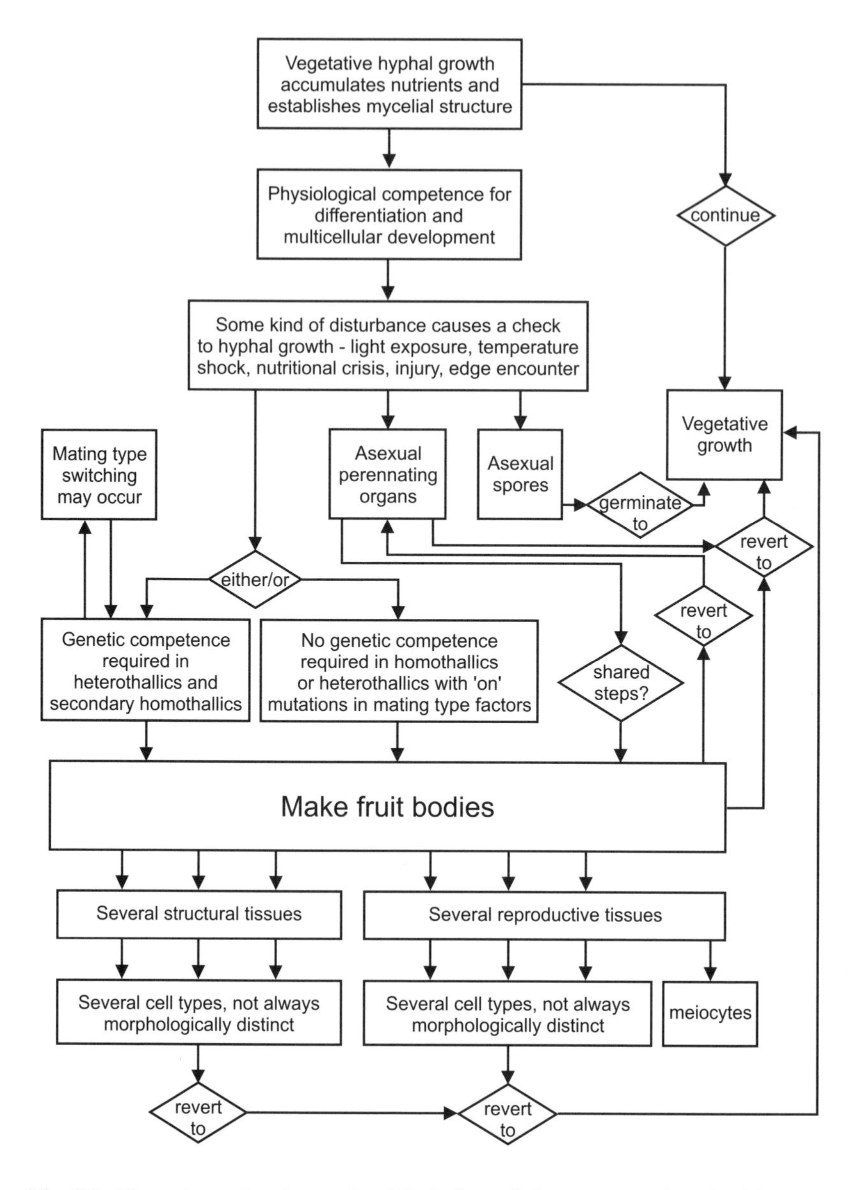

Fig. 7.1 Flow chart showing a simplified view of the processes involved in development of fruit bodies and other multicellular structures in fungi.

(Timberlake, 1993) are not widely used as regulators throughout the higher fungi.

Translational triggering is a mechanism which can relate a morphogenetic pathway to the development of competence on the one hand,

and to initiation in response to environmental cue(s) on the other hand. There are indications from a wide range of physiological studies that nitrogen metabolism may be crucial in regulating morphogenesis. There would certainly be scope for associating particular differentiation pathways with particular aspects of metabolism, so that supply of specific aminoacyl-tRNA molecules might regulate entry into differentiation pathways by affecting translation of a controlling reading frame (trigger-ORF in Fig. 7.2).

The mechanism envisaged is in many ways similar to the attenuation mechanism which regulates several biosynthetic operons in bacteria (Yanofsky and Kolter, 1982). Since translation and transcription are so closely coupled in prokaryotes, attenuation regulates transcription and this cannot apply without modification in eukaryotes. In an operon subject to attenuation, translation of mRNA commences soon after transcription begins. The RNA encodes a short (approx 15 amino acid) leader peptide which contains several adjacent codons for the amino acid product of the operon. When product levels are low, the corresponding aminoacyl-tRNA is limiting and the ribosome stalls at those codons. This allows a secondary structure to form in the mRNA that allows RNA polymerase to continue transcription of the structural genes of the operon. When the product of the operon is readily available, however, translation of the leader proceeds normally and an alternative secondary RNA structure allowing termination of transcription is formed. Attenuation provides a link between cellular levels of the product which an operon is responsible for synthesising and transcription of the operon.

As described here, attenuation depends on transcription and translation occurring simultaneously in time and space as they do in prokaryotes. Attenuation cannot operate in eukaryotes because transcription and translation occur in different places and at different times. Nevertheless, a similar mechanism *could* regulate translation of a messenger transcript which coded for several reading frames. If the trigger-ORF contained adjacent attenuating codons for aminoacyl-tRNAs subject to variation in supply, stalling/non-stalling of translation of trigger-ORF might determine whether the messenger transcript forms secondary structures which permit/do not permit translation of downstream reading frames. Note that either one or both components of the aminoacyl-tRNA may be the limiting factor and the limitation may be imposed by a compartmentalisation. That is, amino acid or a specific tRNA (or, presumably, an aminoacyl-tRNA synthetase) may be compartmentalised, regulated in local concentration, or both.

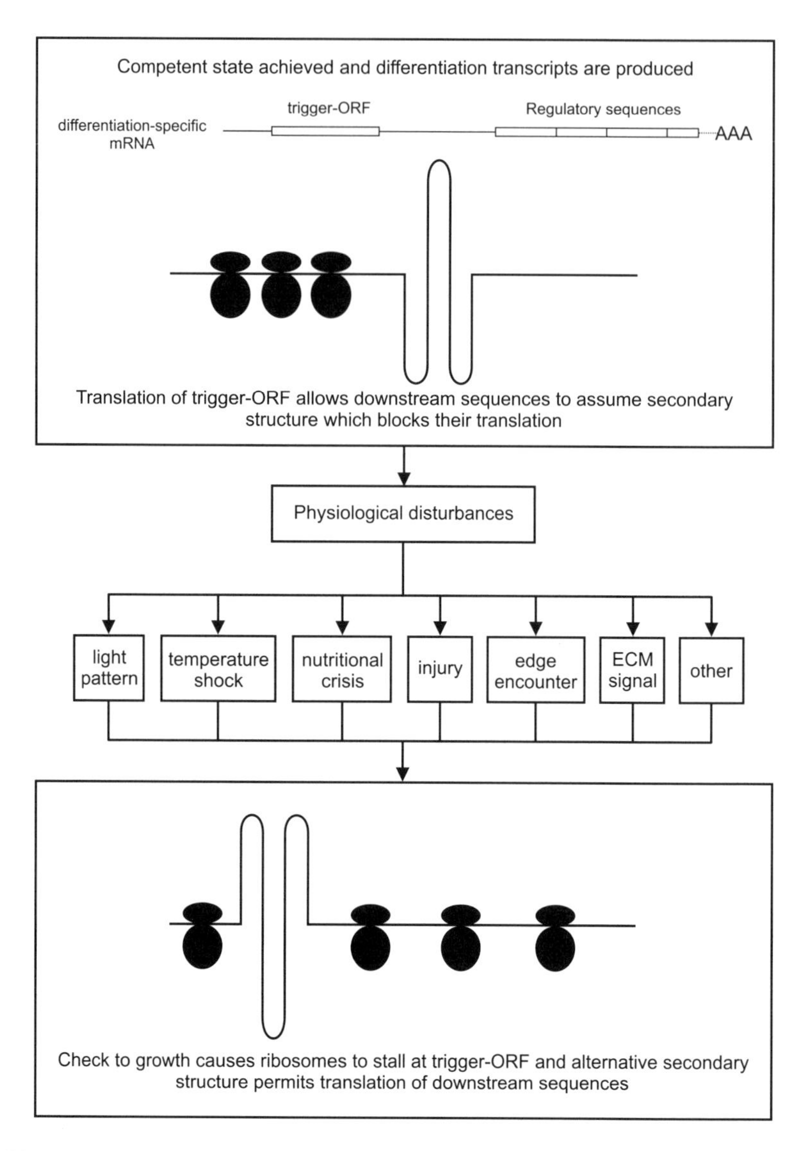

Figure 7.2 Translational triggering adopted as a general model for entry of competent tissues into fungal pathways of differentiation.

The interpretation offers a way by which a competent tissue can be released to undertake differentiation by a range of physiological events. Competence is interpreted to mean that messenger transcripts for the necessary regulators (and perhaps some key structural genes) are produced but not fully translated because an upstream sequence (trigger-

ORF) prevents translation. There may be a number of different such transcripts with regulators corresponding to the different pathways upon which the competent cell can embark, their trigger-ORFs responding to separate physiological events. On the other hand, there may be a number of similar transcripts in different cellular compartments so that the translational trigger can be released by the particular activities of those compartments with the result that one differentiation process may be triggered by different physiological events. It could also be that such a transcript was limited to one compartment, even one type of vesicle, perhaps, from which the trigger molecule can be excluded until some highly specific and/or localised physiological change occurs. Unfortunately, there is no evidence for any of these speculations. The only possibly relevant data is the finding that depletion of carbon sources available to *Saccharomyces cerevisiae* blocks translation but not transcription (Martinez-Pastor and Estruch, 1996). This may not even be relevant, but it does at least indicate that a major physiological crisis can have more immediate impact on translation and it emphasises that carbon metabolism is important, too, even if nitrogen metabolism is more often associated with morphogenetic change. Despite the lack of direct evidence at this time, I suggest that a variety of physiological signals and stresses cause translation-level controls to direct competent fungal tissues to undertake specific differentiation processes. Comparison with the operation of mating type factors (section 5.2.2) makes it reasonable to suggest that the translational trigger could immediately lead to translation of components of highly specific transcription activators and inhibitors which then regulate gene sequences required for the differentiation which has been initiated. These, or their eventual products, may be involved in feedback fixation of the differentiation pathway.

Feedback fixation is the outcome of feedback activation and autoregulation which together reinforce expression of the whole regulatory pathway to make it independent of the external environmental cues which initiated it. Feedback fixation results in developmental determination in the classic embryological sense (section 5.3.2.2). The epigenetic aspects of the network governing fungal morphogenesis starts with feedback fixation, but also includes signals from outside the cell (Fig. 7.3). The fungal extracellular matrix is extensive and complex. Its reaction to, and interaction with the environment can be communicated to the intracellular environment to modify cytoplasmic activity. Since neighbouring cells are components of the external environment, it must be the case that the activity of one hyphal cell is modulated by changes made to the extracellular matrix by a neighbouring hyphal cell.

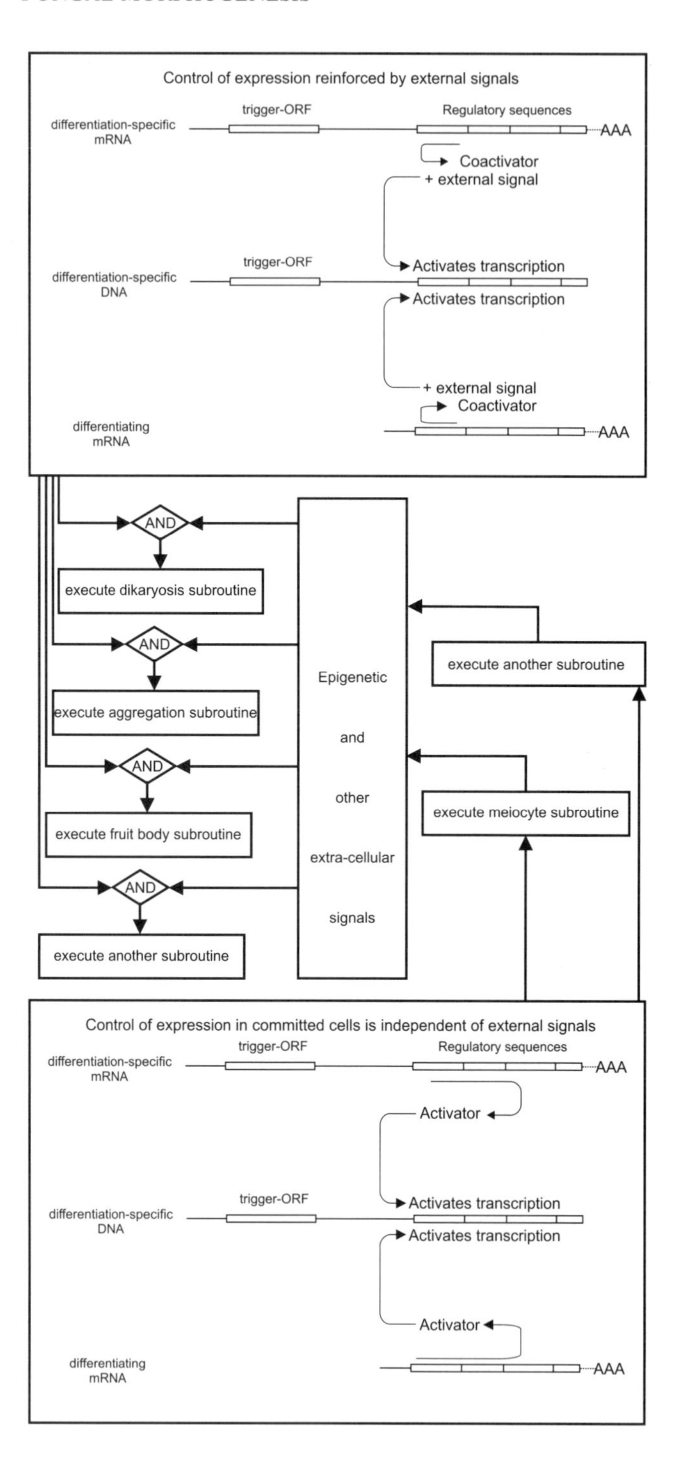
Control of expression reinforced by external signals
trigger-ORF
Regulatory sequences
differentiation-specific mRNA
AAA
Coactivator
+ external signal
trigger-ORF
Activates transcription
differentiation-specific DNA
Activates transcription
+ external signal
Coactivator
differentiating mRNA
AAA
AND
execute dikaryosis subroutine
AND
execute aggregation subroutine
AND
execute fruit body subroutine
AND
execute another subroutine
Epigenetic
and
other
extra-cellular
signals
execute another subroutine
execute meiocyte subroutine
Control of expression in committed cells is independent of external signals
trigger-ORF
Regulatory sequences
differentiation-specific mRNA
AAA
Activator
trigger-ORF
Activates transcription
differentiation-specific DNA
Activates transcription
Activator
differentiating mRNA
AAA

On this interpretation, therefore, continued progress in differentiation for most fungal cells requires continued reinforcement from their local microenvironment. This may involve production of location-specific and/or time-specific extracellular matrix molecules, or any of a range of smaller molecules which might be classed as hormones or growth factors. Smaller molecules might exert their effects by being taken up into the cell. But uptake is not necessary. Any of these molecules may also affect relationships between integrins and the existing extracellular matrix. As a result there could be direct effects on the cytoskeleton which are able to cause immediate metabolic changes in one or more cellular compartments, or directly influence gene transcription.

Connections to the extracellular matrix may also be involved in that other great enigma: the control of hyphal branching. By varying extracellular matrix/membrane or wall/membrane connections external signals may be able to specify branch initiation sites. Similarly, internal cytoskeletal architecture could also arrange specific membrane/wall connections to become branch initiation sites. Branch initiation sites specified in these ways may then become gathering sites for the molecules which create a new hyphal tip. First among these could be molecular chaperones (heat-shock proteins; Hartl *et al.*, 1994) which rearrange existing wall and membrane proteins. The chaperones could arrange for a mature wall region to become juvenile again by altering the conformation of existing polypeptides rather than by severing any covalent bonds. Local activation of resident chitin synthase (perhaps by consequential stress in the membrane, see section 2.2.4.3) and initiation of a localised resumption of wall synthesis would then establish the new hyphal tip. The branch would consequently emerge in a position precisely defined by the stimulation of generalised cytoskeleton/membrane/wall connections by a positional stimulus.

The focus of these hypothetical regulatory activities is, obviously, the hyphal wall, its surface and the immediate extracellular environment. These are features about which we are very ignorant and urgent and

Figure 7.3 Feedback fixation adopted as a general model for maintaining progress through fungal pathways of differentiation. In this flow chart the box at the top shows the type of feedback fixation process envisaged to apply to most developmental subroutines in which epigenetic reinforcement from the local microenvironment is needed to interact with coactivators in order to maintain the feedback activation loop. In the bottom panel, the alternative of direct feedback fixation independently of other signals is shown as being applicable to cell types which show developmental commitment; only meiocytes are known to be committed, but there may be other committed cell types.

extensive research on these topics is necessary. The key to fungal morphogenesis lies in understanding how that which is outside a hypha can influence that which goes on inside the hypha in a time- and place-dependent manner. We are still a long way from reaching that understanding.

References

Abuzinadah, R. A., Finlay, R. D. & Read, D. J. (1986) The role of proteins in the nitrogen nutrition of ectomycorrhizal plants. II. Utilization of protein by mycorrhizal plants of *Pinus contorta*. *New Phytologist* **103**, 495–506.

Abuzinadah, R. A. & Read, D. J. (1986a) The role of proteins in the nitrogen nutrition of ectomycorrhizal plants. I. Utilization of peptides and proteins by ectomycorrhizal fungi. *New Phytologist* **103**, 481–493.

Abuzinadah, R. A. & Read, D. J. (1986b) The role of proteins in the nitrogen nutrition of ectomycorrhizal plants. III. Protein utilization by *Betula*, *Picea* and *Pinus* mycorrhizal association with *Hebeloma crustuliniforme*. *New Phytologist* **103**, 507–514.

Adams, T. H. (1995) Asexual sporulation in higher fungi. In *The Growing Fungus* (ed. N. A. R. Gow & G. M. Gadd), pp. 367–382. Chapman & Hall: London.

Agarwal, M. K. (1993) Receptors for mammalian steroid-hormones in microbes and plants. *FEBS Letters* **322**, 207–210.

Aharonowitz, Y. (1980) Nitrogen metabolite regulation of antibiotic biosynthesis. *Annual Review of Microbiology* **34**, 209–233.

Ainsworth, A. M. & Rayner, A. D. M. (1990) Aerial mycelial transfer by *Hymenochaete corrugata* between stems of hazel and other trees. *Mycological Research* **92**, 263–266.

Ainsworth, G. C. (1961) *Ainsworth & Bisby's Dictionary of the Fungi, 5th edition.* Commonwealth Mycological Institute: Kew, Surrey, UK.

Aist, J. R. (1995) Independent nuclear motility and hyphal tip growth. *Canadian Journal of Botany* **73**, S122–S125.

Aist, J. R. & Wilson, C. L. (1968) Interpretation of nuclear division figures in vegetative hyphae of fungi. *Phytopathology* **58**, 876–877.

Akashi, T., Kanbe, T. & Tanaka, K. (1994) The role of the cytoskeleton in the polarized growth of the germ tube in *Candida albicans*. *Microbiology - UK* **140**, 271–280.

Akimova, Y. D. (1982) [Effects of light and aeration on morphogenesis of *Polyporus ciliatus* Fr. and *Pleurotus ostreatus* (Fr.) Kumm.]. *Mikologiya i Fitopatologiya* **16**, 89–95.

Alexander, M. (1994) *Biodegradation and Bioremediation.* Academic Press: San Diego.

Alfonso, C., Santamaría, F., Nuero, O. M., Prieto, A., Leal, J. A. & Reyes, F. (1995) Biochemical studies on the cell wall degradation of *Fusarium oxysporum* f. sp. *lycopersici* race 2 by its own lytic enzymes for its biocontrol. *Letters in Applied Microbiology* **20**, 105–109.

Al-Gharawi, A. & Moore, D. (1977) Factors affecting the amount and the activity of the glutamate dehydrogenases of *Coprinus cinereus*. *Biochimica et Biophysica Acta* **496**, 95–102.

Allen, J. J., Moore, D. & Elliott, T. J. (1992) Persistent meiotic arrest in basidia of *Agaricus bisporus*. *Mycological Research* **96**, 125–127.

Allen, M. F. (1996) The ecology of arbuscular mycorrhizas: a look back into the 20th century and a peek into the 21st. *Mycological Research* **100**, 769–782.

Ames, A. (1913) A consideration of structure in relation to genera of the Polyporaceae. *Annales Mycologici* **11**, 211–263.

Ander, P. (1994) The cellobiose-oxidizing enzymes CBQ and CbO as related to lignin and cellulose degradation - a review. *FEMS Microbiology Reviews* **13**, 297–312.

Anderson, J. B., Petsche, D. M. & Franklin, A. L. (1984) Nuclear DNA content of benomyl-induced segregants of diploid strains of the phytopathogenic fungus *Armillaria mellea*. *Canadian Journal of Genetics and Cytology* **27**, 47–50.

Andre, B. (1995) An overview of membrane transport proteins in *Saccharomyces cerevisiae*. *Yeast* **11**, 1575–1611.

Andrews, J. H. (1995) Fungi and the evolution of growth form. *Canadian Journal of Botany* **73**, S1206–S1212.

Angiolella, L., Facchin, M., Stringaro, A., Maras, B., Simonetti, N. & Cassone, A. (1996) Identification of a glucan-associated enolase as a main cell wall protein of *Candida albicans* and an indirect target of lipopeptide antimycotics. *Journal of Infectious Diseases* **173**, 684–690.

Armstrong, G. A. & Hearst, J. E. (1996) Carotenoids. 2. Genetics and molecular biology of carotenoid pigment biosynthesis. *FASEB Journal* **10**, 228–237.

Arst, H. N. & Cove, D. J. (1973) Nitrogen metabolite repression in *Aspergillus nidulans*. *Molecular & General Genetics* **126**, 111–141.

Ásgeirsdóttir, S. A., Van Wetter, M. A. & Wessels, J. G. H. (1995) Differential expression of genes under control of the mating-type genes in the secondary mycelium of *Schizophyllum commune*. *Microbiology - UK* **141**, 1281–1288.

Ashby, A. M. & Johnstone, K. (1993) Expression of the *E. coli* ß-glucuronidase gene in the light leaf spot pathogen *Pyrenopeziza brassicae* and its use as a reporter gene to study developmental interactions in fungi. *Mycological Research* **97**, 575–581.

Asina, S., Jain, K. & Cain, R. F. (1977) Factors influencing growth and ascocarp production in three species of *Sporormiella*. *Canadian Journal of Botany* **55**, 1915–1925.

Asselineau, D., Barnet, J. & Labarère, J. (1981) Protoplasmic incompatibility in *Podospora anserina*: possible involvement of the plasma membrane in the trigger mechanism. *Journal of General Microbiology* **125**, 139–146.

Atkins, F. C. (1950) *Mushroom Growing Today*. Faber & Faber: London.

Atkinson, G. F. (1906) The development of *Agaricus campestris*. *Botanical Gazette* **42**, 241–269.

Atkinson, G. F. (1914) The development of *Agaricus arvensis* and *A. comtulus*. *American Journal of Botany* **1**, 3–22.

Atkinson, G. F. (1916) Origin and development of the lamellae in *Coprinus*. *Botanical Gazette* **61**, 89–103.

Ayres, P. G. & Boddy, L. (1986) *Water, Fungi and Plants*. Cambridge University Press: Cambridge, UK.

Badham, E. R. (1980) The effect of light upon basidiocarp initiation in *Psilocybe cubensis*. *Mycologia* **72**, 136–142.

Badham, E. R. (1982) Tropisms in the mushroom *Psilocybe cubensis*. *Mycologia* **74**, 275–279.

Bahn, M. & Hock, B. (1973) Genetic and physiological control of perithecia formation of the ascomycete *Sordaria macrospora*. *Genetics* **74**, 14–15.

Bahn, M. & Hock, B. (1974) Morphogenese von *Sordaria macrospora*: Induktion der Perithezienbildung durch Arginin. *Berichte Deutsche Botanische Gesellschaft* **87**, 433–442.

Bajwa, R. & Read, D. J. (1985) The biology of mycorrhiza in the Ericaceae. IX. Peptides as nitrogen sources for the ericoid endophyte and for mycorrhizal and non-mycorrhizal plants. *New Phytologist* **101**, 459–467.

Baldauf, S. L. & Palmer, J. D. (1993) Animals and fungi are each others closest relatives - congruent evidence from multiple proteins. *Proceedings of the National Academy of Sciences of the USA* **90**, 11558–11562.

Banbury, G. H. (1962) Gravitropism of lower plants. In *Handbuch der Pflanzenphysiologie* (ed. W. Ruhland), pp. 344–377. Springer-Verlag: Berlin.

Banuett, F. & Herskowitz, I. (1994) Morphological transitions in the life cycle of *Ustilago maydis* and their genetic control by the a and b loci. *Experimental Mycology* **18**, 247–266.

Barbesgaard, P. & Wagner, S. (1959) Further studies on the biochemical basis of protoperithecia formation in *Neurospora crassa. Hereditas* **45**, 564–572.

Bard, J. (1990) *Morphogenesis: the Cellular and Molecular Processes of Developmental Anatomy*. Cambridge University Press: Cambridge, UK.

Bardwell, L., Cook, J. G., Inouye, C. J. & Thorner, J. (1994) Signal propagation and regulation in the mating pheromone response pathway of the yeast *Saccharomyces cerevisiae. Developmental Biology* **166**, 363–379.

Barksdale, A. W. (1969) Sexual hormones of *Achlya* and other fungi. *Science* **166**, 831–837.

Barnett, H. L. & Lilly, V. G. (1949) Production of haploid and diploid fruit bodies of *Lenzites trabea* in culture. *Proceedings of the West Virginia Academy of Science* **19**, 34–39.

Barrettbee, K. & Dixon, G. (1995) Ergosterol biosynthesis inhibition - a target for antifungal agents. *Acta Biochimica Polonica* **42**, 465–479.

Bartnicki-Garcia, S. (1973) Fundamental aspects of hyphal morphogenesis. In *Microbial Differentiation* (ed. J. M. Ashworth & J. E. Smith), pp. 245–267. Cambridge University Press: Cambridge, UK.

Bartnicki-Garcia, S. (1990) Role of vesicles in apical growth and a new mathematical model of hyphal morphogenesis. In *Tip Growth in Plant and Fungal Cells* (ed. I. B. Heath), pp. 211–232. Academic Press: San Diego.

Bartnicki-Garcia, S., Bartnicki, D. D., Gierz, G., Lopez-Franco, R. & Bracker, C. E. (1995a) Evidence that Spitzenkörper behavior determines the shape of a fungal hypha - a test of the hyphoid model. *Experimental Mycology* **19**, 153–159.

Bartnicki-Garcia, S., Bartnicki, D. D. & Gierz, G. (1995b) Determinants of fungal cell-wall morphology - the vesicle supply center. *Canadian Journal of Botany* **73**, S372–S378.

Bartnicki-Garcia, S., Hergert, F. & Gierz, G. (1989) Computer-simulation of fungal morphogenesis and the mathematical basis for hyphal (tip) growth. *Protoplasma* **153**, 46–57.

Bartnicki-Garcia, S. & Lippman, E. (1969) Fungal morphogenesis: cell wall construction in *Mucor rouxii. Science* **165**, 302–304.

Bartnicki-Garcia, S. & Lippman, E. (1972) The bursting tendency of hyphal tips of fungi: presumptive evidence for a delicate balance between wall synthesis and wall lysis in apical growth. *Journal of General Microbiology* **73**, 487–500.

Barton, R. C. & Gull, K. (1992) Isolation, characterization and genetic analysis of monosomic, aneuploid mutants of *Candida albicans. Molecular Microbiology* **6**, 171–177.

Bas, C. (1969) Morphology and subdivision of *Amanita* and a monograph of its section Lepidella. *Persoonia* **5**, 285–579.

Basith, M. & Madelin, M. F. (1968) Studies on the production of perithecial stromata by *Cordyceps militaris* in artificial culture. *Canadian Journal of Botany* **46**, 473–480.

Bastouill-Descollonges, Y. & Manachère, G. (1984) Photosporogenesis of *Coprinus congregatus*: correlations between the physiological age of lamellae and the development of their potential for renewed fruiting. *Physiologia Plantarum* **61**, 607–610.

Baxter, C. F. & Roberts, E. (1960) Gamma-aminobutyric acid and cerebral metabolism. In *The Neurochemistry of Nucleotides and Amino Acids* (ed. R. O. Brady & D. B. Tower), pp. 127–145. John Wiley & Son: New York.

Beck-Mannagetta, G. (1923) Versuch einer systematischen Gliederung der Gattung *Boletus* L. *Zeitschrift für Pilzkunde* **2**, 141–149.

Beever, R. E. & Dempsey, G. P. (1978) Function of rodlets on the surface of fungal spores. *Nature* **272**, 608–610.

Beguin, P. & Aubert, J. P. (1994) The biological degradation of cellulose. *FEMS Microbiology Reviews* **13**, 25–58.

Belcour, L. & Bernet. J. (1969) Sur la mise en évidence d'un gene dout la mutation supprime spécifiquement certain manifestations d'incompatibilité chez le *Podospora anserina. Comptes rendu hebdomadaire des Séances de l'Académie des Sciences, Séries D* **269**, 712–714.

Bell, G. (1993) The sexual nature of the eukaryote genome. *Journal of Heredity* **84**, 351–359.

Bell, G. & Mooers, A. O. (1997) Size and complexity among multicellular organisms. *Biological Journal of the Linnean Society* **60**, 345–363.

Bell-Pederson, D., Dunlap, J. C. & Loros, J. J. (1992) The *Neurospora* circadian clock-controlled gene, *ccg-2*, is allelic to *eas* and encodes a fungal hydrophobin required for formation of the conidial rodlet layer. *Genes & Development* **6**, 2382–2394.

Bell-Pedersen, D., Garceau, N. & Loros, J. J. (1996) Circadian rhythms in fungi. *Journal of Genetics* **75**, 387–401.

Belozerskaya, T. A., Kritskii, M. S. & Sokolovskii, V. Y. (1995) Membrane electrogenesis and the problems of cell differentiation in mycelial fungi. *Microbiology* **64**, 243–249.

Benhamou, N. & Ouellete, G. B. (1986) Ultrastructural characterisation of an extracellular sheath on cells of *Ascocalyx abientina*, the scleroderris canker agent of conifers. *Canadian Journal of Botany* **65**, 154–167.

Bennett, J. W. (1995) From molecular genetics and secondary metabolism to molecular metabolites and secondary genetics. *Canadian Journal of Botany* **73**, S917–S924.

Berbee, M. L. & Taylor, J. W. (1992) Detecting morphological convergence in true fungi, using 18S ribosomal-RNA gene sequence data. *BioSystems* **28**, 117–125.

Berbee, M. L. & Taylor, J. W. (1993) Dating the evolutionary radiations of the true fungi. *Canadian Journal of Botany* **71**, 1114–1127.

Berbee, M. L. & Taylor, J. W. (1995) From 18S ribosomal sequence data to evolution of morphology among the fungi. *Canadian Journal of Botany* **73**, S677–S683.

Bergman, K., Eslava, A. P. & Cerdá-Olmedo, E. (1973) Mutants of *Phycomyces* with abnormal phototropism. *Molecular & General Genetics* **123**, 1–16.

Berlin, V. & Yanofsky, C. (1985a) Isolation and characterization of genes differentially expressed during conidiation of *Neurospora crassa. Molecular and Cellular Biology* **5**, 849–855.

Berlin, V. & Yanofsky, C. (1985b) Protein changes during the asexual cycle of *Neurospora crassa. Molecular and Cellular Biology* **5**, 839–848.

Bernet, J. (1971) Sur un cas de suppression de l'incompatibilité cellulaire chez le *Podospora anserina. Comptes rendu hebdomadaire des Séances de l'Académie des Sciences, Séries D* **273**, 1120–1122.

Bernet, J., Bégueret, J. & Labarère, J. (1973) Incompatibility in the fungus *Podospora anserina*: are the mutations abolishing the incompatibility reaction ribosomal mutations? *Molecular & General Genetics* **124**, 35–50.

Bernstein, H., Byerly, H. C., Hopf, F. A. & Michod, R. E. (1985) Genetic damage, mutation and the evolution of sex. *Science* **229**, 1277–1281.

Berry, D. R. (1983) Ascospore formation in yeast. In *Fungal Differentiation: a Contemporary Synthesis* (ed. J. E. Smith), pp. 147–173. Marcel Dekker: New York.

Bestor, T. H., Chandler, V. L. & Feinberg, A. P. (1994) Epigenetic effects in eukaryotic gene expression. *Developmental Genetics* **15**, 458–462.

Bevan, E. A. & Kemp, R. F. O. (1958) Stipe regeneration and fruit body production in *Collybia velutipes* (Curt.) Fr. *Nature* **181**, 1145–1146.

Bianciotto, V., Bandi, C., Minerdi, D., Sironi, M., Tichy, H. V. & Bonfante, P. (1996) An obligately endosymbiotic mycorrhizal fungus itself harbours obligately intracellular bacteria. *Applied and Environmental Microbiology* **62**, 3005–3010.

Bidochka, M. J., St Leger, R. J., Joshi, L. & Roberts, D. W. (1995a) The rodlet layer from aerial and submerged conidia of the entomopathogenic fungus *Beauveria bassiana* contains hydrophobin. *Mycological Research* **99**, 403–406.

Bidochka, M. J., St Leger, R. J., Joshi, L. & Roberts, D. W. (1995b) An inner cell wall protein (*cwp1*) from conidia of the entomopathogenic fungus *Beauveria bassiana. Microbiology - UK* **141**, 1075–1080.

Binks, P. R., Robson, G. D., Goosey, M. W. & Trinci, A. P. J. (1991) Relationships between phosphatidylcholine content, chitin synthesis, growth, and morphology of *Aspergillus nidulans choC*. *FEMS Microbiology Letters* **83**, 159–164.

Bircher, U. & Hohl, H. R. (1997) Surface glycoproteins associated with appressorium formation and adhesion in *Phytophthora palmivora*. *Mycological Research* **101**, 395–402.

Bistis, G. N. (1956) Sexuality in *Ascobolus stercorarius*. I. Morphology of the ascogonium; plasmogamy; evidence for a sexual hormonal mechanism. *American Journal of Botany* **43**, 389–394.

Bistis, G. N. (1957) Sexuality in *Ascobolus stercorarius*. II. Preliminary experiments on various aspects of the sexual process. *American Journal of Botany* **44**, 436–443.

Bistis, G. N. (1983) Evidence for diffusible, mating-type specific trichogyne attractants in *Neurospora crassa*. *Experimental Mycology* **7**, 292–295.

Bistis, G. N. & Georgopoulos, S. G. (1979) Some aspects of sexual reproduction in *Nectria haematococca* var. *cucurbitae*. *Mycologia* **71**, 127–143.

Bistis, G. N. & Raper, J. R. (1963) Heterothallism and sexuality in *Ascobolus stercorarius*. *American Journal of Botany* **50**, 880–891.

Blattn, C., Kasala, B., Pilát, A., Sentilliova-Svobodová, J. & Semerdzieva, M. (1971) Proliferation of *Armillaria mellea* (Vahlin Fl. Dan. ex Fr.) P. Karst. probably caused by a virus. *Ceská Mykologie* **25**, 66–74.

Blattn, C., Kralik, O., Veselsky, J., Kasala, B. & Herzova, H. (1973) Particles resembling virions accompanying the proliferation of agaric mushrooms. *Ceská Mykologie* **27**, 1–5.

Blumenthal, H. J. (1965) Glycolysis. In *The Fungi,* **vol.I** (ed. G. C. Ainsworth & A. S. Sussman), pp. 229–268. Academic Press: New York.

Blumenthal, H. J. (1968) Glucose catabolism in fungi. *Wallernstein Laboratory Communications* **31**, 171–191.

Blumenthal, H. J. (1976) Reserve carbohydrates in fungi. In *The Filamentous Fungi,* **vol II**, Biosynthesis and Metabolism (ed. J. E. Smith & D. R. Berry), pp. 292–307. Edward Arnold: London.

Bobichon, H., Gache, D. & Bouchet, P. (1994) Ultrarapid cryofixation of *Candida albicans*: evidence for a fibrillar reticulated external layer and mannan channels within the cell wall. *Cryo-Letters* **15**, 161–172.

Boddy, L. (1993) Saprotrophic cord-forming fungi: warfare strategies and other ecological aspects. *Mycological Research* **97**, 641–655.

Boddy, L. & Rayner, A. D. M. (1983a) Ecological roles of basidiomycetes forming decay columns in attached oak branches. *New Phytologist* **93**, 77–88.

Boddy, L. & Rayner, A. D. M. (1983b) Mycelial interactions, morphogenesis and ecology of *Phlebia radiata* and *P. rufa* from oak. *Transactions of the British Mycological Society* **80**, 437–448.

Bolker, M., Genin, S., Lehmler, C. & Kahmann, R. (1995) Genetic-regulation of mating and dimorphism in *Ustilago maydis*. *Canadian Journal of Botany* **73**, S320–S325.

Bölker, M. & Kahmann, R. (1993) Sexual pheromones and mating responses in fungi. *Plant Cell* **5**, 1461–1469.

Bond, T. E. T. (1952) A further note on size and form in agarics. *Transactions of the British Mycological Society* **35**, 190–194.

Bondartsev, M. A. (1963) On the anatomical criterion in the taxonomy of Aphyllophorales. *Botanichnyi Zhurnal SSSR* **48**, 362–372.

Bonnen, A. M., Anton, L. H. & Orth, A. B. (1994) Lignin-degrading enzymes of the commercial button mushroom, *Agaricus bisporus*. *Applied and Environmental Microbiology* **60**, 960–965.

Bonner, J. T. (1977) Some aspects of chemotaxis using the cellular slime moulds as an example. *Mycologia* **69**, 443–459.

Bonner, J. T., Hoffman, A. A., Morioka, W. T. & Chiquoine, A. D. (1957) The distribution of polysaccharides and basophilic substances during the development of the mushroom *Coprinus*. *Biological Bulletin, Marine Biological Laboratory, Woods Hole, Massachusetts* **122**, 1–6.

Bonner, J. T., Kane, K. K. & Levey, R. H. (1956) Studies on the mechanics of growth in the common mushroom, *Agaricus campestris*. *Mycologia* **48**, 13–19.

Bonneu, M. & Labarère, J. (1983) Electrophoretic analysis of plasma membrane proteins of wild-type and differentiation-deficient mutant strains of *Podospora anserina. Current Microbiology* **9**, 133–140.

Borriss, H. (1934) Beiträge zur Wachstums- und Entwicklungsphysiologie der Fruchtkörper von *Coprinus lagopus. Planta* **22**, 28–69.

Borrow, A., Jefferys, E. G., Kessell, R. H. J., Lloyd, E. C., Lloyd, P. B. & Nixon, I. S. (1961) The metabolism of *Gibberella fujikuroi* in stirred culture. *Canadian Journal of Microbiology* **7**, 227–276.

Bosch, A., Maronna, R. A. & Yantorno, O. M. (1995) A simple descriptive model of filamentous fungi spore germination. *Process Biochemistry* **30**, 599–606.

Botton, B. & Dexheimer, J. (1977) The ultrastructure of the rhizomorphs of *Sphaerostilbe repens* B. & B. *Zeitschrift für Pflanzenphysiologie* **85**, 429–443.

Boucherie, H., Bégueret, J. & Bernet, J. (1976) The molecular mechanism of protoplasmic incompatibility and its relationship to the formation of protoperithecia in *Podospora anserina. Journal of General Microbiology* **92**, 59–66.

Bougher, N. L., Tommerup, I. C. & Malajczuk, N. (1993) Broad variation in developmental and mature basidiome morphology of the ectomycorrhizal fungus *Hydnangium sublamellatum* sp. nov. bridges morphologically-based generic concepts of *Hydnangium, Podohydnangium* and *Laccaria. Mycological Research* **97**, 613–619.

Bourne, A. N., Chiu, S. W. & Moore, D. (1996) Experimental approaches to the study of pattern formation in *Coprinus cinereus.* In *Patterns in Fungal Development* (ed. S. W. Chiu & D. Moore), pp. 126–155. Cambridge University Press: Cambridge, UK.

Bowman, B. H., Taylor, J. W., Brownlee, A. G., Lee, J., Lu, S. D. & White, T. J. (1992) Molecular evolution of the fungi - relationship of the basidiomycetes, ascomycetes, and chytridiomycetes. *Molecular Biology and Evolution* **9**, 285–296.

Bowman, B. J. & Bowman, E. J. (1996) Mitochondrial and vacuolar ATPases. In *The Mycota,* **vol. III**, Biochemistry and Molecular Biology (ed. R. Brambl & G. A. Marzluf), pp. 57–83. Springer-Verlag: Berlin.

Boylan, M. T., Mirabito, P. M., Willett, C. E., Zimmerman, C. R. & Timberlake, W. E. (1987) Isolation and physical characterization of three essential conidiation genes from *Aspergillus nidulans. Molecular and Cellular Biology* **7**, 3113–3118.

Braaksma, A., Schaap, G. J., de Vrije, T., Jongen, W. M. & Woltering, E. J. (1994) Ageing of the mushroom (*Agaricus bisporus*) under post-harvest conditions. *Postharvest Biology and Technology* **4**, 99–110.

Braaksma, A., Van der Meer, P. & Schaap, D. J. (1996) Polyphosphate accumulation in the senescing mushroom *Agaricus bisporus. Postharvest Biology and Technology* **8**, 121–127.

Brefeld, O. (1877) *Botanische Untersuchungen über Schimmelpilze,* **vol. III**. Felix: Leipzig.

Bremner, J. M. (1967) Nitrogenous compounds. In *Soil Biochemistry* (ed. A. D. McLaren & G. H. Peterson), pp. 19–66. Marcel Dekker: New York.

Bret, J. P. (1977) Respective role of cap and mycelium on stem elongation of *Coprinus congregatus. Transactions of the British Mycological Society* **68**, 262–269.

Brickell, P. M. & Tickle, C. (1989) Morphogens in chick limb development. *BioEssays* **11**, 145–149.

Broda, P., Birch, P. R. J., Brooks, P. R. & Sims, P. F. G. (1996) Lignocellulose degradation by *Phanerochaete chrysosporium* - gene families and gene expression for a complex process. *Molecular Microbiology* **19**, 923–932.

Brown, A. D. (1978) Compatible solutes and extreme water stress in eukaryotic micro-organisms. *Journal of Microbiological Physiology* **17**, 181–242.

Brown, A. D. & Simpson, J. R. (1972) Water relations of sugar-tolerant yeasts: the role of intracellular polyols. *Journal of General Microbiology* **72**, 589–591.

Brown, D. W., Yu, J. H., Kelkar, H. S., Fernandes, M., Nesbitt, T. C., Keller, N. P., Adams, T. H. & Leonard, T. J. (1996) Twenty-five coregulated transcripts define a sterigmatocystin gene cluster in *Aspergillus nidulans. Proceedings of the National Academy of Sciences, USA* **93**, 1418–1422.

Broxholme, S. J., Read, N. D. & Bond, D. J. (1991) Developmental regulation of proteins during fruit-body morphogenesis in *Sordaria brevicollis. Mycological Research* **95**, 958–969.

Brunt, I. C. & Moore, D. (1989) Intracellular glycogen stimulates fruiting in *Coprinus cinereus*. *Mycological Research* **93**, 543–546.

Buades, C. & Moya, A. (1996) Phylogenetic analysis of the isopenicillin-N-synthetase horizontal gene transfer. *Journal of Molecular Evolution* **42**, 537–542.

Buchala, A. J. & Leisola, M. (1987) Structure of β-D-glucan secreted by *P. chrysosporium* in continuous culture. *Carbohydrate Research* **165**, 135–139.

Bull, A. T. & Trinci, A. P. J. (1977) The physiology and metabolic control of fungal growth. *Advances in Microbial Physiology* **15**, 1–84.

Buller, A. H. R. (1905) The reactions of the fruit bodies of *Lentinus lepideus*, Fr., to external stimuli. *Annals of Botany* **19**, 427–438.

Buller, A. H. R. (1909) *Researches on Fungi,* **vol. 1**. Longman, Green & Co.: London.

Buller, A. H. R. (1922) *Researches on Fungi,* **vol. 2**. Longman Green & Co: London.

Buller, A. H. R. (1924) *Researches on Fungi,* **vol. 3**. Longman Green & Co.: London.

Buller, A. H. R. (1930) The biological significance of conjugate nuclei in *Coprinus lagopus* and other hymenomycetes. *Nature* **126**, 686–689.

Buller, A. H. R. (1931) *Researches on Fungi,* **vol. 4**. Longmans, Green & Co.: London.

Buller, A. H. R. (1934) *Researches on Fungi,* **vol. 6**. Longman Green & Co.: London.

Buller, A. H. R. (1941) The diploid cell and the diploidization process in plants and animals, with special reference to the higher fungi, part 1. *The Botanical Review* **7**, 335–387.

Bu'Lock, J. D. (1961) Intermediary metabolism and antibiotic synthesis. *Advances in Applied Microbiology* **3**, 293–342.

Bu'Lock, J. D. (1965) Aspects of secondary metabolism in fungi. In *Biogenesis of Antibiotic Substances* (ed. Z. Vanek & Z. Hostalek), pp. 61–71. Academic Press: New York.

Bu'Lock, J. D. (1967) *Essays in Biosynthesis and Microbial Development*. John Wiley & Sons: New York.

Bu'Lock, J. D. (1975) Cascade expression of the mating type locus in Mucorales. In *Proceedings, 1974 Symposium on the Genetics of Industrial Micro-organisms*. Academic Press: London.

Bu'Lock, J. D. (1976) Hormones in fungi. In *The Filamentous Fungi,* **vol. 2**, Biosynthesis and Metabolism (ed. J. E. Smith & D. R. Berry), pp. 345–368. Edward Arnold: London.

Bu'Lock, J. D., Jones, B. E., Quarrie, S. & Winskill, N. (1973) The biochemical basis of sexuality in Mucorales. *Die Naturwissenschaften* **60**, 550–551.

Bumpus, J. A., Tien, M., Wright, D. & Aust, S. D. (1985) Oxidation of persistent environmental pollutants by a white rot fungus. *Science* **228**, 1434–1436.

Burnett, J. H. & Trinci, A. P. J. (1979) *Fungal Walls and Hyphal Growth*. Cambridge University Press: Cambridge, UK.

Burton, K. S., Partis, M. D., Wood, D. A. & Thurston, C. F. (1997) Accumulation of serine proteinase in senescent sporophores of the cultivated mushroom, *Agaricus bisporus*. *Mycological Research* **101**, 146–152.

Bush, D. A. (1974) Autolysis of *Coprinus comatus* sporophores. *Experientia* **30**, 984–985.

Buston, H. W., Jarbar, A. & Etheridge, D. E. (1953) The influence of hexose phosphates, calcium and jute extract on the formation of perithecia by *Chaetomium globosum*. *Journal of General Microbiology* **8**, 302–306.

Buston, H. W. & King, E. J. (1951) Further observations on the sporulation of *Chaetomium globosum*. *Journal of General Microbiology* **5**, 766–771.

Buston, H. W., Moss, M. O. & Tyrrell, D. (1966) The influence of carbon dioxide on growth and sporulation of *Chaetomium globosum*. *Transactions of the British Mycological Society* **49**, 387–396.

Buston, H. W. & Rickard, B. (1956) The effect of a physical barrier on sporulation of *Chaetomium globosum*. *Journal of General Microbiology* **15**, 194–197.

Buswell, J. A., Ander, P. & Eriksson, K. E. (1982) Ligninolytic actvity and levels of ammonia assimilating enzymes in *Sporotrichum pulverulentum*. *Archives of Microbiology* **133**, 165–171.

Butler, G. M. (1957) The development and behaviour of mycelial strands in *Merulius lacrymans* (Wulf.) Fr. I. Strand development during growth from a food-base through a non-nutrient medium. *Annals of Botany* **21**, 523–537.

Butler, G. M. (1958) The development and behaviour of mycelial strands in *Merulius lacrymans* (Wulf.) Fr. II. Hyphal behaviour during strand formation. *Annals of Botany* **22**, 219–236.

Butler, G. M. (1966) Vegetative structures. In *The Fungi - An Advanced Treatise*, **vol. II** (ed. G. C. Ainsworth & A. S. Sussman), pp. 83–112. Academic Press: New York.

Butler, G. M. (1988) Pattern of pore morphogenesis in the resupinate basidiome of *Phellinus contiguus*. *Transactions of the British Mycological Society* **91**, 677–686.

Butler, G. M. (1992a) Location of hyphal differentiation in the agar pore field of the basidiome of *Phellinus contiguus*. *Mycological Research* **96**, 313–317.

Butler, G. M. (1992b) Capacity for differentiation of setae and other hyphal types of the basidiome in explants from cultures of the polypore *Phellinus contiguus*. *Mycological Research* **96**, 949–955.

Butler, G. M. (1995) Induction of precocious fruiting by a diffusible sex factor in the polypore *Phellinus contiguus*. *Mycological Research* **99**, 325–329.

Butler, G. M. & Wood, A. E. (1988) Effects of environmental factors on basidiome development in the resupinate polypore *Phellinus contiguus*. *Transactions of the British Mycological Society* **90**, 75–83.

Butnick, N. Z., Yager, L. N., Hermann, T. E., Kurtz, M. B. & Champe, S. P. (1984a) Mutants of *Aspergillus nidulans* blocked at an early stage of sporulation secrete an unusual metabolite. *Journal of Bacteriology* **160**, 533–540.

Butnick, N. Z., Yager, L. N., Kurtz, M. B. & Champe, S. P. (1984b) Genetic analysis of mutants of *Aspergillus nidulans* blocked at an early stage of sporulation. *Journal of Bacteriology* **160**, 541–545.

Butt, T. M., Hoch, H. C., Staples, R. C. & St Leger, R. J. (1989) Use of fluorochromes in the study of fungal cytology and differentiation. *Experimental Mycology* **13**, 303–320.

Büttner, P., Koch, F., Voigt, K., Quidde, T., Risch, S., Blaich, R., Brückner, B. & Tudzynski, P. (1994) Variations in ploidy among isolates of *Botrytis cinerea*: implications for genetic and molecular analyses. *Current Genetics* **25**, 445–450.

Byrde, R. J. W. (1982) Fungal pectinases, from ribosome to plant cell wall. *Transactions of the British Mycological Society* **79**, 1–14.

Cairney, J. W. G. (1992) Translocation of solutes in ectomycorrhizal and saprotrophic rhizomorphs. *Mycological Research* **96**, 135–141.

Cairney, J. W. G. & Clipson, N. J. W. (1991) Internal structure of rhizomorphs of *Trechispora vaga*. *Mycological Research* **95**, 764–767.

Cairney, J. W. G., Jennings, D. H. & Veltkamp, C. J. (1989) A scanning electron microscope study of the internal structure of mature linear mycelial organs of four basidiomycete species. *Canadian Journal of Botany* **67**, 2266–2271.

Caldwell, G. A., Naider, F. & Becker, J. M. (1995) Fungal lipopeptide mating pheromones: a model system for the study of protein prenylation. *Microbiological Reviews* **59**, 406–422.

Caldwell, I. Y. & Trinci, A. P. J. (1973) The growth unit of the mould *Geotrichum candidum*. *Archiv für Mikrobiologie* **88**, 1–10.

Calonje, M., Mendoza, C. G., Cabo, A. P. & Novaes-Ledieu, M. (1995) Some significant differences in wall chemistry among four commercial *Agaricus bisporus* strains. *Current Microbiology* **30**, 111–115.

Campbell, A. (1957) Synchronisation of cell division. *Bacteriological Reviews* **21**, 263–272.

Campbell, I. M. (1984) Secondary metabolism and microbial physiology. *Advances in Microbial Physiology* **25**, 1–60.

Campo-Aasen, I. & Albornoz, M. C. (1994) Alkaline phosphatase at the cell wall of the yeast phase of *Paracoccidiodes brasiliensis*. *Mycopathologia* **127**, 69–71.

Cannon, R. D., Timberlake, W. E., Gow, N. A. R., Bailey, D., Brown, A., Gooday, G. W., Hube, B., Monod, M., Nombela, C., Navarro, F., Perez, R., Sanchez, M. & Pla, J. (1994) Molecular biological and biochemical aspects of fungal dimorphism. *Journal of Medical and Veterinary Mycology* **32**, 53–64.

Cantino, E. C. (1966) Morphogenesis in aquatic fungi. In *The Fungi: An Advanced Treatise*, **vol. 2** (ed. G. C. Ainsworth & A. S. Sussman), pp. 283–337. Academic Press: London.

Cantino, E. C. & Lovett, J. S. (1964) Non-filamentous aquatic fungi: model systems for biochemical studies of morphological differentiation. *Advances in Morphogenesis* **3**, 33–93.

Cappellaro, C., Baldermann, C., Rachel, R. & Tanner, W. (1994) Mating type-specific cell-cell recognition of *Saccharomyces cerevisiae*: cell wall attachment and active sites of a- and α-agglutinin. *EMBO Journal* **13**, 4737–4744.

Capy, P., Langin, T., Bigot, Y., Brunet, F., Daboussi, M. J., Periquet, G., David, J. R. & Hartl, D. L. (1994) Horizontal transmission versus ancient origin - mariner in the witness box. *Genetica* **93**, 161–170.

Carlile, M. J. (1970) The photoresponses of fungi. In *Photobiology of Microorganisms* (ed. P. Halldal), pp. 309–344. Wiley Interscience: London.

Carlile, M. J. (1987) Genetic exchange and gene flow: their promotion and prevention. In *Evolutionary Biology of the Fungi* (ed. A. D. M. Rayner, C. M. Brasier & D. Moore), pp. 203–213. Cambridge University Press: Cambridge, UK.

Carlile, M. J. (1995) The success of the hypha and mycelium. In *The Growing Fungus* (ed. N. A. R. Gow & G. M. Gadd), pp. 3-19. Chapman & Hall: London.

Carlile, M. J. & Watkinson, S. C. (1994) *The Fungi.* Academic Press: London.

Carvalho, D. B., Smith, M. L. & Anderson, J. B. (1995) Genetic exchange between diploid and haploid mycelia of *Armillaria gallica. Mycological Research* **99**, 641–647.

Casselton, L. A. & Kües, U. (1994) Mating-type genes in homobasidiomycetes. In *The Mycota,* **vol. I**, Growth, Differentiation and Sexuality (ed. J. G. H. Wessels & F. Meinhardt), pp. 307–321. Springer-Verlag: Berlin.

Casselton, P. J. (1976) Anaplerotic pathways. In *The Filamentous Fungi, vol II, Biosythesis and metabolism* (ed. J. E. Smith & D. R. Berry), pp. 121–136. Edward Arnold: London.

Casselton, P. J., Fawole, M. O. & Casselton, L. A. (1969) Isocitrate lyase in *Coprinus lagopus* (*sensu* Buller) *Canadian Journal of Microbiology* **15**, 637–640.

Castle, E. S. (1942) Spiral growth and reversal of spiralling in *Phycomyces*, and their bearing on primary wall structure. *American Journal of Botany* **29**, 664–672.

Cavalier-Smith, T. (1981) Eukaryote Kingdoms: seven or nine? *BioSystems* **14**, 461–481.

Cavalier-Smith, T. (1987) The origin of fungi and pseudofungi. In *Evolutionary Biology of the Fungi* (ed. A. D. M. Rayner, C. M. Brasier & D. Moore), pp. 339–353. Cambridge University Press: Cambridge, UK.

Cavalier-Smith, T. (1992) Origins of secondary metabolism. *CIBA Foundation Symposia* **171**, 64–87.

Cerdá-Olmedo, E., Fernández-Martín, R. & Avalos, J. (1994) Genetics and gibberellin production in *Gibberella fujikuroi. Antonie Van Leeuwenhoek International Journal of General and Molecular Microbiology* **65**, 217–225.

Chadefaud, M. (1982a) Les principaux types d'ascocarpes: leur organisation et leur évolution. *Cryptogamie Mycologie* **3**, 1–9.

Chadefaud, M. (1982b) Les principaux types d'ascocarpes: leur organisation et leur évolution. Deuxième partie: les discocarpes. *Cryptogamie Mycologie* **3**, 103–144.

Chadefaud, M. (1982c) Les principaux types d'ascocarpes: leur organisation et leur évolution. Troisième partie: les pyrénocarpes. *Cryptogamie Mycologie* **3**, 199–235.

Chambers, S. M., Hardham, A. R. & Scott, E. S. (1995) *In planta* immunolabelling of three types of peripheral vesicles in cells of *Phytophthora cinnamomi* infecting chestnut roots. *Mycological Research* **99**, 1281–1288.

Champe, S. P. & El-Zyat, A. A. E. (1989) Isolation of a sexual sporulation hormone from *Aspergillus nidulans. Journal of Bacteriology* **171**, 3982–3988.

Champe, S. P., Kurtz, M. B., Yager, L. N., Butnick, N. J. & Axelrod, D. E. (1981) Spore formation in *Aspergillus nidulans*: competence and other developmental processes. In *The Fungal Spore: Morphogenetic Controls* (ed. G. Turian & H. R. Hohl), pp. 255–276. Academic Press: London.

Champe, S. P. & Simon, L. D. (1992) Cellular diferentiation and tissue formation in the fungus *Aspergillus nidulans.* In *Morphogenesis: an Analysis of the Development of Biological Form* (ed. E. F. Rossomando & S. Alexander), pp. 63–91. Marcel Dekker: New York.

Chang, S. & Staben, C. (1994) Directed replacement of mtA by mt a-1 effects a mating type switch in *Neurospora crassa. Genetics* **138**, 75–81.

Chang, S. T. & Hayes, W. A. (1978) *Biology and Cultivation of Edible Mushrooms.* Academic Press: New York & London.

Chang, Y. (1967) The fungi of wheat straw compost. II. Biochemical and physiological studies. *Transactions of the British Mycological Society* **50**, 667–677.

Chapman, E. S. & Fergus, C. L. (1973) An investigation of the effects of light on basidiocarp formation of *Coprinus domesticus. Mycopathologia et Mycologia Applicata* **51**, 315–326.

Chaubal, R., Wilmot, V. A. & Wynn, W. K. (1991) Visualisation, adhesiveness, and cytochemistry of the extracellular matrix produced by urediniospore germ tube of *Puccinia sorghi*. *Canadian Journal of Botany* **69**, 2044–2054.

Chebotarev, L. N. & Zemlyanukhin, A. A. (1973) Effect of visible light and ultraviolet rays on the activity of oxidation enzymes in molds. *Mikrobiologia* **42**, 196.

Chet, I. & Henis, Y. (1968) The control mechanism of sclerotial formation in *Sclerotium rolfsii* Sacc. *Journal of General Microbiology* **54**, 231–236.

Chet, I. & Henis, Y. (1975) Sclerotial morphogenesis in fungi. *Annual Review of Phytopathology* **13**, 169–192.

Chet, I., Henis, Y. & Kislev, N. (1969) Ultrastructure of sclerotia and hyphae of *Sclerotium rolfsii* Sacc. *Journal of General Microbiology* **57**, 143–147.

Chet, I., Henis, Y. & Mitchell, R. (1966) The morphogenetic effect of sulphur-containing amino acids, glutathione and iodoacetic acid on *Sclerotium rolfsii* Sacc. *Journal of General Microbiology* **45**, 541–546.

Chiu, S. W. (1993) Evidence for a haploid life-cycle in *Volvariella volvacea* by microspectrophotometric measurements and observation of nuclear behaviour. *Mycological Research* **97**, 1481–1485.

Chiu, S. W. (1996) Nuclear changes during fungal development. In *Patterns in Fungal Development* (ed. S. W. Chiu & D. Moore), pp. 105–125. Cambridge University Press: Cambridge, UK.

Chiu, S. W. & Moore, D. (1988a) Evidence for developmental commitment in the differentiating fruit body of *Coprinus cinereus*. *Transactions of the British Mycological Society* **90**, 247–253.

Chiu, S. W. & Moore, D. (1988b) Ammonium ions and glutamine inhibit sporulation of *Coprinus cinereus* basidia assayed *in vitro*. *Cell Biology International Reports* **12**, 519–526.

Chiu, S. W. & Moore, D. (1990a) Development of the basidiome of *Volvariella bombycina*. *Mycological Research* **94**, 327–337.

Chiu, S. W. & Moore, D. (1990b) A mechanism for gill pattern formation in *Coprinus cinereus*. *Mycological Research* **94**, 320–326.

Chiu, S. W. & Moore, D. (1990c) Sporulation in *Coprinus cinereus*: use of an *in vitro* assay to establish the major landmarks in differentiation. *Mycological Research* **94**, 249–253.

Chiu, S. W. & Moore, D. (1993) Cell form, function and lineage in the hymenia of *Coprinus cinereus* and *Volvariella bombycina*. *Mycological Research* **97**, 221–226.

Chiu, S. W. & Moore, D. (1998) Sexual development of higher fungi. In *Experimental Fungal Biology* (ed. R. Oliver & M. Schweizer), in press. Cambridge University Press: Cambridge, UK.

Chiu, S. W., Moore, D. & Chang, S. T. (1989) Basidiome polymorphism in *Volvariella bombycina*. *Mycological Research* **92**, 69–77.

Chiu, S. W. & To, S. W. (1993) Endogenous glycogen is not a trigger for fruiting in *Pleurotus sajor-caju*. *Mycological Research* **97**, 363–366.

Chow, C. H. (1934) Contribution a l'etude du developpement des coprins. *Le Botaniste* **26**, 89–233.

Clancy, M. J., Smith, L. M. & Magee, P. T. (1982) Developmental regulation of a sporulation-specific enzyme activity in *Saccharomyces cerevisiae*. *Molecular and Cellular Biology* **2**, 171–178.

Claverie-Martin, F., Diaz-Torres, M. R. & Geoghegan, M. J. (1986) Chemical composition and electron microscopy of the rodlet layer of *Aspergillus nidulans* conidia. *Current Microbiology* **14**, 221–225.

Clay, R. P., Enkerli, J. & Fuller, M. S. (1994) Induction and formation of *Cochliobolus sativus* appressoria. *Protoplasma* **178**, 34–47.

Claydon, N. (1985) Secondary metabolic products of selected agarics. In *Developmental Biology of Higher Fungi* (ed. D. Moore, L. A. Casselton, D. A. Wood & J. C. Frankland), pp. 561–579. Cambridge University Press: Cambridge, UK.

Clement, J. A., Porter, R., Butt, T. M. & Beckett, A. (1994) The role of hydrophobicity in attachment of urediniospores and sporelings of *Uromyces viciae-fabae*. *Mycological Research* **98**, 1217–1228.

Clutterbuck, A. J. (1969) A mutational analysis of conidial development in *Aspergillus nidulans*. *Genetics* **63**, 317–327.

Clutterbuck, A. J. (1972) Absence of laccase from yellow-spored mutants of *Aspergillus nidulans*. *Journal of General Microbiology* **70**, 423–435.

Clutterbuck, A. J. (1977) The genetics of conidiation in *Aspergillus nidulans*. In *The Physiology and Genetics of Aspergillus* (ed. J. E. Smith & J. A. Pateman), pp. 305–317. Academic Press: London.

Clutterbuck, A. J. (1978) Genetics of vegetative growth and asexual reproduction. In *The Filamentous Fungi,* **vol III**, Developmental Mycology (ed. J. E. Smith & D. R. Berry), pp. 240–256. Edward Arnold: London.

Clutterbuck, A. J. (1994) Mutants of *Aspergillus nidulans* deficient in nuclear migration during hyphal growth and conidiation. *Microbiology* **140**, 1169–1174.

Clutterbuck, A. J. (1995a) Genetics of fungi. In *The Growing Fungus* (ed. N. A. R. Gow & G. M. Gadd), pp. 239–253. Chapman & Hall: London.

Clutterbuck, A. J. (1995b) Molecular biology. In *The Growing Fungus* (ed. N. A. R. Gow & G. M. Gadd), pp. 255–274. Chapman & Hall: London.

Cochrane, V. W. (1958) *Physiology of the Fungi*. John Wiley & Sons: New York.

Cochrane, V. W. (1976) Glycolysis. In *The Filamentous Fungi, vol II, Biosynthesis and Metabolism* (ed. J. E. Smith & D. R. Berry), pp. 6591. Edward Arnold: London.

Cohen, B. L. (1972) Ammonium repression of extracellular protease in *Aspergillus nidulans*. *Journal of General Microbiology* **71**, 293–299.

Cohen, R. J. (1974) Cyclic AMP Levels in *Phycomyces* during a response to light. *Nature* **251**, 144.

Cohen, S. S. & Barner, H. D. (1954) Studies on unbalanced growth in *Escherichia coli*. *Proceedings of the National Academy of Sciences of the USA* **40**, 885–893.

Cole, G. L. (1986) Models of cell differentiation in conidial fungi. *Microbiological Reviews* **50**, 95–132.

Cole, G. T., Pishko, E. J. & Seshan, K. R. (1995) Possible roles of wall hydrolases in the morphogenesis of *Coccidioides immitis*. *Canadian Journal of Botany* **73**, S1132–S1141.

Coley-Smith, J. R. & Cooke, R. C. (1971) Survival and germination of fungal sclerotia. *Annual Review of Phytopathology* **9**, 65–92.

Colson, B. (1935) The cytology of the mushroom *Psalliota campestris* Quél. *Annals of Botany* **49**, 1–17.

Connolly, J. H., Chen, Y. & Jellison, J. (1995) Environmental SEM observation of the hyphal sheath and mycofibrils in *Postia placenta*. *Canadian Journal of Microbiology* **41**, 433–437.

Connolly, J. H. & Jellison, J. (1995) Calcium translocation, Ca oxalate accumulation and hyphal sheath morphology in the white-rot fungus *Resinicium bicolor*. *Canadian Journal of Botany* **73**, 927–936.

Cooke, R. C. & Rayner, A. D. M. (1984) *Ecology of Saprotrophic Fungi*. Longman: London.

Corda, A., J. C. (1839) *Icones fungorum hucusque cognitorum* **3,** 55pp, 9 pl. Prague.

Corner, E. J. H. (1932a) A *Fomes* with two systems of hyphae. *Transactions of the British Mycological Society* **17**, 51–81.

Corner, E. J. H. (1932b) The fruit-body of *Polystictus xanthopus* Fr. *Annals of Botany* **46**, 71–111.

Corner, E. J. H. (1947) Variation in the size and shape of spores, basidia and cystidia in Basidiomycetes. *New Phytologist* **46**, 195–228.

Corner, E. J. H. (1948) *Asterodon*, a clue to the morphology of fungus fruit-bodies, with notes on *Asterostroma* and *Asterostromella*. *Transactions of the British Mycological Society* **31**, 234–245.

Corner, E. J. H. (1950) *A Monograph of Clavaria and Allied Genera*. *Annals of Botany Memoirs no. 1*. Oxford University Press: London.

Corner, E. J. H. (1953) The construction of polypores 1. Introduction: *Polyporus sulphureus, P. squamosus, P. betulinus* and *Polystictus microcyclus*. *Phytomorphology* 3, 152.

Corner, E. J. H. (1954) Review: Pinto-Lopes, J.(1952) Polyporaceae, contribuiç o para a sua bio-taxonomia. *Transactions of the British Mycological Society* **37**, 92–94.

Corner, E. J. H. (1966) A monograph of the cantharelloid fungi. *Annals of Botany Memoirs* **2**, 1–255.

Corner, E. J. H. (1981) The agaric genera *Lentinus*, *Panus* and *Pleurotus*, with particular reference to the Malaysian species. *Beihefte zur Nova Hedwigia* **69**, 1–169.

Corner, E. J. H. (1991) *Trogia* (Basidiomycetes) *The Garden's Bulletin, Singapore, supplement* **2**, 1–100.

Cove, D. J. (1966) The induction and repression of nitrate reductase in the fungus *Aspergillus nidulans*. *Biochimica et Biophysica Acta* **113**, 51–56.

Cox, R. J. & Niederpruem, D. J. (1975) Differentiation in *Coprinus lagopus*. III. Expansion of excised fruit bodies. *Archives of Microbiology* **105**, 257–260.

Craig, G. D., Gull, K. & Wood, D. A. (1977) Stem elongation in *Agaricus bisporus*. *Journal of General Microbiology* **102**, 337–347.

Cram, W. J. (1976) Negative feedback regulation of transport in cells. The maintenance of turgor, volume and nutrient supply. In *Encyclopedia of Plant Physiology, New Series*, **vol. 2**, Transport in Plants II, Part A, Cells (ed. U. Lüttge & M. G. Pitman), pp. 284–316. Springer-Verlag: Berlin.

Croft, J. H. & Jinks, J. L. (1977) Aspects of the population genetics of *Aspergillus nidulans*. In *The Physiology and Genetics of Aspergillus* (ed. J. E. Smith & J. A. Pateman), pp. 305–317. Academic Press: London.

Cronenberg, C. C. H., Ottengraf, S. P. P., van den Heuvel, J. C., Pottel, F., Sziele, D., Schügerl, K. & Bellgardt, K. H. (1994) Influence of age and structure of *Penicillium chrysogenum* pellets on the internal concentration profiles. *Bioprocess Engineering* **10**, 209–216.

Cullen, D. & Kersten, P. J. (1996) Enzymology and molecular biology of lignin degradation. In *The Mycota*, **vol. III**, Biochemistry and Molecular Biology (ed. R. Brambl & G. A. Marzluf), pp. 295–312. Springer-Verlag: Berlin.

Cunningham, G. H. (1947) Notes on classification of the Polyporaceae. *New Zealand Journal of Science and Technology, ser. A* **28**, 238–251.

Cunningham, G. H. (1954) Hyphal systems as aids in identification of species and genera of the Polyporaceae. *Transactions of the British Mycological Society* **37**, 44–50.

Cunningham, G. H. (1956) Thelephoraceae of New Zealand IX. The genus *Stereum*. *Transactions of the Royal Society of New Zealand* **84**, 201–231.

Cunningham, G. H. (1958) Hydnaceae of New Zealand Part I - The pileate genera *Beenakia, Dentinum, Hercium, Hydnum, Phellodon, Steccherinum*. *Transactions of the Royal Society of New Zealand* **85**, 585–601.

Cunningham, G. H. (1965) Polyporaceae of New Zealand. *New Zealand Department of Scientific and Industrial Research Bulletin* **164**, 1–303.

Curtis, C. R. (1972) Action spectrum of the photoinduced sexual stage in the fungus *Nectria haematococca* Berk. and Br. var. *cucurbitae* (Snyder and Hansen) Dingley. *Plant Physiology* **49**, 235–239.

Damsky, C. H. & Werb, Z. (1992) Signal transduction by integrin receptors for extracellular matrix: cooperative processing of extracellular information. *Current Opinion in Cell Biology* **4**, 772–781.

Daniel, G. (1994) Use of EM for aiding our understanding of wood degradation. *FEMS Microbiology Reviews* **13**, 199–233.

Daniel, J. W. (1977) Photometabolic events programming expression of growth and development in *Physarum polycephalum*. *Abstracts, Second International Mycological Congress, Tampa, Florida*, 126.

Darbyshire, J. (1974) Developmental studies on *Coprinus lagopus*. Ph.D. Thesis, University of Manchester.

Da Silva, S. P., Felipe, M. S. S., Pereira, M., Azevedo, M. O. & Soares, C. M. D. (1994) Phase transition and stage-specific protein synthesis in the dimorphic fungus *Paracoccidioides brasiliensis*. *Experimental Mycology* **18**, 294–299.

Datta, A. (1994) Pathogenicity of *Candida albicans*: quest for a molecular switch. *Brazilian Journal of Medical and Biological Research* **27**, 2721–2732.

Davis, R. H. (1995) Genetics of *Neurospora*. In *The Mycota*, **vol. II**, Genetics and Biotechnology (ed. U. Kück), pp. 3–18. Springer-Verlag: Berlin.

Dawes, I. W. (1983) Genetic control and gene expression during meiosis and sporulation in *Saccharomyces cerevisiae*. In *Yeast Genetics: Fundamental and Applied Aspects* (ed. J. F. T. Spencer, D. M. Spencer & A. R. W. Smith), pp. 29–64. Springer-Verlag: New York.

Dawkins, R. (1976) *The Selfish Gene*. Oxford University Press: Oxford, UK.

Dearnaley, J. D. W., Maleszka, J. & Hardham, A. R. (1996) Synthesis of zoospore peripheral vesicles during sporulation of *Phytophthora cinnamomi*. *Mycological Research* **100**, 39–48.
De Bary, A. (1887) *Comparative Morphology of Fungi, Mycetozoa and Bacteria*. Clarendon Press: Oxford, UK.
Degli-Innocenti, F., Pohl, U. & Russo, V. E. A. (1983) Photoinduction of protoperithecia in *Neurospora crassa* by blue light. *Photochemistry and Photobiology* **37**, 49–51.
De Groot, P. W. J., Schaap, P. J., Sonnenberg, A. S. M., Visser, J. & van Griensven, L. J. L. D. (1996) The *Agaricus bisporus hypA* gene encodes a hydrophobin and specifically accumulates in peel tissue of mushroom caps during fruit body development. *Journal of Molecular Biology* **257**, 1008–1018.
Dehorter, B. & Lacoste, L. (1980) Photoinduction des périthèces du *Nectria galligena*. I. Influence de la lumière blanche. *Canadian Journal of Botany* **58**, 2206–2211.
Dehorter, B. & Perrin, R. (1983) Production *in vitro* de périthèces du *Nectria ditissima*, agent du chancre du hêtre (*Fagus sylvatica*) II. Effets de la composition carbone-azote du milieu nutritif et influence de la lumière. *Canadian Journal of Botany* **61**, 1947–1954.
Delettre, Y. M. & Bernet, J. (1976) Regulation of proteolytic enzymes in *Podospora anserina*: selection and properties of self-lysing mutant strains. *Molecular & General Genetics* **144**, 191–197.
Denis, H. (1995) The origin and early evolution of metazoans. *Bulletin de la Societé Zoologique de France - Evolution et Zoologie* **120**, 171–190.
Dennen, D. W. & Niederpruem, D. J. (1967) Regulation of glutamate dehydrogenases during morphogenesis of *Schizophyllum commune*. *Journal of Bacteriology* **93**, 904–913.
de Silva, L. R., Youatt, J., Gooday, G. W. & Gow, N. A. R. (1992) Inwardly directed ionic currents of *Allomyces macrogynus* and other water moulds indicate sites of proton-driven nutrient transport but are incidental to tip growth. *Mycological Research* **96**, 925–931.
Deutsch, A., Dress, A. & Rensing, L. (1993) Formation of morphological differentiation patterns in the ascomycete *Neurospora crassa*. *Mechanisms of Development* **44**, 17–31.
de Vries, O. M. H., Fekkes, M. P., Wösten, H. A. B. & Wessels, J. G. H. (1993) Insoluble hydrophobin complexes in the walls of *Schizophyllum commune* and other filamentous fungi. *Archives of Microbiology* **159**, 330–335.
de Vries, O. M. H., Hoge, J. H. C. & Wessels, J. G. H. (1980) Translation of RNA from *Schizophyllum commune* in a wheat germ and rabbit reticulocyte cell-free system: comparison of *in vitro* and *in vivo* products after two-dimensional gel electrophoresis. *Biochimica et Biophysica Acta* **607**, 373–378.
de Vries, O. M. H. & Wessels, J. G. H. (1984) Patterns of polypeptide synthesis in non-fruiting monokaryons and a fruiting dikaryon of *Schizophyllum commune*. *Journal of General Microbiology* **133**, 145–154.
Dietert, M. F., Van Etten, H. D. & Matthews, P. S. (1983) *Nectria haematococca* mating population. VI. Cultural parameters affecting growth, conidiation, and perithecial formation. *Canadian Journal of Botany* **61**, 1178–1184.
Dimsterdenk, D., Schafer, W. R. & Rine, J. (1995) Control of Ras mRNA level by the mevalonate pathway. *Molecular Biology of the Cell* **6**, 59–70.
Donk, M. A. (1964) A conspectus of the families of Aphyllophorales. *Persoonia* **3**, 199–324.
Donk, M. A. (1971) Progress in the study of the classification of the higher basidiomycetes. In *Evolution in the Higher Basidiomycetes* (ed. R. H. Petersen), pp. 3–25. University of Tennessee Press: Knoxville.
Dons, J. J. M., Springer, J., de Vries, S. C. & Wessels, J. G. H. (1984) Molecular cloning of a gene abundantly expressed during fruiting body initiation in *Schizophyllum commune*. *Journal of Bacteriology* **157**, 802–808.
Doolittle, R. F., Feng, D. F., Tsang, S., Cho, G. & Little, E. (1996) Determining divergence times of the major kingdoms of living organisms with a protein clock. *Science* **271**, 470–477.
Doonan, J. H. (1992) Cell division in *Aspergillus*. *Journal of Cell Science* **103**, 599–611.
Dormer, K. J. (1980) *Fundamental Tissue Geometry for Biologists*. Cambridge University Press: Cambridge, UK.

Dowson, C. G., Rayner, A. D. M. & Boddy, L. (1986) Outgrowth patterns of mycelial cord-forming basidiomycetes from and between woody resource units in soil. *Journal of General Microbiology* **132**, 203–211.

Drouin, G., Desa, M. M. & Zuker, M. (1995) The *Giardia lamblia* actin gene and the phylogeny of eukaryotes. *Journal of Molecular Evolution* **41**, 841–849.

Dubourdieu, D. & Ribereau-Gayon, P. (1981) Structure of the extracellular β-D-glucan from *Botrytis cinerea. Carbohydrate Research* **93**, 294–299.

Dunn, E. & Pateman, J. A. (1972) Urea and thiourea uptake in *Aspergillus nidulans. Heredity* **29**, 129.

Duntze, W., Betz, R. & Nientiedt, M. (1994) Pheromones in yeasts. In *The Mycota,* **vol. I**, Growth, Differentiation and Sexuality (ed. K. Esser & P. A. Lemke), pp. 381–399. Springer-Verlag: New York.

Durand, R. (1983) Light breaks and fruit-body maturation in *Coprinus congregatus*: dark inhibitory and dark recovery process. *Plant & Cell Physiology* **24**, 899–905.

Durand, R. (1985) Blue-UV-light photoreception in fungi. *Physiologie Végétale* **23**, 935–943.

Durand, R. & Furuya, M. (1985) Action spectrum for stimulatory and inhibitory effects of UV and blue light on fruit-body formation in *Coprinus congregatus. Plant & Cell Physiology* **26**, 1175–1183.

Durrens, P. (1983) *Podospora* mutant defective in glucose-dependent growth control. *Journal of Bacteriology* **154**, 702–707.

Durrens, P. & Bernet, J. (1985) Temporal action of mutations inhibiting the accomplishment of quiescence or disrupting development in the fungus *Podospora anserina. Genetics* **109**, 37–47.

Dütsch, G. A. & Rast, D. (1972) Biochemische beziehung zwischen mannitbildung und hexose-monophosphatzyklus in *Agaricus bisporus. Phytochemistry* **11**, 2677–2681.

Dyer, P. S., Ingram, D. S. & Johnstone, K. (1992) The control of sexual morphogenesis in the ascomycetes. *Biological Reviews* **67**, 421–458.

Eamus, D. & Jennings, D. H. (1986) Water, turgor and osmotic potentials of fungi. In *Water, Fungi and Plants* (ed. P. G. Ayres & L. Boddy), pp. 27–48. Cambridge University Press: Cambridge, UK.

Eaton, D. C. (1985) Mineralization of polychlorinated biphenyls by *Phanerochaete chrysosporium*, a lignolytic fungus. *Enzyme and Microbial Technology* **7**, 194–196.

Ebbole, D. J. (1996) Morphogenesis and vegetative differentiation in filamentous fungi. *Journal of Genetics* **75**, 361–374.

Edelman, G. M. (1992) Morphoregulation. *Developmental Dynamics* **193**, 2–10.

Edelstein, L. (1982) The propagation of fungal colonies: a model for tissue growth. *Journal of Theoretical Biology* **98**, 679–701.

Edelstein, L. & Hadar, Y. (1983) A model for pellet size distributions in submerged cultures. *Journal of Theoretical Biology* **105**, 427–452.

Edelstein, L., Hadar, Y., Chet, I., Henis, Y. & Segel, L. A. (1983) A model for fungal colony growth applied to *Sclerotium rolfsii. Journal of General Microbiology* **129**, 1873–1881.

Edelstein, L. & Segel, L. A. (1983) Growth and metabolism in mycelial fungi. *Journal of Theoretical Biology* **104**, 187–210.

Edmundowicz, J. M. & Wriston, J. C. (1963) Mannitol dehydrogenase from *Agaricus campestris. Journal of Biological Chemistry* **238**, 3539–3541.

Edwardson, J. M. & Marciniak, S. J. (1995) Molecular mechanisms in exocytosis. *Journal of Membrane Biology* **146**, 113–122.

Egorova, V. N. & Lakhchev, K. L. (1994) Genetic control of yeast cell morphology. *Genetika* **30**, 1036–1042.

Eilers, F. I. (1974) Growth regulation in *Coprinus radiatus. Archives of Microbiology* **96**, 353–364.

Elhiti, M. M. Y., Butler, R. D. & Moore, D. (1979) Cytochemical localization of glutamate dehydrogenase during carpophore development in *Coprinus cinereus. New Phytologist* **82**, 153–157.

Elhiti, M. M. Y., Moore, D. & Butler, R. D. (1986) Cytochemical localisation of NADP-linked glutamate dehydrogenase newly induced in nitrogen starved mycelia of *Coprinus cinereus. FEMS Microbiology Letters* **33**, 121–124.

Elhiti, M. M. Y., Moore, D. & Butler, R. D. (1987) Ultrastructural distribution of glutamate dehydrogenases during fruit body development in *Coprinus cinereus. New Phytologist* **107**, 531–539.

Elisashvili, V. I. (1993) Physiological regulation of ligninolytic activity in higher basidium fungi. *Microbiology* **62**, 480–487.

Elliott, C. G. (1994) *Reproduction in Fungi. Genetical and Physiological Aspects*. Chapman & Hall: London.

Elliott, T. J. (1985) Developmental genetics - from spore to sporophore. In *Developmental Biology of Higher Fungi* (ed. D. Moore, L. A. Casselton, D. A. Wood & J. C. Frankland), pp. 451–465. Cambridge University Press: Cambridge, UK.

Elliott, T. J. & Challen, M. P. (1983) Genetic ratio in secondary homothallic basidiomycetes. *Experimental Mycology* **7**, 170–174.

Elliott, T. J. & Challen, M. P. (1984) Effect of temperature on spore number in the cultivated mushroom, *Agaricus bisporus. Transactions of the British Mycological Society* **82**, 293–296.

Ellis, S. W., Grindle, M. & Lewis, D. H. (1991) Effect of osmotic stress on yield and polyol content of dicarboximide-sensitive and -resistant strains of *Neurospora crassa. Mycological Research* **95**, 457–464.

Elorza, M. V., Marcilla, A., Sanjuan, R., Mormeneo, S. & Sentandreu, R. (1994) Incorporation of specific wall proteins during yeast and mycelial protoplast regeneration in *Candida albicans. Archives of Microbiology* **161**, 145–151.

Erikssen, J., Hjortstam, K. & Ryvarden, L. (1978) The Corticiaceae of North Europe. *Fungiflora* **5**, 889–1047.

Eriksson, K. E. L., Habu, N. & Samejima, M. (1993) Recent advances in fungal cellobiose oxidoreductases - review. *Enzyme and Microbial Technology* **15**, 1002–1008.

Erwin, D. H. (1993) The origin of metazoan development: a palaeobiological perspective. *Biological Journal of the Linnean Society* **50**, 255–274.

Esposito, R. E., Frink, N., Bernstein, P. & Esposito, M. S. (1972) The genetic control of sporulation in *Saccharomyces*. II. Dominance and complementation of mutants of meiosis and spore formation. *Molecular & General Genetics* **114**, 241–248.

Esposito, R. E. & Klapholz, S. (1981) Meiosis and ascospore development. In *The Molecular Biology of the Yeast Saccharomyces cerevisiae*. **vol. 1**. Life Cycle and Inheritance (ed. J. N. Strathern, E. W. Jones & J. R. Broach), pp. 211–287. Cold Spring Harbor Laboratory Press: New York.

Esser, K. (1956) Wachstum, fruchtkörper, und pigmentbilding von *Podospora anserina* in synthetischen nährmedien. *Comptes Rendus des Travaux du Laboratoire Carlsberg* **26**, 102–117.

Esser, K. (1968) Phenoloxidases and morphogenesis in *Podospora anserina. Genetics* **60**, 281–288.

Esser, K. (1990) Molecular aspects of ageing: facts and perspectives. In *Frontiers in Mycology* (ed. D. L. Hawksworth), pp. 3–25. CAB International: Wallingford.

Esser, K. & Hoffman, P. (1977) Genetic basis for speciation in higher basidiomycetes with special reference to the genus *Polyporus*. In *The Species Concept in Hymenomycetes* (ed. H. Clémençon), pp. 189–214. J. Cramer: Vaduz.

Esser, K. & Meinhardt, F. (1977) A common genetic control of dikaryotic and monokaryotic fruiting in the basidiomycete *Agrocybe aegerita. Molecular & General Genetics* **155**, 113–115.

Esser, K. & Minuth, W. (1970) The phenoloxidases of the ascomycete *Podospora anserina*. VI. Genetic regulation of the formation of laccase. *Genetics* **64**, 441–458.

Esser, K., Saleh, F. & Meinhardt, F. (1979) Genetics of fruit body production in higher basidiomycetes. II. Monokaryotic and dikaryotic fruiting in *Schizophyllum commune. Current Genetics* **1**, 85–88.

Esser, K., Stahl, U. & Meinhardt, F. (1977) Genetic aspects of differentiation in fungi. In *Biotechnology and Fungal Differentiation* (ed. J. Meyrath & J. D. Bu'Lock), pp. 67–75. Academic Press: London.

Esser, K. & Straub, J. (1958) Genetische Untersuchungen an *Sordaria macrospora* Auersw., Kompensation und induktion bei genbedingten Entwicklungsdefekten. *Zeitschrift für Vererbungslehre* **89**, 729–746.

Evans, H. J. (1959) Nuclear behaviour in the cultivated mushroom. *Chromosoma* **10**, 115–135.
Ewaze, J. O., Moore, D. & Stewart, G. R. (1978) Co-ordinate regulation of enzymes involved in ornithine metabolism and its relation to sporophore morphogenesis in *Coprinus cinereus*. *Journal of General Microbiology* **107**, 343–357.
Fares, H., Goetsch, L. & Pringle, J. R. (1996) Identification of a developmentally regulated septin and involvement of the septins in spore formation in *Saccharomyces cerevisiae*. *Journal of Cell Biology* **132**, 399–411.
Fayod, M. V. (1889) Histoire naturelle des Agaricinés. *Annales des Sciences Naturelles* **9**, 181–411.
Fayret, J. & Parguey-Leduc, A. (1976) Thermo-inhibition du développement du sporophyte au cours de la maturation des périthèces de *Gnomonia leptostyla* (Fr.) Ces. de Not. *Revue de Mycologie* **40**, 245–253.
Fermor, T. R. & Wood, D. A. (1981) Degradation of bacteria by *Agaricus bisporus* and other fungi. *Journal of General Microbiology* **126**, 377–387.
Fiddy, C. & Trinci, A. P. J. (1976) Mitosis, septation, branching and the duplication cycle in *Aspergillus nidulans*. *Journal of General Microbiology* **97**, 169–184.
Fisher, R. A. (1928) The possible modifications of the wild type to recurrent mutations. *American Naturalist* **62**, 115–126.
Fisher, R. A. (1931) The evolution of dominance. *Biological Reviews* **6**, 345–368.
Flegg, P. B. (1972) Response of the mushroom to temperature with particular reference to the control of cropping. *Mushroom Science* **8**, 75–84.
Flegg, P. B. (1978a) Effect of temperature on sporophore initiation and development in *Agaricus bisporus*. *Mushroom Science* **10 (2)**, 595–602.
Flegg, P. B. (1978b) Effect of temperature on the development of sporophores of *Agaricus bisporus* beyond a cap diameter of 2 mm. *Mushroom Science* **10 (2)**, 603–609.
Flegg, P. B. (1983) Response of the sporophores of the cultivated mushroom (*Agaricus bisporus*) to volatile substances. *Scientia Horticulturae* **21**, 301–310.
Flegg, P. B. & Wood, D. A. (1985) Growth and fruiting. In *The Biology and Technology of the Cultivated Mushroom* (ed. P. B. Flegg, D. M. Spencer & D. A. Wood), pp. 141–177. John Wiley & Sons: New York.
Foss, H. M., Roberts, C. J., Claeys, K. M. & Selker, E. U. (1993) Abnormal chromosome behaviour in *Neurospora* mutants defective in DNA methylation. *Science* **262**, 1737–1741.
Fothergill-Gilmore, L. A. (1986) The evolution of the glycolytic pathway. *Trends in Biochemical Sciences* **11**, 47–51.
Frankhauser, C. & Simanis, V. (1994) Cold fission: splitting the pombe cell at room temperature. *Trends in Cell Biology* **4**, 96–101.
Frankland, J. C., Hedger, J. N. & Smith, M. J. (1982) *Decomposer Basidiomycetes*. Cambridge University Press: Cambridge, UK.
Freedman, T. & Pukkila, P. J. (1993) *De novo* methylation of repeated sequences in *Coprinus cinereus*. *Genetics* **135**, 357–366.
Fries, E. (1821) *Systema Mycologicum*, **vol. I**. Gryphiswald: Mauritius.
Fritz, B. J. & Ninnemann, H. (1985) Photoreactivation by triplet flavin and photoinactivation by singlet oxygen of *Neurospora crassa* nitrate reductase. *Photochemistry and Photobiology* **41**, 39–45.
Funtikova, N. S., Katomina, A. A. & Mysyakina, I. S. (1995) Dimorphism and lipid composition of the fungus *Mucor lusitanicus*. *Microbiology* **64**, 238–239.
Gabriel, M. & Kopecka, M. (1995) Disruption of the actin cytoskeleton in budding yeast results in formation of an aberrant cell wall. *Microbiology - UK* **141**, 891–899.
Galbraith, J. C. & Smith, J. E. (1969) Changes in activity of certain enzymes of the tricarboxylic acid cycle and the glyoxylate cycle during initiation of conidiation of *Aspergillus niger*. *Canadian Journal of Microbiology* **15**, 1207–1212.
Galun, E. (1971) Morphogenesis in *Trichoderma*. Induction of conidiation by narrow-beam illumination of restricted areas of the fungal colony. *Plant & Cell Physiology* **12**, 779–783.
Galvagno, M. A., Forchiassin, F., Cantore, M. L. & Passeron, S. (1984) The effect of light and cyclic AMP metabolism on fruiting body formation in *Saccobolus plantensis*. *Experimental Mycology* **8**, 334–341.

Gamow, R. I. & Böttger, B. (1982) Avoidance and rheotropic responses in *Phycomyces*: evidence for an 'avoidance gas' mechanism. *Journal of General Physiology* **79**, 835–848.

Garbaye, J. (1994a) Helper bacteria - a new dimension to the mycorrhizal symbiosis. *New Phytologist* **128**, 197–210.

Garbaye, J. (1994b) Mycorrhization helper bacteria - a new dimension to the mycorrhizal symbiosis. *Acta Botanica Gallica* **141**, 517–521.

Gargas, A., Depriest, P. T., Grube, M. & Tehler, A. (1995) Multiple origins of lichen symbioses in fungi suggested by SSU rDNA phylogeny. *Science* **268**, 1492–1495.

Garraway, M. O., Hüttermann, A. & Wargo, P. M. (1991) Ontogeny and physiology. In *Armillaria Root Disease* (ed. C. Shaw & G. A. Kile), pp. 21–47. United States Department of Agriculture: Washington, DC.

Garrett, S. D. (1953) Rhizomorph behaviour in *Armillaria mellea* (Vahl) Quél. I. Factors controlling rhizomorph initiation by *Armillaria mellea* in pure culture. *Annals of Botany* **17**, 63–79.

Garrett, S. D. (1954) Function of the mycelial strands in substrate colonization by the cultivated mushroom *Psalliota hortensis*. *Transactions of the British Mycological Society* **37**, 51–57.

Garrett, S. D. (1956) *Biology of Root-Infecting Fungi*. Cambridge University Press: Cambridge, UK.

Garrett, S. D. (1960) Inoculum potential. In *Plant Pathology: an Advanced Treatise*, **vol. 3** (ed. J. G. Horsfall & A. E. Dimond), pp. 23–56. Academic Press: New York.

Garrett, S. D. (1970) *Pathogenic Root-Infecting Fungi*. Cambridge University Press: Cambridge, UK.

Garrill, A. (1995) Transport. In *The Growing Fungus* (ed. N. A. R. Gow & G. M. Gadd), pp. 163-181. Chapman & Hall: London.

Garstang, W. (1922) The theory of recapitulation: a critical re-statement of the biogenetic law. *Linnean Society of London, Zoological Journal* **35**, 81–101.

Georgiou, G. & Shuler, M. L. (1986) A computer model for the growth and differentiation of a fungal colony on solid substrate. *Biotechnology and Bioengineering* **28**, 405–416.

Gerin, P. A., Asther, M., Sleytr, U. B. & Rouxhet, P. G. (1994) Detection of rodlets in the outer wall region of conidiospores of *Phanerochaete chrysosporium*. *Canadian Journal of Microbiology* **40**, 412–416.

Ghora, B. K. & Chaudhuri, K. L. (1975a) Comparative studies on the role of nitrogenous compounds in the growth and perithecial development of *Chaetomium aureum*. *Folia Microbiologica* **20**, 157–165.

Ghora, B. K. & Chaudhuri, K. L. (1975b) Effect of pH on carbohydrate utilization and perithecial production in *Chaetomium aureum*. *Bulletin of the Botanical Society of Bengal* **29**, 21–23.

Giovannetti, M., Sbrana, C., Avio, L., Citernesi, A. S. & Logi, C. (1993) Differential hyphal morphogenesis in arbuscular mycorrhizal fungi during preinfection stages. *New Phytologist* **125**, 587–593.

Girbardt, M. (1979) A microfilamentous septal belt (FSB) during induction of cytokinesis in *Trametes versicolor* (L. ex Fr.) *Experimental Mycology* **3**, 215–228.

Gloer, J. B. (1995) The chemistry of fungal antagonism and defense. *Canadian Journal of Botany* **73**, S1265–S1274.

Gold, S., Duncan, G., Barrett, K. & Kronstad, J. (1994) cAMP regulates morphogenesis in the fungal pathogen *Ustilago maydis*. *Genes & Development* **8**, 2805–2816.

Gooday, G. W. (1972a) The role of chitin synthetase in the elongation of fruit bodies of *Coprinus cinereus*. *Journal of General Microbiology* **73**, xxi (abstract).

Gooday, G. W. (1972b) The effect of polyoxin D on morphogenesis in *Coprinus cinereus*. *Biochemical Journal* **129**, 17P (abstract).

Gooday, G. W. (1973a) Activity of chitin synthetase during development of fruit bodies of the toadstool *Coprinus cinereus*. *Biochemical Society Transactions* **1**, 1105.

Gooday, G. W. (1973b) Differentiation in the Mucorales. *Symposia of the Society for General Microbiology* **23**, 269–294.

Gooday, G. W. (1974a) Control of development of excised fruitbodies and stipes of *Coprinus cinereus*. *Transactions of the British Mycological Society* **62**, 391–399.

Gooday, G. W. (1974b) Fungal sex hormones. *Annual Review of Biochemistry* **43**, 35–49.

Gooday, G. W. (1975a) Chemotaxis and chemotropism in fungi and algae. In *Primitive Sensory and Communication Systems* (ed. M. J. Carlile), pp. 155–204. Academic Press: London.
Gooday, G. W. (1975b) The control of differentiation in fruit bodies of *Coprinus cinereus*. *Reports of the Tottori Mycological Institute (Japan)* **12**, 151–160.
Gooday, G. W. (1977) Metabolic control of fruit body morphogenesis in *Coprinus cinereus*. In *Abstracts, Second International Mycological Congress, Tampa Florida*, p. 231.
Gooday, G. W. (1982) Metabolic control of fruit body morphogenesis in *Coprinus cinereus*. In *Basidium and Basidiocarp: Evolution, Cytology, Function and Development* (ed. K. Wells & E. K. Wells), pp. 157–173. Springer-Verlag: New York.
Gooday, G. W. (1985) Elongation of the stipe of *Coprinus cinereus*. In *Developmental Biology of Higher Fungi* (ed. D. Moore, L. A. Casselton, D. A. Wood & J. C. Frankland), pp. 311–331. Cambridge University Press: Cambridge, UK.
Gooday, G. W. (1994) Hormones in mycelial fungi. In *The Mycota,* **vol. I**, Growth, Differentiation and Sexuality (ed. J. G. H. Wessels & F. Meinhardt), pp. 401–411. Springer-Verlag: Berlin.
Gooday, G. W. (1995a) Cell membrane. In *The Growing Fungus* (ed. N. A. R. Gow & G. M. Gadd), pp. 63–74. Chapman & Hall: London.
Gooday, G. W. (1995b) The dynamics of hyphal growth. *Mycological Research* **99**, 385–394.
Gooday, G. W. (1995c) Cell walls. In *The Growing Fungus* (ed. N. A. R. Gow & G. M. Gadd), pp. 43–62. Chapman & Hall: London.
Gooday, G. W. & Adams, D. J. (1993) Sex hormones and fungi. *Advances in Microbial Physiology* **34**, 69–145.
Gooday, G. W., de Rousset-Hall, A. & Hunsley, D. (1976) Effect of polyoxin D on chitin synthesis in *Coprinus cinereus*. *Transactions of the British Mycological Society* **67**, 193–200.
Gooday, G. W. & Schofield, D. A. (1995) Regulation of chitin synthesis during growth of fungal hyphae - the possible participation of membrane stress. *Canadian Journal of Botany* **73**, S114–S121.
Gooday, G. W. & Trinci, A. P. J. (1980) Wall structure and biosynthesis in fungi. *Symposia of the Society for General Microbiology* **30**, 207–251.
Goodwin, B. C. (1984) What are the causes of morphogenesis? *BioEssays* **3**, 32–36.
Goodwin, T. W. (1973) Comparative biochemistry of sterols in eukaryotic micro-organisms. In *Lipids and Biomembranes of Eukaryotic Micro-organisms* (ed. J. A. Erwin), pp. 1–40. Academic Press: New York.
Goodwin, T. W. (1976) Carotenoids. In *The Filamentous Fungi,* **vol II**, Biosynthesis and Metabolism (ed. J. E. Smith & D. R. Berry), pp. 423–444. Edward Arnold: London.
Gorovoj, L. F., Kasatkina, T. B. & Klyushkina, N. S. (1987) [Role of gravitation in the development of carpophores in hymenomycetes]. *Mikologiya i Fitopatologiya* **21**, 301–307, [*in Russian*].
Gorovoj, L. F., Kasatkina, T. B. & Laurinavichius, R. S. (1989) [Morphogenesis of mushrooms in changed gravitation conditions]. *Report of the N. G. Kholodny Institute of Botany, Academy of Sciences of the Ukrainian SSR* [*in Russian*].
Gow, N. A. R. (1994) Growth and guidance of the fungal hypha. *Microbiology - UK* **140**, 3193–3205.
Gow, N. A. R. (1995a) Tip growth and polarity. In *The Growing Fungus* (ed. N. A. R. Gow & G. M. Gadd), pp. 277–299. Chapman & Hall: London.
Gow, N. A. R. (1995b) Yeast-hyphal dimorphism. In *The Growing Fungus* (ed. N. A. R. Gow & G. M. Gadd), pp. 403–422. Chapman & Hall: London.
Gow, N. A. R., Hube, B., Bailey, D. A., Schofield, D. A., Munro, C., Swoboda, R. K., Bertram, G., Westwater, C., Broadbent, I., Smith, R. J., Gooday, G. W. & Brown, A. J. P. (1995) Genes associated with dimorphism and virulence of *Candida albicans*. *Canadian Journal of Botany* **73**, S335–S342.
Goyal, S. & Khuller, G. K. (1994) Structural and functional role of lipids in yeast and mycelial forms of *Candida albicans*. *Lipids* **29**, 793–797.
Gozalbo, D., Elorza, M. V., Sanjuan, R., Marcilla, A., Valentin, E. & Sentandreu, R. (1993) Critical steps in fungal cell wall synthesis: strategies for their inhibition. *Pharmacology & Therapeutics* **60**, 337–345.

Graafmans, W. D. J. (1977) Effect of blue light on metabolism in *Penicillium isariiforme*. *Journal of General Microbiology* **101**, 157.

Granlund, H. I., Jennings, D. H. & Thompson, W. (1985) Translocation of solutes along rhizomorphs of *Armillaria mellea*. *Transactions of the British Mycological Society* **84**, 111–119.

Grant, W. D., Rhodes, L. L., Prosser, B. A. & Asher, R. A. (1986) Production of bacteriolytic enzymes and degradation of bacteria by filamentous fungi. *Journal of General Microbiology* **132**, 2353–2358.

Green, J. B. A. & Smith, J. C. (1991) Growth factors as morphogens: do gradients and thresholds establish body plan? *Trends in Genetics* **7**, 245–250.

Green, J. R., Jones, G. L. & O'Connell, R. J. (1996) Composition and organisation of extracellular matrices around germ tubes and appressoria of *Colletotrichum lindemuthianum*. *Protoplasma* **190**, 119–130.

Green, J. R., Pain, N. A., Cannell, M. E., Jones, G. L., Leckie, C. P., McCready, S., Mendgen, K., Mitchell, A. J., Callow, J. A. & O'Connell, R. J. (1995) Analysis of differentiation and development of the specialized infection structures formed by biotrophic fungal plant pathogens using monoclonal antibodies. *Canadian Journal of Botany* **73**, S408–S417.

Greening, J. P., Holden, J. & Moore, D. (1993) Distribution of mechanical stress is not involved in regulating stem gravitropism in *Coprinus cinereus*. *Mycological Research* **97**, 1001–1004.

Greening, J. P. & Moore, D. (1996) Morphometric analysis of cell size patterning involved in regulating stem gravitropism in *Coprinus cinereus*. *Advances in Space Research* **17**, 83–86.

Greening, J. P., Sánchez, C. & Moore, D. (1997) Coordinated cell elongation alone drives tropic bending in stems of the mushroom fruit body of *Coprinus cinereus*. *Canadian Journal of Botany* **75**, 1174–1181.

Gregory, P. H. (1984) The fungal mycelium - an historical perspective. In *The Ecology and Physiology of the Fungal Mycelium* (ed. D. H. Jennings & A. D. M. Rayner), pp. 1–22. Cambridge University Press: Cambridge, UK.

Gressel, J. & Rau, W. (1983) Photocontrol of fungal development. In *Photomorphogenesis, Encyclopedia of Plant Physiology*, **vol. 16** (ed. W. Shropshire & H. Mohr), pp. 603–639. Springer-Verlag: New York.

Griffin, D. H., Timberlake, W. E. & Cheney, J. C. (1974) Regulation of macromolecular synthesis, colony development and specific growth rate of *Achlya bisexualis* during balanced growth. *Journal of General Microbiology* **80**, 381–388.

Griffiths, A. J. F. (1992) Fungal senescence. *Annual Review of Genetics* **26**, 351–372.

Grove, G. & Marzluf, G. A. (1981) Identification of the product of the major regulatory gene of the nitrogen control circuit of *Neurospora crassa* as a nuclear DNA-binding protein. *Journal of Biological Chemistry* **256**, 463–470.

Grove, S. N. (1978) The cytology of hyphal tip growth. In *The Filamentous Fungi*, **vol III**, Developmental Mycology (ed. J. E. Smith & D. R. Berry), pp. 28–50. Edward Arnold: London.

Grove, S. N. & Sweigard, J. A. (1980) Cytochalasin A inhibits spore germination and hyphal tip growth in *Gilbertella pensicaria*. *Experimental Mycology* **4**, 239–250.

Gruen, H. E. (1963) Endogenous growth regulation in carpophores of *Agaricus bisporus*. *Plant Physiology* **38**, 652–666.

Gruen, H. E. (1969) Growth and rotation of *Flammulina velutipes* fruitbodies and the dependence of stipe elongation on the cap. *Mycologia* **61**, 149–166.

Gruen, H. E. (1976) Promotion of stipe elongation in *Flammulina velutipes* by a diffusate from excised lamellae supplied with nutrients. *Canadian Journal of Botany* **54**, 1306–1315.

Gruen, H. E. (1982) Control of stipe elongation by the pileus and mycelium in fruitbodies of *Flammulina velutipes*. In *Basidium and Basidiocarp: Evolution, Cytology, Function and Development* (ed. K. Wells & E. K. Wells), pp. 125-155. Springer-Verlag: New York.

Gruen, H. (1991) Effects of grafting on stem elongation and pileus expansion in the mushroom *Flammulina velutipes*. *Mycologia* **83**, 480–491.

Grzemski, W. E., Lawrie, A. C. & Grant, B. R. (1994) The role of the plasma membrane in differentiation of *Phytophthora palmivora* zoospores. *Mycological Research* **98**, 1051–1058.

Guerinot, M. L. (1994) Microbial iron transport. *Annual Review of Microbiology* **48**, 743–772.

Guilfoyle, T. J., Hagen, G., Li, Y., Ulmasov, T., Liu, Z-B., Strabala, T. & Gee, M. (1993) Auxin-regulated transcription. *Australian Journal of Plant Physiology* **20**, 489–502.

Guiseppin, M. L. F. (1984) Effects of dissolved oxygen concentration on lipase production by *Rhizopus delemar. Applied Microbiology and Biotechnology* **20**, 161–173.

Gull, K. (1976) Differentiation of septal ultrastructure according to cell type in the basidiomycete *Agrocybe praecox. Journal of Ultrastructure Research* **54**, 89–94.

Gull, K. (1978) Form and function of septa in filamentous fungi. In *The Filamentous Fungi,* **vol III**, Developmental Mycology (ed. J. E. Smith & D. R. Berry), pp. 78–93. Edward Arnold: London.

Gupta, R. S. (1995) Phylogenetic analysis of the 90 kD heat-shock family of protein sequences and an examination of the relationship among animals, plants, and fungi species. *Molecular Biology and Evolution* **12**, 1063–1073.

Gwynne, D. I., Miller, B. L., Miller, K. Y. & Timberlake, W. E. (1984) Structure and regulated expression of the *SpoC1* gene cluster from *Aspergillus nidulans. Journal of Molecular Biology* **180**, 91–109.

Gwynne, D. I. & Timberlake, W. E. (1984) Genetic regulation of conidiation in *Aspergillus nidulans.* In *Control of Cell Growth and Proliferation* (ed. C. M. Venezuale), pp. 47–66. Van Nostrand Reinhold: New York.

Hafner, L. & Thielke, C. (1970) Kernzahl und Zellgrösse im Fruchtkorperstiel von *Coprinus radiatus* (Solt.) Fr. *Berichte Deutsche Botanische Gesellschaft* **83**, 27–31.

Hagimoto, H. (1963) Studies on the growth of fruit body of fungi. IV. The growth of the fruit body of *Agaricus bisporus* and the economy of the mushroom growth hormone. *Botanical Magazine (Tokyo)* **76**, 256–263.

Hagimoto, H. & Konishi, M. (1959) Studies on the growth of fruitbody of fungi. I. Existence of a hormone active to the growth of fruitbody in *Agaricus bisporus* (Lange) Sing. *Botanical Magazine (Tokyo)* **72**, 359–366.

Hagimoto, H. & Konishi, M. (1960) Studies on the growth of fruit body of fungi. II. Activity and stability of the growth hormone in the fruit body of *Agaricus bisporus* (Lange) Sing. *Botanical Magazine (Tokyo)* **73**, 283–287.

Hallenberg, N. (1985) *The Lachnocladiacaeae and Coniophoraceae of North Europe.* Fungiflora: Oslo.

Hammad, F., Ji, J., Watling, R. & Moore, D. (1993a) Cell population dynamics in *Coprinus cinereus*: co-ordination of cell inflation throughout the maturing fruit body. *Mycological Research* **97**, 269–274.

Hammad, F., Watling, R. & Moore, D. (1993b) Cell population dynamics in *Coprinus cinereus*: narrow and inflated hyphae in the basidiome stipe. *Mycological Research* **97**, 275–282.

Hammad, F., Watling, R. & Moore, D. (1993c) Artifacts in video measurements cause growth curves to advance in steps. *Journal of Microbiological Methods* **18**, 113–117.

Hammond, J. B. W. (1977) Carbohydrate metabolism in *Agaricus bisporus*: oxidation pathways in mycelium and sporophore. *Journal of General Microbiology* **102**, 245–248.

Hammond, J. B. W. & Nichols, R. (1976) Carbohydrate metabolism in *Agaricus bisporus* (Lange) Sing.: Changes in soluble carbohydrates during growth of mycelium and sporophore. *Journal of General Microbiology* **93**, 309–320.

Hammond, J. B. W. & Nichols, R. (1977) Carbohydrate metabolism in *Agaricus bisporus* (Lange) Imbach.: metabolism of [^{14}C] labelled sugars by sporophores and mycelium. *New Phytologist* **79**, 315–325.

Hammond, J. B. W. & Wood, D. A. (1985) Metabolism, biochemistry and physiology. In *The Biology and Technology of the Cultivated Mushroom* (ed. P. B. Flegg, D. M. Spencer & D. A. Wood), pp. 63–80. John Wiley & Sons: Chichester.

Han, S., Navarro, J., Greve, R. A. & Adams, T. H. (1993) Translational repression of *brlA* expression prevents premature development in *Aspergillus. EMBO Journal* **12**, 2449–2457.

Harold, F. M. (1990) To shape a cell: an enquiry into the causes of morphogenesis of microorganisms. *Microbiological Reviews* **54**, 381–431.

Harold, F. M. (1991) Biochemical topology - from vectorial metabolism to morphogenesis. *Bioscience Reports* **11**, 347–385.

Harold, F. M. (1994) Ionic and electrical dimensions of hyphal growth. In *The Mycota,* **vol. I**, Growth, Differentiation and Sexuality (ed. J. G. H. Wessels & F. Meinhardt), pp. 89-109. Springer-Verlag: Berlin.

Harold, F. M. (1995) From morphogens to morphogenesis. *Microbiology* **141**, 2765–2778.

Harold, F. M. & Caldwell, J. H. (1990) Tips and currents: electrobiology of apical growth. In *Tip Growth in Plant and Fungal Cells* (ed. I. B. Heath), pp. 59–90. Academic Press: San Diego.

Harold, F. M., Harold, R. L. & Money, N. P. (1995) What forces drive cell-wall expansion. *Canadian Journal of Botany* **73**, S379–S383.

Harper, J. L., Rosen, B. R. & White, J. (1986) *The Growth and Form of Modular Organisms.* The Royal Society: London.

Harris, S. D., Morrell, J. L. & Hamer, J. E. (1994) Identification and characterization of *Aspergillus nidulans* mutants defective in cytokinesis. *Genetics* **136**, 517–532.

Hartl, F.-U., Hlodan, R. & Langer, T. (1994) Molecular chaperones in protein folding: the art of avoiding sticky situations. *Trends in Biochemical Sciences* **19**, 20–25.

Hasebe, K., Murakami, S. & Tsuneda, A. (1991) Cytology and genetics of a sporeless mutant of *Lentinus edodes. Mycologia* **83**, 354–359.

Hasegawa, M., Hashimoto, T., Adachi, J., Iwabe, N. & Miyata, T. (1993) Early branchings in the evolution of eukaryotes - ancient divergence of *Entamoeba* that lacks mitochondria revealed by protein-sequence data. *Journal of Molecular Evolution* **36**, 380–388.

Hass, H., Taylor, T. N. & Remy, W. (1994) Fungi from the Lower Devonian Rhynie Chert - mycoparasitism. *American Journal of Botany* **81**, 29–37.

Hatakka, A. (1994) Lignin-modifying enzymes from selected white-rot fungi - production and role in lignin degradation. *FEMS Microbiology Reviews* **13**, 125–135.

Hawes, C. R. & Beckett, A. (1977a) Conidium ontogeny in the *Chalara* state of *Ceratocystis adiposa.* I. Light microscopy. *Transactions of the British Mycological Society* **68**, 259–265.

Hawes, C. R. & Beckett, A. (1977b) Conidium ontogeny in the *Chalara* state of *Ceratocystis adiposa.* II. Electron microscopy. *Transactions of the British Mycological Society* **68**, 267–276.

Hawker, L. E. (1939) The influence of various sources of carbon on the formation of perithecia by *Melanospora destruens* Shear in the presence of accessory growth factors. *Annals of Botany (NS)* **3**, 455–468.

Hawker, L. E. (1947) Further experiments on growth and fruiting of *Melanospora destruens* Shear in the presence of various carbohydrates, with special reference to the effects of glucose and sucrose. *Annals of Botany (NS)* **11**, 245–259.

Hawker, L. E. (1948) Stimulation of the formation of perithecia of *Melanospora destruens* by small quantities of certain phosphoric esters of glucose and fructose. *Annals of Botany (NS)* **12**, 77–79.

Hawker, L. E. (1950) *Physiology of Fungi.* University of London Press: London.

Hawker, L. E. (1966) Environmental influences on reproduction. In *The Fungi, An Advanced Treatise,* **vol. 2** (ed. G. C. Ainsworth & A. S. Sussman), pp. 435–469. Academic Press: New York.

Hawker, L. E. & Beckett, A. (1971) Fine structure and development of the zygospore of *Rhizopus sexualis* (Smith) Callen. *Philosophical Transactions of the Royal Society of London, Series B* **263**, 71–100.

Hawker, L. E. & Chaudhuri, D. D. (1946) Growth and fruiting of certain ascomycetous fungus as influenced by the nature and concentration of the carbohydrate in the medium. *Annals of Botany (NS)* **9**, 185–194.

Hawkins, A. R., Lamb, H. K., Radford, A. & Moore, J. D. (1994) Evolution of transcription-regulating proteins by enzyme recruitment - molecular-models for nitrogen metabolite repression and ethanol utilization in eukaryotes. *Gene* **146**, 145–158.

Hawksworth, D. L. (1991) The fungal dimension of biodiversity - magnitude, significance, and conservation. *Mycological Research* **95**, 641–655.

Hawksworth, D. L. (1995) Challenges in mycology. *Mycological Research* **99**, 127–128.

Hawksworth, D. L., Kirk, P. M., Sutton, B. C. & Pegler, D. N. (1995) *Ainsworth and Bisby's Dictionary of the Fungi*. CAB International: Wallingford, UK.

Hawley, R. S. & Arbel, T. (1993) Yeast genetics and the fall of the classical view of meiosis. *Cell* **72**, 301–303.

Hazen, K. C. (1990) Cell surface hydrophobicity of medically important fungi, especially *Candida* species. In *Microbial Cell Surface Hydrophobicity* (ed. R. J. Doyle & M. Rosenberg), pp. 249–295. American Society for Microbiology: Washington, DC.

Hazen, K. C. & Glee, P. M. (1994) Hydrophobic cell wall protein glycosylation by the pathogenic fungus *Candida albicans*. *Canadian Journal of Microbiology* **40**, 266–272.

Heath, I. B. (1994) The cytoskeleton in hyphal growth, organelle movements, and mitosis. In *The Mycota*, **vol. I**, Growth, Differentiation and Sexuality (ed. J. G. H. Wessels & F. Meinhardt), pp. 43-65. Springer-Verlag: Berlin.

Heath, I. B. (1995a) Integration and regulation of hyphal tip growth. *Canadian Journal of Botany* **73**, S131–S139.

Heath, I. B. (1995b) The cytoskeleton. In *The Growing Fungus* (ed. N. A. R. Gow & G. M. Gadd), pp. 99–134. Chapman & Hall: London.

Heath, I. B. & van Rensburg, E. J. J. (1996) Critical evaluation of the VSC model for tip growth. *Mycoscience* **37**, 71–80.

Heath, M. C. (1995) Signal exchange between higher plants and rust fungi. *Canadian Journal of Botany* **73**, S616–S623.

Hedger, J. N. (1985) Tropical agarics: resource relations and fruiting periodicity. In *Developmental Biology of Higher Fungi* (ed. D. Moore, L. A. Casselton, D. A. Wood & J. C. Frankland), pp. 41–86. Cambridge University Press: Cambridge, UK.

Hedger, J. N., Lewis, P. & Gitay, H. (1993) Litter-trapping by fungi in moist tropical forest. In *Aspects of Tropical Mycology* (ed. S. Isaac, R. Watling, A. J. S. Whalley & J. C. Frankland), pp. 15–35. Cambridge University Press: Cambridge, UK.

Heim, R. J. (1931) Le genre *Inocybe*. In *Encyclopédie Mycologique*, **vol. I**. Lechevalieret Fils: Paris.

Hemmi, K., Julmanop, C., Hirata, D., Tsuchiya, E., Takemoto, J. Y. & Miyakawa, T. (1995) The physiological roles of membrane ergosterol as revealed by the phenotypes of *syr1/erg3* null mutant of *Saccharomyces cerervisiae*. *Bioscience Biotechnology and Biochemistry* **59**, 482–486.

Hendriks, L., DeBaere, R., Van de Peer, Y., Neefs, J., Goris, A. & Dewachter, R. (1991) The evolutionary position of the rhodophyte *Porphyra umbilicalis* and the basidiomycete *Leucosporidium scottii* among other eukaryotes as deduced from complete sequences of small ribosomal-subunit RNA. *Journal of Molecular Evolution* **32**, 167–177.

Hereward, F. V. & Moore, D. (1979) Polymorphic variation in the structure of aerial sclerotia of *Coprinus cinereus*. *Journal of General Microbiology* **113**, 13–18.

Hermann, T. E., Kurtz, M. B. & Champe, S. P. (1983) Laccase localized in Hülle cells and cleistothecial primordia of *Aspergillus nidulans*. *Journal of Bacteriology* **154**, 955–964.

Hibbett, D. S., Grimaldi, D. & Donoghue, M. J. (1995) Cretaceous mushrooms in amber. *Nature* **377**, 487.

Hibbett, D. S., Murakami, S. & Tsuneda, A. (1993a) Hymenophore development and evolution in *Lentinus*. *Mycologia* **85**, 428–443.

Hibbett, D. S., Murakami, S. & Tsuneda, A. (1993b) Sporocarp ontogeny in *Panus* (Basidiomycotina) - evolution and classification. *American Journal of Botany* **80**, 1336–1348.

Hibbett, D. S., Murakami, S. & Tsuneda, A. (1994a) Postmeiotic nuclear behavior in *Lentinus*, *Panus* and *Neolentinus*. *Mycologia* **86**, 725–732.

Hibbett, D. S., Tsuneda, A. & Murakami, S. (1994b) The secotioid form of *Lentinus tigrinus* - genetics and development of a fungal morphological innovation. *American Journal of Botany* **81**, 466–478.

Hibbett, D. S. & Vilgalys, R. (1991) Evolutionary relationships of *Lentinus* to the Polyporaceae - evidence from restriction analysis of enzymatically amplified ribosomal DNA. *Mycologia* **83**, 425–439.

Hibbett, D. S. & Vilgalys, R. (1993) Phylogenetic relationships of *Lentinus* (Basidiomycotina) inferred from molecular and morphological characters. *Systematic Botany* **18**, 409–433.

Hilgenberg, W. & Sandmann, G. (1977) Light-stimulated carbon dioxide fixation in *Phycomyces blakesleeanus* Bgff. *Experimental Mycology* **1**, 265–270.

Hilt, W. & Wolf, D. H. (1995) Proteasomes of the yeast *S. cerevisiae*: genes, structure and functions. *Molecular Biology Reports* **21**, 3–10.

Hirata, A. & Shimoda, A. (1994) Structural modification of spindle pole bodies during meiosis II is essential for the normal formation of ascospores in *Schizosaccharomyces pombe*: ultrastructural analysis of *spo* mutants. *Yeast* **10**, 173–183.

Hirsch, H. M. (1954) Environmental factors influencing the differentiation of protoperithecia and their relation to tyrosinase and melanin formation in *Neurospora crassa*. *Physiologia Plantarum* **7**, 72–97.

Hitt, A. L. & Luna, E. J. (1994) Membrane interactions with the actin cytoskeleton. *Current Topics in Cell Biology* **6**, 120–130.

Hock, B., Bahn, M., Walk, R. A. & Nitschke, U. (1978) The control of fruiting body formation in the ascomycete *Sordaria macrospora* Auersw. by regulation of hyphal development. An analysis based on scanning electron and light microscopic observations. *Planta* **141**, 93–103.

Hodgkiss, I. J. & Harvey, R. (1972) Effect of carbon dioxide on the growth and sporulation of certain coprophilous Pyrenomycetes. *Transactions of the British Mycological Society* **59**, 409–418.

Hoffman, H. (1860) Bietrage zur Entwickelungsgeschichte und Anatomie der Agaricinen. *Botanische Zeitung* **18**, 389–404.

Holzer, H., Hierholzer, G. & Witt, I. (1965) The role of glutamate dehydrogenases in the linkage and regulation of carbohydrate and nitrogen metabolism in yeast. In *Mecanismes de régulation des activités cullulaires chez les microorganismes, Colloques International du CNRS*, pp. 407–416. Centre National de la Recherche Scientifique: Paris.

Honda, Y. & Yunoki, T. (1975) On spectral dependence for maturation of apothecia in *Sclerotinia trifoliorum* Erik. *Annals of the Phytopathological Society of Japan* **41**, 383–389.

Horgen, P. A. (1977) Steroid induction of differentiation: *Achlya* as a model system. In *Eucaryotic Microbes as Model Developmental Systems* (ed. D. H. O'Day & P. A. Horgen), pp. 272–293. Marcel Dekker: New York.

Horgen, P. A. (1981) The role of the steroid sex pheromone antheridiol in controlling the development of male sex organs in the water mold *Achlya*. In *Sexual Interactions in Eukaryotic Microbes* (ed. D. H. O'Day & P. A. Horgen), pp. 155–178. Academic Press: New York.

Horner, J. & Moore, D. (1987) Cystidial morphogenetic field in the hymenium of *Coprinus cinereus*. *Transactions of the British Mycological Society* **88**, 479–488.

Horowitz, N. H., Fling, M., MacLeod, H. L. & Sueoka, N. (1960) Genetic determination and enzymatic induction of tyrosinase in *Neurospora*. *Journal of Molecular Biology* **2**, 96–104.

Horton, J. S. & Horgen, P. A. (1989) Molecular cloning of cDNAs regulated during steroid-induced sexual differentiation in the aquatic fungus *Achlya*. *Experimental Mycology* **13**, 263–273.

Horton, J. S. & Raper, C. A. (1991) A mushroom-inducing DNA sequence isolated from the basidiomycete, *Schizophyllum commune*. *Genetics* **129**, 707–716.

Hosking, S. L., Robson, G. D. & Trinci, A. P. J. (1995) Phosphoinositides play a role in hyphal extension and branching in *Neurospora crassa*. *Experimental Mycology* **19**, 71–80.

Huang, B.-F. & Staples, R. C. (1982) Synthesis of proteins during differentiation of the Bean Rust fungus. *Experimental Mycology* **6**, 7–14.

Huang, M. Y. & Bungay, H. R. (1973) Microprobe measurements of oxygen concentration in mycelial pellets. *Biotechnology and Bioengineering* **15**, 1193–1197.

Hubert, E. E. (1924) The diagnosis of decay in wood. *Journal of Agricultural Research* **29**, 523–567.

Hughes, S. J. (1953) Conidiophores, conidia, and classification. *Canadian Journal of Botany* **31**, 577–659.

Hughes, S. J. (1971) On conidia of fungi, and gemmae of algae, bryophytes, and pteridophytes. *Canadian Journal of Botany* **49**, 1319–1339.

Hurst, L. D. & Peck, J. R. (1996) Recent advances in understanding of the evolution and maintenance of sex. *Trends in Ecology and Evolution* **11**, 79–82.

Hutchinson, S. A., Sharma, P., Clarke, K. R. & MacDonald, I. (1980) Control of hyphal orientation in colonies of *Mucor hiemalis*. *Transactions of the British Mycological Society* **75**, 177–191.

Hyde, G. J., Gubler, F. & Hardham, A. R. (1991) Ultrastructure of zoosporogenesis in *Phytophthora cinnamomi*. *Mycological Research* **95**, 577–591.

Hynes, M. J. (1975) A *cis*-dominant regulatory mutation affecting enzyme induction in the eukaryote *Aspergillus nidulans*. *Nature* **253**, 210–211.

Hynes, M. J. & Pateman, J. A. (1970) The genetic analysis of regulation of amidase synthesis in *Aspergillus nidulans*. 1. Mutants able to utilize acrylamide. *Molecular & General Genetics* **108**, 97–106.

Hynes, R. O. (1992) Integrins: versatility, modulation, and signalling in cell adhesion. *Cell* **69**, 11–25.

Ilott, T. W., Ingram, D. S. & Rawlinson, C. J. (1986) Evidence of a chemical factor involved in the control of sexual development in the light leaf spot fungus, *Pyrenopeziza brassicae* (Ascomycotina) *Transactions of the British Mycological Society* **87**, 303–308.

Ingold, C. T. (1946) Size and form in agarics. *Transactions of the British Mycological Society* **29**, 108–113.

Ingold, C. T. (1953) *Dispersal in Fungi*. Clarendon Press: Oxford, UK.

Ingold, C. T. & Dring, V. J. (1957) An analysis of spore discharge in *Sordaria*. *Annals of Botany (NS)* **21**, 465–477.

Innocenti, F. G., Pohl, U. & Russo, V. E. A. (1983) Photoinduction of protoperithecia in *Neurospora crassa* by blue light. *Photochemistry and Photobiology* **37**, 49–51.

Inoue, Y. & Furuya, M. (1970) Perithecial formation in *Gelasinospora reticulospora*. I. Effects of light at two different growth states. *Development, Growth and Differentiation* **12**, 141–150.

Inoue, Y. & Furuya, M. (1974) Perithecial formation in *Gelasinospora reticulispora*. III. Inhibitory effects of near-UV and blue light during the inductive dark period. *Plant & Cell Physiology* **15**, 469–475.

Inoue, Y. & Furuya, M. (1975) Perithecial formation in *Gelasinospora reticulispora*. IV. Action spectra for the photoinduction. *Plant Physiology* **55**, 1098–1101.

Inoue, Y. & Furuya, M. (1978) Perithecial formation in *Gelasinospora reticulospora*. VI. Inductive effect of microbeam irradiation with blue light. *Planta* **143**, 255–259.

Inoue, Y. & Watanabe, M. (1984) Perithecial formation in *Gelasinospora reticulispora*. VII. Action spectra in UV region for the photoinduction and the photoinhibition of photoinductive effect brought by blue light. *Plant & Cell Physiology* **25**, 107–113.

Irelan, J. T. & Selker, E. U. (1996) Gene silencing in filamentous fungi: RIP, MIP and quelling. *Journal of Genetics* **75**, 313–324.

Ishikawa, T. & Serizawa, H. (1978) Expression of enzyme activities during the fruiting process in *Coprinus macrorhizus*. *Abstracts, 12th International Congress of Microbiology, Munich, Germany*, p. 82, B11.

Iten, W. (1970) Zur funktion hydrolytischer enzyme bei der autolysate von *Coprinus*. *Berichte Schweitzerische Botanische Gesellschaft* **79**, 175–198.

Iten, W. & Matile, P. (1970) Role of chitinase and other lysosomal enzymes of *Coprinus lagopus* in the autolysis of fruiting bodies. *Journal of General Microbiology* **61**, 301–309.

Jabor, F. N. & Moore, D. (1984) Evidence for synthesis *de novo* of NADP-linked glutamate dehydrogenase in *Coprinus* mycelia grown in nitrogen-free medium. *FEMS Microbiology Letters* **23**, 249–252.

Jackson, S. L. & Heath, I. B. (1990) Evidence that actin reinforces the extensible hyphal apex of the oomycete *Saprolegnia ferax*. *Protoplasma* **157**, 144–153.

Jacobs, M. & Stahl, U. (1995) Gene regulation in mycelial fungi. In *The Mycota*, **vol. II**, Genetics and Biotechnology (ed. U. Kück), pp. 155–167. Springer-Verlag: Berlin.

Jacobs, W. P. (1979) *Plant Hormones and Plant Development*. Cambridge University Press: Cambridge, UK.

Jacobson, K. M. & Miller, O. K., Jr. (1994) Postmeiotic mitosis in the basidia of *Suillus granulatus*: implications for population structure and dispersal biology. *Mycologia* **86**, 511–516.

Jacques-Félix, M. (1967) Recherches morphologiques, anatomiques, morphogénétiques et physiologiques sur des rhizomorphes de champignons supérieurs et sur le déterminisme de leur formation. I. Observations sur les formations 'synnémiques' des champignons supérieurs dans le milieu naturel. *Bulletin Trimestrial de la Societé Mycologique de France* **83**, 5–103.

Janshekar, H. & Feichter, A. (1983) Lignin, biosynthesis, application and biodegradation. *Advances in Biochemical Engineering/Biotechnology* **27**, 119–178.

Jeffreys, D. B. & Greulach, V. A. (1956) The nature of tropisms of *Coprinus sterquilinus*. *Journal of the Elisha Mitchell Scientific Society* **72**, 153–158.

Jelitto, T. C., Page, H. A. & Read, N. D. (1994) Role of external signals in regulating the pre-penetration phase of infection by the rice blast fungus, *Magnaporthe grisea*. *Planta* **194**, 471–477.

Jenkins, C. D. (1993) Selection and the evolution of genetic life cycles. *Genetics* **133**, 401–410.

Jennings, D. H. (1986) Morphological plasticity in fungi. *Symposia of the Society for Experimental Biology* **40**, 329–346.

Jennings, D. H. (1991) The physiology and biochemistry of the vegetative mycelium. In Serpula lacrymans*: Fundamental Biology and Control Strategies* (ed. D. H. Jennings & A. F. Bravery), pp. 55–79. John Wiley & Sons: Chichester.

Jennings, D. H. (1995) *The Physiology of Fungal Nutrition*. Cambridge University Press: Cambridge, UK.

Jennings, D. H. & Austin, S. (1973) The stimulatory effect of the non-metabolized sugar 3-O-methyl glucose on the conversion of mannitol and arabitol to polysaccharide and other insoluble compounds in the fungus *Dendryphiella salina*. *Journal of General Microbiology* **75**, 287–294.

Jennings, D. H. & Rayner, A. D. M. (1984) *The Ecology and Physiology of the Fungal Mycelium*. Cambridge University Press: Cambridge, UK.

Jennings, D. H. & Watkinson, S. C. (1982) Structure and development of mycelial strands in *Serpula lacrimans*. *Transactions of the British Mycological Society* **78**, 465–474.

Ji, J. & Moore, D. (1993) Glycogen metabolism in relation to fruit body maturation in *Coprinus cinereus*. *Mycological Research* **97**, 283–289.

Jimenez-Garcia, L. F., Elizundia, J. M., Lopez-Zamorano, B., Maciel, A., Zavala, G., Echeverria, O. M. & Vazqueznin, G. H. (1989) Implications for evolution of nuclear-structures of animals, plants, fungi and protoctists. *BioSystems* **22**, 103–116.

Jirjis, R. I. & Moore, D. (1976) Involvement of glycogen in morphogenesis of *Coprinus cinereus*. *Journal of General Microbiology* **95**, 348–352.

Jirjis, R. I. & Moore, D. (1979) Comparison of spectrophotometric and electrophoretic measurements of glutamate dehydrogenase activity during mycelial growth and carpophore development in the basidiomycete *Coprinus cinereus*. *FEMS Microbiology Letters* **6**, 115–118.

Johnson, B. F., Calleja, G. B. & Yoo, B. Y. (1996) A new model for hyphal tip extension and its application to differential fungal morphogenesis. In *Patterns in Fungal Development* (ed. S. W. Chiu & D. Moore), pp. 37–69. Cambridge University Press: Cambridge, UK.

Johnson, B. F., Yoo, B. Y. & Calleja, G. B. (1995) Smashed fission yeast walls. Structural discontinuities related to wall growth. *Cell Biophysics* **26**, 57–75.

Johnson, D. L. & Gamow, R. I. (1971) The avoidance response in *Phycomyces*. *Journal of General Physiology* **57**, 41–49.

Johnson, T. E. (1976) Analysis of pattern formation in *Neurospora* perithecial development using genetic mosaics. *Developmental Biology* **54**, 23–36.

Johnson, T. E. (1978) Isolation and characterisation of perithecial development mutants in *Neurospora*. *Genetics* **88**, 27–47.

Johnstone, I. L., Hughes, S. G. & Clutterbuck, A. J. (1985) Cloning an *Aspergillus nidulans* developmental gene by transformation. *EMBO Journal* **4**, 1307–1311.

Jones, B. E., Williamson, I. P. & Gooday, G. W. (1981) Sex pheromones in *Mucor*. In *Sexual Interactions in Eukaryotic Microbes* (ed. D. H. O'Day & P. A. Horgen), pp. 179–198. Academic Press: New York.

Jones, C. L., Lonergan, G. T. & Mainwaring, D. E. (1995) Acid phosphatase positional correlations in solid surface fungal cultivation: a fractal interpretation of biochemical differentiation. *Biochemical and Biophysical Research Communications* **208**, 1159–1165.

Jones, E. B. G. (1994) Fungal adhesion. *Mycological Research* **98**, 961–981.

Jülich, W. (1981) Higher taxa of Basidiomycetes. *Bibliotheca Mycologica* **85**, 1–485.

Kalgutkar, R. M. & Sigler, L. (1995) Some fossil fungal form-taxa from the Maastrichtian and Palaeogene ages. *Mycological Research* **99**, 513–522.

Kalisz, H. M. (1988) Microbial proteinases. *Advances in Biochemical Engineering/Biotechnology* **36**, 1–65.

Kalisz, H. M., Moore, D. & Wood, D. A. (1986) Protein utilization by basidiomycete fungi. *Transactions of the British Mycological Society* **86**, 519–525.

Kalisz, H. M., Wood, D. A. & Moore, D. (1987) Production, regulation and release of extracellular proteinase activity in basidiomycete fungi. *Transactions of the British Mycological Society* **88**, 221–227.

Kalisz, H. M., Wood, D. A. & Moore, D. (1989) Some characteristics of extracellular proteinases from *Coprinus cinereus*. *Mycological Research* **92**, 278–285.

Kamada, T., Hirai, K. & Fujii, M. (1993) The role of the cytoskeleton in the pairing and positioning of the two nuclei in the apical cell of the dikaryon of the basidiomycete *Coprinus cinereus*. *Experimental Mycology* **17**, 338–344.

Kamada, T., Kurita, R. & Takemaru, T. (1978) Effects of light on basidiocarp maturation in *Coprinus macrorhizus*. *Plant & Cell Physiology* **19**, 263–275.

Kamada, T., Miyazaki, S. & Takemaru, T. (1976) Quantitative changes of DNA, RNA and protein during basidiocarp maturation in *Coprinus macrorhizus*. *Transactions of the Mycological Society of Japan* **17**, 451–460.

Kamada, T., Sumiyoshi, T., Shindo, Y. & Takemaru, T. (1989a) Isolation and genetic analysis of resistant mutants to the benzimidazole fungicide benomyl in *Coprinus cinereus*. *Current Microbiology* **18**, 215–218.

Kamada, T., Sumiyoshi, T. & Takemaru, T. (1989b) Mutations in β-tubulin block transhyphal migration of nuclei in dikaryosis in the homobasidiomycete *Coprinus cinereus*. *Plant & Cell Physiology* **30**, 1073–1080.

Kamada, T. & Takemaru, T. (1977a) Stipe elongation during basidiocarp maturation in *Coprinus macrorhizus*: mechanical properties of stipe cell wall. *Plant & Cell Physiology* **18**, 831–840.

Kamada, T. & Takemaru, T. (1977b) Stipe elongation during basidiocarp maturation in *Coprinus macrorhizus*: changes in polysaccharide composition of stipe cell wall during elongation. *Plant & Cell Physiology* **18**, 1291–1300.

Kamada, T. & Takemaru, T. (1983) Modifications of cell wall polysaccharides during stipe elongation in the basidiomycete *Coprinus cinereus*. *Journal of General Microbiology* **129**, 703–709.

Kamada, T., Takemaru, T., Prosser, J. I. & Gooday, G. W. (1991) Right and left handed helicity of chitin microfibrils in stipe cells in *Coprinus cinereus*. *Protoplasma* **165**, 64–70.

Kamada, T. & Tsuru, M. (1993) The onset of the helical arrangement of chitin microfibrils in fruit body development of *Coprinus cinereus*. *Mycological Research* **97**, 884–888.

Kameswar, R. K. V. S. R., Aist, J. R. & Crill, J. P. (1985) Mitosis in the rice blast fungus and its possible implications for pathogenic variability. *Canadian Journal of Botany* **63**, 1129–1134.

Kaminskyj, S. G. W., Garrill, A. & Heath, I. B. (1992) The relation between turgor and tip growth in *Saprolegnia ferax*: turgor is necessary, but not sufficient to explain apical extension rates. *Experimental Mycology* **16**, 64–75.

Kaminskyj, S. G. W. & Heath, I. B. (1995) Integrin and spectrin homologues and cytoplasm-wall adhesion in tip growth. *Journal of Cell Science* **108**, 849–856.

Kanda, T., Arakawa, H., Yasuda, Y. & Takemaru, T. (1990) Basidiospore formation in a mutant of incompatibility factors and in mutants that arrest at meta-anaphase I in *Coprinus cinereus*. *Experimental Mycology* **14**, 218–226.

Kanda, T., Ishihara, H. & Takemaru, T. (1989b) Genetic analysis of recessive primordium-less mutants in the basidiomycete *Coprinus cinereus*. *Botanical Magazine (Tokyo)* **102**, 561–564.
Kanda, T., Goto, A., Sawa, K., Arakawa, H., Yasuda, Y. & Takemaru, T. (1989a) Isolation and characterization of recessive sporeless mutants in the basidiomycete *Coprinus cinereus*. *Molecular & General Genetics* **216**, 526–529.
Kanda, T. & Ishikawa, T. (1986) Isolation of recessive developmental mutants in *Coprinus cinereus*. *Journal of General and Applied Microbiology* **32**, 541–543.
Kanda, T., Uno, I. & Ishikawa, T. (1986) Detection of proteins synthesized in the cap cells of *Coprinus cinereus*. *Journal of General and Applied Microbiology* **32**, 283–291.
Kane, S. M. & Roth, R. (1974) Carbohydrate metabolism during ascospore development in yeast. *Journal of Bacteriology* **118**, 8–14.
Kappeli, O., Muller, M. & Feichter, A. (1978) Chemical and structural alterations at the cell surface of *Candida tropicalis*, induced by hydrocarbon substrate. *Journal of Bacteriology* **133**, 952–958.
Karlin, S. & Ladunga, I. (1994) Comparisons of eukaryotic genomic sequences. *Proceedings of the National Academy of Sciences of the USA* **91**, 12832–12836.
Katohda, S., Ito, H., Takahashi, H. & Kikuchi, H. (1988) Carbohydrate metabolism during sporulation in spheroplasts of yeast, *Saccharomyces cerevisiae*. *Agricultural and Biological Chemistry* **52**, 349–355.
Katz, D. & Rosenberger, R. F. (1971) Hyphal wall synthesis in *Aspergillus nidulans*: effects of protein synthesis inhibition and osmotic shock on chitin insertion and morphogenesis. *Journal of Bacteriology* **108**, 184–190.
Kaur, M. & Despande, K. B. (1980) Physiology of perithecial production in *Cochliobolus miyabeanus* (Ito & Kuribayashi) Dreschler ex Dastur. *Biovigyanam* **6**, 83–85.
Kawai, G. & Ikeda, Y. (1982) Fruiting-inducing activity of cerebrosides observed with *Schizophyllum commune*. *Biochimica et Biophysica Acta* **719**, 612–618.
Kelly, M., Burke, J., Smith, M., Klar, A. & Beach, D. (1988) Four mating-type genes control sexual differentiation in the fission yeast. *EMBO Journal* **7**, 1537–1547.
Kemp, R. F. O. (1977) Oidial homing and the taxonomy and speciation of basidiomycetes with special reference to the genus *Coprinus*. In *The Species Concept in Hymenomycetes* (ed. H. Clemençon), pp. 259–273. J. Cramer: Vaduz.
Kempken, F. (1995) Horizontal transfer of a mitochondrial plasmid. *Molecular & General Genetics* **248**, 89–94.
Keon, J. P. R., Jewitt, S. & Hargreaves, J. A. (1995) A gene encoding gamma-adaptin is required for apical extension growth in *Ustilago maydis*. *Gene* **162**, 141–145.
Kerrigan, R. W., Royer, J. C., Baller, L. M., Kohli, Y., Horgen, P. A. & Anderson, J. B. (1993) Meiotic behavior and linkage relationships in the secondarily homothallic fungus *Agaricus bisporus*. *Genetics* **133**, 225–236.
Key, J. L. (1989) Modulation of gene expression by auxin. *Bioessays* **11**, 52–58.
Keyworth, W. G. (1942) The occurrence of tremelloid outgrowths on the pilei of *Coprinus ephemerus*. *Transactions of the British Mycological Society* **25**, 307–310.
Kher, K., Greening, J. P., Hatton, J. P., Novak Frazer, L. & Moore, D. (1992) Kinetics and mechanics of stem gravitropism in *Coprinus cinereus*. *Mycological Research* **96**, 817–824.
Kihara, J. & Kumagai, T. (1994) Ecotypes of the fungus *Bipolaris oryzae* with various responses of the mycochrome system. *Physiologia Plantarum* **92**, 689–695.
Kirk, K. E. & Morris, N. R. (1991) The *tubB* β-tubulin gene is essential for sexual development in *Aspergillus nidulans*. *Genes & Development* **5**, 2014–2023.
Kirk, T. K., Tien, M., Kersten, P. J., Kalyanaraman, B., Hamel, K. E. & Farrell, R. L. (1990) Lignin peroxidase from fungi: *Phanerochaete chrysosporium*. *Methods in Enzymology* **188**, 159–171.
Kitamoto, Y. & Gruen, H. E. (1976) Distribution of cellular carbohydrates during development of mycelium and fruit bodies of *Flammulina velutipes*. *Plant Physiology* **58**, 485–491.
Kitamoto, Y., Horikoshi, T. & Suzuki, A. (1974a) An action spectrum for photoinduction of pileus formation in a basidiomycete, *Favolus arcularius*. *Planta* **119**, 81–84.
Kitamoto, Y., Horikoshi, T. & Kasai, Z. (1974b) Growth of fruit-bodies of *Favolus arcularius*. *Botanical Magazine (Tokyo)* **87**, 41–49.

Kitamoto, Y., Suzuki, A. & Furukawa, S. (1972) An action spectrum for light-induced primordium formation in a basidiomycete, *Favolus arcularius* (Fr) Ames. *Plant Physiology* **49**, 338–340.

Klebs, G. (1898) Zur Physiologie der Fortpflanzung einiger Pilze. I. *Sporodinia grandis. Jahrbücher für wissenschaftliche Botanik* **32**, 1–69.

Klebs, G. (1899) Zur Physiologie der Fortpflanzung einiger Pilze. I. *Saprolegnia mixta. Jahrbücher für wissenschaftliche Botanik* **33**, 513–593.

Klebs, G. (1900) Zur Physiologie der Fortpflanzung einiger Pilze. III. Allgemeine Betrachtungen. *Jahrbücher für wissenschaftliche Botanik* **35**, 80–203.

Klein, S. (1994) Choose your partner: chromosome pairing in yeast meiosis. *BioEssays* **16**, 869–871.

Klemm, E. & Ninnemann, H. (1978) Correlation between absorbance changes and a physiological response induced by blue light in *Neurospora. Photochemistry and Photobiology* **28**, 227–230.

Klemm, E. & Ninnemann, H. (1979) Nitrate reductase - a key enzyme in blue light-promoted conidiation and absorbance change of *Neurospora. Photochemistry and Photobiology* **29**, 629–632.

Kligman, A. M. (1943) Some cultural and genetic problems in the cultivation of the mushroom *Agaricus campestris. American Journal of Botany* **30**, 745–763.

Kobayashi, H., van Dedeem, G. & Moo-Young, M. (1973) Oxygen transfer into mycelial pellets. *Biotechnology and Bioengineering* **27**, 45–47.

Kobori, H., Sato, M. & Osumi, M. (1992) Relationship of actin organization to growth in the two forms of the dimorphic yeast *Candida tropicalis. Protoplasma* **167**, 193–204.

Koch, A. L. (1994) The problem of hyphal growth in streptomycetes and fungi. *Journal of Theoretical Biology* **171**, 137–150.

Kohlwein, S. D., Daum, G., Schneiter, R. & Paltauf, F. (1996) Phospholipids - synthesis, sorting subcellular traffic - the yeast approach. *Trends in Cell Biology* **6**, 260–266.

Kolattukudy, P. E., Rogers, L. M., Li, D. X., Hwang, C. S. & Flaishman, M. A. (1995) Surface signaling in pathogenesis. *Proceedings of the National Academy of Sciences of the USA* **92**, 4080–4087.

Kollar, R., Petrakova, E., Ashwell, G., Robbins, P. W. & Cabib, E. (1995) Architecture of the yeast cell wall. The linkage between chitin and (1→3)-glucan. *Journal of Biological Chemistry* **270**, 1170–1178.

Konishi, M. (1967) Growth promoting effect of certain amino acids on the *Agaricus* fruit body. *Mushroom Science* **6**, 121–134.

Konishi, M. & Hagimoto, H. (1961) Studies on the growth of fruitbody of fungi. III. Occurrence, formation and destruction of indole acetic acid in the fruitbody of *Agaricus bisporus* (Lange) Sing. *Plant & Cell Physiology* **2**, 425–434.

Konishi, M. & Hagimoto, H. (1962) Growth promoting effect of amino acids in the *Agaricus* fruit body. *Plant Physiology (suppl.)* **37**, x–xi.

Korn, R. W. (1993) Heterogeneous growth of plant tissues. *Botanical Journal of the Linnean Society* **112**, 351–371.

Kosasih, B. D. & Willetts, H. J. (1975) Ontogenetic and histochemical studies of the apothecium of *Sclerotinia sclerotiorum. Annals of Botany (NS)* **39**, 185–191.

Kothe, E. (1996) Tetrapolar fungal mating types - sexes by the thousands. *FEMS Microbiology Reviews* **18**, 65–87.

Kramer, C. L. & Pady, S. M. (1970) Ascospore discharge in *Hypoxylon. Mycologia* **62**, 1170–1186.

Kretschmer, S. (1988) Stepwise increase of elongation rate in individual hyphae of *Streptomyces granaticolor* during outgrowth. *Journal of Basic Microbiology* **28**, 35–43.

Kritsky, M. S. (1977) Nicotinamide dinucleotide coenzymes and regulation of different processes in fungi. *Abstracts, Second International Mycological Congress, Tampa, Florida*, 361.

Kropf, D. L., Caldwell, J. C., Gow, N. A. R. & Harold, F. M. (1984) Transcellular ion currents in the water mould *Achlya.* Amino acid proton symport as a mechanism of current entry. *Journal of Cell Biology* **99**, 486–496.

Kropf, D. L., Lupa, M. D. A., Caldwell, J. C. & Harold, F. M. (1983) Cell polarity: endogenous ion currents precede and predict branching in the water mould *Achlya. Science* **220**, 1385–1387.

Kües, U. & Casselton, L. A. (1992) Fungal mating type genes - regulators of sexual development. *Mycological Research* **96**, 993–1006.

Kuhad, R. C., Rosin, I. V. & Moore, D. (1987) A possible relation between cyclic-AMP levels and glycogen mobilisation in *Coprinus cinereus*. *Transactions of the British Mycological Society* **88**, 229–236.

Kühn, K. & Eble, J. (1994) The structural bases of integrin-ligand interactions. *Trends in Cell Biology* **4**, 256–261.

Kühner, R. (1977) Variation of nuclear behaviour in the homobasidiomycetes. *Transactions of the British Mycological Society* **68**, 1–16.

Kühner, R. (1980) Les Hymenomycètes agaricoides. *Bulletin mensuel de la Société Linnéenne de Lyon numéro special*, 1–1028.

Kumar, S., Mukerji, K. G. & Lal, R. (1996) Molecular aspects of pesticide degradation by microorganisms. *Critical Reviews in Microbiology* **22**, 1–26.

Kumar, S. & Punekar, N. S. (1997) The metabolism of 4-aminobutyrate (GABA) in fungi. *Mycological Research* **101**, 403–409.

Kumar, S. & Rzhetsky, A. (1996) Evolutionary relationships of eukaryotic kingdoms. *Journal of Molecular Evolution* **42**, 183–193.

Kurtz, M. B. & Champe, S. P. (1981) Dominant spore color mutants of *Aspergillus nidulans* defective in germination and sexual development. *Journal of Bacteriology* **148**, 629–638.

Labarère, J. & Bernet, J. (1977) Protoplasmic incompatibility and cell lysis in *Podospora anserina*. I. Genetic investigations on mutations of a novel modifier gene that suppresses cell destruction. *Genetics* **87**, 249–257.

Lafay, J. F., Matricon, J. & Bodere, C. (1975) The avoidance response of *Phycomyces*: distance dependence of the response. *Physiologie Végétale* **13**, 259–263.

Lambert, E. B. (1930) Two new diseases of cultivated mushrooms. *Phytopathology* **20**, 917–919.

Lambert, E. B. (1933) Effect of excess carbon dioxide on growing mushrooms. *Journal of Agricultural Research* **47**, 599–608.

Langeron, M. & Vanbreuseghem, R. (1965) *Outline of Mycology*, **vol 2**, 2nd edn, translated from the French by J. Wilkinson. Sir Isaac Pitman & Sons: London.

Lauter, F. R., Russo, V. E. A. & Yanofsky, C. (1992) Developmental and light regulation of *eas*, the structural gene for the rodlet protein of *Neurospora*. *Genes & Development* **6**, 2373–2381.

Leach, C. M. & Trione, E. J. (1966) Action spectra for light-induced sporulation of the fungi *Pleospora herbarum* and *Alternaria dauci*. *Photochemistry and Photobiology* **5**, 621–630.

Leake, J. R. & Read, D. J. (1990) Chitin as a nitrogen source for mycorrhizal fungi. *Mycological Research* **94**, 993–995.

Leatham, G. F. & Stahmann, M. A. (1981) Studies on the laccase of *Lentinus edodes*: specificity, localization and association with the development of fruiting bodies. *Journal of General Microbiology* **125**, 147–157.

Leatham, G. F. & Stahmann, M. N. (1987) Effect of light and aeration on fruiting of *Lentinula edodes*. *Transactions of the British Mycological Society* **88**, 9–20.

Lee, B. N. & Adams, T. H. (1994) Over-expression of *flbA*, an early regulator of *Aspergillus* asexual sporulation, leads to activation of *brlA* and premature initiation of development. *Molecular Microbiology* **14**, 323–334.

Lee, Y. H. & Dean, R. A. (1993) Stage-specific gene expression during appressorium formation of *Magnaporthe grisea*. *Experimental Mycology* **17**, 215–222.

Leeuw, T., Fourestlieuvin, A., Wu, C. L., Chenevert, J., Clark, K., Whiteway, M., Thomas, D. Y. & Leberer, E. (1995) Pheromone response in yeast: association of Bem1p with proteins of the MAP kinase cascade and actin. *Science* **270**, 1210–1213.

Lemaire, M. (1996) The cellulosome - an exocellular multiprotein complex specialised in cellulose degradation. *Critical Reviews in Biochemistry & Molecular Biology* **31**, 201–236.

Lenard, J. (1992) Mammalian hormones in microbial cells. *Trends in Biochemical Sciences* **17**, 147–150.

Lentz, P. L. (1954) Modified hyphae of hymenomycetes. *The Botanical Review* **20**, 135–199.

Leonard, T. J. (1971) Phenoloxidase activity and fruiting body formation in *Schizophyllum commune*. *Journal of Bacteriology* **106**, 162–167.

Leonard, T. J. (1972) Phenoloxidase activity in mycelia carrying modifier mutations that affect sporocarp development in *Schizophyllum commune*. *Journal of Bacteriology* **111**, 292–293.

Leonard, T. J. & Dick, S. (1968) Chemical induction of haploid fruiting bodies in *Schizophyllum commune*. *Proceedings of the National Academy of Sciences of the USA* **59**, 745–751.

Leonard, T. J. & Dick, S. (1973) Induction of haploid fruiting by mechanical injury in *Schizophyllum commune*. *Mycologia* **65**, 809–822.

Leonard, T. J. & Phillips, L. E. (1973) Study of phenoloxidase activity during the reproductive cycle of *Schizophyllum commune*. *Journal of Bacteriology* **114**, 7–10.

Le Roux, P. (1967) Métabolisme carboné et azoté du champignon cultivé. *Mushroom Science* **6**, 179–190.

Le Roux, P. (1968) Action du gaz carbonique sur le métabolisme du carpophore d'*Agaricus bisporus*. *Mushroom Science* **7**, 31–36.

Leslie, J. F. & Leonard, T. J. (1979a) Monokaryotic fruiting in *Schizophyllum commune*: genetic control of the response to mechanical injury. *Molecular & General Genetics* **175**, 5–12.

Leslie, J. F. & Leonard, T. J. (1979b) Three independent genetic systems that control initiation of a fungal fruiting body. *Molecular & General Genetics* **171**, 257–260.

Letacon, F. & Selosse, M. A. (1984) The mycorrhizal symbiosis in the evolution and the colonization of continents by the life. *Acta Botanica Gallica* **141**, 405–419.

Letham, D. B. (1975) Stimulation by light of apothecial initial development of *Sclerotinia sclerotiorum*. *Transactions of the British Mycological Society* **65**, 333–335.

Levin, D. E. & Errede, B. (1995) The proliferation of MAP kinase signaling pathways in yeast. *Current Opinion in Cell Biology* **7**, 197–202.

Levina, N. N., Lew, R. R., Hyde, G. J. & Heath, I. B. (1995) The roles of Ca^{2+} and plasma membrane ion channels in hyphal tip growth of *Neurospora crassa*. *Journal of Cell Science* **108**, 3405–3417.

Levine, M. (1914) The origin and development of lamellae in *Coprinus micaceus*. *American Journal of Botany* **1**, 343–356.

Lewis, D. H. & Smith, D. C. (1967) Sugar alcohols (polyols) in fungi and green plants. I. Distribution, physiology and metabolism. *New Phytologist* **66**, 143–184.

Lewis, D. & Vakeria, D. (1977) Resistance to *p*-fluorophenylalanine in diploid/haploid dikaryons: dominance modifier gene explained as a controller of hybrid multimer formation. *Genetical Research* **30**, 31–43.

Lewis, F. T. (1931) A comparison between the mosaic of polygons in a film of artificial emulsion and the pattern of simple epithelium in surface view (cucumber epidermis and human amnion) *Anatomical Record*, 235–265.

Lewis, F. T. (1943) The geometry of growth and cell division in epithelial mosaics. *American Journal of Botany* **30**, 766–776.

Lewis, L. A. (1975) Effect of visible light on the partial synchronization of meiosis in *Ascobolus immersus*. *Transactions of the British Mycological Society* **65**, 148–152.

Li, C. Y. (1979) Light and temperature induced sporocarp formation of *Phellinus weirii*. *Canadian Journal of Forestry Research* **9**, 535–538.

Lilly, V. G. & Barnett, H. L. (1951) *Physiology of the Fungi*. McGraw-Hill Book Company: New York.

Lindenmayer, A. (1965) Terminal oxidation and electron transport. In *The Fungi* (ed. G. C. Ainsworth & A. S. Sussman), pp. 301–348. Academic Press: New York.

Löbler, M. & Klämbt, D. (1985) Auxin-binding protein from coleoptile membranes of corn (*Zea mays* L.) II. Localisation of a putative auxin receptor. *Journal of Biological Chemistry* **260**, 9854–9859.

Lodha, B. C. (1974) Decomposition of digested litter. In *Biology of Plant Litter Decomposition* (ed. C. H. Dickinson & J. F. Pugh), pp. 213–241. Academic Press: London.

Lohwag, H. (1941) Anatomie der Asco- und Basidiomyceten. In *Handbuch der Pflanzenanatomie*, **Bd. 4**, Abt. 2 (ed. K. Linsbauer). Borntraeger: Berlin.

Loidl, J. (1995) Meiotic chromosome pairing in triploid and tetraploid *Saccharomyces cerevisiae*. *Genetics* **139**, 1511–1520.

Lombard, F. F. (1990) A cultural study of several species of *Antrodia* (Polyporaceae, Aphyllophorales) *Mycologia* **82**, 185–191.

Long, P. E. & Jacobs, L. (1974) Aseptic fruiting of the cultivated mushroom *Agaricus bisporus*. *Transactions of the British Mycological Society* **63**, 99–107.

Longtine, M. S., Demarini, D. J., Valencik, M. L., Alawar, O. S., Fares, H., de Virgilio, C. & Pringle, J. R. (1996) The septins - roles in cytokinesis and other processes. *Current Opinion in Cell Biology* **8**, 106–119.

Lopez, M. C., Nicaud, J. M., Skinner, H. B., Vergnolle, C., Kader, J. C., Bankaitis, V. A. & Gaillardin, C. (1994) A phosphatidylinositol phosphatidylcholine transfer protein is required for differentiation of the dimorphic yeast *Yarrowia lipolytica* from the yeast to the mycelial form. *Journal of Cell Biology* **125**, 113–127.

López-Franco, R., Bartnicki-Garcia, S. & Bracker, C. E. (1994) Pulsed growth of fungal hyphal tips. *Proceedings of the National Academy of Sciences of the USA* **91**, 12228–12232.

Lopez-Real, J. M. (1975) Formation of pseudosclerotia ("zone lines") in wood decayed by *Armillaria mellea* and *Stereum hirsutum*. *Transactions of the British Mycological Society* **64**, 465–471.

Lopez-Real, J. M. & Swift, M. J. (1975) Formation of pseudosclerotia ("zone lines") in wood decayed by *Armillaria mellea* and *Stereum hirsutum*. II. Formation in relation to the moisture content of the wood. *Transactions of the British Mycological Society* **64**, 473–481.

Lopez-Real, J. M. & Swift, M. J. (1977) Formation of pseudosclerotia ("zone lines") in wood decayed by *Armillaria mellea* and *Stereum hirsutum*. III. Formation in relation to composition of gaseous atmosphere in wood. *Transactions of the British Mycological Society* **68**, 321–325.

Lu, B. C. (1972) Dark dependence of meiosis at elevated temperature in the Basidiomycete, *Coprinus lagopus*. *Journal of Bacteriology* **111**, 833–834.

Lu, B. C. (1974) Meiosis in *Coprinus*. V. The role of light on basidiocarp initiation, mitosis, and hymenium differentiation in *Coprinus lagopus*. *Canadian Journal of Botany* **52**, 299–305.

Lu, B. C. (1982) Replication of deoxyribonucleic acid and crossing over in *Coprinus*. In *Basidium and Basidiocarp: Evolution, Cytology, Function and Development* (ed. K. Wells & E. K. Wells), pp. 93–112. Springer-Verlag: New York.

Lu, B. C. (1991) Cell degeneration and gill remodelling during basidiocarp development in the fungus *Coprinus cinereus*. *Canadian Journal of Botany* **69**, 1161–1169.

Lu, B. C. (1993) Spreading the synaptonemal complex of *Neurospora crassa*. *Chromosoma* **102**, 464–472.

Lu, B. C. (1996) Chromosomes, mitosis and meiosis. In *Fungal Genetics* (ed. C. J. Bos), pp. 119–176. Marcel Dekker,: New York.

Lu, B. C. & Chiu, S. M. (1978) Meiosis in *Coprinus*. IX. The influence of premeiotic S- phase arrest and cold temperature on the meiotic cell cycle. *Journal of Cell Science* **32**, 21–30.

Lu, C. F., Montijn, R. C., Brown, J. L., Klis, F., Kurjan, J., Bussey, H. & Lipke, P. N. (1995) Glycosyl phosphatidylinositol-dependent cross-linking of α-agglutinin and β-1,6-glucan in the *Saccharomyces cerevisiae* cell wall. *Journal of Cell Biology* **128**, 333–340.

Lü, H. & McLaughlin, D. J. (1995) A light and electron microscopic study of mitosis in the clamp connection of *Auricularia auricula-judae*. *Canadian Journal of Botany* **73**, 315–332.

Lugones, L. G., Bosscher, J. S., Scholtmeyer, K., de Vries, O. M. H. & Wessels, J. G. H. (1996) An abundant hydrophobin (ABH1) forms hydrophobic rodlet layers in *Agaricus bisporus* fruiting bodies. *Microbiology - UK* **142**, 1321–1329.

Lysek, G. (1976) Formation of perithecia in colonies of *Podospora anserina*. *Planta* **133**, 81–83.

Lysek, G. (1984) Physiology and ecology of rhythmic growth and sporulation in fungi. In *The Ecology and Physiology of the Fungal Mycelium* (ed. D. H. Jennings & A. D. M. Rayner), pp. 323–342. Cambridge University Press: Cambridge, UK.

Maas Geesteranus, R. A. (1971) *Hydnaceous Fungi of the Eastern Old World*. North-Holland Publishing Co.: Amsterdam.

Maas Geesteranus, R. A. (1975) *The Terrestrial Hydnums of Europe*. North-Holland Publishing Co.: Amsterdam.

MacBride, E. W. (1914) *Textbook of Embryology*, **vol. I**, Invertebrata. Macmillan: London.

MacDonald, D. J. & Bond, D. J. (1976) Genetic and environmental factors influencing the production and distribution of protoperithecia in *Sordaria brevicollis*. *Journal of General Microbiology* **95**, 375–380.

Macfarlane, T. D., Kuo, J. & Hilton, R. N. (1978) Structure of the giant sclerotium of *Polyporus mylittae*. *Transactions of the British Mycological Society* **71**, 359–365.

Machlis, L. (1973) Factors affecting the stability and accuracy of the bioassay for the sperm attractant sirenin. *Plant Physiology* **52**, 524–526.

Madelin, M. F. (1956a) The influence of light and temperature on fruiting of *Coprinus lagopus* Fr. in pure culture. *Annals of Botany (NS)* **20**, 467–480.

Madelin, M. F. (1956b) Studies on the nutrition of *Coprinus lagopus* Fr. especially as affecting fruiting. *Annals of Botany* (*NS*) **20**, 307–330.

Madelin, M. F. (1960) Visible changes in the vegetative mycelium of *Coprinus lagopus* Fr. at the time of fruiting. *Transactions of the British Mycological Society* **43**, 105–110.

Madelin, M. F., Toomer, D. J. & Ryan, J. (1978) Spiral growth of fungus colonies. *Journal of General Microbiology* **106**, 73–80.

Manachère, G. (1970) Recherches physiologiques sur la fructification de *Coprinus congregatus*: action de la lumière; rhythme de production de carpophores. *Annales des Sciences Naturelles, Botanique, 12e série* **11**, 1–96.

Manachère, G. (1971) Research on the fruiting rhythm of a basidiomycete mushroom *Coprinus congregatus* Bull. ex Fr. *Journal of Interdisciplinary Cycle Research* **2**, 199–209.

Manachère, G. (1985) Sporophore differentiation of higher fungi: a survey of some actual problems. *Physiologie Végétale* **23**, 221–230.

Manachère, G. (1988) Regulation of sporophore differentiation in some macromycetes, particularly in coprini: an overview of some experimental studies, from fruiting initiation to sporogenesis. *Cryptogamie Mycologie* **9**, 291–323.

Manachère, G. & Bastouill-Descollonges, Y. (1983) Effect of 'reduced' photoperiods on development of carpophores of *Coprinus congregatus*. *Transactions of the British Mycological Society* **81**, 630–634.

Manachère, G., Robert, J. C., Durand, R., Bret, J. P. & Fevre, M. (1983) Differentiation in the Basidiomycetes. In *Fungal Differentiation: a Contemporary Synthesis* (ed. J. E. Smith), pp. 481–514. Marcel Dekker: New York.

Manavathu, A. K. & Thomas, D. D. S. (1985) Chemotropism of *Achlya ambisexualis* to methionine and methionyl compounds. *Journal of General Microbiology* **131**, 751–756.

Manning, J. T. (1976) Is sex maintained to facilitate or minimise mutational advance? *Heredity* **36**, 351–357.

Manning, J. T. (1977) The evolution of dominance: Haldane v Fisher revisited. *Heredity* **38**, 117–119.

Manocha, M. S. (1965) Fine structure of the *Agaricus* carpophore. *Canadian Journal of Botany* **43**, 1329–1333.

Margulis, L. (1974) Five-Kingdom classification and the origin and evolution of cells. *Evolutionary Biology* **7**, 45–78.

Margulis, L. (1992) Biodiversity - molecular biological domains, symbiosis and Kingdom origins. *BioSystems* **27**, 39–51.

Markham, P. (1992) Stress management - filamentous fungi as exemplary survivors. *FEMS Microbiology Letters* **100**, 379–385.

Markham, P. (1994) Occlusions of septal pores in filamentous fungi. *Mycological Research* **98**, 1089–1106.

Markham, P. (1995) Organelles of filamentous fungi. In *The Growing Fungus* (ed. N. A. R. Gow & G. M. Gadd), pp. 75–98. Chapman & Hall: London.

Marks, G. C. (1991) Causal morphology and evolution of mycorrhizas. *Agriculture Ecosystems & Environment* **35**, 89–104.

Marks, J., Hagan, I. M. & Hyams, J. S. (1987) Spatial association of F-actin with growth polarity and septation in the fission yeast *Schizosaccharomyces pombe*. *Special Publication of the Society for General Microbiology* **23**, 119–135.

Marshall, K. C. (1991) The importance of studying microbial cell surfaces. In *Microbial Cell Surface Analysis* (ed. N. Mozes, P. S. Handley, H. J. Busscher & P. G. Rouxhet), pp. 3–19. VCH Publishers: New York.

Martin, F. N. (1995) Electrophoretic karyotype polymorphisms in the genus *Pythium*. *Mycologia* **87**, 333–353.

Martinelli, S. D. (1979) Phenotypes of double conidiation mutants of *Aspergillus nidulans*. *Journal of General Microbiology* **114**, 277–287.

Martinelli, S. D. & Clutterbuck, A. J. (1971) A quantitative survey of conidiation mutants in *Aspergillus nidulans*. *Journal of General Microbiology* **69**, 261–268.

Martinez-Pastor, M. T. & Estruch, F. (1996) Sudden depletion of carbon source blocks translation, but not transcription, in the yeast *Saccharomyces cerevisiae*. *FEBS Letters* **390**, 319–322.

Marzluf, G. A. (1996) Regulation of nitrogen metabolism in mycelial fungi. In *The Mycota*, **vol. III**, Biochemistry and Molecular Biology (ed. R. Brambl & G. A. Marzluf), pp. 357–368. Springer-Verlag: Berlin.

Mathew, K. T. (1961) Morphogenesis of mycelial strands in the cultivated mushroom *Agaricus bisporus*. *Transactions of the British Mycological Society* **44**, 285–290.

Matsuyama, S. S., Nelson, R. E. & Siegel, R. W. (1974) Mutations specifically blocking differentiation of macroconidia in *Neurospora crassa*. *Developmental Biology* **41**, 278–287.

Matthews, T. R. & Niederpruem, D. J. (1972) Differentiation in *Coprinus lagopus*. I. Control of fruiting and cytology of initial events. *Archiv für Mikrobiologie* **87**, 257–268.

Matthews, T. R. & Niederpruem, D. J. (1973) Differentiation in *Coprinus lagopus*. II. Histology and ultrastructural aspects of developing primordia. *Archiv für Mikrobiologie* **88**, 169–180.

Mau, J. L., Beelman, R. B. & Ziegler, G. R. (1992) Effect of 10-oxo-*trans*-8-decenoic acid on growth of *Agaricus bisporus*. *Phytochemistry* **31**, 4059–4064.

Maynard Smith, J. (1978) *The Evolution of Sex*. Cambridge University Press: Cambridge, UK.

Mays, R. W., Beck, K. A. & Nelson, W. J. (1994) Organization and function of the cytoskeleton in polarized epithelial cells: a component of the protein sorting machinery. *Current Topics in Cell Biology* **6**, 16–24.

McCorkindale, N. J. (1976) The biosynthesis of terpenes and steroids. In *The Filamentous Fungi*, **vol II**, Biosynthesis and Metabolism (ed. J. E. Smith & D. R. Berry), pp. 369–422. Edward Arnold: London.

McGarry, A. & Burton, K. S. (1994) Mechanical properties of the mushroom *Agaricus bisporus*. *Mycological Research* **98**, 241–245.

McGillivray, A. M. & Gow, N. A. R. (1986) Applied electrical fields polarize the growth of mycelial fungi. *Journal of General Microbiology* **132**, 2515–2525.

McKerracher, L. J. & Heath, I. B. (1986) Fungal nuclear behaviour analysed by ultraviolet microbeam irradiation. *Cell Motility & Cytoskeleton* **6**, 35–47.

McLaughlin, D. J. (1974) Ultrastructural localization of carbohydrate in the hymenium and subhymenium of *Coprinus*: evidence for the function of the Golgi apparatus. *Protoplasma* **82**, 341–364.

McLaughlin, D. J. (1977) Basidiospore initiation and early development in *Coprinus cinereus*. *American Journal of Botany* **64**, 1–16.

McLaughlin, D. J. (1982) Ultrastructure and cytochemistry of basidial and basidiospore development. In *Basidium and Basidiocarp: Evolution, Cytology, Function and Development* (ed. K. Wells & E. K. Wells), pp. 37–74. Springer-Verlag: New York.

McLaughlin, D. J., Berres, M. E. & Szabo, L. J. (1995) Molecules and morphology in basidiomycete phylogeny. *Canadian Journal of Botany* **73**, S684–S692.

McLean, K. M. & Prosser, J. I. (1987) Development of vegetative mycelium during colony growth of *Neurospora crassa*. *Transactions of the British Mycological Society* **88**, 489–495.

McMorris, T. C. (1978a) Antheridiol and the oogoniols, steroid hormones which control sexual reproduction in *Achlya*. *Philosophical Transactions of the Royal Society of London, Series B* **284**, 459–470.

McMorris, T. C. (1978b) Sex hormones of the aquatic fungus *Achlya*. *Lipids* **13**, 716–722.

Megnet, R. (1965) Effect of 2-deoxyglucose on *Schizosaccharomyces pombe. Journal of Bacteriology* **90**, 1032–1035.

Meinhardt, F. & Esser, K. (1983) Genetic aspects of sexual differentiation in fungi. In *Fungal Differentiation* (ed. J. E. Smith), pp. 537–557. Marcel Dekker: New York.

Meinhardt, H. (1976) Morphogenesis of lines and nets. *Differentiation* **6**, 117–123.

Meinhardt, H. (1984) Models of pattern formation and their application to plant development. In *Positional Controls in Plant Development* (ed. P. W. Barlow & D. J. Carr), pp. 1–32. Cambridge University Press: Cambridge, UK.

Meinhardt, H. & Gierer, A. (1974) Applications of a theory of biological pattern formation based on lateral inhibition. *Journal of Cell Science* **15**, 321–346.

Metzenberg, R. L. (1990) The role of similarity and difference in fungal mating. *Genetics* **125**, 457–462.

Metzenberg, R. L. (1991) The impact of molecular biology on mycology. *Mycological Research* **95**, 9–13.

Metzenberg, R. L. & Glass, N. L. (1990) Mating type and mating strategies in *Neurospora. BioEssays* **12**, 53–60.

Meyer, P. (1996) Repeat-induced gene silencing - common mechanisms in plants and fungi. *Hoppe-Seyler's Biological Chemistry* **377**, 87–95.

Meyerhoff, J., Tiller, V. & Bellgardt, K. H. (1995) Two mathematical models for the development of a single microbial pellet. 1. Detailed morphological model based on the description of individual hyphae. *Bioprocess Engineering* **12**, 305–313.

Michel, F. C., Grulke, E. A. & Reddy, C. A. (1992) A kinetic model for the fungal pellet lifecycle. *AIChE Journal* **38**, 1449–1460.

Miller, B. L. (1990) The developmental genetics of asexual reproduction in *Aspergillus nidulans. Seminars in Developmental Biology* **1**, 207–219.

Miller, J. H. (1980) Orientation of the plane of cell division in fern gametophytes: the roles of cell shape and stress. *American Journal of Botany* **67**, 534–542.

Miller, K. Y., Wu, J. & Miller, B. L. (1992) *StuA* is required for cell pattern formation in *Aspergillus. Genes & Development* **6**, 1770–1782.

Mims, C. W., Richardson, E. A. & Timberlake, W. E. (1988) Ultrastructural analysis of conidiophore development in the fungus *Aspergillus nidulans* using freeze substitution. *Protoplasma* **144**, 132–141.

Mims, C. W. & Seabury, F. (1989) Ultrastructure of tube formation and basidiospore development in *Ganoderma lucidum. Mycologia* **81**, 754–764.

Minter, D. W., Kirk, P. M. & Sutton, B. C. (1983) Thallic phialides. *Transactions of the British Mycological Society* **80**, 39–66.

Mirabito, P. M., Adams, T. H. & Timberlake, W. E. (1989) Interactions of three sequentially expressed genes control temporal and spatial specificity in *Aspergillus* development. *Cell* **57**, 859–868.

Mirocha, C. J. & DeVay, J. E. (1971) Growth of fungi on an inorganic medium. *Canadian Journal of Microbiology* **17**, 1373–1378.

Miyake, H., Takemaru, T. & Ishikawa, T. (1980a) Sequential production of enzymes and basidiospore formation in fruiting bodies of *Coprinus macrorhizus. Archives of Microbiology* **126**, 201–205.

Miyake, H., Tanaka, K. & Ishikawa, T. (1980b) Basidiospore formation in monokaryotic fruiting bodies of a mutant strain of *Coprinus macrorhizus. Archives of Microbiology* **126**, 207–212.

Miyata, M., Kanbe, T. & Tanaka, K. (1985) Morphological alterations of the fission yeast *Schizosaccharomyces pombe* in the presence of aculeacin A: spherical wall formation. *Journal of General Microbiology* **131**, 611–621.

Miyata, M., Kitamura, J. & Miyata H. (1980) Lysis of growing fission-yeast cells induced by aculeacin A, a new antifungal antibiotic. *Archives of Microbiology* **127**, 11–16.

Miyata, M., Miyata, H. & Johnson, B. F. (1986a) Asymmetric location of the septum in morphologically altered cells of the fission yeast *Schizosaccharomyces pombe. Journal of General Microbiology* **132**, 883–891.

Miyata, M., Miyata, H. & Johnson, B. F. (1986b) Establishment of septum orientation in a morphologically altered fission yeast *Schizosaccharomyces pombe. Journal of General Microbiology* **132**, 2535–2540.

Moens, P. B. (1994) Molecular perspectives of chromosome pairing at meiosis. *BioEssays* **16**, 101–106.

Mol, P. C., Park, H. M., Mullins, J. T. & Cabib, E. (1994) A GTP-binding protein regulates the activity of (1→3)-β-glucan synthase, an enzyme directly involved in yeast cell wall morphogenesis. *Journal of Biological Chemistry* **269**, 31267–31274.

Mollenhauer, H. H. & Morre, D. J. (1994) Structure of Golgi-apparatus. *Protoplasma* **180**, 1428.

Moller, A. P. (1993) A fungus infecting domestic flies manipulates sexual behavior of its host. *Behavioral Ecology and Sociobiology* **33**, 403–407.

Molowitz, R., Bahn, M. & Hock, B. (1976) The control of fruiting body formation in the ascomycete *Sordaria macrospora* Auersw. by arginine and biotin: a two-factor analysis. *Planta* **128**, 143–148.

Momany, M. & Hamer, J. E. (1996) Actin and chitin ring formation in septation. In *Abstracts, American Phytopathological Society & Mycological Society of America joint annual meeting, Indianapolis, Indiana, USA, July 27–31, 1996* abstract 175M.

Momany, M., Morrell, J. L., Harris, S. D. & Hamer, J. E. (1995) Septum formation in *Aspergillus nidulans*. *Canadian Journal of Botany* **73**, S396–S399.

Money, N. P. (1994) Osmotic adjustment and the role of turgor in mycelial fungi. In *The Mycota,* **vol. I**, Growth, Differentiation and Sexuality (ed. J. G. H. Wessels & F. Meinhardt), pp. 67-88. Springer-Verlag: Berlin.

Money, N. P. & Harold, F. M. (1992) Extension growth of the water mold *Achlya*: interplay of turgor and wall strength. *Proceedings of the National Academy of Sciences of the USA* **89**, 4245–4249.

Money, N. P. & Harold, F. M. (1993) Two water molds can grow without measurable turgor pressure. *Planta* **190**, 426–430.

Monzer, J. (1995) Actin filaments are involved in cellular graviperception of the basidiomycete *Flammulina velutipes*. *European Journal of Cell Biology* **66**, 151–156.

Moore, D. (1969a) Effect of 2-deoxy-D-glucose on mycelial growth of filamentous fungi. *Transactions of the British Mycological Society* **53**, 139–141.

Moore, D. (1969b) Sources of carbon and energy used by *Coprinus lagopus sensu* Buller. *Journal of General Microbiology* **58**, 49–56.

Moore, D. (1981a) Developmental genetics of *Coprinus cinereus*: genetic evidence that carpophores and sclerotia share a common pathway of initiation. *Current Genetics* **3**, 145–150.

Moore, D. (1981b) Effects of hexose analogues on fungi: mechanisms of inhibition and of resistance. *New Phytologist* **87**, 487–515.

Moore, D. (1981c) Evidence that the NADP-linked glutamate dehydrogenase of *Coprinus cinereus* is regulated by acetyl-CoA and ammonium levels. *Biochimica et Biophysica Acta* **661**, 247–254.

Moore, D. (1984a) Developmental biology of the *Coprinus cinereus* carpophore: metabolic regulation in relation to cap morphogenesis. *Experimental Mycology* **8**, 283–297.

Moore, D. (1984b) Positional control of development in fungi. In *Positional Controls in Plant Development* (ed. P. W. Barlow & D. J. Carr), pp. 107–135. Cambridge University Press: Cambridge, UK.

Moore, D. (1987) The formation of agaric gills. *Transactions of the British Mycological Society* **89**, 105–108.

Moore, D. (1991) Perception and response to gravity in higher fungi - a critical appraisal. *New Phytologist* **117**, 3–23.

Moore, D. (1995) Tissue formation. In *The Growing Fungus* (ed. N. A. R. Gow & G. M. Gadd), pp. 423–465. Chapman & Hall: London.

Moore, D. (1996) Graviresponses in fungi. *Advances in Space Research* **17**, 73–82.

Moore, D. & Devadatham, M. S. (1979) Sugar transport in *Coprinus cinereus*. *Biochimica et Biophysica Acta* **550**, 515–526.

Moore, D., Elhiti, M. M. Y. & Butler, R. D. (1979) Morphogenesis of the carpophore of *Coprinus cinereus*. *New Phytologist* **83**, 695–722.

Moore, D. & Ewaze, J. O. (1976) Activities of some enzymes involved in metabolism of carbohydrate during sporophore development in *Coprinus cinereus*. *Journal of General Microbiology* **97**, 313–322.

Moore, D., Hammad, F. & Ji, J. (1994) The stage in sporulation between the end of meiosis and emergence of sterigmata is most sensitive to ammonium inhibition in *Coprinus cinereus*. *Microbios* **76**, 197–201.

Moore, D., Hock, B., Greening, J. P., Kern, V. D., Novak Frazer, L. & Monzer, J. (1996) Centenary review. Gravimorphogenesis in agarics. *Mycological Research* **100**, 257–273.

Moore, D., Horner, J. & Liu, M. (1987) Co-ordinate control of ammonium-scavenging enzymes in the fruit body cap of *Coprinus cinereus* avoids inhibition of sporulation by ammonium. *FEMS Microbiology Letters* **44**, 239–242.

Moore, D. & Jirjis, R. I. (1976) Regulation of sclerotium production by primary metabolites in *Coprinus cinereus* (= *C. lagopus sensu* Lewis) *Transactions of the British Mycological Society* **66**, 377–382.

Moore, D. & Jirjis, R. I. (1981) Electrophoretic studies of carpophore development in the basidiomycete *Coprinus cinereus*. *New Phytologist* **87**, 101–113.

Moore, D., Liu, M. & Kuhad, R. C. (1987) Karyogamy-dependent enzyme derepression in the basidiomycete *Coprinus*. *Cell Biology International Reports* **11**, 335–341.

Moore, D. & Novak Frazer, L. (1993) Controlling fungal growth. Patent Application (to Victoria University of Manchester): UK. 9303073.2, 3 Feb 1993.

Moore, D. & Novak Frazer, L. (1994) Controlling fungal growth. Patent Application (to Victoria University of Manchester): UK. 9403007.9, 3 Feb 1994.

Moore, D. & Stewart, G. R. (1972) Effects of 2-deoxy-D-glucose, D-glucosamine, and L-sorbose on the growth of *Coprinus lagopus* hyphae. *Journal of General Microbiology* **71**, 333–342.

Moore, J. A. (1993) *Science as a Way of Knowing. The Foundations of Modern Biology*. Harvard University Press: Cambridge, Massachusetts.

Moore-Landecker, E. (1975) Effect of cultural conditions on apothecial morphogenesis in *Pyronema domesticum*. *Canadian Journal of Botany* **53**, 2759–2769.

Moore-Landecker, E. (1979a) Effect of cultural age and a single photoperiod on morphogenesis of the discomycete *Pyronema domesticum*. *Canadian Journal of Botany* **57**, 1541–1549.

Moore-Landecker, E. (1979b) Effect of light regimens and intensities on morphogenesis of the discomycete *Pyronema domesticum*. *Mycologia* **71**, 699–712.

Moore-Landecker, E. (1981a) Histochemical observations on the discomycete, *Pyronema domesticum*, with special reference to apothecial ontogeny. *Mycologia* **73**, 301–320.

Moore-Landecker, E. (1981b) Histochemical observations on the discomycete, *Pyronema domesticum*, with special reference to light. *Canadian Journal of Botany* **59**, 1726–1737.

Moore-Landecker, E. (1983) Chemical induction of *Pyronema domesticum* apothecia in the dark. *Mycologia* **75**, 832–838.

Moore-Landecker, E. (1987) Effects of medium composition and light on formation of apothecia and sclerotia by *Pyronema domesticum*. *Canadian Journal of Botany* **65**, 2276–2279.

Moore-Landecker, E. (1993) Physiology and biochemistry of ascocarp induction and development. *Mycological Research* **96**, 705–716.

Moore-Landecker, E. & Shropshire, W. (1982) Effects of aeration and light on apothecia, sclerotia, and mycelial growth in the discomycete *Pyronema domesticum*. *Mycologia* **74**, 1000–1013.

Moore-Landecker, E. & Shropshire, W. (1984) Effects of ultraviolet A radiation and inhibitory volatile substances on the discomycete, *Pyronema domesticum*. *Mycologia* **76**, 820–829.

Morimoto, N. & Oda, Y. (1973) Effects of light on fruit-body formation in a basidiomycete, *Coprinus macrorhizus*. *Plant & Cell Physiology* **14**, 217–225.

Morimoto, N. & Oda, Y. (1974) Photo-induced karyogamy in a basidiomycete, *Coprinus macrorhizus*. *Plant & Cell Physiology* **15**, 183–186.

Morimoto, N., Suda, S. & Sagara, N. (1981) Effect of ammonia on fruit-body induction of *Coprinus cinereus* in darkness. *Plant & Cell Physiology* **22**, 247–254.

Morris, N. R. & Enos, A. P. (1992) Mitotic gold in a mold: *Aspergillus* genetics and the biology of mitosis. *Trends in Genetics* **8**, 32–37.

Moss, M. O. (1996) Mycotoxins. *Mycological Research* **100**, 513–523.

Motoyama, T., Sudoh, M., Horiuchi, H., Ohta, A. & Takagi, M. (1994) Isolation and characterization of two chitin synthase genes of *Rhizopus oligosporus*. *Bioscience Biotechnology and Biochemistry* **58**, 1685–1693.

Motta, J. J. (1967) A note on the mitotic apparatus in the rhizomorph meristem of *Armillaria mellea*. *Mycologia* **59**, 370–375.

Motta, J. J. (1969) Cytology and morphogenesis in the rhizomorph of *Armillaria mellea*. *American Journal of Botany* **56**, 610–619.

Motta, J. J. (1971) Histochemistry of the rhizomorph meristem of *Armillaria mellea*. *American Journal of Botany* **58**, 80–87.

Motta, J. J. & Peabody, D. C. (1982) Rhizomorph cytology and morphogenesis in *Armillaria tabescens*. *Mycologia* **74**, 671–674.

Mukhamedzhanova, T. G. & Bezborodov, A. M. (1982) Effect of nitrogen and carbon sources on lipase accumulation by the fungus *Rhizopus oryzae* 1414. *Mikrobiologia* **18**, 16–22.

Mulder, G. H. & Wessels, J. G. H. (1986) Molecular cloning of RNAs differentially expressed in monokaryons and dikaryons of *Schizophyllum commune* in relation to fruiting. *Experimental Mycology* **10**, 214–227.

Mulholland, J., Preuss, D., Moon, A., Wong, A., Drubin, D. & Botstein, D. (1994) Ultrastructure of the yeast actin cytoskeleton and its association with the plasma membrane. *Journal of Cell Biology* **125**, 381–391.

Muller, M. M., Kantola, R. & Kitunen, V. (1994) Combining sterol and fatty acid profiles for the characterization of fungi. *Mycological Research* **98**, 593–603.

Mullins, J. T. (1994) Hormonal control of sexual dimorphism. In *The Mycota,* **vol. I**, Growth, Differentiation and Sexuality (ed. K. Esser & P. A. Lemke), pp. 413–421. Springer-Verlag: Berlin.

Mullins, J. T. & Ellis, E. A. (1974) Sexual morphogenesis in *Achlya*: ultrastructural basis for the hormonal induction of antheridial hyphae. *Proceedings of the National Academy of Sciences of the USA* **71**, 1347–1350.

Muñoz, V. & Butler, W. L. (1975) Photoreceptor pigment for blue light in *Neurospora crassa*. *Plant Physiology* **55**, 421–426.

Murakami, S. & Takemaru, T. (1980) Nuclear number in stipe cells of some hymenomycetes. *Reports of the Tottori Mycological Institute (Japan)* **18**, 143–148.

Murakami, S. & Takemaru, T. (1984) Cytological studies on basidiospore formation in polyploid fruitbodies of *Coprinus cinereus*. *Reports of the Tottori Mycological Institute (Japan)* **22**, 67–68.

Musgrave, A., Ero, L., Scheffer, R. & Oehlers, E. (1977) Chemotropism of *Achlya bisexualis* germ tube hyphae to casein hydrolysate and amino acids. *Journal of General Microbiology* **101**, 65–70.

Nakasone, K. K. (1990) Cultural studies and identification of wood-inhabiting Corticiaceae and selected Hymenomycetes from North America. *Mycologia Memoirs* **15**, 1–412.

Nasrallah, J. B. & Srb, A. M. (1973) Genetically related protein variants specifically associated with fruiting body maturation in *Neurospora*. *Proceedings of the National Academy of Sciences of the USA* **70**, 1891–1893.

Nasrallah, J. B. & Srb, A. M. (1977) Occurrence of a major protein associated with fruiting body development in *Neurospora* and related Ascomycetes. *Proceedings of the National Academy of Sciences of the USA* **74**, 3831–3834.

Nasrallah, J. B. & Srb, A. M. (1978) Immunofluorescent localization of a phase-specific protein in *Neurospora tetrasperma* perithecia. *Experimental Mycology* **2**, 211–215.

Nathan, C. & Sporn, M. (1991) Cytokines in context. *Journal of Cell Biology* **113**, 981–986.

Nauta, M. J. & Hoekstra, R. F. (1992a) Evolution of reproductive systems in filamentous ascomycetes. 1. Evolution of mating types. *Heredity* **68**, 405–410.

Nauta, M. J. & Hoekstra, R. F. (1992b) Evolution of reproductive systems in filamentous ascomycetes. 2. Evolution of hermaphroditism and other reproductive strategies. *Heredity* **68**, 537–546.

Nauta, M. J. & Hoekstra, R. F. (1994) Evolution of vegetative incompatibility in filamentous ascomycetes. 1. Deterministic models. *Evolution* **48**, 979–995.

Navarro, E., Sandmann, G. & Torresmartinez, S. (1995) Mutants of the carotenoid biosynthetic pathway of *Mucor circinelloides*. *Experimental Mycology* **19**, 186–190.

Navarro-Bordonaba, J. & Adams, T. H. (1994) Development of conidia and fruiting bodies in ascomycetes. In *The Mycota,* **vol. I**, Growth, Differentiation and Sexuality (ed. J. G. H. Wessels & F. Meinhardt), pp. 333–349. Springer-Verlag: Berlin.

Nelson, M. A. (1996) Mating systems in ascomycetes - a romp in the sac. *Trends in Genetics* **12**, 69–74.

Nelson, R. R. (1971) Hormonal involvement in sexual reproduction in the fungi, with special reference to F-2, a fungal estrogen. In *Morphological and Biochemical Events in Plant-Parasite Interactions* (ed. S. Akai & S. Ouchi), pp. 181–205. Phytopathological Society of Japan: Tokyo.

Neville, N. M., Suskind, S. R. & Roseman, S. (1971) A derepressible active transport system for glucose in *Neurospora crassa. Journal of Biological Chemistry* **246**, 1294–1301.

Newell, P. C. (1978) Cellular communication during aggregation of *Dictyostelium. Journal of General Microbiology* **104**, 1–13.

Nichols, R. (1985) Post-harvest physiology and storage. In *The Biology and Technology of the Cultivated Mushroom* (ed. P. B. Flegg, D. M. Spencer & D. A. Wood), pp. 195–210. John Wiley & Sons: Chichester.

Nicholson, R. (1996) Adhesion of fungal propagules. In *Histology, Ultrastructure and Molecular Cytology of Plant-Microorganism Interactions* (ed. M. Nicole & V. Gianinazzi-Pearson), pp. 117–134. Kluwer Academic Publishers: Dordrecht, The Netherlands.

Nicole, M., Chamberland, H., Rioux, D., Lecours, N., Rio, B., Geiger, J. P. & Ouellete, G. B. (1993) A cytochemical study of extracellular sheaths associated with *Rigidoporus lignosus* during wood decay. *Applied and Environmental Microbiology* **59**, 2578–2588.

Nicole, M., Chamberland, H., Rioux, D., Xixuan, X., Blanchette, R. A., Geiger, J. P. & Ouellete, G. B. (1995) Wood degradation by *Phellinus noxius*: ultrastructure and cytochemistry. *Canadian Journal of Microbiology* **41**, 553–565.

Niederpruem, D. J. (1965) Tricarboxylic acid cycle. In *The Fungi* (ed. G. C. Ainsworth & A. S. Sussman), pp. 269–300. Academic Press: New York.

Niederpruem, D. J., Jersild, R. A. & Lane, P. L. (1971) Direct microscopic studies of clamp connection formation in growing hyphae of *Schizophyllum commune.* I. The dikaryon. *Archives of Microbiology* **78**, 268–280.

Niederpruem, D. J. & Jersild, R. A. (1972) Cellular aspects of morphogenesis in the mushroom *Schizophyllum commune. CRC Critical Reviews in Microbiology* **1**, 545–576.

Niederpruem, D. J. & Wessels, J. G. H. (1969) Cytodifferentiation and morphogenesis in *Schizophyllum commune. Bacteriological Reviews* **33**, 505–535.

Nielsen, J. (1993) A simple morphologically structured model describing the growth of filamentous microorganisms. *Biotechnology and Bioengineering* **41**, 715–727.

Nielsen, O. & Davey, J. (1995) Pheromone communication in the fission yeast *Schizosaccharomyces pombe. Seminars in Cell Biology* **6**, 95–104.

Nikoh, N., Hayase, N., Iwabe, N., Kuma, K. & Miyata, T. (1994) Phylogenetic relationship of the Kingdoms Animalia, Plantae, and Fungi, inferred from 23 different protein species. *Molecular Biology and Evolution* **11**, 762–768.

Nishida, H. (1991) Diversity and significance of Late Cretaceous permineralized plant remains from Hokkaido, Japan. *Botanical Magazine (Tokyo)* **104**, 253–273.

Nobles, M. K. (1958) Cultural characters as a guide to the taxonomy and phylogeny of the Polyporaceae. *Canadian Journal of Botany* **36**, 883–926.

Nobles, M. K. (1965) Identification of cultures of wood-inhabiting Hymenomycetes. *Canadian Journal of Botany* **43**, 1097–1139.

Nobles, M. K. (1971) Cultural characters as a guide to the taxonomy of the Polyporaceae. In *Evolution in the Higher Basidiomycetes* (ed. R. H. Petersen), pp. 169–196. University of Tennessee Press: Knoxville.

Noel, T., Axiotis, S. & Labarère, J. (1995) Study of nutritional and environmental-factors affecting the fruiting competence of *Rhizoctonia solani* AG-1. *Journal of Phytopathology - Phytopathologische Zeitschrift* **143**, 173–177.

Novak Frazer, L. (1996) Control of growth and patterning in the fungal fruiting structure. A case for the involvement of hormones. In *Patterns in Fungal Development* (ed. S. W. Chiu & D. Moore), pp. 156–181. Cambridge University Press: Cambridge, UK.

Nuss, I., Jennings, D. H. & Veltkamp, C. J. (1991) Morphology of *Serpula lacrymans*. In *Serpula lacrymans: Fundamental Biology and Control Strategies* (ed. D. H. Jennings & A. F. Bravery), pp. 9–38. John Wiley & Sons: Chichester.

Nutting, W. H., Rapoport, H. & Machlis, L. (1968) The structure of sirenin. *Journal of the American Chemical Society* **90**, 6434–6438.

Nyunoya, H. & Ishikawa, T. (1979) Control of unusual hyphal morphology in a mutant of *Coprinus macrorhizus*. *Japanese Journal of Genetics* **54**, 11–20.

Oberwinkler, F. (1982) The significance of the morphology of the basidium in the phylogeny of basidiomycetes. In *Basidium and Basidiocarp: Evolution, Cytology, Function and Development* (ed. K. Wells & E. K. Wells), pp. 9-35. Springer-Verlag: New York.

O'Brien, B. L. & Khuu, L. (1996) A fatal Sunday brunch - *Amanita* mushroom poisoning in a gulf coast family. *American Journal of Gastroenterology* **91**, 581–583.

O'Donnell, K. (1994) A re-evaluation of the mitotic spindle pole body cycle in *Tilletia caries* based on freeze-substitution techniques. *Canadian Journal of Botany* **72**, 1412–1423.

Ohkura, H., Hagan, I. M. & Glover, D. M. (1995) The conserved *Schizosaccharomyces pombe* kinase *plo1*, required to form a bipolar spindle, the actin ring, and septum, can drive septum formation in G1 and G2 cells. *Genes & Development* **9**, 1059–1073.

Olsson, S. & Hansson, B. S. (1995) Action potential-like activity found in fungal mycelia is sensitive to stimulation. *Die Naturwissenschaften* **82**, 30–31.

Orlowski, M. (1994) Yeast/mycelial dimporphism. In *The Mycota,* **vol. I**, Growth, Differentiation and Sexuality (ed. J. G. H. Wessels & F. Meinhardt), pp. 143–162. Springer-Verlag: Berlin.

Orlowski, M. (1995) Gene-expression in *Mucor dimorphism*. *Canadian Journal of Botany* **73**, S326–S334.

Osmani, A. H., Osmani, S. A. & Morris, N. R. (1990) The molecular cloning and identification of a gene product specifically required for nuclear movement in *Aspergillus nidulans*. *Journal of Cell Biology* **111**, 543–552.

Padmore, R., Cao, L. & Klechkner, N. (1991) Temporal comparison of recombination and synaptonemal complex formation during meiosis in *Saccharomyces cerevisiae*. *Cell* **66**, 1239–1256.

Pady, S. M. & Kramer, C. L. (1969) Periodicity in ascospore discharge in *Bombardia*. *Transactions of the British Mycological Society* **53**, 449–454.

Pagga, U., Beimborn, D. B., Boelens, J. & Dewilde, B. (1995) Determination of the aerobic biodegradability of polymeric material in a laboratory controlled composting test. *Chemosphere* **31**, 4475–4487.

Pall, M. L. & Robertson, C. K. (1986) Cyclic AMP control of hierarchical growth pattern of hyphae in *Neurospora crassa*. *Experimental Mycology* **10**, 161–165.

Palmer, J. G., Murmanis, L. & Highley, T. L. (1983) Visualisation of hyphal sheath in wood-decay hymenomycetes. I. Brown-rotters. *Mycologia* **75**, 995–1004.

Paquin, B., Roewer, I., Wang, Z. & Lang, B. F. (1995) A robust fungal phylogeny using the mitochondrially encoded NAD5 protein-sequence. *Canadian Journal of Botany* **73**, S180–S185.

Park, D. (1985a) Application of Horton's first and second laws of branching to fungi. *Transactions of the British Mycological Society* **84**, 577–584.

Park, D. (1985b) Does Horton's law of branch length apply to open branching systems? *Journal of Theoretical Biology* **112**, 299–313.

Parks, L. W. & Casey, W. M. (1995) Physiological implications of sterol biosynthesis in yeast. *Annual Review of Microbiology* **49**, 95–116.

Parks, L. W., Smith, S. J. & Crowley, J. H. (1995) Biochemical and physiological effects of sterol alterations in yeast - a review. *Lipids* **30**, 227–230.

Pateman, J. A. & Cove, D. J. (1967) Regulation of nitrate reduction in *Aspergillus nidulans*. *Nature* **215**, 1234–1239.

Pateman, J. A., Kinghorn, J. R., Dunn, E. & Forres, E. (1973) Ammonium regulation in *Aspergillus nidulans*. *Journal of Bacteriology* **114**, 943–950.

Pateman, J. A., Kinghorn, J. R. & Dunn, E. (1974) Regulatory aspects of L-glutamate transport in *Aspergillus nidulans*. *Journal of Bacteriology* **119**, 534–542.

Patouillard, N. (1887) *Les Hyménomycètes d'Europe*. Paul Klincksieck: Paris.

Patouillard, N. (1900) *Essai Taxonomique sur les Familles et les Genres des Hyménomycètes*. Lucian Declume: Lous-le-Saunier.

Paul, G. C. & Thomas, C. R. (1996) A structured model for hyphal differentiation and penicillin production using *Penicillium chrysogenum. Biotechnology and Bioengineering* **51**, 558–572.

Peberdy, J. F. & Ferenczy, L. (1985) *Fungal Protoplasts: Applications in Biochemistry and Genetics*. Marcel Dekker: New York.

Pegg, G. F. (1973) Occurrence of gibberellin-like growth substances in basidiomycete sporophores. *Transactions of the British Mycological Society* **61**, 277–286.

Pegler, D. N. (1983) The genus *Lentinus*. A world monograph. *Kew Bulletin, Additional Series* **10**, 1–281.

Pegler, D. N. (1996) Hyphal analysis of basidiomata. *Mycological Research* **100**, 129–142.

Pegler, D. N., Læssøe, T. & Spooner, B. M. (1995) *British Puffballs, Earthstars and Stinkhorns: an account of the British Gasteroid Fungi*. Royal Botanic Garden: Kew.

Perasso, R. & Barointourancheau, A. (1992) The eukaryogenesis - a model derived from ribosomal-RNA molecular phylogenies. *Comptes Rendus des Séances de la Societé de Biologie et de ses Filiales* **186**, 656–665.

Perkins, J. H. (1969) Morphogenesis in *Schizophyllum commune*. I. Effects of white light. *Plant Physiology* **44**, 1706–1711.

Perkins, J. H. & Gordon, S. A. (1969) Morphogenesis in *Schizophyllum commune*. II. Effects of monochromatic light. *Plant Physiology* **44**, 1712–1716.

Persoon, D. C. H. (1801) *Synopsis Methodica Fungorum*. Henricum Dieterich: Gottingae.

Petersen, R. H. (1995) There's more to a mushroom than meets the eye: mating studies in the agaricales. *Mycologia* **87**, 1–17.

Philley, M. L. & Staben, C. (1994) Functional analyses of the *Neurospora crassa* MT a-1 mating type polypeptide. *Genetics* **137**, 715–722.

Pinto-Lopes, J. (1952) Polyporaceae, contribuiça o para a sua bio-taxonomia. *Memórias da Sociedade Broteriana* **8**, 1–195.

Piquemal, M., Latché, J. C. & Baldy, P. (1972) Importance de l'acide glutamique dans la nutrition et le metabolisme d'*Agaricus bisporus* Lge. *Mushroom Science* **8**, 413–424.

Pirozynski, K. A. (1976) Fungal spores in fossil record. *Biological Memoirs* **1**, 104–120.

Pirt, S. J. (1966) A theory of the mode of growth of fungi in the form of pellets in submerged culture. *Proceedings of the Royal Society, London, Series B* **166**, 369–373.

Plesofsky-Vig, N. (1996) The heat shock proteins and the stress response. In *The Mycota,* **vol. III**, Biochemistry and Molecular Biology (ed. R. Brambl & G. A. Marzluf), pp. 171–190. Springer-Verlag: Berlin.

Plunkett, B. E. (1956) The influence of factors of the aeration complex and light upon fruit-body form in pure cultures of an agaric and a polypore. *Annals of Botany* **20**, 563–586.

Plunkett, B. E. (1961) The change of tropism in *Polyporus brumalis* stipes and the effect of directional stimuli on pileus differentiation. *Annals of Botany* **25**, 206–223.

Pöder, R. (1992) Phylogenetical aspects of gill development and proportions in basidiocarps. *Mycologia Helvetica* **5**, 39–46.

Pöder, R. & Kirchmair, M. (1995) Gills and pores: the impact of geometrical constraints on form, size and number of basidia. *Documents Mycologiques* **100**, 337–348.

Podila, G. K., Rogers, L. M. & Kolattukudy, P. E. (1993) Chemical signals from avocado surface wax trigger germination and appressorium formation in *Colletotrichum gloeosporioides. Plant Physiology* **103**, 267–272.

Poinar, G. O., Waggoner, B. M. & Bauer, U. C. (1993) Terrestrial soft-bodied protists and other microorganisms in Triassic amber. *Science* **259**, 222–224.

Polizeli, M. D. T. M., Noventa-Jordao, M. A., DaSilva, M. M., Jorge, J. A. & Terenzi, H. F. (1995) (1,3)-β-D-glucan synthase activity in mycelial and cell wall-less phenotypes of the *fz*, *sg*, *os-1* ("slime") mutant strain of *Neurospora crassa. Experimental Mycology* **19**, 35–47.

Pollock, R. T. (1975) Control of the pattern of perithecium development in *Sordaria fimicola* on agar medium. *Growth* **39**, 223–232.

Pommerville, J. C. (1981) The role of sexual pheromones in *Allomyces*. In *Sexual Interactions in Eukaryotic Microbes* (ed. D. H. O'Day & P. A. Horgen), pp. 53–92. Academic Press: New York.

Pommerville, J. C., Strickland, J. B., Romo, D. & Harding, K. E. (1988) Effects of analogs of the fungal sex pheromone sirenin on male gamete motility in *Allomyces macrogynus*. *Plant Physiology* **88**, 139–142.

Pontefract, R. D. & Miller, J. J. (1962) The metabolism of yeast sporulation. IV. Cytological and physiological changes in sporulating cells. *Canadian Journal of Microbiology* **8**, 573–584.

Pouzar, Z. (1966) Studies on the taxonomy of polypores 1. *Ceská Mykologie* **20**, 171–177.

Powell, K. A. & Rayner, A. D. M. (1983) Ultrastructure of the rhizomorph apex in *Armillaria bulbosa* in relation to mucilage production. *Transactions of the British Mycological Society* **81**, 529–534.

Price, I. P. (1973) A study of cystidia in effused Aphyllophorales. *Nova Hedwigia* **24**, 515–618.

Prillinger, H., Deml, G., Dorfler, C., Laaser, G. & Lockau, W. (1991) A contribution to the systematics and evolution of higher fungi - yeast-types in the basidiomycetes. 2. *Microbotryum*-type. *Botanica Acta* **104**, 5–17.

Prillinger, H. & Esser, K. (1977) The phenoloxidases of the ascomycete *Podospora anserina*. XIII. Action and interaction of genes controlling the formation of laccase. *Molecular & General Genetics* **156**, 333–346.

Prosser, J. I. (1982) Growth of fungi. In *Mathematical Models in Microbiology* (ed. M. J. Bazin), pp. 125-166. CRC Press: Boca Raton, Florida.

Prosser, J. I. (1983) Hyphal growth patterns. In *Fungal Differentiation* (ed. J. E. Smith), pp. 357–396. Marcel Dekker: New York.

Prosser, J. I. (1993) Growth kinetics of mycelial colonies and aggregates of ascomycetes. *Mycological Research* **97**, 513–528.

Prosser, J. I. (1995a) Kinetics of filamentous growth and branching. In *The Growing Fungus* (ed. N. A. R. Gow & G. M. Gadd), pp. 301–318. Chapman & Hall: London.

Prosser, J. I. (1995b) Mathematical modelling of fungal growth. In *The Growing Fungus* (ed. N. A. R. Gow & G. M. Gadd), pp. 319–335. Chapman & Hall: London.

Prosser, J. I. & Tough, A. J. (1991) Growth mechanisms and growth-kinetics of filamentous microorganisms. *Critical Reviews in Biotechnology* **10**, 253–274.

Prosser, J. I. & Trinci, A. P. J. (1979) A model for hyphal growth and branching. *Journal of General Microbiology* **111**, 153–164.

Prusinkiewicz, P. & Lindenmayer, A. (1996) *The Algorithmic Beauty of Plants*. Springer: New York.

Pukkila, P. J. (1994) Meiosis in mycelial fungi. In *The Mycota,* **vol. I**, Growth, Differentiation and Sexuality (ed. J. G. H. Wessels & F. Meinhardt), pp. 267–281. Springer-Verlag: Berlin.

Pukkila, P. J. & Casselton, L. A. (1991) Molecular genetics of the agaric *Coprinus cinereus*. In *More Gene Manipulations in Fungi* (ed. J. W. Bennett & L. A. Lasure), pp. 126–150. Academic Press: New York.

Pukkila, P. J. & Skrzynia, C. (1993) Frequent changes in the number of reiterated ribosomal RNA genes throughout the life cycle of the basidiomycete *Coprinus cinereus*. *Genetics* **133**, 203–211.

Pukkila, P. J., Yashar, B. M. & Binninger, D. M. (1984) Analysis of meiotic development in *Coprinus cinereus*. *Symposia of the Society for Experimental Biology* **38**, 177–194.

Radford, A., Stone, P. J. & Taleb, F. (1996) Cellulase and amylase complexes. In *The Mycota,* **vol. III**, Biochemistry and Molecular Biology (ed. R. Brambl & G. A. Marzluf), pp. 269–294. Springer-Verlag: Berlin.

Rajchenberg, M. (1983) Cultural studies of resupinate polypores. *Mycotaxon* **17**, 275–293.

Raju, N. B. & Lu, B. C. (1970) Meiosis in *Coprinus*. III. Timing of meiotic events in *C. lagopus* (*sensu* Buller) *Canadian Journal of Botany* **48**, 2183–2186.

Raju, N. B. & Lu, B. C. (1973) Meiosis in *Coprinus*. IV. Morphology and behavior of spindle pole bodies. *Journal of Cell Science* **12**, 131–141.

Raju, N. B. & Perkins, D. D. (1994) Diverse programs of ascus development in pseudohomothallic species of *Neurospora*, *Gelasinospora* and *Podospora*. *Developmental Genetics* **15**, 104–118.

Ramsdale, M. & Rayner, A. D. M. (1994) Distribution patterns of number of nuclei in conidia from heterokaryons of *Heterobasidion annosum* (Fr.) Bref and their interpretation in terms of genomic conflict. *New Phytologist* **128**, 123–134.

Ramsdale, M. & Rayner, A. D. M. (1996) Imbalanced nuclear ratios, postgermination mortality and phenotype-genotype relationships in allopatrically-derived heterokaryons of *Heterobasidion annosum*. *New Phytologist* **133**, 303–319.

Rand, R. D. (1917) The production of spores by *Alternaria solani* in pure culture. *Phytopathology* **7**, 316–317.
Rao, P. S. & Niederpruem, D. J. (1969) Carbohydrate metabolism during morphogenesis of *Coprinus lagopus* (*sensu* Buller) *Journal of Bacteriology* **100**, 1222–1228.
Raper, J. R. (1952) Chemical regulation of sexual processes in the Thallophytes. *Botanical Reviews* **18**, 447–545.
Raper, J. R. (1966) Life cycles, basic patterns of sexuality and sexual mechanisms. In *The Fungi: An Advanced Treatise, vol. 2* (ed. G. C. Ainsworth & A. S. Sussman), pp. 473–511. Academic Press: London.
Raper, J. R. & Krongelb, G. S. (1958) Genetic and environmental aspects of fruiting in *Schizophyllum commune* Fr. *Mycologia* **50**, 707–740.
Rast, D. M. (1965) Zur stoffwechselphysiologischen Bedeutung von Mannit und Trehalose in *Agaricus bisporus* (eine gaschromatographische Studie) *Planta* **64**, 81–93.
Rast, D. M., Stauble, E. J. & Zobrist, P. (1976) The Krebs cycle and control of dormancy in the *Agaricus bisporus* spore. *New Phytologist* **76**, 469–477.
Rath, J., Messner, R., Kosma, P., Altmann, F., Marz, L. & Kubicek, C. P. (1995) The α-D-mannan core of a complex cell-wall heteroglycan of *Trichoderma reesei* is responsible for β-glucosidase activation. *Archives of Microbiology* **164**, 414–419.
Raudaskoski, M. & Lu, B. C. (1980) The effect of hydroxyurea on meiosis and genetic recombination in the fungus *Coprinus lagopus*. *Canadian Journal of Genetics and Cytology* **22**, 41–50.
Raudaskoski, M., Mao, W. Z. & Ylimattila, T. (1994) Microtubule cytoskeleton in hyphal growth. Response to nocodazole in a sensitive and a tolerant strain of the homobasidiomycete *Schizophyllum commune*. *European Journal of Cell Biology* **64**, 131–141.
Raudaskoski, M. & Salonen, M. (1984) Interrelationships between vegetative development and basidiocarp initiation. In *The Ecology and Physiology of the Fungal Mycelium* (ed. D. H. Jennings & A. D. M. Rayner), pp. 291–322. Cambridge University Press: Cambridge, UK.
Raudaskoski, M. & Viitanen, H. (1982) Effect of aeration and light on fruit body induction in *Schizophyllum commune*. *Transactions of the British Mycological Society* **78**, 89–96.
Ray, P. M., Green, P. B. & Cleland, R. (1972) Role of turgor in plant cell growth. *Nature* **239**, 163–164.
Rayner, A. D. M. (1992) Conflicting flows: the dynamics of mycelial territoriality. *McIlvainea* **10**, 24–35.
Rayner, A. D. M. (1997) *Degrees of Freedom. Living in Dynamic Boundaries*. Imperial College Press: London.
Rayner, A. D. M. & Boddy, L. (1988) *Fungal Decomposition of Wood*. John Wiley & Sons: Chichester.
Rayner, A. D. M., Brasier, C. M. & Moore, D. (1987) *Evolutionary Biology of the Fungi*. Cambridge University Press: Cambridge, UK.
Rayner, A. D. M. & Coates, D. (1987) Regulation of mycelial organisation and responses. In *Evolutionary Biology of the Fungi* (ed. A. D. M. Rayner, C. M. Brasier & D. Moore), pp. 115–136. Cambridge University Press: Cambridge, UK.
Rayner, A. D. M., Griffith, G. S. & Wildman, H. G. (1994) Differential insulation and the generation of mycelial patterns. In *Shape and Form in Plants and Fungi* (ed. D. S. Ingram & A. Hudson), pp. 291–310. Academic Press: London.
Rayner, A. D. M., Griffith, G. S. & Ainsworth, A. M. (1995a) Mycelial interconnectedness. In *The Growing Fungus* (ed. N. A. R. Gow & G. M. Gadd), pp. 21–40. Chapman & Hall: London.
Rayner, A. D. M., Powell, K. A., Thompson, W. & Jennings, D. H. (1985a) Morphogenesis of vegetative organs. In *Developmental Biology of Higher Fungi* (ed. D. Moore, L. A. Casselton, D. A. Wood & J. C. Frankland), pp. 249–279. Cambridge University Press: Cambridge, UK.
Rayner, A. D. M., Ramsdale, M. & Watkins, Z. R. (1995b) Origins and significance of genetic and epigenetic instability in mycelial systems. *Canadian Journal of Botany* **73**, S1241–S1248.
Rayner, A. D. M. & Todd, N. K. (1979) Population and community structure and dynamics of fungi in decaying wood. *Advances in Botanical Research* **7**, 333–420.

Rayner, A. D. M., Watling, R. & Frankland, J. C. (1985b) Resource relations - an overview. In *Developmental Biology of Higher Fungi* (ed. D. Moore, L. A. Casselton, D. A. Wood & J. C. Frankland), pp. 1–40. Cambridge University Press: Cambridge, UK.

Rayner, A. D. M. & Webber, J. F. (1984) Interspecific mycelial interactions – an overview. In *The Ecology and Physiology of the Fungal Mycelium* (ed. D. H. Jennings & A. D. M. Rayner), pp. 383–417. Cambridge University Press: Cambridge, UK.

Rea, C. (1922) *British Basidiomycetae*. Cambridge University Press: Cambridge, UK.

Read, D. J. (1991) Mycorrhizas in ecosystems - Nature's response to the "Law of the Minimum". In *Frontiers of Mycology* (ed. D. L. Hawksworth), pp. 101–130. CAB International: Wallingford, UK.

Read, D. J. & Armstrong, W. (1972) A relationship between oxygen transport and the formation of the ectotrophic mycorrhizal sheath in conifer seedlings. *New Phytologist* **71**, 49–53.

Read, D. J., Leake, J. R. & Langdale, A. R. (1989) The nitrogen nutrition of mycorrhizal fungi and their host plants. In *Nitrogen, Phosphorus and Sulphur Utilization by Fungi* (ed. L. Boddy, R. Marchant & D. J. Read), pp. 181–204. Cambridge University Press: Cambridge, UK.

Read, N. D. (1983) A scanning electron microscopic study of the external features of perithecium development in *Sordaria humana. Canadian Journal of Botany* **61**, 3217–3229.

Read, N. D. (1994) Cellular nature and multicellular morphogenesis of higher fungi. In *Shape and Form in Plants and Fungi* (ed. D. S. Ingram & A. Hudson), pp. 251–269. Academic Press: London.

Read, N. D. & Beckett, A. (1985) The anatomy of the mature perithecium in *Sordaria humana*; and its significance for fungal multicellular development. *Canadian Journal of Botany* **63**, 281–296.

Read, N. D. & Beckett, A. (1996) Centenary review. Ascus and ascospore morphogenesis. *Mycological Research* **100**, 1281–1314.

Reddy, C. A. & Dsouza, T. M. (1994) Physiology and molecular biology of the lignin peroxidases of *Phanerochaete chrysosporium. FEMS Microbiology Reviews* **13**, 137–152.

Redhead, S. A. (1987) The Xerulaceae (Basidiomycetes), a family with sarcodimitic tissues. *Canadian Journal of Botany* **65**, 1551–1562.

Rees, B., Shepherd, V. A. & Ashford, A. E. (1994) Presence of a motile tubular vacuole system in different phyla of fungi. *Mycological Research* **98**, 985–992.

Reeves, R. J. (1975) Behaviour of *Phytophthora cinnamomi* in different soils and water regimes. *Soil Biology & Biochemistry* **7**, 19–24.

Reeves, R. J. & Jackson, R. M. (1974) Stimulation of sexual reproduction in *Phytophthora* by damage. *Journal of General Microbiology* **84**, 303–310.

Regalado, C. M., Crawford, J. W., Ritz, K. & Sleeman, B. D. (1996) The origins of spatial heterogeneity in vegetative fungal mycelia: a reaction-diffusion model. *Mycological Research* **100**, 1473–1480.

Reid, I. D. (1995) Biodegradation of lignin. *Canadian Journal of Botany* **73**, S1011–S1018.

Reijnders, A. F. M. (1948) Etudes sur le developpement et l'organisation histologique des carpophores dans les Agaricales. *Recueil des Travaux botanique de la Néerlande,* **41**, 213–396.

Reijnders, A. F. M. (1963) *Les problèmes du développement des carpophores des Agaricales et de quelques groupes voisins.* Dr W. Junk: The Hague.

Reijnders, A. F. M. (1976) Recherches sur le développement et l'histogenèse dans les Asterérosporales. *Persoonia* **9**, 65–83.

Reijnders, A. F. M. (1977) The histogenesis of bulb and trama tissue of the higher Basidiomycetes and its phylogenetic implications. *Persoonia* **9**, 329–362.

Reijnders, A. F. M. (1979) Developmental anatomy of *Coprinus. Persoonia* **10**, 383–424.

Reijnders, A. F. M. (1993) On the origin of specialised trama types in the Agaricales. *Mycological Research* **97**, 257–268.

Reijnders, A. F. M. & Moore, D. (1985) Developmental biology of agarics - an overview. In *Developmental Biology of Higher Fungi* (ed. D. Moore, L. A. Casselton, D. A. Wood & J. C. Frankland), pp. 581–595. Cambridge University Press: Cambridge, UK.

Reijnders, A. F. M. & Stalpers, J. A. (1992) The development of the hymenophoral trama in the Aphyllophorales and the Agaricales. In *Studies in Mycology*, **No. 34**. Centraalbureau voor Schimmelcultures: Baarn, The Netherlands.

Reinbothe, H. (1964) Urea metabolism in basidiomycetes. II. Formation of τ-guanidobutyric acid in fruit bodies of *Lycoperdon*. *Phytochemistry* **3**, 327–333.

Reinbothe, H., Tintemann, H. & Schlee, D. (1969) Harnstoff-metabolismus bei Basidiomyceten. V. Der Ornithin-cyclus als Hautweg der Harnstoffsynthese in Fruchtkörpern von *Agaricus bisporus* Lange. *Flora* **160**, 422–437.

Reinbothe, H. & Tschiersch, B. (1962) Harnstoff-metabolismus bei Basidiomyceten. I. Zur Harnstoffbiosynthese in *Agaricus bisporus* Lange und *Lycoperdon perlatum* Pers. *Flora* **152**, 423–446.

Reinbothe, H., Wasternack, C. & Miersch, J. (1967) Harnstoff-metabolismus bei Basidiomyceten. IV. Untersuchungen zur physiologie des harnstoffs. *Flora* **158**, 27–57.

Reiser, V. & Gasperik, J. (1995) Purification and characterization of the cell-wall-associated and extracellular α-glucosidases from *Saccharomycopsis fibuligera*. *Biochemical Journal* **308**, 753–760.

Reizer, J., Finley, K., Kakuda, D., Macleod, C. L., Reizer, A. & Saier, M. H. (1993) Mammalian integral membrane-receptors are homologous to facilitators and antiporters of yeast, fungi, and eubacteria. *Protein Science* **2**, 20–30.

Remy, W., Taylor, T. N., Hass, H. & Kerp, H. (1994) 4-Hundred-million-year-old vesicular-arbuscular mycorrhizae. *Proceedings of the National Academy of Sciences of the USA* **91**, 11841–11843.

Retallack, G. J. (1994) Were the ediacaran fossils lichens? *Paleobiology* **20**, 523–544.

Reyna Lopez, G. E., Simpson, J. & Ruiz-Herrera, J. (1997) Differences in DNA methylation patterns are detectable during the dimorphic transition of fungi by amplification of restriction polymorphisms. *Molecular & General Genetics* **253**, 703–710.

Richartz, G. & MacLellan, A. (1987) Action spectra for hyphal aggregation, the first stage of fruiting, in the basidiomycete *Pleurotus ostreatus*. *Photochemistry and Photobiology* **45**, 815–820.

Riehl, R. M. & Toft, D. O. (1984) Analysis of the steroid receptor of *Achlya ambisexualis*. *Journal of Biological Chemistry* **259**, 15324–15330.

Riehl, R. M., Toft, D. O., Meyer, M. D., Carlson, G. L. & McMorris, T. C. (1984) Detection of a pheromone binding protein in the aquatic fungus *Achlya ambisexualis*. *Experimental Cell Research* **153**, 544–549.

Riley, D. E. & Krieger, J. N. (1995) Molecular and phylogenetic analysis of PCR-amplified cyclin-dependent kinase (CDK) family sequences from representatives of the earliest available lineages of eukaryotes. *Journal of Molecular Evolution* **41**, 407–413.

Rishbeth, J. (1985) *Armillaria*: resources and hosts. In *Developmental Biology of Higher Fungi* (ed. D. Moore, L. A. Casselton, D. A. Wood & J. C. Frankland), pp. 87–101. Cambridge University Press: Cambridge, UK.

Ritchie, E. D. (1960) Spiral growth of fungus colonies. *Growth* **24**, 391–400.

Roach, P. & Larner, J. (1976) Regulation of glycogen synthase. *Trends in Biochemical Sciences* **1**, 110.

Robert, J. C. (1977a) Fruiting of *Coprinus congregatus*: biochemical changes in fruit bodies during morphogenesis. *Transactions of the British Mycological Society* **68**, 379–387.

Robert, J. C. (1977b) Fruiting of *Coprinus congregatus*: relationship to biochemical changes in the whole culture. *Transactions of the British Mycological Society* **68**, 389–395.

Robert, J. C. (1990) On the light control of fruit-body development in *Coprinus congregatus*. In *Abstracts, Fourth International Mycological Congress* (ed. A. Reissinger & A. Bresinsky), p. 91. University of Regensberg: Regensberg.

Robert, J. C. & Bret, J. P. (1987) Release of an inhibitor of stipe elongation from illuminated caps of *Coprinus congregatus* mushrooms. *Canadian Journal of Botany* **65**, 505–508.

Roberts, A. V., Berlin, V., Hager, K. M. & Yanofsky, C. (1988) Molecular analysis of a *Neurospora crassa* gene expressed during conidiation. *Molecular Cell Biology* **8**, 2411–2418.

Roberts, A. V. & Yanofsky, C. (1989) Genes expressed during conidiation in *Neurospora crassa*: characterization of *con-8*. *Nucleic Acids Research* **17**, 197–213.

Roberts, R. L. & Fink, G. R. (1994) Elements of a single MAP kinase cascade in *Saccharomyces cerevisiae* mediate two developmental programs in the same cell type: mating and invasive growth. *Genes & Development* **8**, 2974–2985.

Robertson, N. F. (1965a) The fungal hypha. *Transactions of the British Mycological Society* **48**, 1–8.

Robertson, N. F. (1965b) The mechanism of cellular extension and branching. In *The Fungi,* **vol. 1**, The Fungal Cell (ed. G. C. Ainsworth & A. S. Sussman), pp. 613–623. Academic Press: New York.

Robertson, N. F. (1968) The growth process in fungi. *Annual Review of Phytopathology* **6**, 115–136.

Robinow, C. F. & Hyams, J. S. (1989) General cytology of fission yeasts. In *Molecular Biology of the Fission Yeast* (ed. A. Nasim, P. R. Young & B. F. Johnson), pp. 273–330. Academic Press: San Diego.

Robinson, P. M. (1973a) Autotropism in fungal spores and hyphae. *The Botanical Review* **39**, 367–384.

Robinson, P. M. (1973b) Chemotropism in fungi. *Transactions of the British Mycological Society* **61**, 303–313.

Robinson, P. M. & Bolton, S. K. (1984) Autotropism in hyphae of *Saprolegnia ferax*. *Transactions of the British Mycological Society* **83**, 257–263.

Robinson, P. M. & Smith, J. M. (1979) Development of cells and hyphae of *Geotrichum candidum* in chemostat and batch culture. *Transactions of the British Mycological Society* **72**, 39–47.

Robinson, W. (1926) The conditions of growth and development of *Pyronema confluens*. *Annals of Botany* **40**, 245–272.

Robson, G. D., Bell, S. D., Kuhn, P. J. & Trinci, A. P. J. (1987) Glucose and penicillin concentrations in agar medium below fungal colonies. *Journal of General Microbiology* **133**, 361–367.

Robson, G. D., Wiebe, M. G. & Trinci, A. P. J. (1991a) Exogenous cAMP and cGMP modulate branching in *Fusarium graminearum*. *Journal of General Microbiology* **137**, 963–969.

Robson, G. D., Wiebe, M. G. & Trinci, A. P. J. (1991b) Involvement of Ca^{2+} in the regulation of hyphal extension and branching in *Fusarium graminearum* A-3/5. *Experimental Mycology* **15**, 263–272.

Rosin, I. V., Horner, J. & Moore, D. (1985) Differentiation and pattern formation in the fruit body cap of *Coprinus cinereus*. In *Developmental Biology of Higher Fungi* (ed. D. Moore, L. A. Casselton, D. A. Wood & J. C. Frankland), pp. 333–351. Cambridge University Press: Cambridge, UK.

Rosin, I. V. & Moore, D. (1985a) Origin of the hymenophore and establishment of major tissue domains during fruit body development in *Coprinus cinereus*. *Transactions of the British Mycological Society* **84**, 609–619.

Rosin, I. V. & Moore, D. (1985b) Differentiation of the hymenium in *Coprinus cinereus*. *Transactions of the British Mycological Society* **84**, 621–628.

Ross, I. K. (1979) *Biology of the Fungi. Their Development Regulation and Associations*. McGraw-Hill: New York.

Ross, I. K. (1982a) Localization of carpophore initiation in *Coprinus congregatus*. *Journal of General Microbiology* **128**, 2755–2762.

Ross, I. K. (1982b) The role of laccase in carpophore initiation in *Coprinus congregatus*. *Journal of General Microbiology* **128**, 2763–2770.

Ross, I. K. (1985) Determination of the initial steps in differentiation in *Coprinus congregatus*. In *Developmental Biology of Higher Fungi* (ed. D. Moore, L. A. Casselton, D. A. Wood & J. C. Frankland), pp. 353–373. Cambridge University Press: Cambridge, UK.

Ross, I. K. & Margalith, P. (1987) Nuclear behaviour in the basidia of the secondarily homothallic *Coprinus bilanatus*. *Mycologia* **79**, 595–602.

Ross, R. G. & Bremner, D. J. (1971) Effect of ammonium nitrogen and amino acids on perithecial formation of *Venturia inaequalis*. *Canadian Journal of Plant Science* **51**, 29–33.

Ross, R. G. & Hamlin, S. A. (1965) Influence of nutrients on perithecial production of *Venturia inaequalis* (Cke.) Wint. *Canadian Journal of Botany* **43**, 959–965.

Roth, R. & Lusnak, K. (1970) DNA synthesis during yeast sporulation: genetic control of an early developmental event. *Science* **168**, 493–494.

Ruiters, M. H. J. & Wessels, J. G. H. (1989a) *In situ* localization of specific RNAs in developing fruit bodies of the basidiomycete *Schizophyllum commune. Experimental Mycology* **13**, 212–222.

Ruiters, M. H. J. & Wessels, J. G. H. (1989b) *In situ* localization of specific RNAs in whole fruiting colonies of *Schizophyllum commune. Journal of General Microbiology* **135**, 1747–1754.

Ruiz-Herrera, J. (1994) Polyamines, DNA methylation, and fungal differentiation. *Critical Reviews in Microbiology* **20**, 143–150.

Ruiz-Herrera, J., Iranzo, M., Elorza, M. V., Sentandreu, R. & Mormeneo, S. (1995a) Involvement of transglutaminase in the formation of covalent cross-links in the cell wall of *Candida albicans. Archives of Microbiology* **164**, 186–193.

Ruiz-Herrera, J., Leon, C. G., Guevara-Olvera, L. & Carabez-Trejo, A. (1995b) Yeast-mycelial dimorphism of haploid and diploid strains of *Ustilago maydis. Microbiology - UK* **141**, 695–703.

Ruiz-Herrera, J., Mormeneo, S., Vanaclocha, P., Font-de-Mora, J., Iranzo, M., Puertes, I. & Sentandreu, R. (1994) Structural organization of the components of the cell wall from *Candida albicans. Microbiology - UK* **140**, 1513–1523.

Rusmin, S. & Leonard, T. J. (1978) Biochemical induction of fruiting in *Schizophyllum. Plant Physiology* **61**, 538–543.

Russo, G. M., Dahlberg, K. R. & Van Etten, J. L. (1982) Identification of a development-specific protein in sclerotia of *Sclerotinia sclerotiorum. Experimental Mycology* **6**, 259–267.

Ryvarden, L. (1992) Genera of polypores, nomenclature and taxonomy. *Synopsis Fungorum* **5**, 1–363.

Sachs, M. S. & Yanofsky, C. (1991) Developmental expression of genes involved in conidiation and amino acid biosynthesis in *Neurospora crassa. Developmental Biology* **148**, 117–128.

Sachs, T. (1991) Cell polarity and tissue patterning in plants. *Development* **1**, (Supplement) 83–93.

Sainsard-Chanet, A., Sellem, C., Silar, P., Belcour, L., Dequard-Chablat, M. & Picard, M. (1994) Sénescence chez les champignons filamenteux. *Médecine Sciences* **10**, 574–576.

Salisbury, F. B. & Ross, C. W. (1985) *Plant Physiology*. Wadsworth Publishing Company: Belmont.

Samuels, A. L., Giddings, T. H. & Staehelin, L. A. (1995) Cytokinesis in tobacco BY-2 and root tip cells - a new model of cell plate formation in higher plants. *Journal of Cell Biology* **130**, 1345–1357.

Samuels, R. I. & Paterson, I. C. (1995) Cuticle degrading proteases from insect moulting fluid and culture filtrates of entomopathogenic fungi. *Comparative Biochemistry and Physiology B - Biochemistry & Molecular Biology* **110**, 661–669.

San Antonio, J. & Thomas, R. L. (1971) Carbon dioxide stimulation of hyphal growth of the cultivated mushroom *Agaricus bisporus* (Lange) Sing. *Mushroom Science* **8**, 623–629.

Sanders, S. L. & Field, C. M. (1994) Cell division - Septins in common? *Current Biology* **4**, 907–910.

Sanwal, B. D. & Lata, M. (1961) The occurrence of two different glutamic dehydrogenases in *Neurospora. Canadian Journal of Microbiology* **7**, 319–328.

Sanwal, B. D. & Lata, M. (1962) Concurrent regulation of glutamic acid dehydrogenases of *Neurospora. Archives of Biochemistry and Biophysics* **97**, 582–588.

Saupe, S., Stenberg, L., Shiu, K. T., Griffiths, A. J. F. & Glass, N. L. (1996) The molecular nature of mutations in the mt *A-1* gene of the *Neurospora crassa A* idiomorph and their relation to mating-type function. *Molecular & General Genetics* **250**, 115–122.

Scazzocchio, C. & Darlington, A. J. (1968) The induction and repression of the enzymes of purine breakdown in *Aspergillus nidulans. Biochimica et Biophysica Acta* **166**, 557–568.

Schaeffer, H. P., Hildreth, S. W. & Drake, F. H. (1977) Glucan and glucanase involvement in autolysis of *Coprinus cinereus* gills. In *Abstracts, Second International Mycological Congress*, p. 594. IMA: Tampa, Florida.

Schatz, G. & Dobberstein, B. (1996) Common principles of protein translocation across membranes. *Science* **271**, 1519–1526.

Scherthan, H., Loidl, J., Schuster, T. & Schweizer, D. (1992) Meiotic chromosome condensation and pairing in *Saccharomyces cerevisiae* studied by chromosome painting. *Chromosoma* **101**, 590–595.

Schmidt, H. & Gutz, H. (1994) The mating type switch in yeasts. In *The Mycota,* **vol. I**, Growth, Differentiation and Sexuality (ed. J. G. H. Wessels & F. Meinhardt), pp. 283–294. Springer-Verlag: Berlin.

Schmidt, W. (1980) Artificial flavin/membrane systems; a possible model for physiological blue light action. In *The Blue Light Syndrome* (ed. H. Senger), pp. 210–220. Springer-Verlag: New York.

Schmitz, J. (1842) Mykologische Beobachtungen, als Beitrage zur lebensund Entwicklungsgeschichte einiger Schwamme aus der Klasse der Gasteromyceten und Hymenomyceten. *Linnaea* **16**, 141–215.

Schuren, F. H. J., van der Lende, T. R. & Wessels, J. G. H. (1993) Fruiting genes of *Schizophyllum commune* are transcriptionally regulated. *Mycological Research* **97**, 538–542.

Schwalb, M. N. (1974) Changes in activity of enzymes metabolizing glucose 6-phosphate during development of the basidiomycete *Schizophyllum commune. Developmental Biology* **40**, 84–89.

Schwalb, M. N. (1978) Regulation of fruiting. In *Genetics and Morphogenesis in the Basidiomycetes* (ed. M. N. Schwalb & P. G. Miles), pp. 135–165. Academic Press: New York.

Schwantes, H. O. & Barsuhn, E. (1971) Tropische Reaktionen der Fruchtkörperbildung von *Lentinus tigrinus* (Bull.: Fr.) *Zeitschrift für Pilzkunde* **37**, 169–182.

Selitrennikoff, C. P. (1974) Phase-specific genes for macroconidiation in *Neurospora crassa. Genetics* **78**, 679–690.

Sen, S. (1992) Programmed cell death: concept, mechanism and control. *Biological Reviews* **67**, 287–319.

Senathirajah, S. & Lewis, D. (1975) Resistance to amino acid analogues in *Coprinus*: dominance modifier genes and dominance reversal in dikaryons and diploids. *Genetical Research* **25**, 95–107.

Sentandreu, R., Mormeneo, S. & Riuz-Herrera, J. (1994) Biogenesis of the fungal cell wall. In *The Mycota,* **vol. I**, Growth, Differentiation and Sexuality (ed. J. G. H. Wessels & F. Meinhardt), pp. 111–123. Springer-Verlag: Berlin.

Sentandreu, R., Sentandreu, M., Elorza, M. V., Iranzo, M. & Mormeneo, S. (1995) Interactions of proteins with other wall components - a pivotal step in fungal cell-wall construction. *Canadian Journal of Botany* **73**, S384–S387.

Serlupi-Crescenzi, O., Kurtz, M. B. & Champe, S. P. (1983) Developmental defects resulting from arginine auxotrophy in *Aspergillus nidulans. Journal of General Microbiology* **129**, 3535–3544.

Sewall, T. C., Mims, C. W. & Timberlake, W. E. (1990) Conidial differentiation in wild type and *wetA*⁻ strains of *Aspergillus nidulans. Developmental Biology* **138**, 499–508.

Shah, M. M., Barr, D. P., Chung, N. & Aust, S. D. (1992) Use of white rot fungi in the degradation of environmental chemicals. *Toxicology Letters* **64/5**, 493–501.

Sharland, P. R., Burton, J. L. & Rayner, A. D. M. (1986) Mycelial dimorphism, interactions and pseudosclerotial plate formation in *Hymenochaete corrugata. Transactions of the British Mycological Society* **86**, 158–163.

Sharland, P. R. & Rayner, A. D. M. (1986) Mycelial interactions in *Daldinia concentrica. Transactions of the British Mycological Society* **86**, 643–649.

Shearer, C. A. (1995) Fungal competition. *Canadian Journal of Botany* **73**, S1259–S1264.

Shepherd, V. A., Orlovich, D. A. & Ashford, A. E. (1993a) Cell-to-cell transport via motile tubules in growing hyphae of a fungus. *Journal of Cell Science* **105**, 1173–1178.

Shepherd, V. A., Orlovich, D. A. & Ashford, A. E. (1993b) A dynamic continuum of pleiomorphic tubules and vacuoles in growing hyphae of a fungus. *Journal of Cell Science* **104**, 495–507.

Sheppard, P. M. (1967) *Natural Selection and Heredity*. Hutchinson: London.

Sherwood-Higham, J., Zhu, W. Y., Devine, C. A., Gooday, G. W., Gow, N. A. R. & Gregory, D. W. (1994) Helical growth of hyphae of *Candida albicans*. *Journal of Medical and Veterinary Mycology* **32**, 437–445.

Sherwoodpike, M. (1991) Fossils as keys to evolution in fungi. *BioSystems* **25**, 121–129.

Siddiq, A. A., Ingram, D. S., Johnstone, K., Friend, J. & Ashby, A. A. (1989) The control of asexual and sexual development by morphogens in fungal pathogens. *Aspects of Applied Biology* **23**, 417–426.

Sidow, A. & Thomas, W. K. A. (1994) Molecular evolutionary framework for eukaryotic model organisms. *Current Biology* **4**, 596–603.

Sietsma, J. H., Din, A. B., Ziv, V., Sjollema, K. A. & Yarden, O. (1996) The localization of chitin synthase in membranous vesicles (chitosomes) in *Neurospora crassa*. *Microbiology - UK* **142**, 1591–1596.

Sietsma, J. H. & Wessels, J. G. H. (1994) Apical wall biogenesis. In *The Mycota,* **vol. I**, Growth, Differentiation and Sexuality (ed. J. G. H. Wessels & F. Meinhardt), pp. 125–141. Springer-Verlag: Berlin.

Sietsma, J. H., Wosten, H. A. B. & Wessels, J. G. H. (1995) Cell-wall growth and protein secretion in fungi. *Canadian Journal of Botany* **73**, S388–S395.

Silver, J. C., Brunt, S. A., Kyriakopoulou, G., Borkar, M. & Nazarianarmavil, V. (1993) Heat-shock proteins in hyphal branching and secretion in steroid-hormone induced fungal development. *Journal of Cellular Biochemistry* **S17C**, 136 (meeting abstract).

Simanis, V. (1995) The control of septum formation and cytokinesis in fission yeast. *Seminars in Cell Biology* **6**, 79–87.

Singer, R. (1936) Studien zur Systematik der Basidiomyceten. *Beihefte zum Botanischen Centralblatt* **56B**, 137–174.

Singer, R. (1945) The *Laschia* complex. *Lloydia* **8**, 170–230.

Singer, R. (1947) Coscinoids and coscino-cystidia in *Linderomyces lateritius*. *Farlowia* **3**, 155–157.

Singer, R. (1951) The Agaricales in modern taxonomy. *Lilloa* **22**, 1-830.

Singer, R. (1955) New and interesting species of Basidiomycetes IV. *Mycologia* **47**, 763–777.

Singer, R. (1963) Notes on secotiaceous fungi: *Galeropsis* and *Brauniella*. *Koninklijke Nederlandse Akademie van Wetenschappen - Proceedings Series C: Biological and Medical Sciences* **66**, 106–117.

Singer, R. (1975) *The Agaricales in Modern Taxonomy, 3rd edn*. J. Cramer: Vaduz.

Singer, R. (1986) *The Agaricales in Modern Taxonomy*. Koeltz Scientific Books: Koenigstein.

Singh, R. A. & Shankar, G. (1972) Effect of physical factors and nutrition on growth and perithecial development of *Glomerella cingulata in vitro*. *Indian Phytopathology* **24**, 728–734.

Singleton, I. (1994) Microbial-metabolism of xenobiotics - fundamental and applied research. *Journal of Chemical Technology and Biotechnology* **59**, 9–23.

Slack, J. M. W. (1991) *From Egg to Embryo: Regional Specification in Early Development*. Cambridge University Press: Cambridge, UK.

Smith, A. H. (1966) The hyphal structure of the basidiocarp. In *The Fungi: an Advanced Treatise, vol II* (ed. G. C. Ainsworth & A. S. Sussman), pp. 151–177. Academic Press: New York.

Smith, J. E., Anderson, J. G., Deans, S. G. & Davis, B. (1977) Asexual development in *Aspergillus*. In *Genetics and Physiology of Aspergillus* (ed. J. E. Smith & J. A. Pateman), pp. 23–58. Academic Press: New York.

Smith, M. L., Bruhn, J. N. & Anderson, J. B. (1992) The fungus *Armillaria bulbosa* is among the largest and oldest living organisms. *Nature* **356**, 428–431.

Snider, P. J. (1959) Stages of development in rhizomorphic thalli of *Armillaria mellea*. *Mycologia* **51**, 693–707.

Soderling, T. R. & Park, C. R. (1974) Recent advances in glycogen metabolism. *Advances in Cyclic Nucleotide Research* **4**, 283.

Sommer, T., Degli-Innocenti, F. & Russo, V. E. A. (1987) Role of nitrogen in the photoinduction of protoperithecia and carotenoids in *Neurospora crassa*. *Planta* **170**, 205–208.

Song, P. S. (1980) Spectroscopic and photochemical characterization of flavoproteins and carotenoproteins as blue light photoreceptors. In *The Blue Light Syndrome* (ed. H. Senger), pp. 157–171. Springer-Verlag: New York.
Spatafora, J. W. (1995) Ascomal evolution of filamentous ascomycetes - evidence from molecular data. *Canadian Journal of Botany* **73**, S811–S815.
Spinner, S. & Haselwandter, K. (1985) Proteins as nitrogen sources for *Hymenoscyphus* (= *Pezizella*) *ericae*. In *Proceedings of the 6th North American Conference on Mycorrhizae* (ed. R. Molina), p. 422. Forest Research Laboratory, Oregon State University: Corvallis, Oregon.
Springer, M. L., Hager, K. M., Garrett-Eagele, C. & Yanofsky, C. (1992) Timing of synthesis and cellular localization of two conidiation-specific proteins of *Neurospora crassa*. *Developmental Biology* **152**, 255–262.
Springer, M. L. & Yanofsky, C. (1989) A morphological and genetic analysis of conidiophore development in *Neurospora crassa*. *Genes & Development* **3**, 559–571.
Springer, M. L. & Yanofsky, C. (1992) Expression of *con* genes along the three sporulation pathways of *Neurospora crassa*. *Genes & Development* **6**, 1052–1057.
Staben, C. (1995) Sexual reproduction in higher fungi. In *The Growing Fungus* (ed. N. A. R. Gow & G. M. Gadd), pp. 383–402. Chapman & Hall: London.
Stachelhaus, T., Schneider, A. & Marahiel, M. A. (1995) Rational design of peptide antibiotics by targeted replacement of bacterial and fungal domains. *Science* **269**, 69–72.
Stahl, U. & Esser, K. (1976) Genetics of fruit body production in higher basidiomycetes. I. Monokaryotic fruiting and its correlation with dikaryotic fruiting in *Polyporus ciliatus*. *Molecular & General Genetics* **148**, 183–197.
Stalpers, J. A. (1978) Identification of wood-inhabiting Aphyllophorales in pure culture. *Studies in Mycology* **16**, 1–248.
Steinberg, G. & Schliwa, M. (1993) Organelle movements in the wild type and wall-less *fz; sg; os-1* mutants of *Neurospora crassa* are mediated by cytoplasmic microtubules. *Journal of Cell Science* **106**, 555–564.
Steinberg, G. & Schliwa, M. (1995) The *Neurospora* organelle motor - a distant relative of conventional kinesin with unconventional properties. *Molecular Biology of the Cell* **6**, 1605–1618.
Stephenson, N. A. & Gooday, G. W. (1984) Nuclear numbers in the stipe cells of *Coprinus cinereus*. *Transactions of the British Mycological Society* **82**, 531–534.
Sternlicht, E., Katz, D. & Rosenberger, R. F. (1973) Subapical wall synthesis and wall thickening induced by cycloheximide in hyphae of *Aspergillus nidulans*. *Journal of Bacteriology* **114**, 819–823.
Stewart, G. R. & Moore, D. (1971) Factors affecting the level and activity of pyruvate kinase from *Coprinus lagopus sensu* Buller. *Journal of General Microbiology* **66**, 361–370.
Stewart, G. R. & Moore, D. (1974) The activities of glutamate dehydrogenases during mycelial growth and sporophore development in *Coprinus lagopus* (*sensu* Lewis) *Journal of General Microbiology* **83**, 73–81.
Stine, G. J. (1968) Enzyme activities during the asexual cycle of *Neurospora crassa*. II. NAD- and NADP-dependent glutamic dehydrogenases and nicotinamide adenine dinucleotidase. *Journal of Cell Biology* **37**, 81–88.
St Leger, R. J. (1995) The role of cuticle-degrading proteases in fungal pathogenesis of insects. *Canadian Journal of Botany* **73**, S1119–S1125.
St Leger, R. J., Staples, R. C. & Roberts, D. W. (1992) Cloning and regulatory analysis of starvation stress gene, *ssgA*, encoding a hydrophobin-like protein from the entomopathogenic fungus, *Metarhizium anisopliae*. *Gene* **120**, 119–124.
Stočkus, A. & Moore, D. (1996) Comparison of plant and fungal gravitropic responses using imitational modelling. *Plant, Cell & Environment* **19**, 787–800.
Stoller, B. B. (1952) Abnormal growth and fructification of the cultivated mushroom. *Science* **116**, 320–322.
Stone, K. J. & Hemming, F. W. (1967) The stereochemistry of hexahydroprenol, ubiquinone and ergosterol biosynthesis in the mycelium of *Aspergillus fumigatus*. *Biochemical Journal* **104**, 43–56.
Stowe, M. K., Tumlinson, J. H. & Heath, J. H. (1987) Chemical mimicry: bolas spiders emit components of moth prey species sex pheromones. *Science* **236**, 964–967.

Stratford, M. (1994) Another brick in the wall? Recent developments concerning the yeast cell envelope. *Yeast* **10**, 1741–1752.

Streuli, C. H. (1993) Extracellular matrix and gene expression in mammary epithelium. *Seminars in Cell Biology* **4**, 203–212.

Stringer, M. A., Dean, R. A., Sewall, T. C. & Timberlake, W. E. (1991) Rodletless, a new *Aspergillus* developmental mutant induced by directed gene inactivation. *Genes & Development* **5**, 1161–1171.

Sussman, A. S. (1968) Longevity and survivability of fungi. In *The Fungi: An Advanced Treatise*, **vol. III** (ed. G. C. Ainsworth & A. S. Sussman), pp. 447–486. Academic Press: New York.

Swamy, S., Uno, I. & Ishikawa, T. (1984) Morphogenetic effects of mutations at the *A* and *B* incompatibility factors in *Coprinus cinereus*. *Journal of General Microbiology* **130**, 3219–3224.

Swamy, S., Uno, I. & Ishikawa, T. (1985) Regulation of cyclic AMP metabolism by the incompatibility factors in *Coprinus cinereus*. *Journal of General Microbiology* **131**, 3211–3217.

Swann, E. C. & Taylor, J. W. (1993) Higher taxa of basidiomycetes - an 18S ribosomal-RNA gene perspective. *Mycologia* **85**, 923–936.

Swann, E. C. & Taylor, J. W. (1995) Phylogenetic perspectives on basidiomycete systematics - evidence from the 18S ribosomal RNA gene. *Canadian Journal of Botany* **73**, S862–S868.

Swift, M. J. (1977) The ecology of wood decomposition. *Science Progress (Oxford)* **64**, 175–199.

Tabak, H. H. & Bridge-Cooke, W. (1968) The effect of gaseous environments on the growth and metabolism of fungi. *The Botanical Review* **34**, 126–252.

Tagu, D., Nasse, B. & Martin, F. (1996) Cloning and characterization of hydrophobins-encoding cDNAs from the ectomycorrhizal basidiomycete *Pisolithus tinctorius*. *Gene* **168**, 93–97.

Taj-Aldeen, S. J. & Moore, D. (1982) The *ftr* cistron of *Coprinus cinereus* is the structural gene for a multifunctional transport molecule. *Current Genetics* **5**, 209–213.

Taj-Aldeen, S. J. & Moore, D. (1983) Evidence for a single hexokinase in *Coprinus cinereus*. *FEMS Microbiology Letters* **17**, 73–75.

Takemaru, T. & Kamada, T. (1969) A bibliography on the basidiocarp morphogenesis in the hymenomycetes and gasteromycetes. *Reports of the Tottori Mycological Institute (Japan)* **7**, 127–140.

Takemaru, T. & Kamada, T. (1971) Gene control of basidiocarp development in *Coprinus macrorhizus*. *Reports of the Tottori Mycological Institute (Japan)* **9**, 21–35.

Takemaru, T. & Kamada, T. (1972) Basidiocarp development in *Coprinus macrorhizus*. I. Induction of developmental variations. *Botanical Magazine (Tokyo)* **85**, 51–57.

Talbot, N. J., Ebbole, D. J. & Hamer, J. E. (1993) Identification and characterization of *MPG1*, a gene involved in pathogenicity from the rice blast fungus *Magnaporthe grisea*. *Plant Cell* **5**, 1575–1590.

Talbot, P. H. B. (1968) Fossilized pre-Patouillardian taxonomy? *Taxon* **17**, 622–628.

Tan, K. K. (1978) Light-induced fungal development. In *The Filamentous Fungi*, **vol. III**, Developmental Mycology (ed. J. E. Smith & D. R. Berry), pp. 334–357. Edward Arnold: London.

Tan, Y. H. & Moore, D. (1994) High concentrations of mannitol in the shiitake mushroom *Lentinula edodes*. *Microbios* **79**, 31–35.

Tan, Y. H. & Moore, D. (1995) Glucose catabolic pathways in *Lentinula edodes* determined with radiorespirometry and enzymic analysis. *Mycological Research* **99**, 859–866.

Taylor, J. W. (1985) Mitosis in the basidiomycete fungus *Tulasnella araneosa*. *Protoplasma* **126**, 1–18.

Taylor, J. W., Bowman, B. H., Berbee, M. L. & White, T. J. (1993) Fungal model organisms - phylogenetics of *Saccharomyces*, *Aspergillus*, and *Neurospora*. *Systematic Biology* **42**, 440–457.

Taylor, T. N. & Osborn, J. M. (1996) The importance of fungi in shaping the paleoecosystem. *Review of Palaeobotany and Palynology* **90**, 249–262.

Taylor, T. N., Remy, W., Hass, H. & Kerp, H. (1995) Fossil arbuscular mycorrhizae from the early Devonian. *Mycologia* **87**, 560–573.

Teston, D. (1953a) Étude de la différentiation des hyphes chez les polypores dimidiés de la flore française. I. Historique. *Annales de l Université de Lyon, Sect. C* **7**, 11–23.

Teston, D. (1953b) Étude de la différentiation des hyphes chez les polypores dimidiés de la flore française. II. Descriptions et figures. *Bulletin de la Société des Naturalistes d'Oyannax* **7**, 80–110.

Thompson, K. S. (1988) *Morphogenesis and Evolution.* Oxford University Press: Oxford, UK.

Thompson, W. (1984) Distribution, development and functioning of mycelial cord systems of decomposer basidiomycetes of the deciduous woodland floor. In *The Ecology and Physiology of the Fungal Mycelium* (ed. D. H. Jennings & A. D. M. Rayner), pp. 185–214. Cambridge University Press: Cambridge, UK.

Tickle, C. (1991) Retinoic acid and chick limb development. *Development Supplement* **1**, 113–121.

Timberlake, W. E. (1980) Developmental gene regulation in *Aspergillus nidulans. Developmental Biology* **78**, 497–510.

Timberlake, W. E. (1986) Isolation of stage- and cell-specific genes from fungi. In *Biology and Molecular Biology of Plant-Pathogen Interactions, NATO ASI Series,* **vol. H1** (ed. J. Bailey), pp. 343–357. Springer-Verlag: Berlin.

Timberlake, W. E. (1987) Molecular genetic analysis of development in *Aspergillus nidulans.* In *Genetic Regulation of Development, 45th Symposium of the Society for Developmental Biology* (ed. W. F. Loomis), pp. 63–82. Alan R. Liss: New York.

Timberlake, W. E. (1990) Molecular genetics of *Aspergillus* development. *Annual Review of Genetics* **24**, 5–36.

Timberlake, W. E. (1993) Translational triggering and feedback fixation in the control of fungal development. *Plant Cell* **5**, 1453–1460.

Timberlake, W. E., Boylan, M. T., Cooley, M. B., Mirabito, P. M., O'Hara, E. B. & Willett, C. E. (1985) Rapid identification of mutation-complementing restriction fragments from *Aspergillus nidulans* cosmids. *Experimental Mycology* **9**, 351–355.

Timberlake, W. E. & Clutterbuck, A. J. (1993) Genetic regulation of conidiation. In *Physiology and Genetics of Aspergillus* (ed. D. Martinelli & J. R. Kinghorn). Chapman & Hall: London.

Timberlake, W. E., Gwynne, D. I., Hamer, J. E., Miller, B. L., Miller, K. Y., Mullaney, E. M., de Souza, R., Yelton, M. M. & Zimmerman, C. R. (1983) Gene regulation during conidiation in *Aspergillus nidulans.* In *Plant Molecular Biology* (ed. R. B. Goldberg), pp. 179–199. Alan R. Liss: New York.

Timberlake, W. E. & Hamer, J. E. (1986) Regulation of gene activity during conidiophore development in *Aspergillus nidulans.* In *Genetic Engineering,* **vol.8** (ed. J. K. Setlow & A. Hollaender), pp. 1–29. Plenum Publishing Corp.: New York.

Timberlake, W. E. & Marshall, M. A. (1988) Genetic regulation of development in *Aspergillus nidulans. Trends in Genetics* **4**, 162–169.

Timberlake, W. E. & Orr, W. C. (1984) Steroid hormone regulation of sexual reproduction in *Achlya.* In *Biological Regulation and Development,* **Vol. 3B** (ed. R. F. Goldberg & K. Yamamoto), pp. 255–283. Plenum Press: New York.

Todd, N. K. & Aylmore, R. C. (1985) Cytology of hyphal interactions and reactions in *Schizophyllum commune.* In *Developmental Biology of Higher Fungi* (ed. D. Moore, L. A. Casselton, D. A. Wood & J. C. Frankland), pp. 231–248. Cambridge University Press: Cambridge, UK.

Todd, N. K. & Rayner, A. D. M. (1980) Fungal individualism. *Science Progress (Oxford)* **66**, 331–354.

Tooley, P. W. & Carras, M. M. (1992) Separation of chromosomes of *Phytophthora megasperma* species using gel electrophoresis. *Experimental Mycology* **16**, 188–196.

Tough, A. J., Pulham, J. & Prosser, J. I. (1995) A mathematical model for the growth of mycelial pellet populations. *Biotechnology and Bioengineering* **46**, 561–572.

Towers, G. H. N. (1976) Secondary metabolites derived through the shikimate-chorismate pathway. In *The Filamentous Fungi,* **vol II**, Biosynthesis and Metabolism (ed. J. E. Smith & D. R. Berry), pp. 460–474. Edward Arnold: London.

Townsend, B. B. (1954) Morphology and development of fungal rhizomorphs. *Transactions of the British Mycological Society* **37**, 222–233.

Townsend, B. B. & Willetts, H. J. (1954) The development of sclerotia of certain fungi. *Transactions of the British Mycological Society* **37**, 213–221.

Tressl, R., Bahri, D. & Engel, K. H. (1982) Formation of eight-carbon and ten-carbon components in mushrooms (*Agaricus campestris*) *Journal of Agricultural and Food Chemistry* **30**, 89–93.

Trewavas, A. (1986) Understanding the control of plant development and the role of growth substances. *Australian Journal of Plant Physiology* **13**, 447–457.

Tribe, H. T. & Mabadeje, S. A. (1972) Growth of moulds on media prepared without organic nutrients. *Transactions of the British Mycological Society* **58**, 127–137.

Trinci, A. P. J. (1971) Influence of the peripheral growth zone on the radial growth rate of fungal colonies. *Journal of General Microbiology* **67**, 325–344.

Trinci, A. P. J. (1973) Growth of wild-type and spreading colonial mutants of *Neurospora crassa* in batch culture and on agar medium. *Archiv für Mikrobiologie* **91**, 113–126.

Trinci, A. P. J. (1974) A study of the kinetics of hyphal extension and branch initiation of fungal mycelia. *Journal of General Microbiology* **81**, 225–236.

Trinci, A. P. J. (1978) The duplication cycle and vegetative development in moulds. In *The Filamentous Fungi,* **vol III**, Developmental Mycology (ed. J. E. Smith & D. R. Berry), pp. 132–163. Edward Arnold: London.

Trinci, A. P. J. (1979) The duplication cycle and branching in fungi. In *Fungal Walls and Hyphal Growth* (ed. J. H. Burnett & A. P. J. Trinci), pp. 319–358. Cambridge University Press: Cambridge, UK.

Trinci, A. P. J. (1984) Regulation of hyphal branching and hyphal orientation. In *The Ecology and Physiology of the Fungal Mycelium* (ed. D. H. Jennings & A. D. M. Rayner), pp. 23–52. Cambridge University Press: Cambridge, UK.

Trinci, A. P. J. (1994) Evolution of the Quorn$^{(R)}$ myco-protein fungus, *Fusarium graminearum* A3/5. *Microbiology - UK* **140**, 2181–2188.

Trinci, A. P. J. & Banbury, G. H. (1967) A study of the tall conidiophores of *Aspergillus giganteus. Transactions of the British Mycological Society* **50**, 525–538.

Trinci, A. P. J. & Collinge, A. J. (1973) Influence of L-sorbose on the growth and morphology of *Neurospora crassa. Journal of General Microbiology* **78**, 179–192.

Trinci, A. P. J. & Cutter, E. G. (1986) Growth and form in lower plants and the occurrence of meristems. *Philosophical Transactions of the Royal Society of London, Series B* **313**, 95–113.

Trinci, A. P. J. & Halford, E. A. (1975) The extension zone of stage I sporangiophores of *Phycomyces blakesleeanus. New Phytologist* **74**, 81–83.

Trinci, A. P. J., Saunders, P. T., Gosrani, R. & Campbell, K. A. S. (1979) Spiral growth of mycelial and reproductive hyphae. *Transactions of the British Mycological Society* **73**, 283–292.

Trinci, A. P. J., Wiebe, M. G. & Robson, G. D. (1994) The mycelium as an integrated entity. In *The Mycota,* **vol. I**, Growth, Differentiation and Sexuality (ed. J. G. H. Wessels & F. Meinhardt), pp. 175–193. Springer-Verlag: Berlin.

Trione, E. J. & Leach, C. M. (1969) Light-induced sporulation and sporogenic substances in fungi. *Phytopathology* **59**, 1077–1083.

Tronchin, G., Bouchara, J. P., Ferron, M., Larcher, G. & Chabasse, D. (1995) Cell surface properties of *Aspergillus fumigatus* conidia: correlation between adherence, agglutination, and rearrangements of the cell wall. *Canadian Journal of Microbiology* **41**, 714–721.

Tschabold, E. (1967) Physiology of sexual reproduction in *Hypomyces solani* f. *cucurbitae.* IV. Influence of flavin inhibitors on perithecium formation. *Phytopathology* **57**, 1140–1141.

Tschierpe, H. J. (1959) Die Bedeutung des Kohlendioxyd für den Kulturchampignon. *Gartenbauwissenschaft* **24**, 18–75.

Tsuda, M., Waki, T., Taga, M. & Ueyama, A. (1982) Ascocarp production of *Magnaporthe salvinii* in culture. *Transactions of the British Mycological Society* **78**, 515–519.

Tsusue, Y. M. (1969) Experimental control of fruitbody formation in *Coprinus macrorhizus. Development, Growth and Differentiation* **11**, 164–178.

Turian, G. (1966) Morphogenesis in ascomycetes. In *The Fungi: An Advanced Treatise,* **vol. 2** (ed. G. C. Ainsworth & A. S. Sussman), pp. 339–385. Academic Press: New York.

Turian, G. (1978) Sexual morphogenesis in the Ascomycetes. In *The Filamentous Fungi*, **vol III**, Developmental Mycology (ed. J. E. Smith & D. R. Berry), pp. 315–333. Edward Arnold: London.

Turian, G. & Combépine, G. (1963) Activité comparée de la transaminase alanine-glyoxylate de *Neurospora* cultivé en présence de saccharose ou d'acétate. *Helvetica Chimica Acta* **46**, 2453–2457.

Turing, A. (1952) The chemical basis of morphogenesis. *Philosophical Transactions of the Royal Society, Series B* **237**, 37–72.

Turner, E. M. (1974) Phenoloxidase activity in relation to substrate and development stage in the mushroom *Agaricus bisporus*. *Transactions of the British Mycological Society* **63**, 541–547.

Turner, E. M. (1977) Development of excised sporocarps of *Agaricus bisporus* and its control by CO_2. *Transactions of the British Mycological Society* **69**, 183–186.

Turner, E. M., Wright, M., Ward, T., Osborne, D. J. & Self, R. (1975) Production of ethylene and other volatiles and changes in cellulase and laccase activities during the life cycle of the cultivated mushroom, *Agaricus bisporus*. *Journal of General Microbiology* **91**, 167–176.

Turner, G. (1992) Genes for the biosynthesis of penicillin in microorganisms. *CIBA Foundation Symposia* **171**, 113–128.

Turner, W. B. (1971) *Fungal Metabolites*. Academic Press: London.

Turner, W. B. (1976) Polyketides and related metabolites. In *The Filamentous Fungi*, **vol II**, Biosynthesis and Metabolism (ed. J. E. Smith & D. R. Berry), pp. 445–459. Edward Arnold: London.

Turner, W. B. & Aldridge, D. C. (1983) *Fungal Metabolites II*. Academic Press: London.

Ullrich, R. C. (1973) Sexuality, incompatibility, and intersterility in the biology of the *Sistotrema brinkmannii* aggregate. *Mycologia* **65**, 1234–1249.

Ulmasov, T., Liu, Z-B., Hagen, G. & Guilfoyle, T. J. (1995) Composite structure of auxin responsive elements. *Plant Cell* **7**, 1611–1623.

Umar, M. H. & Van Griensven, L. J. L. D. (1997a) Hyphal regeneration and histogenesis in *Agaricus bisporus*. *Mycological Research* **101**, 1025–1032.

Umar, M. H. & Van Griensven, L. J. L. D. (1997b) Morphological studies on the life span, developmental stages, senescence and death of *Agaricus bisporus*. *Mycological Research* **101**, 1409–1422.

Umar, M. H. & Van Griensven, L. J. L. D. (1997c) Morphogenetic cell death in developing primordia of *Agaricus bisporus*. *Mycologia* **89**, 274–277.

Uno, I. & Ishikawa, T. (1971) Chemical and genetical control of induction of monokaryotic fruiting bodies in *Coprinus macrorhizus*. *Molecular & General Genetics* **113**, 228–239.

Uno, I. & Ishikawa, T. (1973a) Purification and identification of the fruiting-inducing substances in *Coprinus macrorhizus*. *Journal of Bacteriology* **113**, 1240–1248.

Uno, I. & Ishikawa, T. (1973b) Metabolism of adenosine 3′, 5′-cyclic monophosphate and induction of fruiting bodies in *Coprinus macrorhizus*. *Journal of Bacteriology* **113**, 1249–1255.

Uno, I. & Ishikawa, T. (1974) Effect of glucose on the fruiting body formation and adenosine 3′,5′-cyclic monophosphate levels in *Coprinus macrorhizus*. *Journal of Bacteriology* **120**, 96–100.

Uno, I. & Ishikawa, T. (1975) Subcellular localization of adenylate cyclase during fruiting body formation in *Coprinus macrorhizus*. *Journal of General and Applied Microbiology* **21**, 335–343.

Uno, I. & Ishikawa, T. (1976) Effect of cyclic AMP on glycogen phosphorylase in *Coprinus macrorhizus*. *Biochimica et Biophysica Acta* **452**, 112–120.

Uno, I. & Ishikawa, T. (1978) Effect of cyclic AMP on glycogen synthetase in *Coprinus macrorhizus*. *Journal of General and Applied Microbiology* **24**, 193–197.

Uno, I. & Ishikawa, T. (1982) Biochemical and genetic studies on the initial events of fruitbody formation. In *Basidium and Basidiocarp: Evolution, Cytology, Function and Development* (ed. K. Wells & E. K. Wells), pp. 113–123. Springer-Verlag: New York.

Uno, I., Yamaguchi, M. & Ishikawa, T. (1974) The effect of light on fruiting body formation and adenosine 3′:5′-cyclic monophosphate metabolism in *Coprinus macrorhizus*. *Proceedings of the National Academy of Sciences of the USA* **71**, 479–483.

Urayama, T. (1969) Stimulative effect of extracts from fruit bodies of *Agaricus bisporus* and some other hymenomycetes on primordium formation in *Marasmius* sp. *Transactions of the Japanese Mycological Society* **10**, 73–78.

Valadon, L. R. G. (1976) Carotenoids as additional taxonomic characters in fungi. *Transactions of the British Mycological Society* **67**, 1–15.

Valla, G. (1984) Changes in DNA content of nuclei in apical and intercalary compartments of *Polyporus arcularius* during hyphal growth. *Transactions of the British Mycological Society* **83**, 265–273.

van den Ende, H. (1984) Sexual interactions in the lower filamentous fungi. In *Encyclopedia of Plant Physiology*, **vol. 17** (ed. H. F. Linskens & J. Heslop-Harison), pp. 333–349. Springer-Verlag: Berlin.

Van de Peer, Y., Neefs, J. M., Derijk, P. & Dewachter, R. (1993) Evolution of eukaryotes as deduced from small ribosomal-subunit RNA sequences. *Biochemical Systematics and Ecology* **21**, 43–55.

Van der Valk, P. & Marchant, R. (1978) Hyphal ultrastructure in fruit body primordia of the basidiomycetes *Schizophyllum commune* and *Coprinus cinereus*. *Protoplasma* **95**, 57–72.

van der Westhuizen, C. G. A. (1958) Studies of wood-rotting fungi 1. Cultural characters of some common species. *Bothalia* **7**, 83–107.

van der Westhuizen, C. G. A. (1963) The cultural characters, structure of the fruitbody, and the type of interfertility of *Cerrena unicolor* (Bull. ex Fr.) Murr. *Canadian Journal of Botany* **41**, 1487–1499.

van der Westhuizen, C. G. A. (1971) Cultural characters and carpophore construction of some poroid Hymenomycetes. *Bothalia* **10**, 137–328.

Van Essen, D. C. (1997) A tension-based theory of morphogenesis and compact wiring in the central nervous system. *Nature* **385**, 313–318.

Van Laere, A. (1995) Intermediary metabolism. In *The Growing Fungus* (ed. N. A. R. Gow & G. M. Gadd), pp. 211–238. Chapman & Hall: London.

Varjú, D., Edgar, L. & Delbrück, M. (1961) Interplay between the reactions to light and to gravity in *Phycomyces*. *Journal of General Physiology* **45**, 47–58.

Vaughn, J. C., Mason, M. T., Sperwhitis, G. L., Kuhlman, P. & Palmer, J. D. (1995) Fungal origin by horizontal transfer of a plant mitochondrial group-I intron in the chimeric coxI gene of *Peperomia*. *Journal of Molecular Evolution* **41**, 563–572.

Verrinder Gibbins, A. M. & Lu, B. C. (1984) Induction of normal fruiting on originally monokaryotic cultures of *Coprinus cinereus*. *Transactions of the British Mycological Society* **83**, 331–335.

Viniegra-Gonzalez, G., Saucedo-Castaneda, G., Lopez-Isunza, F. & Favelatorres, E. (1993) Symmetrical branching model for the kinetics of mycelial growth. *Biotechnology and Bioengineering* **42**, 1–10.

Viswanath-Reddy, M. & Turian, G. (1975) Physiological changes during protoperithecial differentiation in *Neurospora tetrasperma*. *Physiologia Plantarum* **35**, 166–174.

Voderberg, K. (1949) Zur Fruchtkörperbildung von *Coprinus lagopus*. *Planta* **37**, 149–160.

Volz, P. A. & Niederpruem, D. J. (1970) The sclerotia of *Coprinus lagopus*. *Archiv für Mikrobiologie* **70**, 369–377.

Vraná, D. (1983a) The fission yeast *Schizosaccharomyces pombe* in continuous culture. *Biotechnology and Bioengineering* **25**, 1989–1994.

Vraná, D. (1983b) Morphological properties of *Schizosaccharomyces pombe* in a continuous culture. *Folia Microbiologica* **28**, 414–419.

Wainright, P. O., Hinkle, G., Sogin, M. L. & Stickel, S. K. (1993) Monophyletic origins of the Metazoa - an evolutionary link with fungi. *Science* **260**, 340–342.

Wainwright, M. (1988) Metabolic diversity of fungi in relation to growth and mineral cycling in soil - a review. *Transactions of the British Mycological Society* **90**, 159–170.

Wainwright, M., Ali, T. A. & Killham, K. (1994) Anaerobic growth of fungal mycelium from soil particles onto nutrient-free silica gel. *Mycological Research* **98**, 761–762.

Walker, P. & Woodbine, M. (1976) The biosynthesis of fatty acids. In *The Filamentous Fungi*, **vol II**, Biosynthesis and Metabolism (ed. J. E. Smith & D. R. Berry), pp. 137–158. Edward Arnold: London.

Walkey, D. G. A. & Harvey, R. (1967a) Spore discharge rhythms in Pyrenomycetes. II. Endogenous and exogenous rhythms of spore discharge. *Transactions of the British Mycological Society* **50**, 229–240.

Walkey, D. G. A. & Harvey, R. (1967b) Spore discharge rhythms in Pyrenomycetes. III. Ascospore production and the quantitative and qualitative influences of light on spore discharge. *Transactions of the British Mycological Society* **50**, 241–249.

Wardle, K. C. & Schisler, L. C. (1969) The effects of various lipids on growth of mycelium of *Agaricus bisporus*. *Mycologia* **61**, 305–314.

Wasternack, C. & Reinbothe, H. (1967) Harnstoff-metabolismus bei Basidiomyceten. III. Untersuchungen zur Harnstoffsynthese aus Pyrimidinen in Fruchtkorpern von *Agaricus bisporus* Lange. *Flora* **158**, 1–26.

Waters, H., Butler, R. D. & Moore, D. (1972) Thick-walled sclerotial medullary cells in *Coprinus lagopus*. *Transactions of the British Mycological Society* **59**, 167–169.

Waters, H., Butler, R. D. & Moore, D. (1975a) Structure of aerial and submerged sclerotia of *Coprinus lagopus*. *New Phytologist* **74**, 199–205.

Waters, H., Moore, D. & Butler, R. D. (1975b) Morphogenesis of aerial sclerotia of *Coprinus lagopus*. *New Phytologist* **74**, 207–213.

Watkins, N. K., Fitter, A. H., Graves, J. D. & Robinson, D. (1996) Carbon transfer between C-3 and C-4 plants linked by a common mycorrhizal network, quantified using stable carbon isotopes. *Soil Biology & Biochemistry* **28**, 471–477.

Watkinson, S. C. (1971) The mechanism of mycelial strand induction in *Serpula lacrimans*: a possible effect of nutrient distribution. *New Phytologist* **70**, 1079–1088.

Watkinson, S. C. (1975) The relation between nitrogen nutrition and the formation of mycelial strands in *Serpula lacrimans*. *Transactions of the British Mycological Society* **64**, 195–200.

Watkinson, S. C. (1979) Growth of rhizomorphs, mycelial strands coremia and sclerotia. In *Fungal Walls and Hyphal Growth* (ed. J. H. Burnett & A. P. J. Trinci), pp. 93–113. Cambridge University Press: Cambridge, UK.

Watling, R. (1971) Polymorphism in *Psilocybe merdaria*. *New Phytologist* **70**, 307–326.

Watling, R. (1974) Dimorphism in *Entoloma abortivum*. *Bulletin bimensuel de la Société Linnéenne de Lyon (numéro spécial dédiés à R. Kühner)*, 449–470.

Watling, R. (1978) From infancy to adolescence: advances in the study of higher fungi. *Transactions of the Botanical Society of Edinburgh* **42 (supplement)**, 61–73.

Watling, R. (1985) Developmental characters of agarics. In *Developmental Biology of Higher Fungi* (ed. D. Moore, L. A. Casselton, D. A. Wood & J. C. Frankland), pp. 281–310. Cambridge University Press: Cambridge, UK.

Watling, R. (1996) Patterns in fungal development fruiting patterns in nature. In *Patterns in Fungal Development* (ed. S. W. Chiu & D. Moore), pp. 182–222. Cambridge University Press, Cambridge, UK.

Watling, R. & Moore, D. (1994) Moulding moulds into mushrooms: shape and form in the higher fungi. In *Shape and Form in Plants and Fungi* (ed. D. S. Ingram & A. Hudson), pp. 270–290. Academic Press: London.

Watling, R. & Nicoll, H. (1980) Sphaerocysts in *Lactarius rufus*. *Transactions of the British Mycological Society* **75**, 331–333.

Watson, K. (1976) The biochemistry and biogenesis of mitochondria. In *The Filamentous Fungi,* **vol II**, Biosynthesis and Metabolism (ed. J. E. Smith & D. R. Berry), pp. 92–120. Edward Arnold: London.

Webster, J. (1980) *Introduction to Fungi*, 2nd ed. Cambridge University Press, Cambridge, UK.

Weete, J. D. & Gandhi, S. R. (1996) Biochemistry and molecular biology of fungal sterols. In *The Mycota,* **vol. III**, Biochemistry and Molecular Biology (ed. R. Brambl & G. A. Marzluf), pp. 421–438. Springer-Verlag: Berlin.

Weil, J., Miramonti, J. & Ladisch, M. R. (1995a) Cephalosporin C: mode of action and biosynthetic pathway. *Enzyme and Microbial Technology* **17**, 85–87.

Weil, J., Miramonti, J. & Ladisch, M. R. (1995b) Biosynthesis of cephalosporin C: regulation and recombinant technology. *Enzyme and Microbial Technology* **17**, 88–90.

Wellman, C. H. (1995) Phytodebris from Scottish Silurian and Lower Devonian continental deposits. *Review of Palaeobotany and Palynology* **84**, 255–279.

Wells, K. (1977) Meiotic and mitotic divisions in the Basidiomycotina. In *Mechanisms and Control of Cell Division* (ed. T. L. Rost & E. M. Gifford), pp. 337-374. Dowden, Hutchinson & Ross: Stroudsburg, Pa, USA.

Wendland, J., Vaillancourt, L. J., Hegner, J., Lengeler, K. B., Laddison, K. J., Specht, C. A., Raper, C. A. & Kothe, E. (1995) The mating-type locus Ba1 of *Schizophyllum commune* contains a pheromone receptor gene and putative pheromone genes. *EMBO Journal* **14**, 5271–5278.

Wessels, J. G. H. (1965) Morphogenesis and biochemical processes in *Schizophyllum commune* Fr. *Wentia* **13**, 1–113.

Wessels, J. G. H. (1966) Control of cell-wall glucan degradation during development in *Schizophyllum commune. Antonie Van Leeuwenhoek International Journal of General and Molecular Microbiology* **32**, 341–355.

Wessels, J. G. H. (1969) A β-1,6-glucan glucanohydrolase involved in hydrolysis of cell-wall glucan in *Schizophyllum commune. Biochimica et Biophysica Acta* **178**, 191–193.

Wessels, J. G. H. (1992) Gene expression during fruiting in *Schizophyllum commune. Mycological Research* **96**, 609–620.

Wessels, J. G. H. (1993) Wall growth, protein excretion and morphogenesis in fungi. *New Phytologist* **123**, 397–413.

Wessels, J. G. H. (1994a) Development of fruit bodies in Homobasidiomycetes. In *The Mycota, I, Growth, Differentiation and Sexuality* (ed. J. G. H. Wessels & F. Meinhardt), pp. 351–366. Springer-Verlag: Berlin.

Wessels, J. G. H. (1994b) Developmental regulation of fungal cell wall formation. *Annual Review of Phytopathology* **32**, 413–437.

Wessels, J. G. H. (1996) Fungal hydrophobins: proteins that function at an interface. *Trends in Plant Science* **1**, 9–15.

Wessels, J. G. H., de Vries, O. M. H., Asgeirsdóttir, S. A. & Springer, J. (1991) The *thn* mutations of *Schizophyllum commune*, which supresses formation of aerial hyphae, affects expression of the *Sc3* hydrophobin gene. *Journal of General Microbiology* **137**, 2439–2445.

Wessels, J. G. H., Mulder, G. H. & Springer, J. (1987) Expression of dikaryon-specific and non-specific mRNAs of *Schizophyllum commune* in relation to environmental conditions and fruiting. *Journal of General Microbiology* **133**, 2557–2561.

Wessels, J. G. H. & Niederpruem, D. J. (1967) Role of cell wall glucan-degrading enzyme in mating of *Schizophyllum commune. Journal of Bacteriology* **94**, 1594–1602.

Wessels, J. G. H. & Sietsma, J. H. (1979) Wall structure and growth in *Schizophyllum commune.* In *Fungal Walls and Hyphal Growth* (ed. J. H. Burnett & A. P. J. Trinci), pp. 29–48. Cambridge University Press: Cambridge, UK.

Wessels, J. G. H. & Sietsma, J. H. (1981) Fungal cell walls: a survey. In *Encyclopedia of Plant Physiology, New Series* **vol. 13B**, Plant Carbohydrates II (ed. W. Tanner & F. A. Loewus), pp. 352–394. Springer-Verlag: Berlin.

Weste, G. (1970a) Factors affecting vegetative growth and the production of perithecia in culture by *Ophiobolus graminis.* I. Variations in media and age of mycelium. *Australian Journal of Botany* **18**, 1–10.

Weste, G. (1970b) Factors affecting vegetative growth and the production of perithecia in culture by *Ophiobolus graminis.* II. Variations in light and temperature. *Australian Journal of Botany* **18**, 11–28.

Westergaard, M. & von Wettstein, D. (1970) Studies on the mechanism of crossing over. IV. The molecular organization of the synaptinemal complex in *Neottiella* (Cooke) Saccardo (Ascomycetes) *Comptes Rendus des Travaux du Laboratoire Carlsberg* **37**, 239–268.

Whalley, A. J. S. & Edwards, R. L. (1987) Xylariaceous fungi: use of secondary metabolites. In *Evolutionary Biology of the Fungi* (ed. A. D. M. Rayner, C. M. Brasier & D. Moore), pp. 423–434. Cambridge University Press: Cambridge, UK.

Whetzel, H. H. (1945) A synopsis of the genera and species of the Sclerotiniaceae, a family of stromatic inoperculate Discomycetes. *Mycologia* **37**, 648–714.

Whittaker, R. H. (1969) New concepts of Kingdoms of organisms. *Science* **163**, 150–160.

Wiebe, M. G., Robson, G. D., Oliver, S. G. & Trinci, A. P. J. (1996) pH oscillations and constant low pH delay the appearance of highly branched (colonial) mutants in che-

mostat cultures of the Quorn$^{(R)}$ myco-protein fungus, *Fusarium graminearum* A3/5. *Biotechnology and Bioengineering* **51**, 61–68.
Wiebe, M. G., Robson, G. D. & Trinci, A. P. J. (1992) Evidence for the independent regulation of hyphal extension and branch initiation in *Fusarium graminearum* A-3/5. *FEMS Microbiology Letters* **90**, 179–184.
Wiebe, M. G., Robson, G. D., Cunliffe, B., Oliver, S. G. & Trinci, A. P. J. (1993) Periodic selection in long-term continuous-flow cultures of the filamentous fungus *Fusarium graminearum*. *Journal of General Microbiology* **139**, 2811–2817.
Wilke, N. W. & Wach, M. P. (1993) Detection of methylated mushroom DNA by restriction enzyme analysis. *Mycologia* **85**, 585–591.
Willetts, H. J. (1969) Structure of the outer surfaces of sclerotia of certain fungi. *Archiv für Mikrobiologie* **69**, 48–53.
Willetts, H. J. (1971) The survival of fungal sclerotia under adverse environmental conditions. *Biological Reviews* **46**, 387–407.
Willetts, H. J. (1972) The morphogenesis and possible evolutionary origins of fungal sclerotia. *Biological Reviews* **47**, 515–536.
Willetts, H. J. (1997) Morphology, development and evolution of stromata/sclerotia and macroconidia of the Sclerotiniaceae. *Mycological Research* **101**, 939–952.
Willetts, H. J. & Bullock, S. (1992) Developmental biology of sclerotia. *Mycological Research* **96**, 801–816.
Willetts, H. J. & Calonge, F. D. (1969) Spore development in the brown rot fungi (*Sclerotinia* spp.) *New Phytologist* **68**, 123–131.
Willetts, H. J. & Wong, A. L. (1971) Ontogenetic diversity of sclerotia of *Sclerotinia sclerotiorum* and related species. *Transactions of the British Mycological Society* **57**, 515–524.
Williams, M. A. J., Beckett, A. & Read, N. D. (1985) Ultrastructural aspects of fruit body differentiation in *Flammulina velutipes*. In *Developmental Biology of Higher Fungi* (ed. D. Moore, L. A. Casselton, D. A. Wood & J. C. Frankland), pp. 429–450. Cambridge University Press: Cambridge, UK.
Williams, M. J., Hughes, P. E., O'Toole, T. E. & Ginsberg, M. H. (1994) The inner world of cell adhesion: integrin cytoplasmic domains. *Trends in Cell Biology* **4**, 109–112.
Wilmotte, A., Van de Peer, Y., Goris, A., Chapelle, S., Debaere, R., Nelissen, B., Neefs, J. M., Hennebert, G. L. & Dewachter, R. (1993) Evolutionary relationships among higher fungi inferred from small ribosomal-subunit RNA sequence analysis. *Systematic and Applied Microbiology* **16**, 436–444.
Winkelmann, G. (1986) Iron complex products (siderophores) In *Biotechnology 4* (ed. H. J. Rehn & G. Reed), pp. 215–243. VCH Verlagsgesellschaft: Weinheim.
Wittler, R., Baumgartl, H., Lubbers, D. W. & Schugerl, K. (1986) Investigations of oxygen transfer into *Penicillium chrysogenum* pellets by microprobe measurements. *Biotechnology and Bioengineering* **28**, 1024–1036.
Wolf, C. & Mirocha, C. J. (1973) Regulation of sexual reproduction in *Gibberella zeae* (*Fusarium roseum* 'Graminearum') by F-2 (zearalenone) *Canadian Journal of Microbiology* **19**, 725–734.
Wong, W. M. & Gruen, H. E. (1977) Changes in cell size and nuclear number during elongation of *Flammulina velutipes* fruitbodies. *Mycologia* **69**, 899–913.
Wood, D. A. (1980a) Inactivation of extracellular laccase of *Agaricus bisporus*. *Journal of General Microbiology* **117**, 339–345.
Wood, D. A. (1980b) Production, purification and properties of extracellular laccase of *Agaricus bisporus*. *Journal of General Microbiology* **177**, 327–338.
Wood, D. A. (1980c) Production, purification and properties of laccase of *Agaricus bisporus*. *Journal of General Microbiology* **117**, 327–338.
Wood, D. A., Craig, G. D., Atkey, P. T., Newsam, R. J. & Gull, K. (1985) Ultrastructural studies on the cultivation processes and growth and development of the cultivated mushroom *Agaricus bisporus*. *Food Microstructure* **4**, 143–164.
Worsdell, W. C. (1915) *The Principles of Plant Teratology*, **vol 1**. The Ray Society: London.
Wösten, H. A. B., Schuren, F. H. J. & Wessels, J. G. H. (1994) Interfacial self-assembly of a hydrophobin into and amphipathic membrane mediates fungal attachment to hydrophobic surfaces. *EMBO Journal* **13**, 5848–5854.

Wright, J. L. C. & Vining, L. C. (1976) Secondary metabolites derived from non-aromatic amino acids. In *The Filamentous Fungi,* **vol II**, Biosynthesis and Metabolism (ed. J. E. Smith & D. R. Berry), pp. 475–502. Edward Arnold: London.

Wurster, B., Pan, P., Tyan, G. G. & Bonner, J. T. (1976) Preliminary characterization of the acrasin of the cellular slime mould *Polysphondylium violaceum. Proceedings of the National Academy of Sciences of the USA* **73**, 795–799.

Xiang, X., Roghi, C. & Morris, N. R. (1995) Characterization and localization of the cytoplasmic dynein heavy chain in *Aspergillus nidulans. Proceedings of the National Academy of Sciences of the USA* **92**, 9890–9894.

Yabe, T., Yamadaokabe, T., Kasahara, S., Furuichi, Y., Nakajima, T., Ichishima, E., Arisawa, M. & Yamadaokabe, H. (1996) HKR1 encodes a cell surface protein that regulates both cell wall β-glucan synthesis and budding pattern in the yeast *Saccharomyces cerevisiae. Journal of Bacteriology* **178**, 477–483.

Yager, L. N. (1992) Early developmental events during asexual and sexual sporulation in *Aspergillus nidulans.* In *Aspergillus Biology and Industrial Applications* (ed. J. W. Bennett & M. A. Klich), pp. 19–41. Butterworth-Heinemann: Boston.

Yang, H., King, R., Reichl, U. & Gilles, E. D. (1992) Mathematical model for apical growth, septation, and branching of mycelial microorganisms. *Biotechnology and Bioengineering* **39**, 49–58.

Yano, T., Kodama, T. & Yamada, K. (1961) Fundamental studies on the aerobic fermentation. *Agricultural and Biological Chemistry* **25**, 580–584.

Yanofsky, C. & Kolter, R. (1982) Attenuation in amino acid biosynthesis operons. *Annual Review of Genetics* **16**, 113–134.

Yarden, O. & Russo, V. E. A. (1996) Changes in chitin deposition accompany runner hypha branching of *Gaeumannomyces graminis* in culture. *Mycological Research* **100**, 444–448.

Yashar, B. M. & Pukkila, P. J. (1985) Changes in polyadenylated RNA sequences associated with fruiting body morphogenesis in *Coprinus cinereus. Transactions of the British Mycological Society* **84**, 215–226.

Yli-Mattila, T. (1985) Action spectrum for fruiting in the basidiomycete *Schizophyllum commune. Physiologia Plantarum* **65**, 287–293.

Yoon, K. S. & Kim, Y. S. (1994) Ultrastructure of mitosis and clamp formation in the somatic hyphae of *Pleurotus ostreatus. Mycologia* **86**, 593–601.

Zadeh, L. A. (1996) Fuzzy logic = computing with words. *IEEE Transactions on Fuzzy Systems* **4**, 103–111.

Zantinge, B., Dons, H. & Wessels, J. G. H. (1979) Comparison of poly(A)-containing RNAs in different cell types of the lower eukaryote *Schizophyllum commune. European Journal of Biochemistry* **101**, 251–260.

Zickler, H. (1952) Zur Entwicklungsgechichte des Askomyceten *Bombardia lunata. Archiv für Protistenkunde* **98**, 1–70.

Zimmermann, C. R., Orr, W. C., Leclerc, R. F., Barnard, E. C. & Timberlake, W. E. (1980) Molecular cloning and selection of genes regulated in *Aspergillus* development. *Cell* **21**, 709–715.

Zolan, M. E. (1995) Chromosome-length polymorphism in fungi. *Microbiological Reviews* **59**, 686–698.

Zolan, M. E., Heyler, N. K. & Stassen N. Y. (1994) Inheritance of chromosome-length polymorphisms in *Coprinus cinereus. Genetics* **137**, 87–94.

Zolan, M. E., Tremel, C. J. & Pukkila, P. J. (1988) Production and characterization of radiation-sensitive meiotic mutants of *Coprinus cinereus. Genetics* **120**, 379–387.

Zonneveld, B. J. M. (1977) Biochemistry and ultrastructure of sexual development in *Aspergillus.* In *Genetics and Physiology of Aspergillus* (ed. J. E. Smith & J. A. Pateman), pp. 59–80. Academic Press: London.

Zonneveld, B. J. M. (1980) Protease levels in relation to cAMP and a reserve polymer during growth and sexual differentiation in *Aspergillus nidulans. Experimental Mycology* **4**, 140–146.

Index